The Handbook of Environmental Chemistry

Volume 116

Founding Editor: Otto Hutzinger

Series Editors: Damià Barceló • Andrey G. Kostianoy

In over four decades, *The Handbook of Environmental Chemistry* has established itself as the premier reference source, providing sound and solid knowledge about environmental topics from a chemical perspective. Written by leading experts with practical experience in the field, the series continues to be essential reading for environmental scientists as well as for environmental managers and decision-makers in industry, government, agencies and public-interest groups.

Two distinguished Series Editors, internationally renowned volume editors as well as a prestigious Editorial Board safeguard publication of volumes according to high scientific standards.

Presenting a wide spectrum of viewpoints and approaches in topical volumes, the scope of the series covers topics such as

- local and global changes of natural environment and climate
- anthropogenic impact on the environment
- water, air and soil pollution
- remediation and waste characterization
- environmental contaminants
- biogeochemistry and geoecology
- chemical reactions and processes
- chemical and biological transformations as well as physical transport of chemicals in the environment
- environmental modeling

A particular focus of the series lies on methodological advances in environmental analytical chemistry.

The Handbook of Environmental Chemistry is available both in print and online via https://link.springer.com/bookseries/698. Articles are published online as soon as they have been reviewed and approved for publication.

Meeting the needs of the scientific community, publication of volumes in subseries has been discontinued to achieve a broader scope for the series as a whole.

Chemical Oceanography of Frontal Zones

Volume Editor: Igor M. Belkin

With contributions by

S. Aliani · S. Aliani · M. B. Alkire · T. H. Badewien ·
I. M. Belkin · M. Berta · M. Berta · G. S. Durán Gómez ·
S. K. Eliasen · J. Elken · M.-Z. Fu · A. Griffa · N. Gruber · C. Guéguen ·
H. Hátún · R. Karri · P. Kowalczuk · K. M. H. Larsen · X.-S. Li ·
I. Marinov · M. Mathis · J. Meyerjürgens · A. Molcard · T. Nagai ·
T. M. Özgökmen · J. B. Palter · I. Polyakov · R. Rember · M. Ricker ·
J. L. Sarmiento · E. V. Stanev · G. Suaria · J.-C. Sun · Ü. Suursaar ·
S. Takahashi · S. Tanabe · B.-D. Wang · Q.-S. Wei · Z.-G. Yu ·
E. Zambianchi

Springer

Editor
Igor M. Belkin
College of Marine Science and Technology
Zhejiang Ocean University
Zhoushan, Zhejiang, China

ISSN 1867-979X ISSN 1616-864X (electronic)
The Handbook of Environmental Chemistry
ISBN 978-3-662-65837-6 ISBN 978-3-662-65839-0 (eBook)
https://doi.org/10.1007/978-3-662-65839-0

This Springer imprint is published by the registered company Springer-Verlag GmbH, DE, part of Springer Nature.
The registered company address is: Heidelberger Platz 3, 14197 Berlin, Germany

Series Preface

With remarkable vision, Prof. Otto Hutzinger initiated *The Handbook of Environmental Chemistry* in 1980 and became the founding Editor-in-Chief. At that time, environmental chemistry was an emerging field, aiming at a complete description of the Earth's environment, encompassing the physical, chemical, biological, and geological transformations of chemical substances occurring on a local as well as a global scale. Environmental chemistry was intended to provide an account of the impact of man's activities on the natural environment by describing observed changes.

While a considerable amount of knowledge has been accumulated over the last four decades, as reflected in the more than 150 volumes of *The Handbook of Environmental Chemistry*, there are still many scientific and policy challenges ahead due to the complexity and interdisciplinary nature of the field. The series will therefore continue to provide compilations of current knowledge. Contributions are written by leading experts with practical experience in their fields. *The Handbook of Environmental Chemistry* grows with the increases in our scientific understanding, and provides a valuable source not only for scientists but also for environmental managers and decision-makers. Today, the series covers a broad range of environmental topics from a chemical perspective, including methodological advances in environmental analytical chemistry.

In recent years, there has been a growing tendency to include subject matter of societal relevance in the broad view of environmental chemistry. Topics include life cycle analysis, environmental management, sustainable development, and socio-economic, legal and even political problems, among others. While these topics are of great importance for the development and acceptance of *The Handbook of Environmental Chemistry*, the publisher and Editors-in-Chief have decided to keep the handbook essentially a source of information on "hard sciences" with a particular emphasis on chemistry, but also covering biology, geology, hydrology and engineering as applied to environmental sciences.

The volumes of the series are written at an advanced level, addressing the needs of both researchers and graduate students, as well as of people outside the field of "pure" chemistry, including those in industry, business, government, research establishments, and public interest groups. It would be very satisfying to see these volumes used as a basis for graduate courses in environmental chemistry. With its high standards of scientific quality and clarity, *The Handbook of Environmental Chemistry* provides a solid basis from which scientists can share their knowledge on the different aspects of environmental problems, presenting a wide spectrum of viewpoints and approaches.

The Handbook of Environmental Chemistry is available both in print and online via https://link.springer.com/bookseries/698. Articles are published online as soon as they have been approved for publication. Authors, Volume Editors and Editors-in-Chief are rewarded by the broad acceptance of *The Handbook of Environmental Chemistry* by the scientific community, from whom suggestions for new topics to the Editors-in-Chief are always very welcome.

Damià Barceló
Andrey G. Kostianoy
Series Editors

Preface

This book was conceived in 2008 as a review of chemical and biogeochemical oceanography of fronts. This idea seemed natural despite the fact that nobody has done it before. Perhaps, many chemical oceanographers avoided fronts because their presence complicates data analysis. As Tom Trull wrote to me in 2020, "I've spent the last decade trying to place my field programs far from fronts to avoid their complexities!" Nonetheless, and against all odds, a worldwide search for potential contributors and rigorous peer-review process have produced 11 chapters that comprised the first monograph on chemical and biogeochemical aspects of vastly different physical fronts in all major oceans.

I am grateful to three editors at Springer Nature for their unwavering support. The idea of this book was suggested by Andrey Kostianoy, the long-time Editor-in-Chief of the *Handbook of Environmental Chemistry Series*, who together with Andrea Schlitzberger was always supportive despite countless delays on our part. Ramya Venkitachalam ably and cheerfully assisted during the last 2 years. My most sincere thanks go to 38 authors from 28 organizations in 14 countries who made this book possible. Hopefully, it will stimulate further studies of chemical and biogeochemical processes at ocean fronts.

Zhoushan, China
June 2022

Igor M. Belkin

Contents

Introduction to the *Chemical Oceanography of Frontal Zones*

Igor M. Belkin, Stefano Aliani, Matthew B. Alkire, Thomas H. Badewien, Maristella Berta, Gloria Silvana Durán Gómez, Sólvá Káradóttir Eliasen, Jüri Elken, Annalisa Griffa, Nicolas Gruber, Céline Guéguen, Hjálmar Hátún, Ramu Karri, Piotr Kowalczuk, Karin Margretha H. Larsen, Irina Marinov, Moritz Mathis, Jens Meyerjürgens, Anne Molcard, Takeyoshi Nagai, Tamay M. Özgökmen, Jaime B. Palter, Igor Polyakov, Robert Rember, Marcel Ricker, Jorge L. Sarmiento, Emil V. Stanev, Giuseppe Suaria, Ülo Suursaar, Shin Takahashi, Shinsuke Tanabe, Qin-Sheng Wei, and Enrico Zambianchi

Contents

I. M. Belkin (✉)
College of Marine Science and Technology, Zhejiang Ocean University, Zhoushan, China
e-mail: igormbelkin@gmail.com

S. Aliani, M. Berta, A. Griffa, and G. Suaria
Institute of Marine Sciences, National Research Council, Lerici, Italy

M. B. Alkire
School of Oceanography, University of Washington, Seattle, WA, USA

T. H. Badewien and J. Meyerjürgens
Institute for Chemistry and Biology of the Marine Environment, University of Oldenburg, Oldenburg, Germany

G. S. Durán Gómez and T. Nagai
Department of Ocean Sciences, Tokyo University of Marine Science and Technology, Tokyo, Japan

S. K. Eliasen, H. Hátún, and K. M. H. Larsen
Faroe Marine Research Institute, Tórshavn, Faroe Islands

J. Elken
Department of Marine Systems, Tallinn University of Technology, Tallinn, Estonia

Igor M. Belkin (ed.), *Chemical Oceanography of Frontal Zones*,
Hdb Env Chem (2022) 116: 1–24, DOI 10.1007/698_2022_894,

Published online: 25 May 2022

N. Gruber
Environmental Physics, Institute of Biogeochemistry and Pollutant Dynamics, ETH Zurich, Zurich, Switzerland

C. Guéguen
Department of Chemistry, Université de Sherbrooke, Sherbrooke, QC, Canada

R. Karri
National Centre for Coastal Research, Ministry of Earth Sciences, Government of India, Chennai, India

Center for Marine Environmental Studies, Ehime University, Matsuyama, Japan

P. Kowalczuk
Institute of Oceanology, Polish Academy of Sciences, Sopot, Poland

I. Marinov
Department of Earth and Environmental Science, University of Pennsylvania, Philadelphia, PA, USA

M. Mathis
Max Planck Institute for Meteorology, Hamburg, Germany

A. Molcard
Mediterranean Institute of Oceanography, Université de Toulon, Toulon, France

T. M. Özgökmen
Rosenstiel School of Marine and Atmospheric Science, University of Miami, Coral Gables, FL, USA

J. B. Palter
Graduate School of Oceanography, University of Rhode Island, Narragansett, RI, USA

I. Polyakov and R. Rember
International Arctic Research Center, University of Alaska Fairbanks, Fairbanks, AK, USA

M. Ricker
Institute for Chemistry and Biology of the Marine Environment, University of Oldenburg, Oldenburg, Germany

Institute of Coastal Systems – Analysis and Modeling, Helmholtz-Zentrum Hereon, Geesthacht, Germany

Abstract This chapter provides a concise overview of the entire monograph by assembling summaries of 10 individual chapters starting with a global review of large-scale, persistent nutrient fronts of the World Ocean followed by regional chapters on the Arctic Ocean, North Atlantic, Baltic Sea, Kuroshio Current, and the Yellow Sea, a global review of CDOM dynamics at fronts, a chapter on persistent organic pollutants and marine organisms in the Kuroshio-Oyashio frontal zone, and two chapters on marine litter and its dynamics in frontal zones.

Keywords Diapycnal mixing processes, Frontal zone, Geostrophic currents, Halocline waters, Phytoplankton

Preface

This chapter consists of 10 summaries of individual chapters that together provide a concise overview of the entire book starting with a chapter on large-scale, persistent nutrient fronts of the World Ocean [1] and followed by regional chapters on the Arctic Ocean [2], North Atlantic [3], Baltic Sea [4], Kuroshio Current [5], Yellow Sea [6], a global review of CDOM dynamics at fronts [7], a chapter on persistent

J. L. Sarmiento
Atmospheric and Oceanic Sciences Program, Princeton University, Princeton, NJ, USA

E. V. Stanev
Institute of Coastal Systems – Analysis and Modeling, Helmholtz-Zentrum Hereon, Geesthacht, Germany

Department of Meteorology and Geophysics, Faculty of Physics, Sofia University “St. Kliment Ohridski”, Sofia, Bulgaria

Ü. Suursaar
Estonian Marine Institute, University of Tartu, Tallinn, Estonia

S. Takahashi
Center for Marine Environmental Studies, Ehime University, Matsuyama, Japan

Center of Advanced Technology for the Environment, Graduate School of Agriculture, Ehime University, Matsuyama, Japan

S. Tanabe
Center for Marine Environmental Studies, Ehime University, Matsuyama, Japan

Q.-S. Wei
Key Laboratory of Marine Eco-Environmental Science and Technology, First Institute of Oceanography, Ministry of Natural Resources, Qingdao, China

Laboratory for Marine Ecology and Environmental Science, Qingdao National Laboratory for Marine Science and Technology, Qingdao, China

E. Zambianchi
Department of Science and Technology, Parthenope University of Naples and CoNISMA, Naples, Italy

organic pollutants and their accumulation in deep-sea ecosystems of the Kuroshio-Oyashio frontal zone [8], and two chapters on marine litter (especially macro- and microplastics) and its dynamics in frontal zones [9, 10]. The summaries below are numbered as listed above. Authors of individual summaries are listed after the titles of respective summaries.

1 Large-Scale, Persistent Nutrient Fronts of the World Ocean: Impacts on Biogeochemistry [1]

J. B. Palter, I. Marinov, J. L. Sarmiento, and N. Gruber

Palter et al. [1] describe the ocean's large-scale nutrient fronts, where sharp gradients extending horizontally thousands of kilometers separate dynamically and biogeochemically distinct regimes. One intriguing feature of these fronts is that the exchange across them has enormous implications for biogeochemistry and ecosystems, yet a resistance to mixing is what maintains their sharp frontal structure. Palter et al. [1] aim to reconcile the view of fronts as barriers vs. gateways to property exchange. For instance, nutrient transport across fronts of the Antarctic Circumpolar Current (ACC), accompanied by phytoplankton uptake at different nitrogen:phosphorus:silicic acid ratios, is thought to ultimately endow much of the global thermocline with its characteristic nutrient stoichiometry [11, 12]. Palter et al. [1] showed that down-front winds drive cross-frontal Ekman transport that moves nutrients across the ACC, with nitrate transport extending into the Subantarctic Mode Water formation zone, while silicic acid is depleted poleward of this region, in accord with previous studies. Though Ekman transport is a huge term in the nutrient transport across the ACC fronts, where winds are at their global maximum and surface concentrations of macronutrients are held high by iron limitation, the same is not true of many other large-scale nutrient fronts. For instance, at the edges of the tropical oxygen minimum zones, sluggish mesoscale turbulent mixing is the main mechanism transporting properties (including nutrients) across their bounding fronts [13]. Across the Western boundary currents that divide the subtropical and subpolar gyres, Ekman transport plays a role in supplying nutrients to the surface layer in the subtropics, while turbulent mesoscale eddies play an important role in transporting nutrients deeper in the water column. Figure 1 schematizes these fluxes across a front aligned with a geostrophic jet. These insights continue to play a role in our understanding of the nutrient budgets and ecology in the vast regions surrounded by nutrient fronts (e.g., [14–16]).

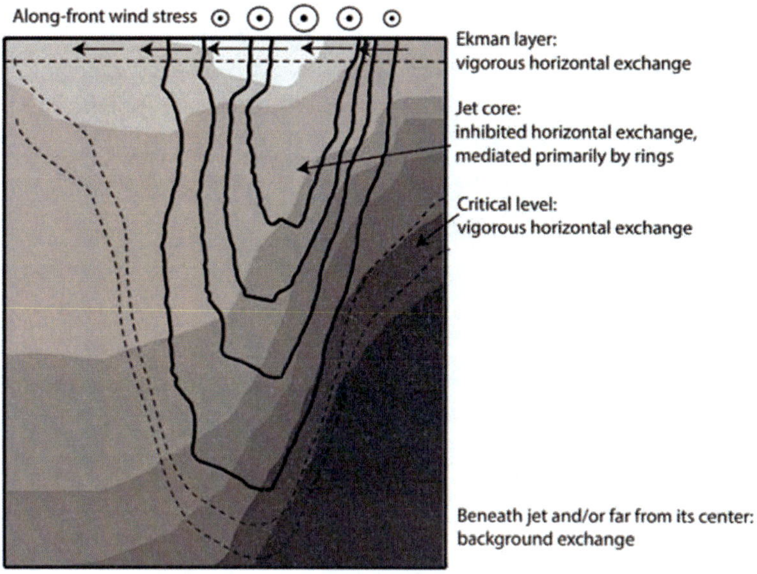

Fig. 1 Schematic of exchange across a front that aligns with a geostrophic jet [1]. Transport of biogeochemical properties across fronts can strongly influence the ecosystem in the regions that it separates, so the schematic explores the mechanisms that facilitate such exchange. The solid lines represent isotachs of positive velocity (flow out of the page) and show the core of the jet; grayscale contours represent density, with higher densities in darker colors. The horizontal dotted line at the top represents the base of the Ekman layer. The dotted curves outline the schematized critical level which descends from the near-surface at the front's outer edges to depths beneath the jet's core. An along-front wind stress drives a cross-front Ekman transport. Exchange across the jet is enhanced beneath the jet core at the depth of the critical level and in the Ekman layer but is inhibited in the jet's core between the Ekman depth and critical level. The limited exchange that is permitted across the center of these jets is mediated primarily by rings shed from the jet

2 The Pacific-Atlantic Front in the East Siberian Sea of the Arctic Ocean [2]

M. B. Alkire, R. Rember, and I. Polyakov

During the summer of 2015, five different research cruises operated within the Arctic Ocean, providing unusually good spatial coverage of observations in a relatively short period of time (Fig. 2). Alkire et al. [2] combined these data to map out the physical and chemical characteristics of Arctic surface and halocline waters. Specifically, by combining sensor-based measurements of nitrate and dissolved oxygen, fronts were identified in the central East Siberian Sea that delineated spreading pathways of different water masses that ventilate the upper and lower regions of the Arctic halocline. For example, a front was observed roughly along the 165°E longitude line that separated shelf waters originating from the Chukchi and East Siberian Seas. The associated spreading pathways indicated East Siberian Sea shelf

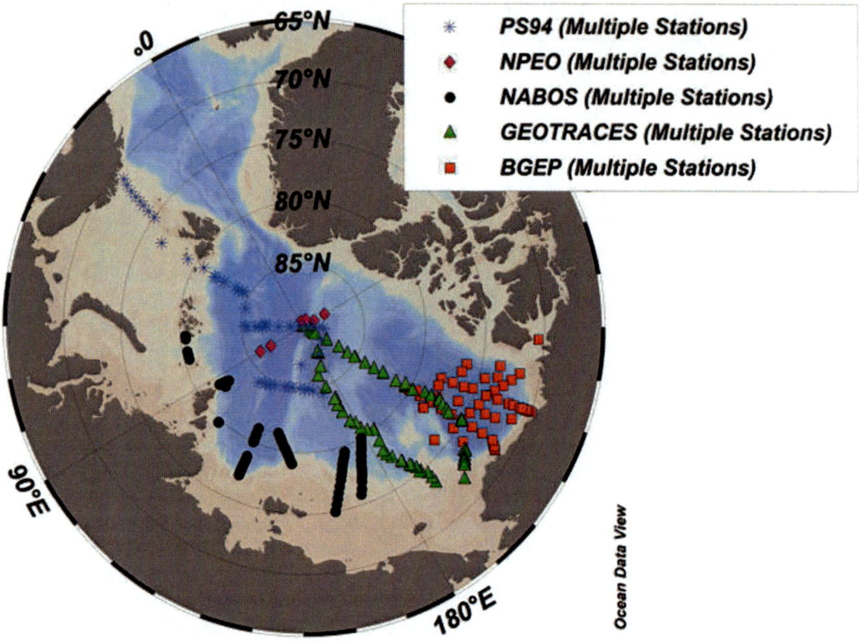

Fig. 2 Map of stations occupied during spring and summer of 2015 as part of the PS94 expedition (blue stars), North Pole Environmental Observatory (purple diamonds), Nansen and Amundsen Basin Observation System (black circles), GEOTRACES (green triangles), and Beaufort Gyre Exploration Program (red squares) [2]

waters ventilated the halocline of the Makarov Basin whereas Chukchi Sea waters were generally restricted to the Canada Basin. In addition, cold and oxygen-rich waters from the Eurasian Basin that ventilate the lower portion of the halocline were observed to spread offshore from the East Siberian Sea continental slope into the Makarov and northern Canada Basins. These observations highlight the benefits of combining physical and chemical observations across multiple research cruises to improve our understanding of circulation patterns in the Arctic Ocean. Increased international coordination and cooperation would enhance opportunities for repeating and/or improving upon the basin-scale coverage achieved during 2015; such campaigns are necessary to further understand and predict future environmental changes across the Arctic Ocean.

3 Major Nutrient Fronts in the Northeastern Atlantic: From the Subpolar Gyre to Adjacent Shelves [3]

H. Hátún, K. M. H. Larsen, S. K. Eliasen, and M. Mathis

The North Atlantic subpolar gyre and its variability are known to regulate the temperature and salinity variability of the North Atlantic Current and its northward branches. The subpolar gyre also has the potential to impact nutrient concentrations in the NE Atlantic. When subtropical nutrient-poor waters are carried northward by ocean currents, these waters mix with nutrient-rich subarctic waters. This mixing occurs mainly along and across the frontal zones between the Atlantic and subarctic water masses (Fig. 3; [3]). When the subpolar gyre is intensified, its cyclonic (counterclockwise) circulation and the attendant upwelling bring nutrient-rich subsurface waters to the upper layers. When the gyre is weak, nutrient-poor subtropical waters become more prevalent in this region. The observed decrease in silicate concentrations in the upper layer of Atlantic waters since the late 1980s was tentatively linked to the declining strength of the subpolar gyre and a weaker winter overturning in the Labrador and Irminger Seas. The declining trend was most pronounced near the Rockall-Hatton Plateau, extending downstream to the Nordic Seas and Labrador Sea. Hátún et al. [3] suggested potential fertilization hotspots along the northeastern periphery of the subpolar gyre and links between these hotspots and feeding migrations of pelagic fish. They also reviewed the ocean-to-shelf nutrient fluxes across shelf-slope fronts and tidal mixing fronts—a process called nutrient *onwelling*. Variations in this *onwelling* likely propagate up the food chain and impact ecosystems in the Northwest European shelf seas as well as the Faroe shelf. Hátún et al. [3] hypothesize that ecosystems on the south Iceland shelf, Faroe shelf, Northwest European shelf, and potentially the Norwegian shelf are influenced by pre-bloom nutrient concentrations in the Atlantic water masses and currents that connect these regions. The results obtained by Hátún et al. [3] from the Faroe shelf case study could help understand other shelf ecosystems as well.

4 Fronts in the Baltic Sea: A Review with a Focus on Its North-Eastern Part [4]

Ü. Suursaar, J. Elken, and I. M. Belkin

The Baltic Sea, a brackish semi-enclosed sea with rugged coastline and bathymetry, hosts numerous fronts—narrow zones of enhanced horizontal gradients of seawater properties such as salinity gradients (Fig. 4a) caused by the Atlantic inflow of oceanic saline water and freshwater river discharges, or temperature gradients caused by wind-driven coastal upwelling. Numerous rivers carry substantial amounts of nutrients and polluting substances from the catchment area to the sea.

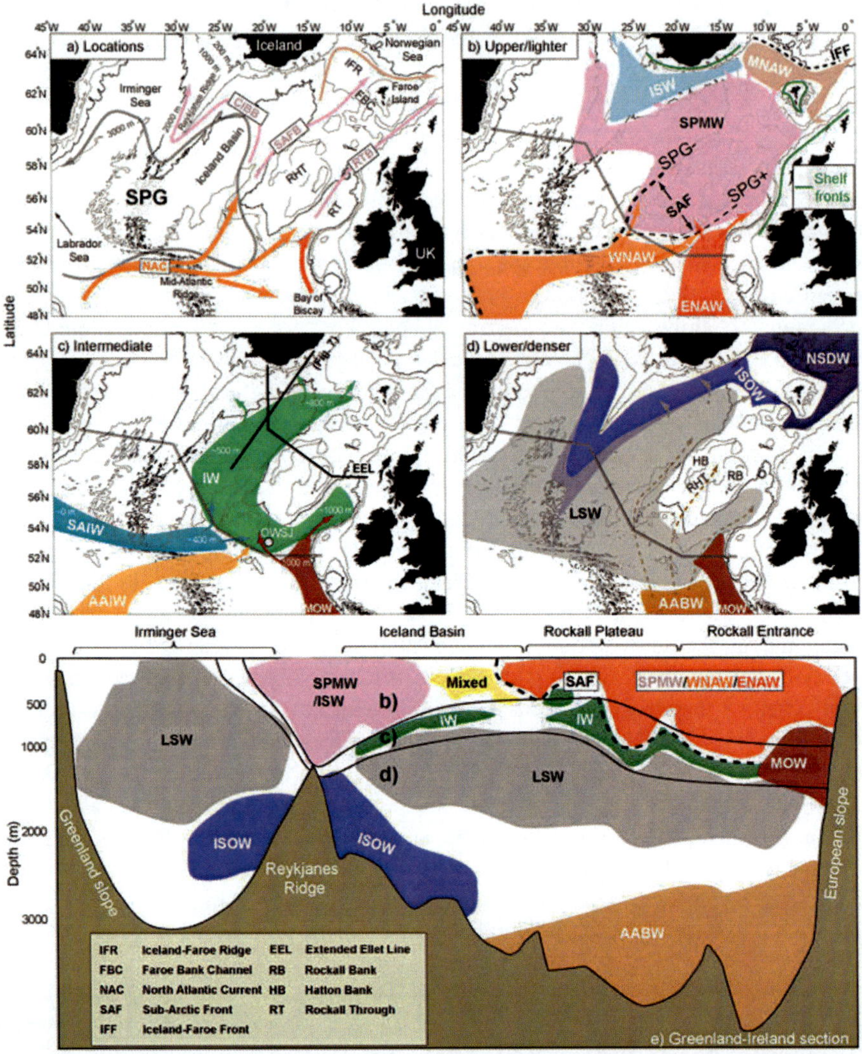

Fig. 3 (After Fig. 1 in [3]). **(a)** Main currents (lines with arrows), including the subpolar gyre (SPG, gray), North Atlantic Current (NAC, orange), Rockall Trough Branch (RTB), Subarctic Front Branch (SAFB), and Central Iceland Basin Branch (CIBB) (after [17]). The location of the WOCE Section A1E from Greenland to Ireland is shown with gray lines in **(b)**, **(c)**, and **(d)**. Main water masses along the A1E line based on its occupation in 1991 (presented by Bersch et al. [18]) are shown in **(e)**, where labels **b**, **c**, and **d** mark, respectively, the upper/lighter layer, intermediate (main thermocline) layer, and deeper/denser layer. The Subarctic Front (SAF, also called Subpolar Front, SPF) is shown with black dashed lines in **(b)** and **(e)** corresponding, respectively, to strong and weak subpolar gyre states (SPG+ and SPG-, dashed lines in **(b)**). Shelf-slope (shelf-break) fronts are shown with green lines in **(b)**. The black lines in **(c)** show the sections presented in Figs. 5, 7, and 8 in [3]. The water mass acronyms are summarized in Table 1 in [3]

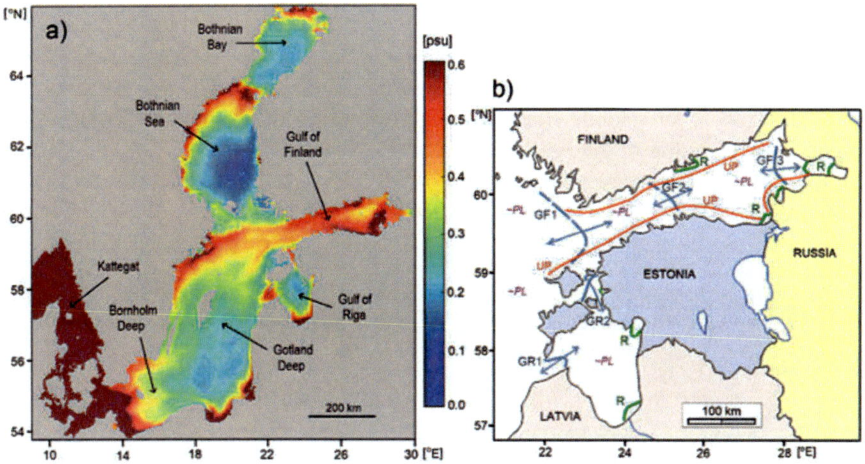

Fig. 4 (**a**) Variability of sea surface salinity of the Baltic Sea shown as standard deviations of the daily CMEMS reanalysis data for the period 1993–2019. Zones of higher variability mostly indicate salinity fronts. (**b**) A schematic of fronts in the north-eastern Baltic Sea. Shown are quasi-permanent thermohaline fronts (blue) in the Gulf of Finland (GF1, GF2, GF3) and the Gulf of Riga (GR1, GR2), upwelling fronts in the Gulf of Finland (UP; red), river plume fronts (R; green), and occasional summertime plankton (~PL) bloom patterns all over the sea. Both figures reproduced from [4]

The Baltic Sea fronts are similar to oceanic fronts at smaller scales since the deformation radius is 10–20 times smaller than that in the ocean. Suursaar et al. [4] provide an overview of various types of fronts, with a spatial focus on the north-eastern Baltic Sea (Fig. 4b). In the nearly 400 km long Gulf of Finland, the mean east-west salinity gradient with a sea surface salinity difference of 1–7 psu is split into several fronts. The easternmost front features distinct cross-front differences in salinity, nutrients, turbidity, and pollutants due to the vicinity of the Neva River, by far the largest river that discharges into the Baltic Sea. Narrow straits that connect the Gulf of Riga with the Baltic Proper host well-defined fronts that enclose the less saline and more nutrient-rich waters of the Gulf. These fronts are rather persistent, although they exhibit back-and-forth excursions of several tens of km depending on meteorological conditions. Fronts are also evident in the Baltic Proper. Although their salinity and density differences are small compared to those of the strait and river plume fronts, meso- and submesoscale features of frontal dynamics are evident. Driven by a divergent Ekman drift, coastal upwelling is a relatively frequent phenomenon in the Baltic Sea. The upwelling-related cross-frontal SST steps may reach up to 20°C in summer, with cold, deep water outcropping nearshore. Therefore, such fronts are easily detected in SST images. Winter upwelling of relatively warm, deep water occurs along certain coastal sections. In winter, the surface water is usually fresher, cooler, and lighter than the deep water below, so that during winter upwelling the saltier and warmer subsurface water outcrops. The winter upwelling of relatively warm water prolongs the ice-free conditions along the southern coast of the Gulf of Finland. Discharges of large rivers such as Neva, Daugava, Oder, and Vistula create plumes that expand during high discharge. Such river plumes differ sharply from ambient water masses. Therefore, the associated plume fronts are readily observed in satellite imagery. High-resolution in situ and remote sensing observational studies of the Baltic Sea since the 1980s provided a wealth of data for frontal studies. These data and new data services are currently available from CMEMS, EMODNET, and other initiatives.

5 The Kuroshio Nutrient Stream: Where Diapycnal Mixing Matters [5]

T. Nagai and G. S. Durán Gómez

Understanding how inorganic nutrients are transported and supplied to phytoplankton is essential for better predictions on how the biological pump can respond, especially in a changing climate. This is particularly important in the western boundary currents since their subsurface layers transport a large amount of nutrients as "nutrient streams" to downstream regions [19–23] that feature major net CO_2 sinks [24]. Along these currents, nutrient concentrations on density surfaces below the euphotic layer are elevated. This is particularly important since the nutrients in the upper layer are susceptible to surface mixing processes, thereby contributing to phytoplankton photosynthesis in the euphotic layer. Several studies of the Gulf Stream have shown that diapycnal mixing in the ocean interior is too slow to account for the observed elevated nutrient concentrations in the Gulf Stream [25, 26]. The Kuroshio Current encounters more topographic features (seamounts, Tokara Strait, and Izu Ridge) than does the Gulf Stream. The interaction of these features with the Kuroshio Current enhances diapycnal mixing (Fig. 5) to a greater degree than similar flow-topography interaction along the Gulf Stream, which only strongly interacts with the New England Seamount Chain. Using state-of-the-art turbulence and biogeochemical profilers and an eddy-resolving numerical simulation in the Kuroshio region, Nagai and Durán Gómez [5] elucidated how inorganic nutrients are transported by the Kuroshio, with special emphasis on diapycnal mixing which maintains the elevated nitrate concentrations on density surfaces along the Kuroshio. In contrast to recent studies of the Gulf Stream, Nagai and Durán Gómez [5] found that diapycnal mixing can account for the observed elevated nitrate concentrations along the Kuroshio. When the Kuroshio Extension approaches the Oyashio Current east of Japan, the proximity of two water masses on density surfaces is conducive to double-diffusive convection, which is another diapycnal mixing process in addition to turbulence. The Kuroshio thus could be very efficient in bringing up subsurface nutrients to the euphotic layer (Fig. 5).

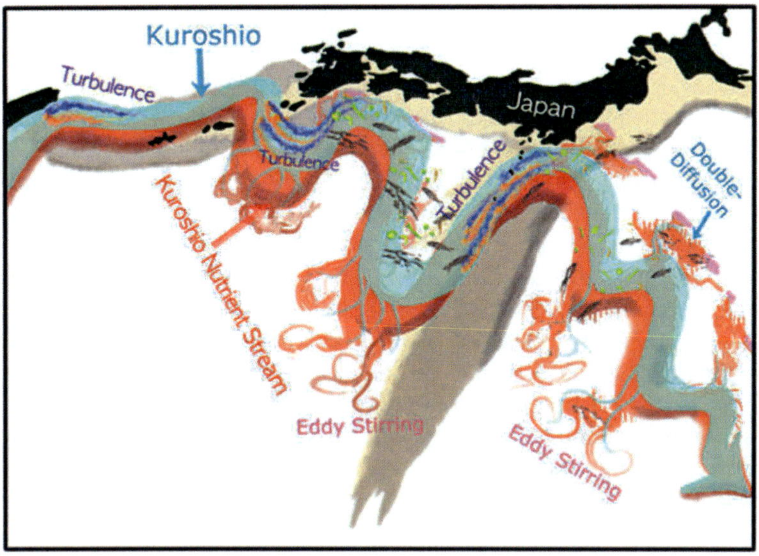

Fig. 5 The Kuroshio Current transports a large amount of nutrients in subsurface layers. The meso- and submesoscale eddies spawned by the Kuroshio front stir these subsurface nutrients along isopycnal surfaces, promoting lateral nutrient transport across the front. As the Kuroshio flows over topographic features south of Japan, its interaction with the topography generates vigorous turbulence and diapycnal mixing that diffuses subsurface nutrients upward, where these nutrients can enrich waters of adjacent continental shelves. As the Kuroshio flows eastward away from Japan as the Kuroshio Extension, its warm and salty water approaches cold and fresh Oyashio water. The spatial juxtaposition of the two contrasting water masses separated by this front is conducive to cross-frontal interleaving, which leads to double-diffusive convection (another diapycnal mixing process). With these multiple diapycnal mixing processes, the Kuroshio brings up nutrients from dark subsurface to sunlit surface layers off the Japanese coast and along the Kuroshio Extension and Kuroshio-Oyashio frontal zone [5]

6 Front-Driven Physical–Biogeochemical–Ecological Interactions in the Yellow Sea Large Marine Ecosystem [6]

Q. S. Wei, M. Z. Fu, X. S. Li, J. C. Sun, B. D. Wang, and Z. G. Yu

Wei et al. [6] describe fronts in the western part of the southern Yellow Sea (SYS) and provide an integrative synthesis of physical–biogeochemical–ecological interactions in the Yellow Sea Large Marine Ecosystem (YSLME), mainly addressing (1) thermohaline, density, turbidity, and nutrient fronts in relation to adjacent water masses; (2) nutrient transport and light conditions associated with the fronts; (3) front-driven primary production regime; and (4) anchovy ecology in the frontal zone. The interaction of coastal water/currents, river discharge, the Yellow Sea Warm Current (YSWC), and the Yellow Sea Cold Water Mass (YSCWM) maintains robust thermohaline and density fronts in the western SYS. Three types of nutrient fronts—estuarine, coastal, and offshore—exist in this region. The nutrient fronts are generally collocated with respective physical fronts. During warm seasons, the water column is stratified; thus, nutrients can accumulate within the YSCWM due to decomposition of organic matter under the pycnocline, resulting in a bottom nutrient pool and a bottom nutrient front around the pool. Upwelling near the bottom YSCWM front in summer leads to surface cold patches, nutricline shoaling, and upward transport of nutrients. A thermohaline and density front along the ~30-m isobath largely prevents the offshore transport of suspended sediments; as a result, the turbid water becomes coastally trapped, leading to a turbidity front. Across this front, turbidity sharply decreases, while light conditions rapidly improve seaward.

The SYS-wide frontal system largely regulates the primary production regime in the YSLME. The front between the southward cold water belt and eastern YSWC shapes the western boundary of the central phytoplankton-blooming area in spring. During the stratified warm seasons, upwelling brings nutrients from the YSCWM to the euphotic layer, resulting in a spatial shift in phytoplankton blooms from the central SYS in spring to the YSCWM frontal region in summer. Phytoplankton biomass also tends to peak near the autumn YSCWM front and winter YSWC front as a result of convergence and improved light conditions. Anchovy distribution is linked to the fronts. In winter, the front adjacent to the YSWC core area at the Yellow Sea entrance acts as an overwintering ground for anchovy. In spring, anchovy migrates northward for feeding, and a high biomass forms in the YSWC-affected phytoplankton bloom area. In summer and autumn, the YSCWM frontal region is an important spawning and nursery ground for anchovy due to food availability and suitable temperature.

The study by Wei et al. [6] reveals that fronts shape physical–biogeochemical processes in the Yellow Sea and determine the complexity of ecosystem dynamics (Fig. 6). The strong front-ecosystem linkages established in this work thus warrant front-related studies of marginal sea ecosystems.

7 Colored Dissolved Organic Matter in Frontal Zones [7]

C. Guéguen and P. Kowalczuk

Guéguen and Kowalczuk [7] reviewed the dynamics of dissolved organic matter (DOM) across frontal zones. DOM is the largest reservoir of organic carbon in the global ocean, with an estimated 662 Pg C [27]. The dominant source of organic matter in the world's ocean is the in situ or autochthonous production, accounting for more than 95% of the total organic matter. The input of terrestrial DOM represents only 2–3% of the total oceanic DOM pool, although it may be a dominant source of DOM in coastal zones [28]. Guéguen and Kowalczuk [7] reviewed how the optical measurements of absorbance and fluorescence are used to track the source, composition, and dynamics of DOM, emphasizing field and satellite studies showing strong

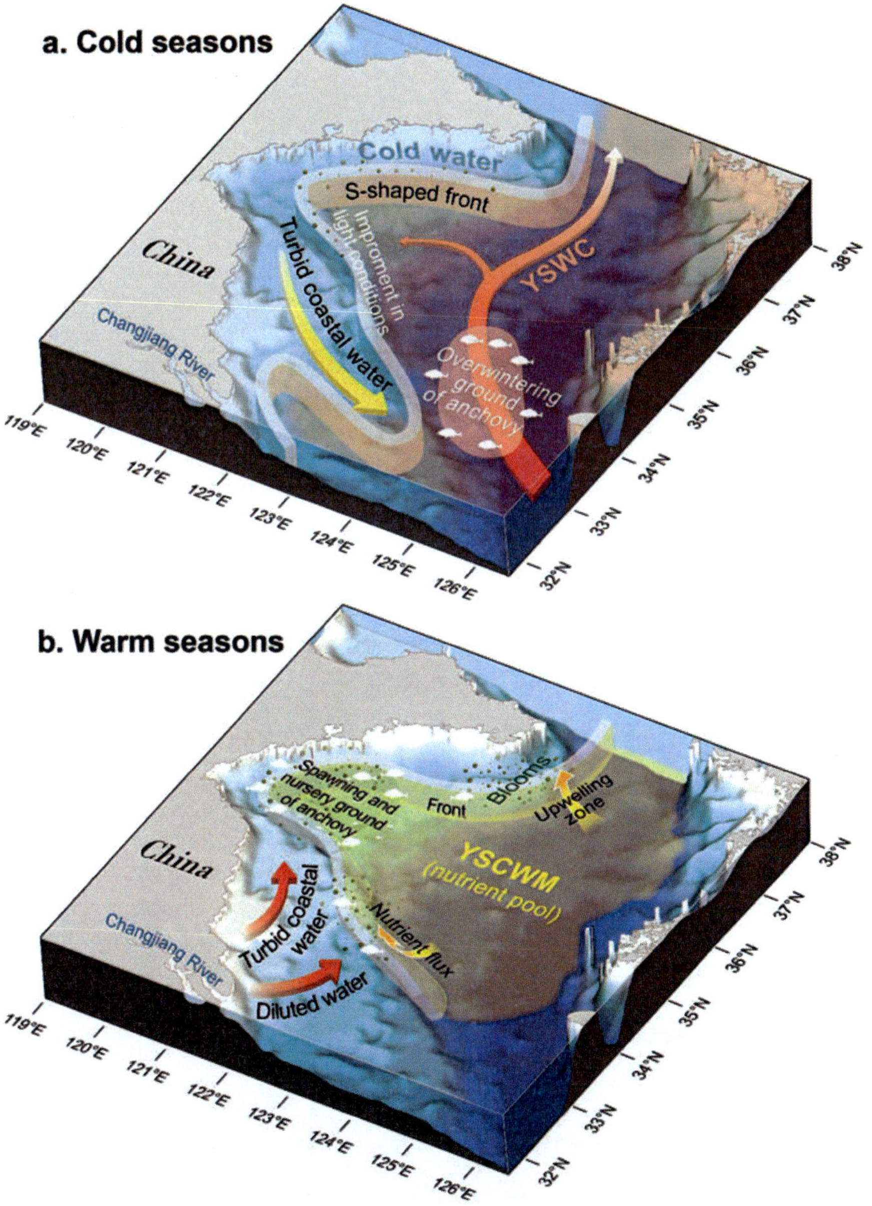

Fig. 6 Schematic diagram illustrating the front-associated, physical–biogeochemical–ecological interactions in the Yellow Sea Large Marine Ecosystem (YSLME) [6]. Robust thermohaline and density fronts are generally formed near the boundary of the Yellow Sea Warm Current (YSWC) during cold seasons (**a**) and the Yellow Sea Cold Water Mass (YSCWM) during warm seasons (**b**). Light conditions, nutrient transport, primary production regime, and the seasonal migration of anchovy (one of the most critical fish species in the YSLME that plays a key "trophic linkage role" in the ecosystem structure) are closely related to the fronts. The turbid water becomes coastally

evidence of DOM dynamics across frontal zones in the Atlantic, Pacific and Arctic Oceans, and in marginal seas such as the Baltic Sea. The frontal transitions featured marked changes in DOM optical properties and anomalies in physical (e.g., salinity and temperature), chemical (e.g., nutrients, $\delta^{18}O$, and dissolved oxygen), and biological properties (e.g., chlorophyll-*a* and primary production). An example of a front that separates two distinct water masses with contrasting DOM optical properties is in Fig. 7. The low temperature and lower salinity water flowing from the Arctic Ocean with the East Greenland Current via the western Fram Strait is rich in CDOM and humic-like FDOM [29–31], while warm and saline Atlantic Water flowing to the Arctic Ocean with the West Spitsbergen Current via the eastern Fram Strait contains low CDOM and low humic-like FDOM. The Atlantic water is more productive than the Polar Water. High chlorophyll-*a* concentrations correspond to a higher intensity of freshly produced protein-like FDOM [32]. The sea ice meltwater (lower salinity) west of the Polar Front dilutes both $a_{CDOM}(350)$ and the humic-like FDOM, I_{CH1} in the surface layer. Details on fluorometric in situ measurements of different fractions of FDOM can be found in [32].

8 Contamination by Persistent Organic Pollutants and Related Compounds in Deep-Sea Ecosystems Along Frontal Zones Around Japan [8]

S. Takahashi, K. Ramu, and S. Tanabe

To protect human health and the environment, national and international control measures on the production and usage of persistent organic pollutants (POPs) have been (or are being) introduced to reduce their emissions to the environment. Though these efforts have resulted in the decline of POPs environmental levels, reservoirs of these pollutants remain in the world's oceans [33]. The oceans play an important role in the cycling and removal of POPs [34]. Pollutants entering the deep oceans are deposited in sediments and can readily accumulate in the food chain. There is a considerable amount of scientific literature indicating that frontal zones are relatively

Fig. 6 (continued) trapped under the frontal system, and the sharp reduction in turbidity across the front provides a rapid seaward improvement in light conditions. Nutrients accumulate within the YSCWM due to decomposition of organic matter under the pycnocline during the stratified warm seasons, resulting in a notable nutrient pool in this region and a pronounced bottom nutrient front around the pool and along the collocated YSCWM boundary. Meanwhile, upwelling exists near the bottom YSCWM front, leading to upward transport of nutrients and a spatial shift in phytoplankton blooms to the YSCWM frontal region in summer. Phytoplankton biomass also tends to peak near the front in winter as a result of convergence and improved light conditions. The front adjacent to the YSWC core area at the Yellow Sea entrance acts as an overwintering ground for anchovy in winter. During warm seasons, the YSCWM frontal region is an important spawning and nursery ground for anchovy due to food availability and suitable temperature

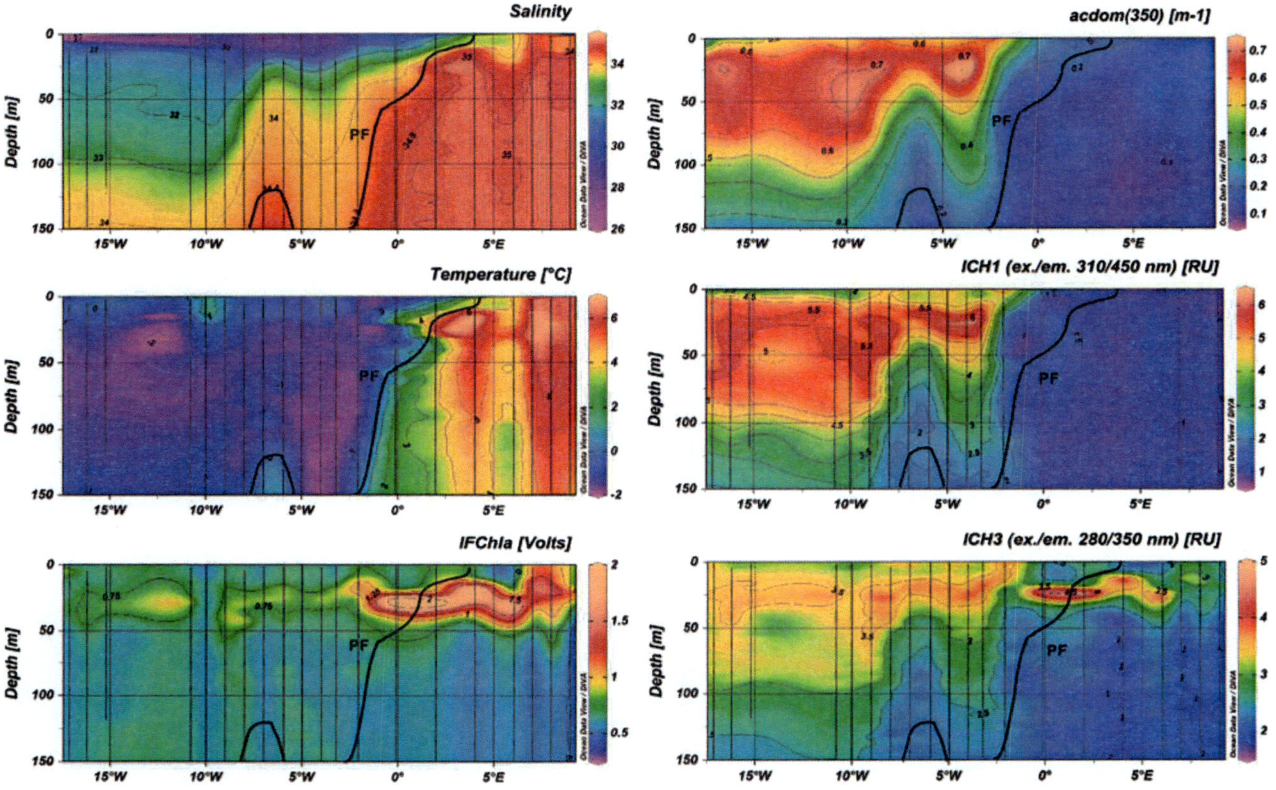

Fig. 7 Section plots across the Fram Strait at 79.9°N in August 2014. Shown are vertical distributions of salinity, temperature, chlorophyll-*a* fluorescence intensity, I_{FChla} (relative instrument readings in volts), optical properties of the dissolved organic matter: CDOM absorption coefficient at 350 nm, $a_{CDOM}(350)$, fluorescence intensity of the humic-like fraction of fluorescent dissolved organic matter, I_{CH1} (in Raman units), and fluorescence intensity of the protein-like fraction of FDOM, I_{CH3} (in Raman units). The Polar Front (PF, bold black lines) is delineated by isotherm 0°C and isohaline 34.4

productive and ecologically significant across the oceans. Nutrient enrichment in frontal zones enhances primary production, which eventually becomes a forage ground for grazers such as copepods, planktivorous fish, and their predators. Convergences associated with frontal zones can accumulate pollutants that strongly adsorb onto man-made neutrally buoyant plastic particles [35], thereby increasing the chances of exposure to these pollutants of marine organisms that forage in these zones.

Studies reviewed by Takahashi et al. [8] suggest vertical transport of POPs along oceanic fronts (Fig. 8). Jamieson et al. [36] and Dasgupta et al. [37] reported high concentrations of POPs in the benthic fauna from the Mariana Trench (11 km deep) and Kermadec Trench (10 km deep) and higher concentrations of PCBs in the Mariana Trench sediments, respectively. Thus, these pollutants penetrate the greatest depths of the oceans.

The Finely Advanced Transboundary Environmental (FATE), 3D dynamic multimedia model for prediction of the global fate and sink of PCBs [38, 39] is used to predict pollution hotspots and critical exposure levels of PCBs in high trophic consumers. Currently, a web interface to visualize and extract selected simulation outputs from FATE is available (https://saviour.create.niigata-u.ac.jp/projects/fov/fov.wsgi/). The model output shows hotspots of PCBs along oceanic frontal zones. Consistent with earlier studies of POPs in skipjack tuna by Ueno et al. [40–42], a recent global study of yellowfin tuna by Nicklisch et al. [43] reported significant variations of POPs, >36-fold on a mass basis, depending on the fish capture locations. Further observations and modeling of POPs' global distribution and fate should identify appropriate and inappropriate locations for harvesting seafood and assess the vulnerability of marine animals regarding POPs, particularly in oceanic frontal zones.

9 Dynamics of Transport, Accumulation, and Export of Plastics at Oceanic Fronts [9]

G. Suaria, M. Berta, A. Griffa, A. Molcard, T. M. Özgökmen, E. Zambianchi, and S. Aliani

The first reports about the accumulation of buoyant plastic material on the ocean surface date back to the early 1970s (e.g., [44]). Since then, the global production of plastics has been steadily growing, together with the massive release of these synthetic materials into the ocean. Synthetic polymers of various sizes, shapes, and typologies are now widespread in the marine environment, from the highly urbanized Mediterranean Sea to the most remote polar waters. What only a few decades ago was an innovative material, prerogative of a few, has now become a widespread solid contaminant, the ubiquitous presence of which has been reported from most marine habitats worldwide. In the last few years, the number of scientific publications on this issue has been skyrocketing [45], and we now have a much

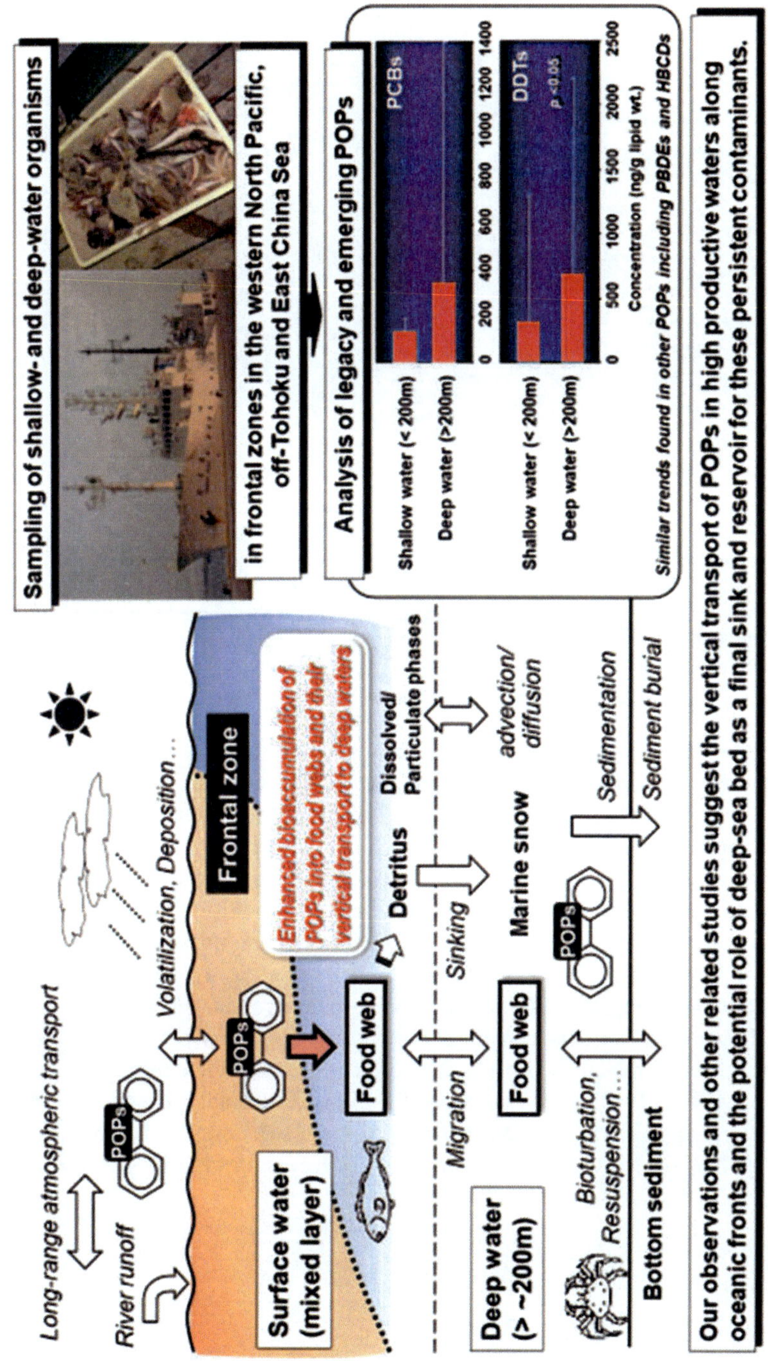

Fig. 8 Transport processes of POPs in deep-sea frontal ecosystems around Japan [8]

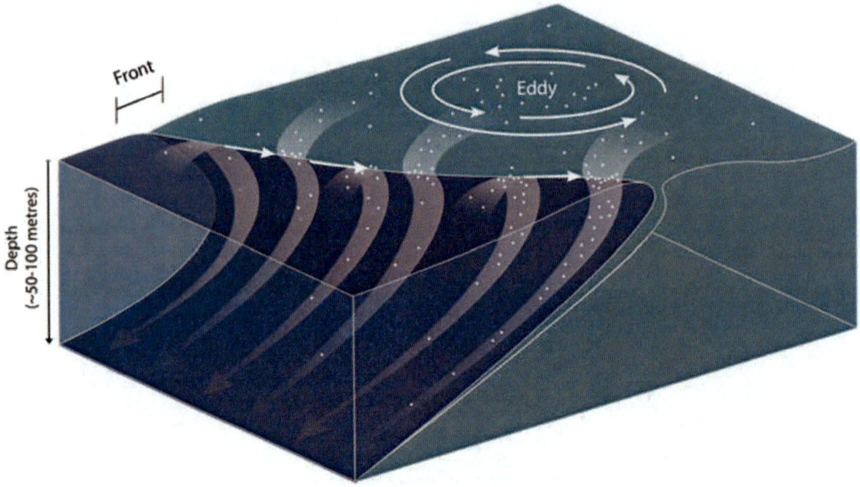

Fig. 9 The processes summarized by Suaria et al. [9] show the potential dynamics of transport, accumulation, and export of small plastic particles in the vicinity of oceanic fronts. In this case, ocean surface currents converge and sink at a density front separating light and heavy water, sweeping small floating particles to the front line where they accumulate. These convergence zones can also create significant downward velocities that are capable of subducting smaller plastic particles, thereby potentially exporting neutrally buoyant plastic particles to the base of the mixed layer. Modified after Fig. 2 of D'Asaro et al. [46]

better understanding on what are the main sources, fates, and impacts of this emerging contaminant. At the same time, we still know relatively little about the physical processes governing the transport of plastics in the marine environment. Even less is known about the sinking dynamics of these particles in the deep sea. Research has shown that frontal systems and their associated convergence zones can accumulate considerable quantities of plastic materials at the ocean surface. However, the complex interactions between oceanic fronts and microplastic subduction mechanisms are still unclear. Suaria et al. [9] review the dynamics of transport, accumulation, and export of plastics at oceanic fronts, providing an accurate description of how oceanic fronts can influence the accumulation of plastic materials, shaping their distribution over the ocean surface, and possibly contributing to their final sinking (Fig. 9). Given the widespread presence of frontal systems in the World Ocean at any spatial scale, the contribution of these structures to the global plastic cycle is potentially very important and clearly deserves further investigation.

10 Lagrangian Methods for Visualizing and Assessing Frontal Dynamics of Floating Marine Litter with a Focus on Tidal Basins [10]

M. Ricker, J. Meyerjürgens, T. H. Badewien, and E. V. Stanev

Marine litter that passively drifts in the ocean is a Lagrangian tracer. It is crucial to understand the litter pathways in the oceans in order to develop appropriate abatement strategies. Studies of the dynamics of marine litter focused largely on the formation and stability of the subtropical garbage patches and the driving physics, such as Stokes drift, geostrophic currents, and Ekman drift. Ocean fronts have been investigated with respect to Lagrangian dynamics. In contrast, marginal seas have more complex dynamics due to riverine freshwater discharge, bottom stress, complex coastlines, and high diffusion rates. Thermohaline fronts can dynamically affect the trajectories of floating matter. Ricker et al. [10] present Lagrangian methods, which enable the visualization and assessment of frontal dynamics with a focus on tidal basins. The main emphasis of this chapter is on providing an overview of suitable methods and demonstration of their specific characteristics by applying some selected methods to the North Sea. The sources of the Lagrangian data are numerical models (Fig. 10a) and satellite-tracked drifters. It is demonstrated that surface-intensified fronts caused by freshwater discharges act as transport barriers leading to the accumulation of virtual particles (Fig. 10b). This accumulation is mainly driven by the so-called finite-domain Lagrangian divergence, the along-path divergence of the virtual particles. Similarly, clusters of satellite-tracked drifters reveal a separation distance decrease while crossing a thermohaline front. In

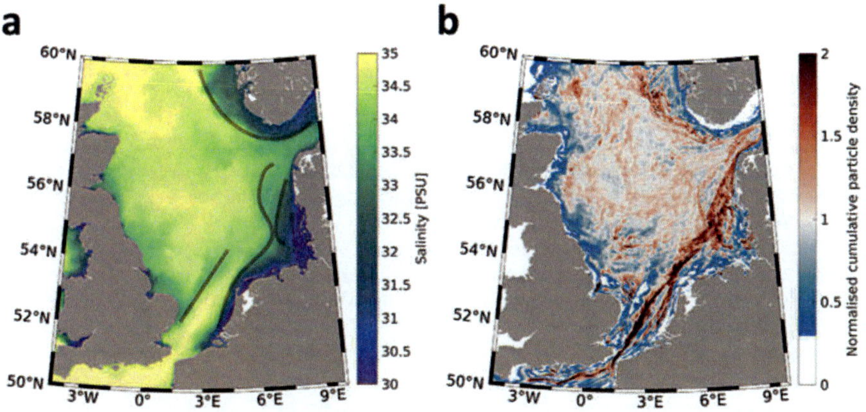

Fig. 10 (**a**) Sea surface salinity in summer 2019 of the North Sea averaged over 1 month. The data are obtained from the operational ocean model AMM15 of the European northwest shelf. Double lines mark the approximate positions of haline fronts. (**b**) Normalized cumulative particle density at the sea surface of the same period represents the accumulation (>1) and dispersal (<1) of virtual particles. Technical details about the data and methods are given in [10]

addition, the importance of tides on the (3-D) Lagrangian particle propagation is shown by analyzing numerical simulations. Although some Lagrangian methods have already been applied to tidal basins, other Lagrangian methods initially developed for other purposes are potentially applicable to fronts in tidal basins. The strengths and weaknesses of each method should be evaluated, and region-dependent simulations should be conducted in order to separate the influence of each process on fronts by taking into account typical spatial and temporal scales. Furthermore, the advantages of exploring merged Lagrangian in situ and model data is emphasized. Despite improved analytics, future experiments could also benefit from such a symbiosis by providing information about interesting regions for in situ drifter experiments and, conversely, confirming model results. As the number of satellite-tracked drifters is often limited in marginal seas, specific methods for sparse trajectory data should be developed.

Acknowledgments Authors of individual chapters express their acknowledgments at the end of their respective chapters. While working on this chapter, Igor Belkin was supported by the Zhejiang Ocean University. The entire chapter was reviewed by all authors. All figures have been either created anew by respective authors of individual chapters or modified after original figures. This chapter was meticulously edited by Daphne Johnson, whose dedication and attention to detail are truly and greatly appreciated.

References

1. Palter JB, Marinov I, Sarmiento JL, Gruber N (2021; online-first in 2013) Large-scale, persistent nutrient fronts of the World Ocean: impacts on biogeochemistry. In: Belkin IM (ed) Chemical oceanography of frontal zones. Handbook of environmental chemistry series, this volume, Springer-Verlag GmbH Germany, Berlin. https://doi.org/10.1007/698_2013_241
2. Alkire MB, Rember R, Polyakov I (2021) The Pacific-Atlantic front in the East Siberian Sea of the Arctic Ocean. In: Belkin IM (ed) Chemical oceanography of frontal zones. Handbook of environmental chemistry series, this volume. Springer-Verlag GmbH Germany, Berlin. https://doi.org/10.1007/698_2021_795
3. Hátún H, Larsen KMH, Eliasen SK, Mathis M (2021) Major nutrient fronts in the northeastern Atlantic: from the subpolar gyre to adjacent shelves. In: Belkin IM (ed) Chemical oceanography of frontal zones. Handbook of environmental chemistry series, this volume. Springer-Verlag GmbH Germany, Berlin. https://doi.org/10.1007/698_2021_794
4. Suursaar Ü, Elken J, Belkin IM (2021) Fronts in the Baltic Sea: a review with a focus on its north-eastern part. In: Belkin IM (ed) Chemical oceanography of frontal zones. Handbook of environmental chemistry series, this volume. Springer-Verlag GmbH Germany, Berlin. https://doi.org/10.1007/698_2021_813
5. Nagai T, Durán Gómez GS (2021) The Kuroshio nutrient stream: where diapycnal mixing matters. In: Belkin IM (ed) Chemical oceanography of frontal zones. Handbook of environmental chemistry series, this volume. Springer-Verlag GmbH Germany, Berlin. https://doi.org/10.1007/698_2021_833
6. Wei QS, Fu MZ, Li XS, Sun JC, Wang BD, Yu ZG (2021) Front-driven physical-biogeochemical-ecological interactions in the southwestern Yellow Sea large marine ecosystem. In: Belkin IM (ed) Chemical oceanography of frontal zones. Handbook of environmental chemistry series, this volume. Springer-Verlag GmbH Germany, Berlin. https://doi.org/10.1007/698_2021_832

7. Guéguen C, Kowalczuk P (2021; online first in 2013) Colored dissolved organic matter in frontal zones. In: Belkin IM (ed) Chemical oceanography of frontal zones, Handbook of environmental chemistry series, this volume, Springer-Verlag GmbH Germany, Berlin. https://doi.org/10.1007/698_2013_244
8. Takahashi S, Ramu K, Tanabe S (2021; online-first in 2014) Contamination by persistent organic pollutants and related compounds in deep-sea ecosystems along frontal zones around Japan. In: Belkin IM (ed) Chemical oceanography of frontal zones, Handbook of environmental chemistry series, this volume, Springer-Verlag GmbH Germany, Berlin. https://doi.org/10.1007/698_2013_252
9. Suaria G, Berta M, Griffa A, Molcard A, Özgökmen TM, Zambianchi E, Aliani S (2021) Dynamics of transport, accumulation, and export of plastics at oceanic fronts. In: Belkin IM (ed) Chemical oceanography of frontal zones. Handbook of environmental chemistry series, this volume. Springer-Verlag GmbH Germany, Berlin. https://doi.org/10.1007/698_2021_814
10. Ricker M, Meyerjürgens J, Badewien TH, Stanev EV (2021) Lagrangian methods for visualizing and assessing frontal dynamics of floating marine litter with a focus on tidal basins. In: Belkin IM (ed) Chemical oceanography of frontal zones. Handbook of environmental chemistry series, this volume. Springer-Verlag GmbH Germany, Berlin. https://doi.org/10.1007/698_2021_812
11. Sarmiento JL, Gruber N, Brzezinski MA, Dunne JP (2004) High-latitude controls of thermocline nutrients and low latitude biological productivity. Nature 427(6969):56–60. https://doi.org/10.1038/nature02127
12. Weber TS, Deutsch C (2010) Ocean nutrient ratios governed by plankton biogeography. Nature 467(7315):550–554. https://doi.org/10.1038/nature09403
13. Lévy M, Resplandy L, Palter JB, Couespel D, Lachkar Z (2021) The crucial contribution of mixing to present and future ocean oxygen distribution. In: Meredith M, Garabato AN (eds) Ocean mixing: drivers, mechanisms and impacts, Chapter 13. Elsevier, Amsterdam, pp 329–344. https://doi.org/10.1016/B978-0-12-821512-8.00020-7
14. Letscher RT, Primeau F, Moore JK (2016) Nutrient budgets in the subtropical ocean gyres dominated by lateral transport. Nat Geosci 9(11):815–819. https://doi.org/10.1038/ngeo2812
15. Palter JB, Ames EJ, Benavides M, Gonçalves Neto A, Granger J, Moisander PH, Watkins-Brandt KS, White AE (2020) High N2 fixation in and near the Gulf Stream consistent with a circulation control on diazotrophy. Geophys Res Lett 47(16):e2020GL089103. https://doi.org/10.1029/2020GL089103
16. Yamamoto A, Palter JB, Dufour CO, Griffies SM, Bianchi D, Claret M, Dunne JP, Frenger I, Galbraith ED (2018) Roles of the ocean mesoscale in the lateral supply of mass, heat, carbon and nutrients to the northern hemisphere subtropical gyres. J Geophys Res Oceans 123(10): 7016–7036. https://doi.org/10.1029/2018JC013969
17. Brambilla E, Talley LD (2008) Subpolar Mode Water in the northeastern Atlantic: 1. Averaged properties and mean circulation. J Geophys Res Oceans 113(4):C04025. https://doi.org/10.1029/2006JC00406
18. Bersch M, Meincke J, Sy A (1999) Interannual thermohaline changes in the northern North Atlantic 1991-1996. Deep-Sea Res II 46(1–2):55–75. https://doi.org/10.1016/S0967-0645(98)00114-3
19. Pelegri JL, Csanady GT (1991) Nutrient transport and mixing in the Gulf stream. J Geophys Res Oceans 96(C2):2577–2583. https://doi.org/10.1029/90JC02535
20. Chen CTA, Liu CT, Pai SC (1995) Variations in oxygen, nutrient and carbonate fluxes of the Kuroshio current. La Mer 33(3):161–176
21. Chen CTA (1996) The Kuroshio intermediate water is the major source of nutrients on the East China Sea continental shelf. Oceanol Acta 19(5):523–527
22. Guo XY, Zhu XH, Wu QS, Huang DJ (2012) The Kuroshio nutrient stream and its temporal variation in the East China Sea. Global Biogeochem Cycles 117(1):C01026. https://doi.org/10.1029/2011JC007292

23. Guo XY, Zhu XH, Long Y, Huang DJ (2013) Spatial variations in the Kuroshio nutrient transport from the East China Sea to south of Japan. Biogeosciences 10(10):6403–6417. https://doi.org/10.5194/bg-10-6403-2013
24. Takahashi T, Sutherland SC, Wanninkhof R, Sweeney C, Feely RA, Chipman DW, Hales B, Friederich G, Chavez F, Sabine C, Watson A, Bakker DCE, Schuster U, Metzl N, Yoshikawa-Inoue H, Ishii M, Midorikawa T, Nojiri Y, Körtzinger A, Steinhoff T, Hoppema M, Olafsson J, Arnarson TS, Tilbrook B, Johannessen T, Olsen A, Bellerby R, Wong CS, Delille B, Bates NR, de Baar HJW (2009) Climatological mean and decadal change in surface ocean pCO_2, and net sea-air CO_2 flux over the global oceans. Deep-Sea Res II 56(8–10):554–577. https://doi.org/10.1016/j.dsr2.2008.12.009
25. Palter JB, Lozier MS (2008) On the source of Gulf Stream nutrients. J Geophys Res Oceans 113 (C6):C06018. https://doi.org/10.1029/2007JC004611
26. Whitt DB (2019) On the role of the Gulf stream in the changing Atlantic nutrient circulation during the 21st century. In: Nagai T, Saito H, Suzuki K, Takahashi M (eds) Kuroshio current: physical, biogeochemical, and ecosystem dynamics, Chapter 4. American Geophysical Union, Wiley, Washington, Hoboken, pp 51–82. https://doi.org/10.1002/9781119428428.ch4
27. Hansell DA, Carlson CA, Repeta DJ, Schlitzer R (2009) Dissolved organic matter in the ocean: a controversy stimulates new insights. Oceanography 22(4):52–61. https://doi.org/10.5670/oceanog.2009.109
28. Opsahl S, Benner R (1997) Distribution and cycling of terrigenous dissolved organic matter in the ocean. Nature 386(6624):480–482. https://doi.org/10.1038/386480a0
29. Granskog MA, Stedmon CA, Dodd PA, Amon RMW, Pavlov AK, de Steur L, Hansen E (2012) Characteristics of colored dissolved organic matter (CDOM) in the Arctic outflow in the Fram Strait: assessing the changes and fate of terrigenous CDOM in the Arctic Ocean. J Geophys Res Oceans 117(12):C12021. https://doi.org/10.1029/2012JC008075
30. Granskog MA, Pavlov AK, Sagan S, Kowalczuk P, Raczkowska A, Stedmon CA (2015) Effect of sea-ice melt on inherent optical properties and vertical distribution of solar radiant heating in Arctic surface waters. J Geophys Res Oceans 120(10):7028–7039. https://doi.org/10.1002/2015JC011087
31. Pavlov AK, Granskog MA, Stedmon CA, Ivanov BV, Hudson SR, Falk-Petersen S (2015) Contrasting optical properties of surface waters across the Fram Strait and its potential biological implications. J Mar Syst 143:62–72. https://doi.org/10.1016/j.jmarsys.2014.11.001
32. Makarewicz A, Kowalczuk P, Sagan S, Granskog MA, Pavlov AK, Zdun A, Borzycka K, Zabłocka M (2018) Characteristics of chromophoric and fluorescent dissolved organic matter in the Nordic seas. Ocean Sci 14(3):543–562. https://doi.org/10.5194/os-14-543-2018
33. Lohmann R, Jurado E, Pilson MEQ, Dachs J (2006) Oceanic deep water formation as a sink of persistent organic pollutants. Geophys Res Lett 33(12):L12607. https://doi.org/10.1029/2006GL025953
34. Palm A, Cousins I, Gustafsson O, Axelman J, Grunder K, Broman D, Brorström-Lundén E (2004) Evaluation of sequentially-coupled POP fluxes estimated from simultaneous measurements in multiple compartments of an air-water-sediment system. Environ Pollut 128(1–2): 85–97. https://doi.org/10.1016/j.envpol.2003.08.023
35. Lohmann R, Belkin IM (2014) Organic pollutants and ocean fronts across the Atlantic Ocean: a review. Prog Oceanogr 128:172–184. https://doi.org/10.1016/j.pocean.2014.08.013
36. Jamieson AJ, Malkocs T, Piertney SB, Fujii T, Zhang Z (2017) Bioaccumulation of persistent organic pollutants in the deepest ocean fauna. Nat Ecol Evol 1(3):0051. https://doi.org/10.1038/s41559-016-0051
37. Dasgupta S, Peng X, Chen S, Li J, Du M, Zhou YH, Zhong G, Xu H, Ta K (2018) Toxic anthropogenic pollutants reach the deepest ocean on earth. Geochem Perspectives Lett 7:22–26. https://doi.org/10.7185/geochemlet.1814
38. Kawai T, Handoh IC, Takahashi S (2009) The rise of the finely-advanced transboundary environmental model (FATE): a state-of-the-art model prediction of the global sink of persistent organic pollutants. Organohalogen Compd 71(1610):e1615

39. Handoh IC, Kawai T (2014) Modelling exposure of oceanic higher trophic-level consumers to polychlorinated biphenyls: pollution 'hotspots' in relation to mass mortality events of marine mammals. Mar Pollut Bull 85(2):824–830. https://doi.org/10.1016/j.marpolbul.2014.06.031
40. Ueno D, Takahashi S, Tanaka H, Subramanian A, Fillmann G, Nakata H, Lam PKS, Zheng J, Muchtar M, Prudente M, Chung KH, Tanabe S (2003) Global pollution monitoring of PCBs and organochlorine pesticides using skipjack tuna as a bioindicator. Arch Environ Contam Toxicol 45(3):378–389. https://doi.org/10.1007/s00244-002-0131-9
41. Ueno D, Kajiwara N, Tanaka H, Subramanian A, Fillmann G, Lam PKS, Zheng GJ, Muchtar M, Razak H, Prudente M, Chung KH, Tanabe S (2004) Global pollution monitoring of polybrominated diphenyl ethers using skipjack tuna as a bioindicator. Environ Sci Technol 38(8):2312–2316. https://doi.org/10.1021/es035323k
42. Ueno D, Alaee M, Marvin C, Muir DCG, Macinnis G, Reiner E, Crozier P, Furdui VI, Subramanian A, Fillmann G, Lam PKS, Zheng GJ, Muchtar M, Razak H, Prudente M, Chung KH, Tanabe S (2006) Distribution and transportability of hexabromocyclododecane (HBCD) in the Asia-Pacific region using skipjack tuna as a bioindicator. Environ Pollut 144(1): 238–247. https://doi.org/10.1016/j.envpol.2005.12.024
43. Nicklisch SCT, Bonito LT, Sandin S, Hamdoun A (2017) Mercury levels of yellowfin tuna (*Thunnus albacares*) are associated with capture location. Environ Pollut 229:89–93. https://doi.org/10.1016/j.envpol.2017.05.070
44. Carpenter EJ, Smith KL (1972) Plastics on the Sargasso Sea surface. Science 175(4027): 1240–1241. https://doi.org/10.1126/science.175.4027.1240
45. Ryan PG (2015) A brief history of marine litter research. In: Bergmann M, Gutow L, Klages M (eds) Marine anthropogenic litter, Chapter 1. Springer International Publishing, Cham, pp 1–25. https://doi.org/10.1007/978-3-319-16510-3_1
46. D'Asaro EA, Shcherbina AY, Klymak JM, Molemaker J, Novelli G, Guigand CM, Haza AC, Haus BK, Ryan EH, Jacobs GA, Huntley HS, Laxague NJM, Chen SY, Just F, McWilliams JC, Barkan R, Kirwan Jr AD, Poje AC, Özgökmen TM (2018) Ocean convergence and the dispersion of flotsam. Proc Natl Acad Sci U S A 115(6):1162–1167. https://doi.org/10.1073/pnas.1718453115

Large-Scale, Persistent Nutrient Fronts of the World Ocean: Impacts on Biogeochemistry

Jaime B. Palter, Irina Marinov, Jorge L. Sarmiento, and Nicolas Gruber

Abstract This chapter identifies and describes the large-scale nutrient fronts that span the width of basins and explores the processes that maintain these fronts and those that act against them. In particular, we investigate the nutrient fronts that ring the subtropical gyres and propose that exchange across these fronts represents a critical pathway for nutrients to enter the gyres. However, these biogeochemical fronts most often coincide with dynamical fronts or jets, which are often considered barriers to exchange. Therefore, our view of ocean fronts as nutrient gateways must be reconciled with their tendency to act as barriers to exchange. Ekman transport is one mechanism that allows for nutrient transport across the surface of the fronts and is shown to be a leading term in the subtropical nutrient budgets. Ring formation and mixing beneath the core of jets are other mechanisms that can mediate cross-frontal exchange and have intriguing implications for nutrient budgets and their variability.

Keywords Lateral exchange, Nutrients, Ocean fronts, Productivity

---

J.B. Palter (✉)
Department of Atmospheric and Oceanic Sciences, McGill University, Montreal, QC, Canada
e-mail: jaime.palter@mcgill.ca

I. Marinov
Department of Earth and Environmental Science, University of Pennsylvania, Philadelphia, PA, USA
e-mail: imarinov@sas.upenn.edu

J.L. Sarmiento
Atmospheric and Oceanic Sciences Program, Princeton University, Princeton, NJ, USA
e-mail: jls@princeton.edu

N. Gruber
Environmental Physics, Institute of Biogeochemistry and Pollutant Dynamics, ETH Zurich, Zurich, Switzerland
e-mail: Nicolas.gruber@env.ethz.ch

Igor M. Belkin (ed.), *Chemical Oceanography of Frontal Zones*,
Hdb Env Chem (2022) 116: 25–62, DOI 10.1007/698_2013_241,

Published online: 12 April 2013

Contents

1 Introduction

Ocean fronts separate dynamically distinct regions of the ocean and are characterized by sharp lateral property gradients (Figs. 1 and 2). The sharpness of these fronts arises as an inherent feature of fluid motion in the presence of stratification and rotation, and narrow fronts are common features of the ocean and atmosphere [1, 2]. Most, but not all, ocean fronts of biogeochemically important constituents, such as nutrients and carbon, coincide with density fronts, narrow regions where density surfaces slope towards the ocean's surface. Such spatial coincidence is most simply explained by the action of the biological pump, which strips nutrients and carbon from the low-density surface waters and transports them as organic matter to denser waters at depth. Thus, just as density increases with ocean depth, so too do nutrients and carbon, and where density surfaces rise in depth, the isosurfaces of these biogeochemical properties rise with them. Indeed, it is difficult to visually distinguish between maps of horizontal gradients of density and nutrients on a depth surface (Figs. 1 and 2).

On either side of these sloping density surfaces or isopycnals, different dynamical and biogeochemical regimes create water masses with dissimilar properties. Therefore, a biogeochemical front does not arise solely from sloping isopycnals but also from property gradients on the isopycnals themselves. These isopycnal property gradients generally reflect the boundaries between different water masses with distinct biogeochemical histories. Differences between density fronts and nutrient fronts also arise because density is a dynamical tracer that influences the velocity field. In situations where temperature and salinity changes are density compensating, nutrients and other tracers can have sharp gradients while density remains essentially constant [e.g., 3, 4]. An example of a nutrient front on a nearly constant density surface is found on the $\sigma_\theta = 26.5$ isopycnal (equivalent to a seawater density of 1,026.5 kg m^{-3}), where strong lateral gradients in nitrate and

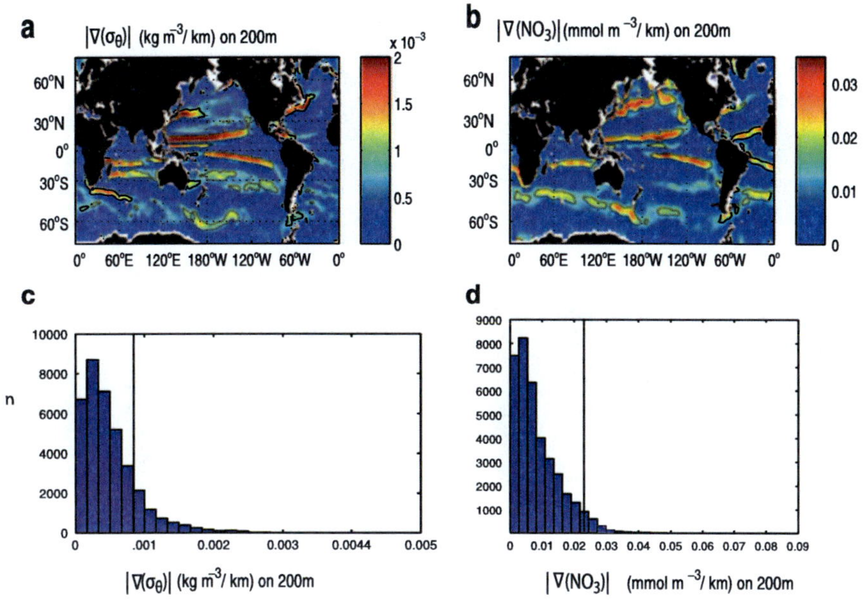

Fig. 1 The lateral gradients of potential density (**a**) and nitrate (**b**) at 200 m constructed from World Ocean Atlas 2009 annual means (WOA09 [87], found online at: http://www.nodc.noaa.gov/OC5/WOA09/pr_woa09.html). The contour that corresponds with the 85th percentile of the gradient (i.e., 85% of the gradients are smaller than this contour) is outlined in black. Panels (**c**) and (**d**) are histograms of the lateral gradients at 200 m with the 85th percentile shown as a black line. The histograms demonstrate that over most of the ocean the lateral density and nitrate gradients are very small, but their distributions are strongly skewed. The high gradients that form the right tail of the histograms in the lower panels are clearly organized into narrow fronts, as shown in the maps of the *upper panels*

oxygen mark the water mass boundaries that separate the inner tropics from the mid latitudes at 20°S and 20°N (Fig. 3).

Property fronts at the boundaries of water masses are often intensified because steep isopycnal slopes are associated with strong geostrophic jets, which are known to create dynamic barriers to exchange and further isolate the water masses on either side of a front [5–7]. The low- to mid-latitude subtropical gyres are regions where this barrier effect might be particularly important. These regions are laterally homogeneous, well stratified, and bounded by jets that tend to act as barriers to exchange. In the hypothetical extreme case of a subtropical gyre bounded by fronts that inhibit all lateral exchange, property budgets within the gyre would have to balance vertically. This requirement becomes clear when we consider a hypothetical Lagrangian parcel circuiting this gyre, as depicted schematically in Fig. 4. In the absence of lateral exchange across the streamlines that define this gyre, any net downward property flux across the bottom of that parcel's maximum mixed layer depth must be restored by a corresponding upward flux to maintain steady-state concentrations on timescales longer than a few years [8].

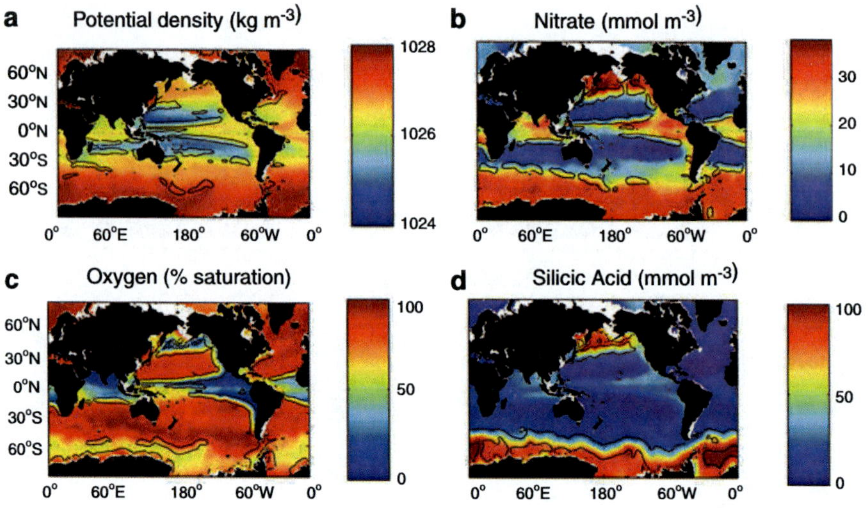

Fig. 2 Properties on 200 m in color with the 85th percentile of their gradients contoured in *black* (i.e., 85% of the gradients are smaller than this contour). (**a**) Potential density referenced to the surface (kg m^{-3}); (**b**) nitrate (mmol m^{-3}); (**c**) oxygen (percent saturation); (**d**) silicic acid (mmol m^{-3})

Such a requirement for a vertical property balance would be problematic for near-surface biology to maintain its productivity, as biology typically represents a one-way street for nutrients: phytoplankton transform nutrients from the dissolved inorganic form into organic particulate forms, which sink from the illuminated upper ocean constituting the euphotic zone. Sediment trap studies suggest that 10–20% of the organic particles sinking from the base of the euphotic zone continue past a depth of a thousand meters in the tropics and subtropics [9–11], leading to a net loss of these nutrient elements from the subtropical gyre system. Local vertical processes in the stratified, downwelling subtropical gyres cannot restore these lost nutrients. Instead, we suggest a two-step process for returning these nutrients to the euphotic zone of the gyres. First, upwelling and vertical mixing in the regions surrounding the gyre bring the nutrients to shallower depths. Second, advection and mixing across the fronts then resupply these nutrients to the gyre regions where they were originally lost by vertical export. Indeed, cross-frontal exchange of nutrients has been shown to be critically important in fueling primary productivity in low latitudes. In particular, the equatorward Ekman transport of nutrients across the northern subtropical boundaries is an important source of nutrients to the subtropical North Atlantic and North Pacific [8, 12, 13].

It has also been hypothesized that exchange across the Antarctic Circumpolar Current (ACC) plays a critical role in restoring nutrients lost to depths beneath the global pycnocline and maintaining the biological productivity of much of the mid- and low-latitude ocean [14–17]. It is argued that without this Southern Ocean nutrient source, the sum of the vertical processes that transport nutrients upwards across the thermocline everywhere outside the Southern Ocean would be insufficient

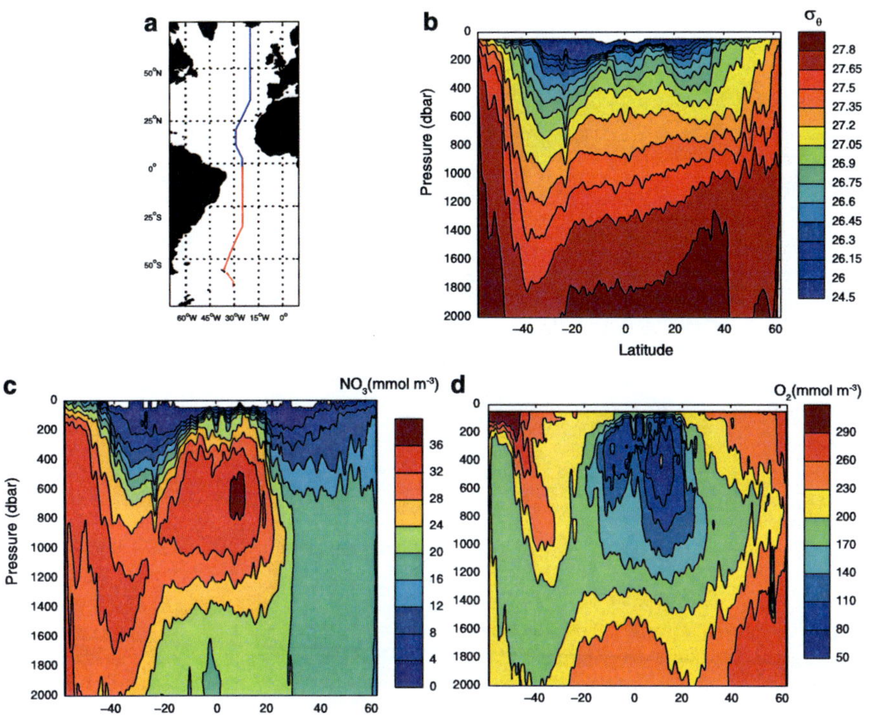

Fig. 3 Meridional section of density nutrients and oxygen through the Atlantic. (**a**) The station locations of WOCE section A16n occupied in June–August 2003 (*blue*) and A16s occupied January–February 2005 (*red*). These stations have been stitched together to plot a continuous section as a function of depth and latitude: (**b**) Potential density anomaly referenced to the surface. (**c**) Nitrate (mmol m^{-3}), (**d**) oxygen (mmol m^{-3}). Nitrate and oxygen fronts generally coincide with density fronts (as is also apparent in Fig. 2). In addition, water mass boundaries give rise to biogeochemical fronts along isopycnals, most apparent near 20°N and 20°S where gradients in nitrate and oxygen are much stronger than in potential density

to restore all of the nutrients that slowly leak across its base. Instead, wind-driven upwelling in the Southern Ocean and the cross-ACC Ekman transport of nutrients is invoked as a restoring mechanism for nutrients to the global thermocline [14]. Therefore, cross-frontal exchange appears to be critical in determining the transport of nutrients back to the near-surface ocean, controlling the productivity of the mid- and low latitudes to a substantial degree.

Here we attempt to reconcile these two views of fronts as the barriers and the gateways to property exchange. By synthesizing a variety of hydrographic data in a dynamical context, we address the questions: Where are the ocean's most important biogeochemical fronts? What are the dynamical and biogeochemical processes that maintain these fronts? How does exchange across them shape biogeochemical processes in the relatively homogenous ocean gyres bounded by such fronts? In exploring these questions, we first define the ocean's large-scale biogeochemical

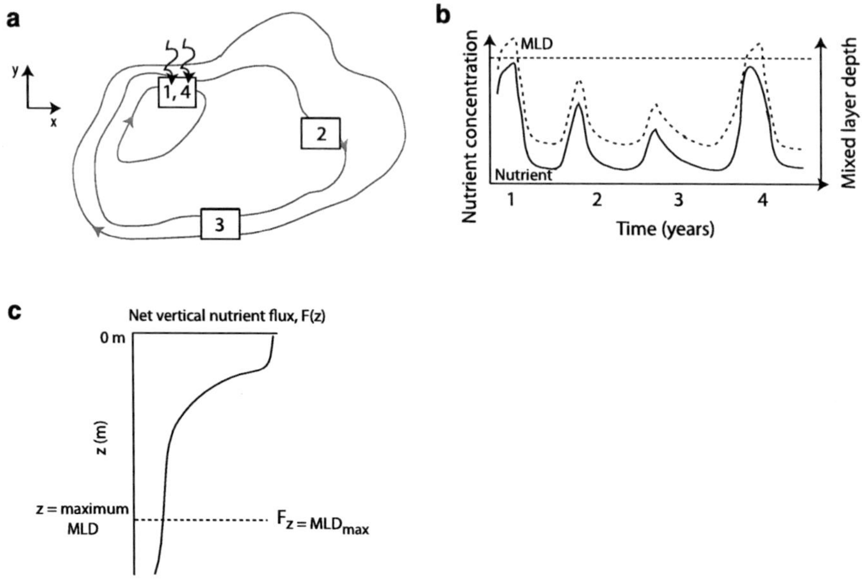

Fig. 4 A schematic for a hypothetical parcel transiting a subtropical gyre in the ocean mixed layer, inspired by Williams and Follows [13]. (**a**) The parcel transits a closed loop in a subtropical gyre in 4 years, with each year labeled in the box representing the parcel. (**b**) It starts in its most poleward position, where mixed layers (*dotted line*) are at a maximum along its trajectory and the seasonal vertical entrainment flux creates high mixed layer nutrient concentrations (*solid line*). In subsequent winters, shoaling of the mixed layer depth along the parcel's trajectory reduces the entrainment flux from below and gives rise to a decline in mixed layer nutrient concentrations. (**c**) There is a net vertical flux of nutrients (i.e., the downward export of nutrients exceeds the upward vertical mixing and entrainment of nutrients) to depths beneath the maximum mixed layer along the parcel's trajectory. Thus, a lateral supply of nutrients is required to restore nutrient concentrations to their initial concentrations at the parcel's initial position. We propose that a cross-frontal nutrient flux (*curvy arrows* in panel *a*) is a critical supply pathway by which nutrients are restored to the mixed layer

fronts and explore the dynamical and biogeochemical regimes that maintain them. Here, we look primarily at the macronutrient fronts of nitrate, phosphate, and silicic acid. Iron is also an important limiting nutrient, but observations are still too sparse to identify large-scale fronts in iron. Next, we investigate the role of cross-frontal exchange in providing nutrients to the homogenous regions bounded by fronts. Williams and Follows [8] provide an excellent reference for the mechanisms by which cross-frontal exchange helps maintain biological production. Our exploration is therefore focused on identifying the specific locales in the ocean where fronts separate water masses with strongly contrasting biogeochemical signatures and on providing simple diagnostics of their potential impact on biogeochemical processes.

2 A Description of the Ocean's Biogeochemical Fronts

The vast majority of the ocean's volume is characterized by strong vertical stratification and weak lateral property gradients, at least when averaged over several mesoscale eddy time and length scales, i.e., timescales of months and length scales of several hundreds of kilometers. In contrast, fronts have lateral property gradients an order of magnitude greater than their surrounding regions (Fig. 1). Our focus is on the large-scale, persistent, biogeochemical fronts of the open ocean. Before launching our discussion of these fronts, we recognize the important biogeochemical fronts in the coastal ocean that fall beyond the scope of this chapter. These include the fronts that arise at the edges of coastal upwelling regions, where the thermocline and nutricline break the ocean's surface under the influence of an alongshore wind stress [18]. Three additional coastal fronts can be found at the edges of shallow shelf seas: a freshwater front, influenced by the input of fresher estuarine water into the coastal zone; the shelf sea tidal mixing front, caused by the competition between surface buoyancy forcing and tidal mixing; and the shelf-break front, a site of enhanced tidal and wind-driven mixing where the relatively flat continental shelf transitions to the steeper incline of the continental slope [19, 20]. All of the shelf sea fronts divide coastal waters from the open ocean. Because coastal waters are generally more nutrient rich than the offshore waters, the shelf sea fronts often mark stark biogeochemical boundaries. Furthermore, strong mixing at these fronts is thought to provide nutrients to the euphotic zone and enhance productivity.

Our discussion of basin-scale biogeochemical fronts is guided by the maps of density, nutrients, and oxygen at 200 m (Figs. 1 and 2). We examine property gradients at 200 m to focus on persistent fronts that have a near-surface expression: the fronts at the edges of the ocean's oxygen minimum zones, those at the subtropical-subpolar boundaries, and those circumnavigating the globe in the Southern Ocean. It is across these fronts that mixing and advection bring nutrients into the euphotic zone to directly influence primary productivity in the neighboring regions. By looking at 200 m instead of directly at the surface, we avoid dealing with spatial variability driven exclusively by surface processes, such as biological drawdown of nutrients and air-sea heat and freshwater exchange. Shelf-break fronts are typically found at depths shallower than 200 m, so the maps of Figs. 1 and 2 will not capture these features. Vertical sections suggest that there are also biogeochemical fronts deeper in the ocean interior that fall outside the scope of this chapter (e.g., at the boundary separating oxygen-poor AAIW from oxygen-rich Mediterranean waters at approximately 20°N in the Atlantic as seen in Fig. 3d).

In many regions, notable features in satellite-derived chlorophyll concentration are coincident with the nutrient fronts, providing an additional motivation to understand what sets the locations of these fronts. In particular, the transition region between the low-chlorophyll North Atlantic and North Pacific subtropical gyres and their neighboring chlorophyll-rich regions coincide with strong nutrient fronts (Fig. 5). In the open waters of the Indian and Atlantic sectors of the Southern Ocean, the chlorophyll concentration maxima overlap with the nutrient fronts (Fig. 5).

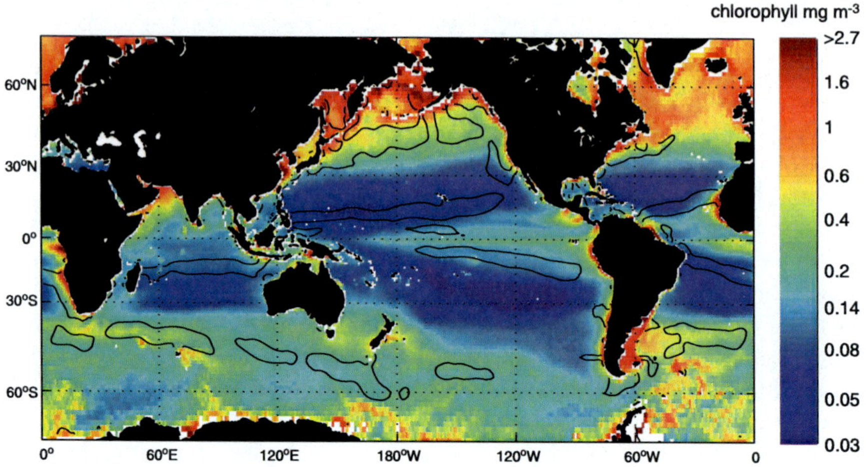

Fig. 5 Satellite-observed chlorophyll from SeaWifs Level 3 data (*colors*) averaged over the period 1997–2009. Because chlorophyll concentrations in the global ocean are distributed lognormally [111], concentrations are plotted on a log scale. Black contours show the 85th percentile of the nitrate gradient (as in Fig. 2b)

We begin in the next section by describing several of the large-scale, persistent biogeochemical fronts of the global ocean: at the edges of oxygen minimum zones (Sect. 2.1), across western boundary currents of the North Pacific and North Atlantic (Sect. 2.2), and across the ACC (Sect. 2.3). In Sect. 2, we also briefly discuss the balance between the mechanisms that maintain these sharp gradient regions and the processes that transport biogeochemical constituents across them. In Sect. 3, we explore the far-reaching consequences on nutrient and carbon budgets of cross-frontal exchange in the ACC region. Section 4 attempts to reconcile the two views of fronts, as both barriers and gateways to property exchange, and quantifies cross-frontal nutrient supply relative to other supply pathways of nutrients in several ocean regions. Finally, we summarize and conclude in Sect. 5.

2.1 *The Edges of Oxygen Minimum Zones*

Among the most visually striking biogeochemical fronts are those found at the edges of the ocean's oxygen minimum zones (OMZs) in the tropical oceans and North Pacific (Figs. 2c and 6). These fronts are produced by the interplay of biology and physics. In the ocean interior, oxygen is consumed by the remineralization of organic matter and it is replenished by the advection and mixing of oxygen into the domain. It is no coincidence that oxygen minimum zones coincide with the ocean's so-called shadow zones where subsurface streamlines recirculate without a direct connection to the surface ocean. Whereas the subsurface layers of the subtropical gyre are oxygenated by the formation and spreading of mode waters, the shadow zone is found in the space created between the subtropical gyre circulation and the

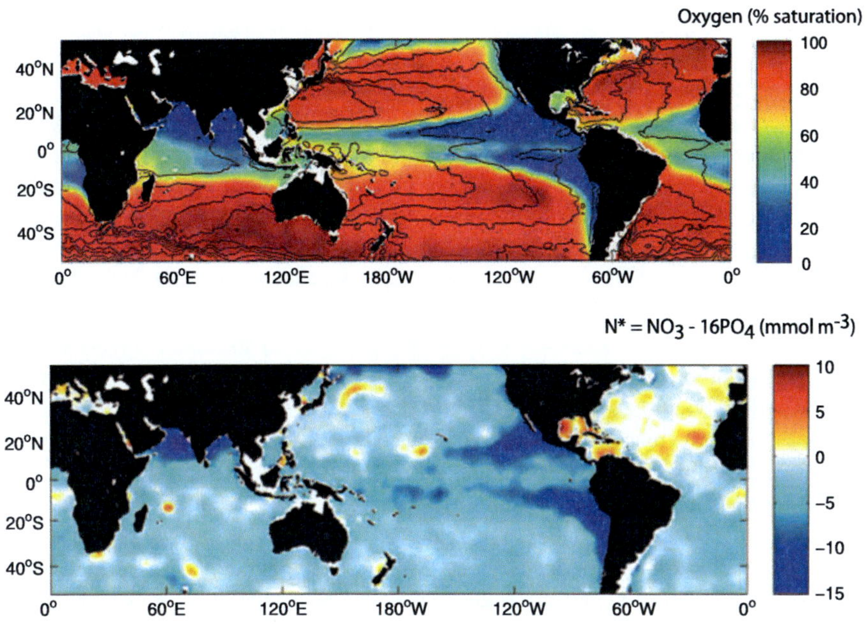

Fig. 6 Oxygen (percent saturation) and N* (mmol m^{-3}) on 200 m with sea surface height (mean dynamic topography) from the altimetry-based climatology of Rio and Hernandez [23]. Contours of the mean dynamic topography resemble geostrophic streamlines at the ocean's surface

basin's eastern boundary. Here, exchange with the neighboring subtropical gyre is slow and the subsurface layers are shielded from exchange with the overlying atmosphere by a shallow surface mixed layer.

The shadow zone circulation arises from constraints on the potential vorticity (PV) budget, which cause the geostrophic streamlines that close the subtropical circulation to pull away from the basins' eastern boundaries [21, 22]. For flows with low Rossby number (i.e., in flows where the Coriolis term is much larger than the relative vorticity), PV is approximately equal to f/H, where f is the planetary vorticity and H is the thickness of an isopycnal layer. In the nearly inviscid ocean interior, potential vorticity is conserved. Therefore, along the trajectory of a fluid parcel, the quotient f/H remains nearly constant and changes only through slow mixing with surrounding fluids. Applying this PV conservation constraint to a fluid parcel following the eastern equatorward-flowing branch of the subtropical gyre circulation requires that the layer thickness (H) decreases in order to compensate for the decrease in the planetary vorticity (f). However, any meridional change in layer thickness creates a meridional pressure gradient and a corresponding zonal geostrophic flow. To avoid flow into or out of a basin's eastern boundary, the eastern subtropical circulation is deflected westward [22]. We see manifestations of this physical constraint in ocean observations: contours of sea surface height [23] are clearly deflected to the west at the eastern boundary of the subtropical oceans (Fig. 6). This westward deflection shows that the surface geostrophic flow closing the anticyclonic, subtropical gyre circulation is found at a distance from the basin's eastern boundary. The shadow zone extends eastwards from the edge of the gyre below a shallow, oxygenated surface layer.

Between the westward deflected streamlines and the continents, the residence time in the thermocline is at a maximum for each basin [24–26]. Overlying and adjacent to these shadow zones are highly productive eastern boundary upwelling systems, where the upwelling of nutrients fuels photosynthetic production of organic matter, a fraction of which is exported to the thermocline. Thus, the oxygen minimum zones result from the combined influence of organic matter export and remineralization in the thermocline and a lack of a direct advective connection with the surface ocean. The situation is quite the contrary in the upper layers of the neighboring subtropical gyres. Here well-oxygenated water masses are formed by wintertime convection at the gyres' poleward edges. These water masses, called subtropical mode waters [27], are transferred into the thermocline, efficiently ventilating its upper layers [21, 28]. Furthermore, low-nutrient conditions in the surface of the subtropical gyres support lower rates of organic matter export than in the upwelling regions [29]. The boundary between these dissimilar physical and biogeochemical regimes marks the ocean's strongest oxygen fronts.

The edges of the oxygen minimum zones are also locales of strong fronts in nitrate and phosphate (Fig. 1 and 2). A basic explanation for these nutrient fronts relies on the same processes as the oxygen fronts. In the subsurface shadow zones, the remineralization of organic matter produces large pools of inorganic nutrients whose principal sink is realized through mixing with lower-nutrient waters in the neighboring gyre [30, 31]. Whereas nutrients in the subtropical gyre are kept low via exchange with the productive surface layer in mode water formation regions [32], their accumulation in the OMZ reflects the lack of physical connection between the subsurface shadow zone and the productive surface layer. In the OMZs of the eastern tropical Pacific Ocean and the northernmost Indian Ocean, there is a sink for nitrate: the microbial respiration of nitrate and/or ammonium to gaseous nitrogen, a process known as denitrification. Because denitrification is slightly less energetically favorable than aerobic respiration, it occurs in the eastern tropical Pacific Ocean and the northern tropical Indian Ocean where dissolved oxygen concentrations fall below 10 μmol kg^{-1}, but not in the eastern tropical Atlantic Ocean where the minimum oxygen concentrations typically exceed 30 μmol kg^{-1} [33, 34].

Whereas nitrate (NO_3^-) and phosphate (PO_4^{3-}) are typically utilized and remineralized in a ratio that averages 16:1, denitrification removes only nitrate, leaving phosphate concentrations unchanged. Thus, a useful tracer that reflects the occurrence of denitrification is $N^* = NO_3^- - 16PO_4^{3-}$ [35]. The depressed N* at 200 m in the OMZs reflects the denitrification that occurs in ocean's lowest oxygen waters (Fig. 6b). At the edges of the OMZ, an N* front reflects the boundary between the nitrate-depleted water from the center of the OMZs with the relatively NO_3-rich neighboring regions. Leakage of low-N* water from the Pacific and Indian OMZs clearly influences the mid-depths of the basins far beyond the edges of the ocean's shadow zones. In the North Atlantic a front divides the N* maximum waters of the subtropical gyre from the tropical regions. This front is thought to divide the salty subtropical waters where N* is added via nitrogen fixation from the fresher tropical waters in the North Equatorial Current [35], which are influenced by the low-N* waters formed in the Southern Ocean [36, 37].

2.2 *Subtropical-Subpolar Gyre Boundaries*

Several aspects of the ocean's subpolar gyre circulation maintain high nutrient concentrations near the surface. First, wind-induced upwelling provides a flux of nutrients from intermediate depths in the ocean to the surface [38]. Second, nutrients flowing poleward in the stratified western boundary currents are advected into the subpolar seasonal boundary layer, a process called induction [37]. Finally, deep wintertime mixing mines these nutrients from the subsurface, returning seasonally lost nutrients to the euphotic zone [38]. This relatively high availability of nutrients together with deep surface mixed layers tend to create an environment where light, rather than nutrients, limits photosynthesis throughout much of the year [e.g., 39, 40]. In stark contrast, the subtropical gyres are downwelling domains with shallower convective mixing. Furthermore, in the Northern Hemisphere subtropical gyres, nutrients are consumed at the gyre's northern flank at the time of mode water formation. This water mass is then subducted with a low-nutrient signature, further endowing the subtopics with a nutrient-depleted upper thermocline [e.g., 32, 37]. Given these dynamical differences, it is unsurprising that the subtropical-subpolar boundaries constitute strong biogeochemical frontal regions (recall Fig. 2). In both the North Atlantic and North Pacific surface and thermocline waters, this boundary marks a transition from low nitrate concentrations in the subtropics to high concentrations in the subpolar gyres and a related transition from low to high chlorophyll in the surface ocean (Fig. 5).

2.2.1 The Kuroshio

The Kuroshio jet marks the transition between the nutrient-depleted surface waters of the subtropical North Pacific and the nutrient-rich surface waters of the region north of the current (Fig. 7). In addition to being a strong nitrate and phosphate front, the Kuroshio is also the only front north of the ACC where high silicic acid concentrations outcrop. Outside of the subpolar North Pacific and the ACC, the global thermocline is severely silicic acid-depleted, a feature that has long fascinated oceanographers [41]. Recent work suggests that subduction on the equatorward edge of the ACC floods the lower thermocline with water that is depleted in silicic acid, setting this property in the global thermocline [14, 15], a concept we will return to in Sect. 3. Only in the ACC and in the subpolar North Pacific and its marginal seas do mixing and upwelling return deep silicic acid to the upper thermocline and surface ocean. The sharp surface gradients prompt the questions: Is there an important cross-Kuroshio

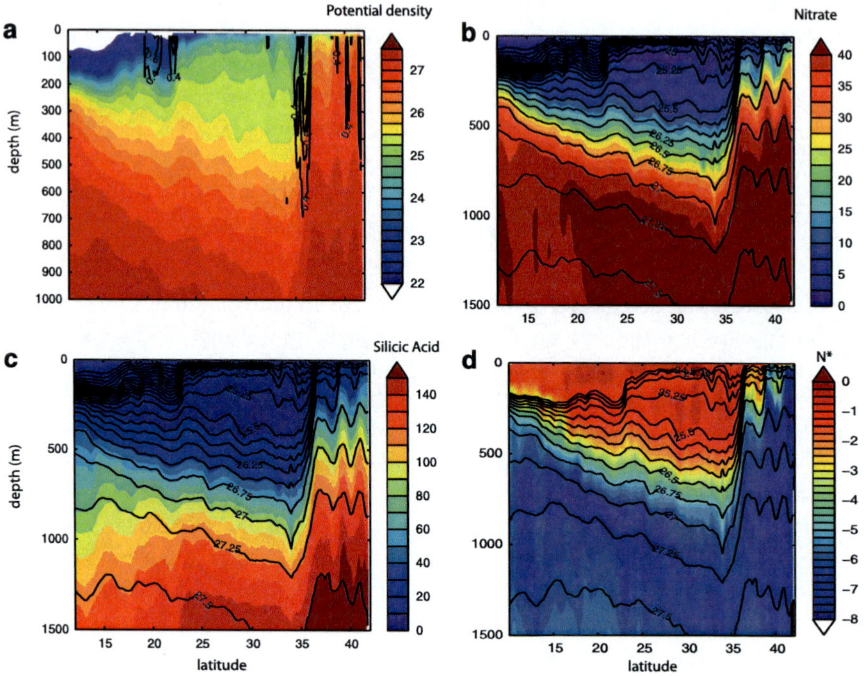

Fig. 7 WOCE section P10 occupied in 2005 from New Guinea to Japan crosses the Kuroshio and the North Pacific subtropical mode water region. (**a**) Potential density referenced to the sea surface (*color*) with velocity (contours in m s^{-1}) measured by a shipboard ADCP. (**b**) Nitrate (mmol m^{-3}). (**c**) Silicic acid (mmol m^{-3}). (**d**) N* (mmol m^{-3}). The Kuroshio appears as a strong jet at approximately 35°N and clearly divides the domain between the relatively nitrate and silicic acid poor subtropical gyre and a nutrient rich region north of the Kuroshio. The thermocline shows strong nitrate depletion relative to phosphate (negative N*), a signal that rises to the surface in the Kuroshio

exchange of nitrate and silicic acid? What are the physical processes that facilitate such exchange?

The importance of cross-Kuroshio transport has recently been highlighted by several studies [12, 42, 43]. Ayers and Lozier [12] showed that seasonal variability in the position of the North Pacific chlorophyll front was tied to the Ekman advection of nutrients across the Kuroshio. Qiu and coauthors [42, 43] suggest that variability in eddy-mediated cross-Kuroshio exchange is reflected in the temporal variability of the Subtropical Mode Water (STMW). The STMW is a pool of weakly stratified water between the $\sigma_\theta = 25.25$ and 25.5 isopycnals that is formed by convection each winter at the poleward edge of the subtropical gyre [27] and extends from the equatorward edge of the Kuroshio to the southern limit of the subtropical gyre. The temperature, salinity and thickness of the STMW appear to be strongly influenced by the number of rings formed by the Kuroshio [42, 43]. These rings are thought to be one of the principal mechanisms for exchanging fluid across jets [44].

Similar to what has been shown in the subtropical North Atlantic [32], the North Pacific STMW appears to inject a wedge of relatively nutrient-poor water above the nutricline (Fig. 7b, c). The nutrient concentrations in the mode water are set by a competition between biological drawdown on one hand and the supply of nutrients via vertical entrainment from the underlying nutricline and cross-frontal nutrient transport on the other. Given that the stratified waters crossing from the region north of the Kuroshio into the subtropics are high in nitrate and silicic acid and low in N* (Fig. 7) and that these rings are thought to set the physical properties of the North Pacific STMW [42, 43], we speculate that the rings shed by the Kuroshio may exert an important influence on the temporal variability of nutrient concentrations and ratios in the STMW. The Kuroshio has two dynamic states, one with a relatively stable path and the other with a highly variable path and vigorous ring formation [42]. Therefore, we hypothesize that variability in the nutrient composition of the gyre may be linked to the dynamic state of the Kuroshio, the number of rings shed from the current, and the Ekman flux across it. Because the STMW comprises the nutrient reservoir at the base of the euphotic zone for much of the western subtropical gyre, such variability could possibly exert a critical control on primary productivity and perhaps species composition in the gyre.

2.2.2 The Gulf Stream

Similar to the Kuroshio, the Gulf Stream forms the boundary between the nutrient-rich subpolar North Atlantic and the depleted waters of the subtropics (Fig. 8). However, a comparison of silicic acid concentrations in the North Atlantic (Fig. 8c) with those of the North Pacific (Fig. 7c) reveals that the Gulf Stream region has roughly 15% the silicic acid found in the Kuroshio region. The low Gulf Stream silicic acid concentrations are consistent with the view that silicic acid-depleted water is advected in the shallow limb of the Meridional Overturning Circulation (MOC) [14], as the Gulf Stream is a principal pathway by which the shallow MOC enters the North Atlantic. The low-silicic acid signature of the Southern Ocean water masses is discussed in detail in Sect. 2.3 on Southern Ocean Fronts.

A number of observational and modeling studies have confirmed that the Gulf Stream is a conduit of nutrients imported from outside the North Atlantic [36, 37, 45–47]. These imported nutrients in the Gulf Stream are suggested by the cross-stream nutrient distributions when plotted as a function of density (Fig. 8d), which show elevated nutrient concentrations that coincide with the high-velocity core in the Gulf Stream. The temperature and salinity signature of the Gulf Stream's high-nutrient waters has been used to link these nutrients to water masses imported in the shallow limb of the MOC [45]. Along the length of the Western Boundary Current, the elevated nutrient concentrations generally decline, especially for the densest isopycnals [45]. This along-stream decline in nutrient concentrations suggests that the Gulf Stream is a source of nutrients to the northern recirculation gyre and/or the subtropical gyre via isopycnal mixing. Thus, the exchange of waters between the Gulf Stream and its neighboring regions brings nutrients from distant locales to

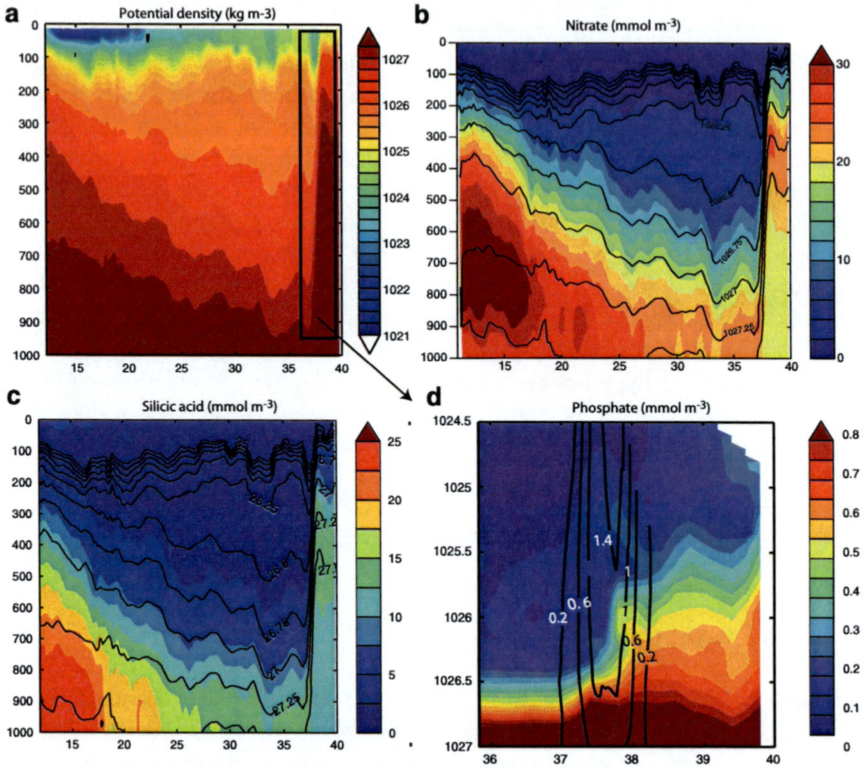

Fig. 8 WOCE Section A22 occupied in 2003 through the Gulf Stream along 64°W. (**a**) Potential density referenced to the sea surface (*color*). (**b**) Nitrate (mmol m^{-3}). (**c**) Silicic acid (mmol m^{-3}). (**d**) Phosphate (mmol m^{-3}) in color and velocity (m/s) in black contours as a function of density and latitude in a narrow region encompassing the Gulf Stream (outlined in panel a). The high-velocity core of the Gulf Stream corresponds with a bullet of elevated phosphate that can be traced to waters imported in the shallow limb of the MOC [112]

the subtropical gyre. Two important consequences of such exchange are that phosphate advected northward in the shallow MOC may help sustain North Atlantic primary productivity [14, 37] and biological nitrogen fixation [36].

While the recent work by Qiu and coauthors [42, 43] suggests a critical role for Kuroshio rings in setting the PV of the North Pacific STMW, rings shed from the Gulf Stream are thought to be a relatively minor player in tracer exchange in the Atlantic. Bower et al. [5] evaluated the role of ring formation in the subtropical oxygen budget with a scale analysis that used typical properties of Gulf Stream rings and the frequency of their formation. The Kuroshio studies combined satellite and in situ data to study the influence of particular rings on the PV budget of the North Pacific subtropical mode water [42, 43]. The question remains whether the difference in the importance of rings for the North Atlantic and North Pacific is due to differences in the tracers studied (i.e. PV or oxygen), differences in the dynamics of exchange across the gyres' western boundary currents, or a difference primarily in the techniques used to study the rings. In any case, variability in the Gulf Stream is not known to manifest as variability in ring formation as it does in the Kuroshio. Thus, North Atlantic subtropical gyre nutrient variability may be governed by a distinct set of processes, with possibilities including wind-driven Ekman exchange, eddy exchange that does not result in ring formation, and/or variability in the quantity of nutrients imported in the shallow limb of the MOC and recirculated in the subtropical gyre.

2.3 Biogeochemical Fronts of the Southern Ocean

More than any other region, the Southern Ocean has been defined by its many fronts [e.g., 48–50]. Sharp meridional gradients in temperature and salinity have long been used to divide the Southern Ocean into distinct dynamical provinces. In this section we focus on the Southern Ocean Fronts within the ACC, where more than 100 Sv (1 Sv $=$ 10^6 m^3 s^{-1}) of eastward transport circumnavigates the globe [50]. The current is organized into a series of narrow jets, which have been historically grouped into three circumpolar fronts. These fronts, listed from south to north, are the Southern ACC Front (SACCF), the Antarctic Polar Front (APF), and the Subantarctic Front (SAF). In addition to the fronts in the ACC, a Subtropical Frontal Zone (STFZ) bounded by fronts to the north and south can be found north of the ACC (Fig. 9). Traditionally, hydrographic properties and water mass features have been used to define only three ACC fronts, but recent research has revealed that they can split into more than 9 narrower jets or frontal filaments [51], each associated with strong lateral property gradients, swift currents, and the rapid shoaling of subsurface water masses. These fronts are not necessarily continuous, as they can merge and split and are temporally variable [51–53]. Chlorophyll concentrations follow these streamlines even at the scale of mesoscale eddies, a dependence traced in part to the upwelling of nutrients at many of the fronts [54]. Grouping these narrow, discontinuous jets into three ACC fronts remains a useful organizing framework to understand the outcropping of nutrient properties and the cross-front exchange of these nutrients.

Many of the biogeochemical fronts in the Southern Ocean can be linked to the presence of various water masses with distinct properties. As a consequence of wind-driven upwelling and mixing, dense Circumpolar Deep Water (CDW) rises in depth south of the APF (Fig. 10). The most voluminous water mass found in the ACC, CDW is divided into Upper Circumpolar Deep Water (UCDW) and Lower Circumpolar Deep Water (LCDW) based on a variety of contrasting properties. The LCDW is characterized by a salinity maximum ($S > 34.73$) and a subsurface nutrient minimum, both of which can be traced to its origins as North Atlantic Deep Water [55]. In contrast, the UCDW is characterized by an oxygen minimum and nutrient maximum, reflecting deep waters being returned from the lower latitudes of the Indian and Pacific Ocean [56]. Antarctic Intermediate Water (AAIW), characterized by a salinity minimum, is found above the UCDW. The vertical property gradients caused by the stacking of UCDW, LCDW, and AAIW become horizontal gradients

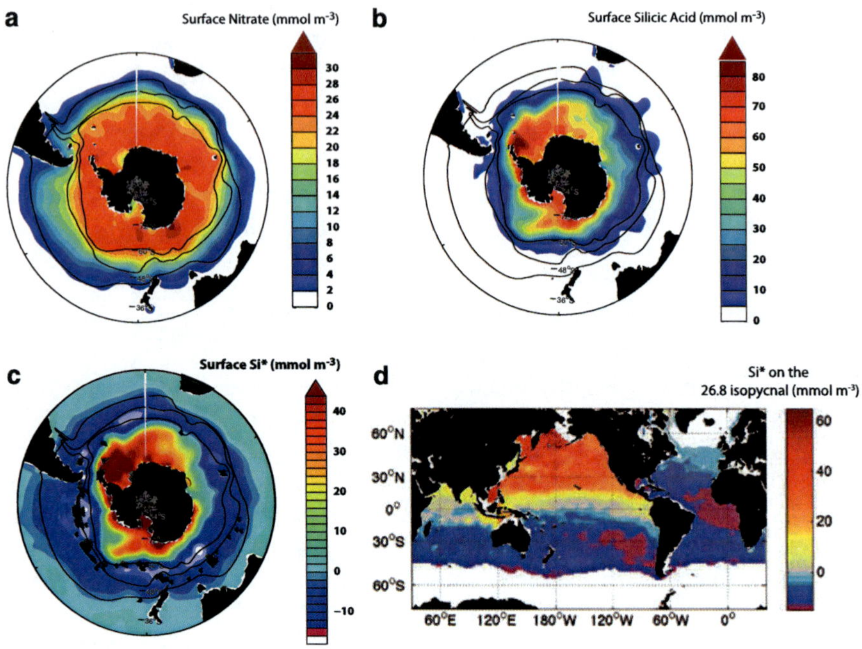

Fig. 9 Polar stereographic maps of the Southern Ocean (South Pole to 30°N) surface nutrient concentrations and the global spreading of its low Si* signal on the $\sigma_\theta = 26.8$ isopycnal, after Sarmiento et al 2004. (**a**) Surface nitrate (mmol m^{-3}). (**b**) Surface silicate (mmol m^{-3}). (**c**) Surface Si* = silicate–nitrate (mmol m^{-3}) in color. The black dots mark the locations of winter maximum mixed layer depth greater than 400 m, from the climatology of de Boyer Montégut et al. [82]. (**d**) Si* on the $\sigma_\theta = 26.8$ isopycnal, where nutrient concentrations were interpolated from pressure surfaces to the isopycnal. In the Polar Stereographic maps, the *black contours* show the position of the major circumpolar fronts of the Southern Ocean defined according to Belkin and Gordon [48]. From south to north they are the Antarctic Polar Front (APF), the Subantarctic Front (SAF), and the Southern Subtropical Front (STF). The nutrient data in panels *a–c* are taken from WOA09 [87]. Panel *d* is reproduced from Palter et al. [76]

by the action of upwelling in the ACC; to first order, the tilting of these vertical gradients is what produces the biogeochemical fronts across the southern ACC. Next, we briefly introduce the properties of the three traditional ACC fronts from south to north. The goal of this terse introduction is to provide a context for the discussion of nutrient and carbon return pathways from the deep ocean (Sect. 3); the interested reader is directed to Artamonova and Belkin (this volume) for a more thorough discussion of the chemical fronts of the Southern Ocean.

2.3.1 The Southern ACC Front (SACCF)

The Southern ACC Front (SACCF) is marked by large meridional temperature gradients associated with the southward shoaling of the UCDW [50]. The UCDW oxygen minimum and nutrient maximum reside on the $\sigma_\theta = 27.6$ isopycnal, which

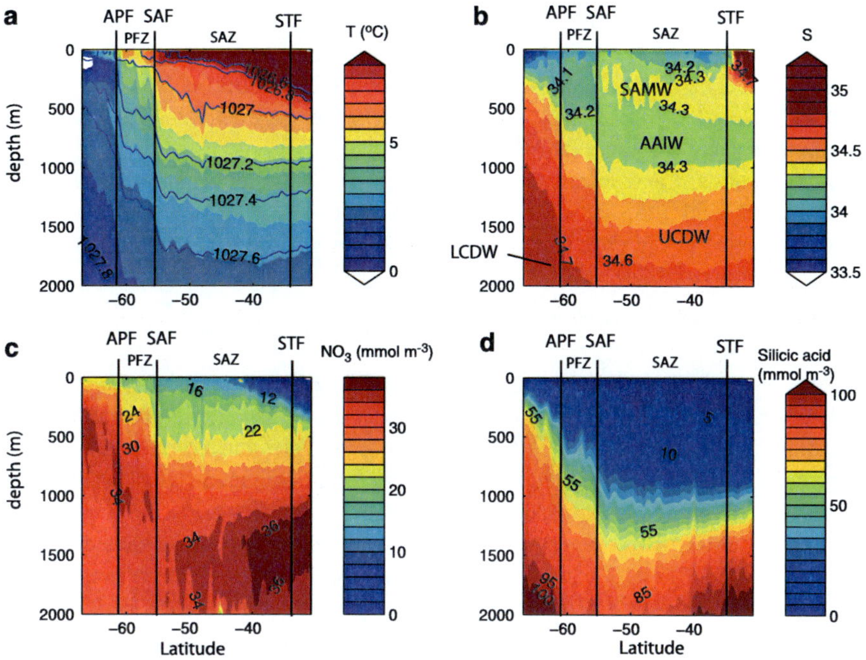

Fig. 10 The P18 WOCE section along 103°W in the Pacific Sector of the Southern Ocean. Properties are a) potential temperature (*colors*) and potential density (contours), (**b**) salinity, (**c**) nitrate, and (**d**) silicic acid. The approximate locations of the Antarctic Polar Front (APF), Subantarctic Front (SAF), and Subtropical Front (STF) are marked by vertical lines, according to the water mass property definitions of Orsi et al. [50] and Belkin and Gordon [48], as synthesized in Pollard et al. [55]. The Subantarctic Zone (SAZ) lies between the SAF and STF; the Polar Frontal Zone (PFZ) lies between the PF and SAF; both are labeled above panel *a*. The Southern ACC Front (SACCF) is very near the southern edge of this section. Water masses are labeled in panel b: Subantarctic Mode Water (SAMW), Antarctic Intermediate Water (AAIW), Upper Circumpolar Deep Water (UCDW), and Lower Circumpolar Deep Water (LCDW)

rises to 500 m at the SACCF. In contrast, the denser and silicic acid-rich LCDW penetrates south of the SACCF and beyond the southern boundary of the ACC, the southernmost circumpolar contour of the ACC found in Drake Passage. The LCDW reaches the Antarctic continental shelves where it mixes with shelf waters to form dense Antarctic Bottom Water (AABW) that sinks downslope and spreads northward. There is also a shelf-break front surrounding the Antarctic Continent called the Antarctic Slope Front, which has been shown to play a critical role in the formation of AABW [56, 57].

2.3.2 The Antarctic Polar Front (APF)

The Antarctic Polar Front (APF) marks a change in the balance of stratification, from salinity-dominated stratification everywhere south of the APF to approximately equal contributions to stratification by temperature and salinity north of the APF [55]. The APF is also the northernmost extension of the temperature minimum waters (approximately 2°C), a remnant of winter convection commonly called the Antarctic Winter Water after Mosby [58]. Thus, the zone between the APF and the SACCF is characterized by the presence of the upwelling UCDW, along with its signature oxygen minimum and nitrate maximum. The APF also represents the Southern Ocean silicic acid front, separating silicic acid-rich waters in the upwelling Antarctic Zone from silicic acid-poor waters in the Subantarctic Zone (Figs. 9 and 10).

2.3.3 The Subantarctic Front (SAF)

The Subantarctic Front (SAF) is the northernmost frontal jet that passes through Drake Passage. Like the APF, the SAF also marks a change in the balance of stratification, this time from the Antarctic Polar Frontal Zone where temperature and salinity contribute equally to stratification to a temperature-dominated Subantarctic Zone north of the SAF [55]. The SAF is identified by the northward subduction of the salinity minimum (salinity less than 34) associated with the well-oxygenated AAIW. South of the SAF in the Polar Frontal Zone, the lowest salinity water is found at the surface; this minimum descends at the SAF to depths greater than 400 m [50]. North of the SAF, nitrate-rich, silicic acid-depleted Subantarctic Mode Water is subducted from the surface to the thermocline [14].

2.3.4 The Subtropical Frontal Zone (STFZ)

The Southern Ocean's natural equatorward boundary is the Subtropical Frontal Zone, a 400–500 km wide region bounded by two sharp fronts: the North and South Subtropical Fronts [48]. These fronts separate the relatively saline subtropical gyre waters (surface salinities greater than 35.5 g kg^{-1}) from the fresher subantarctic waters and are often coincident with strong gradients at the ocean's surface in chlorophyll [59].

3 The Fronts of the ACC and Their Role in Nutrient and Carbon Return Pathways from the Deep Ocean

In the introduction, we proposed a two-step process for returning nutrients to the euphotic zone: (1) upwelling and vertical mixing in the regions surrounding the gyre bring the nutrients to shallower depths and (2) advection and mixing across the frontal region then redistribute these nutrients back into the gyres. The frontal regions of the Southern Ocean are thought to be pivotal to the first step of this process, providing a locale where nutrients return from abyssal waters to the pycnocline [14, 17, 60]. As a consequence of its long residence time in the ocean interior, the UCDW is rich in nutrients. The upwelling and mixing of this water mass to the surface ocean in the Antarctic Zone south of the APF create the high surface concentrations of nitrate and

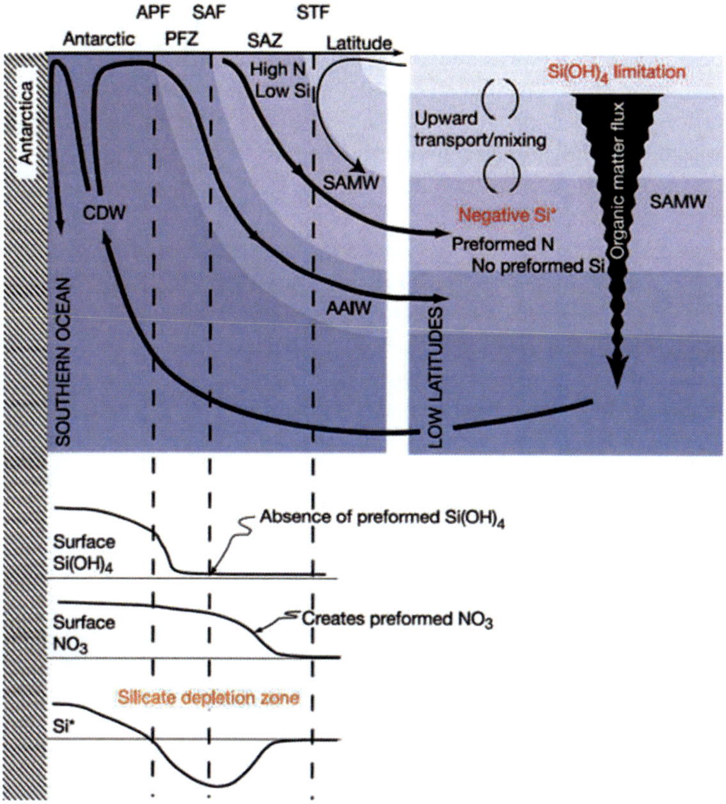

Fig. 11 Southern Ocean control on thermocline nutrient concentrations. Conceptual diagram depicting the Southern Ocean physical and biological processes that form low-Si* waters and feed them into the global thermocline. *Top*, water pathways; bottom, details of surface processes. Upper Circumpolar Deep Water (UCDW) upwells to the surface in the Southern Ocean and is transported to the north across the Antarctic Polar Front (APF) into the Polar Frontal Zone (PFZ), where Antarctic Intermediate Water (AAIW) forms, and then across the Subantarctic Front (SAF) into the Subantarctic Zone (SAZ), which is bounded to the north by the Subtropical Front (STF). Silicic acid is stripped out preferentially over nitrate as the water moves to the north, thus generating negative Si* values. This negative-Si* water is a signature of Subantarctic Mode Water (SAMW), which sinks into the base of the main thermocline and feeds biological production in the low latitudes. Lightly modified from Sarmiento et al. [14] to reflect that a portion of the SAMW source water is supplied from the subtropics as in Iudicone [113] and Talley [114]

silicic acid apparent in the surface maps (Fig. 9) and the meridional section in the Southern Ocean's Pacific Sector (Fig. 10). A portion of the upwelled UCDW is advected northward with the Ekman transport into the Polar Frontal Zone and Subantarctic Zone, as depicted schematically in Fig. 11. There, it is mixed with subtropical waters and subducted as an important constituent of Subantarctic Mode Water (SAMW).

As the upwelled UCDW moves northward across the APF, phytoplankton consume its various nutrients. Diatoms, the dominant phytoplankton group in the biome, take up silicic acid and nitrate in a 1:1 ratio under adequate light and nutrient conditions. However, under the stress of scarce iron or light, diatoms preferentially remove silicic acid over nitrate from the water column [61]. Because of severe iron and light limitation of photosynthesis in the Southern Ocean [62, 63], silicic acid is stripped from the northward moving surface waters in the Polar Frontal Zone north of the APF, while elevated nitrate concentrations persist well to the north of the APF. This signature of low silicic acid and high nitrate concentrations at the ocean's surface is found exclusively in the Polar Frontal Zone and Subantarctic Zone [14]. Thus, a useful tracer of the water masses that are formed in this region is their low value of $Si* = Si(OH)_4 - NO_3^-$ [14]. A band of negative Si* coincides with the deep mixed layers where SAMW is formed (Figs. 9c and 11). Sarmiento et al. [14] argue that the subsurface pool of negative Si* is an indication of the spreading of Subantarctic Mode Water (SAMW) by advective and diffusive transport processes. The strength of this argument depends on Si* being a conserved property; nonconservation of the tracer can arise from differences in remineralization length scales of nitrate and silicic acid, with this effect shown to be of the order of 5 mmol m^{-3} at the depth of the SAMW layer [14]. Because of its high nitrate and low silicic acid concentrations, it is believed that the spreading of SAMW in the thermocline helps sustain low-latitude primary productivity but restricts the abundance of diatoms (Fig. 11) [14]. Related to these ideas, it has further been hypothesized that a change in iron input to the Southern Ocean on glacial-interglacial timescales might have led to changes in the relative leakage of nitrate and silicic acid to the low-latitude pycnocline, causing floral shifts in the Equatorial Pacific and changes in the export of organic matter to the deep ocean [64–66].

Model studies also support the idea that advection across the ACC and subduction of nutrients in the SAMW layer provide a major supply of nutrients supporting low-latitude productivity. Depending on model winds and parameterizations of subgridscale processes, the nutrients returned to the pycnocline in the Southern Ocean frontal regions sustain 33–75% of the productivity at low latitudes [14, 17]. The Southern Ocean control on low-latitude productivity can be traced even more specifically to the region immediately north of the APF. Marinov et al. [60] divided the Southern Ocean into Antarctic and Subantarctic regions roughly equivalent to the zones south and north of the APF and forced modeled surface nutrients to zero in these individual regions. As expected, the removal of nutrients north of the APF in the SAMW and AAIW formation region sharply reduced low-latitude productivity relative to the control model. In vivid contrast, the drawdown of nutrients poleward of the APF had almost no influence on low-latitude productivity.

While these model experiments suggest that the region poleward of the APF is of little consequence for low-latitude productivity, it appears to be critical in setting the ocean-atmosphere carbon balance [60]. In this region, the upwelling of LCDW brings nutrients to the surface ocean where they are inefficiently drawn down by biology. Thus, the newly formed AABW returns to the deep ocean interior with a large load of *preformed* nutrients. Preformed nutrients are those inorganic nutrients that enter the ocean interior without having fueled primary productivity at the surface. In the ocean interior, the total nutrient concentration is the sum of preformed and remineralized nutrient concentrations. Remineralized nutrients are added to the inorganic nutrient pool of the ocean interior by the remineralization of organic matter. Preformed nutrients are the signature of inefficiency in the biological pump. A more efficient biological pump utilizes a greater portion of the surface preformed nutrients, converts them to organic material, and increases the remineralized nutrient and carbon concentration deeper in the water column. Therefore, a lower global preformed nutrient inventory signifies a higher remineralized nutrient inventory, greater ocean biological carbon storage, and lower atmospheric carbon dioxide [66–69]. Preformed surface nutrients increase with southward distance across the ACC and are highest south of the Polar Front (Fig. 9) where biological uptake is least efficient. In the model simulations, the intensified drawdown of nutrients in the AABW formation region south of the Antarctic PF reduced the ocean's preformed nutrient inventory and removed a significant quantity of carbon dioxide from the atmosphere. On the other hand, drawing down nutrients in the intermediate and mode water formation regions north of the PF had far less impact on atmospheric CO_2 on a timescale of a few thousand years, as these preformed nutrients are already efficiently utilized and converted to remineralized nutrients downstream of the water mass formation region on decadal timescales [17]. The conclusion drawn from these model experiments is that on the long ocean equilibrium timescales (thousands of years), the air-sea balance of carbon dioxide is strongly influenced by the quantity of preformed nutrients that sink on the poleward side of the Antarctic Polar Front, while global biological productivity is highly sensitive to the rate at which nutrients are subducted on the equatorward side of the Subantarctic Front (Fig. 10).

4 A Closer Look at Cross-Frontal Exchange

So far, we have identified several of the ocean's strongest biogeochemical fronts, namely, those at the edges of the subtropical oceans that comprise the majority of the ocean's surface and in the ACC. We have qualitatively described the importance of exchange across these fronts in returning nutrients from the deep ocean to the pycnocline in the Southern Ocean and from the thermocline to the euphotic zone across the subtropical-subpolar boundaries. Now, we take a closer look at mechanisms of exchange and return to the question of how fronts may serve as both barriers to exchange and gateways for the transport of biogeochemical properties.

4.1 Fronts as Barriers: Reviewing Results from Kinematic Analysis

A kinematic framework for studying exchange across jets, introduced by Bower [6] and recently synthesized by Wiggins [44], has proven to be a powerful tool in this line

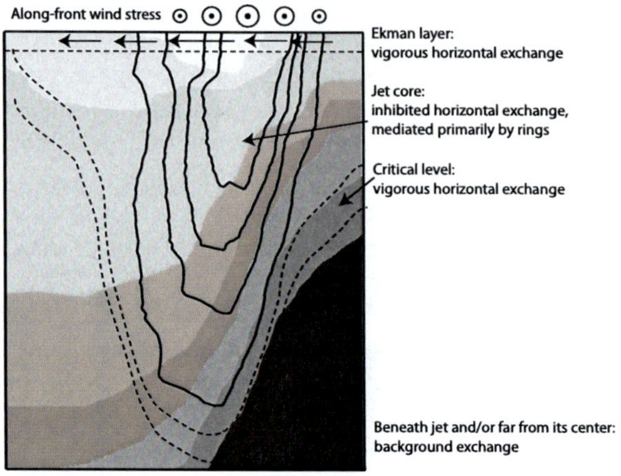

Fig. 12 Schematic of exchange across a front. The *solid lines* are isotachs of positive velocity (flow out of the page) and show the core of the jet; grayscale contours represent density, with higher densities in darker colors. The horizontal dotted line at the top represents the base of the Ekman layer. The *dotted curves* outline the schematized critical level which descends from the near-surface at the front's outer edges to depths beneath the jet's core. An along-front wind stress drives a cross-front Ekman transport. Exchange across the jet is enhanced beneath the jet core at the depth of the critical level and in the Ekman layer, but is inhibited in the jet's core between the Ekman depth and critical level. The limited exchange that is permitted across the center of these jets is mediated primarily by rings shed from the jet [44], with important consequences discussed with regard to North Pacific STMW formation (see Sect. 2.2)

of inquiry and a source of our understanding of ocean jets as barriers to exchange. In the kinematic approach, a flow field is defined from either a model or observations and the trajectories of Lagrangian parcels diagnosed by integrating their position in the flow. Several of these studies have usefully defined the flow as a straight jet perturbed by propagating waves, which force fluid parcels to cross jet streamlines. Such an approach has revealed that cross-frontal exchange is strongly inhibited across the core of jets [6, 70, 71], a view confirmed by the trajectories of floats deployed in the Gulf Stream [72]. On the other hand, lateral exchange is favored where the speed of the jet and the propagation speed of its meanders are approximately equal, locations called steering lines or critical levels. At these critical levels, the motion of a fluid parcel originally in the jet can deviate substantially from the streamlines at the jet's center, permitting exchange between the jet and its surroundings [6, 7, 73].

Near the surface of the ocean, the translation speed at the center of a jet is typically much faster than the propagation speed of its meanders. In contrast, a jet's translation speed and its meander speed are approximately equal at the jet's edges and at depths below the jet's maximum velocity. Thus, there is a three-dimensional surface where turbulent exchange is favored; in a two-dimensional cross section, this surface is defined as a steering line which typically descends from the jet's edges at the surface to depths beneath jet's core (Fig. 12). In the Gulf Stream where swift velocities persist to a depth of 700 m (shown previously in Fig. 7) cross-frontal transport by mesoscale motions is inhibited between the base of the Ekman layer and the $\sigma_\theta = 27.1$ isopycnal [6]. A similar structure is expected for the Kuroshio. Recent results from an eddy-resolving GCM of the Southern Ocean agree with the results of the kinematic studies: contours of high effective diffusivities driven by eddies follow steering lines as they descend from the northern edge of the ACC to a maximum at 1,500 m beneath the ACC's center [73]. Because the strength of mixing influences the strength of tracer transport, this gradient in horizontal diffusion can produce a tracer flux divergence, a fundamental physical process that is currently unresolved and unparameterized in the coarse-resolution models typically used to simulate climate and biogeochemical cycles. Therefore, how a cross-frontal diffusivity gradient might impact the exchange of biogeochemical tracers remains largely an open question.

4.2 *Fronts as Regions of Exchange: Ekman Fluxes*

Because it has been used primarily for studying frontal dynamics in the ocean interior, the kinematic framework has been most often deployed without consideration of wind-driven Ekman transport. However, Ekman advection is hypothesized to be a critical route by which nutrients enter into the subtropical gyres [12, 13] and cross the ACC to enrich the SAMW [14]. In the surface Ekman layer, a wind stress, τ, drives an Ekman transport, U_{Ek}, according to the Ekman relation:

$$U_{\mathrm{Ek}} = -\widehat{k} \times \frac{\tau}{\rho_o f} \tag{1}$$

where $\widehat{k}$ is the vertical coordinate, ρ_o is a reference density, and f is the Coriolis parameter. It is natural to speculate that Ekman transport across the subtropical-subpolar boundaries and the ACC should provide an important flux of nutrients to the equatorward side of these fronts. The location of these fronts is set by the zero contour of the wind stress curl. Where the wind stress curl vanishes, the along-front wind stress is maximized, as is the equatorward cross-stream Ekman transport it supports. Given the sharp increase in nutrients across the subtropical-subpolar and ACC fronts, described in Sect. 2, such Ekman transport is expected to provide a significant source of nutrients to the subtropics. A complication to this expectation is due to the effect of ocean eddies. Eddies provide a transfer of nutrients along isopycnals through familiar down-gradient eddy-induced diffusion, which augments the Ekman supply. However, eddies also transport mass due to the temporal correlation between anomalies of velocity on an isopycnal layer and the thickness of that layer [74]. The mass flux due to mesoscale motions is often called an "eddy bolus transport" and it can transfer nutrients either down-gradient or up-gradient [75]. This bolus transport has been shown to partially offset the Ekman transfer of nutrients across the ACC [17] and Gulf Stream [76] but enhance the offshore export of nutrients from eastern boundary upwelling regions [77]. In this section, we explore the strength of the global Ekman supply of nutrients alone and return to the question of the impact of eddies on nutrient transport in Sect. 4.3.

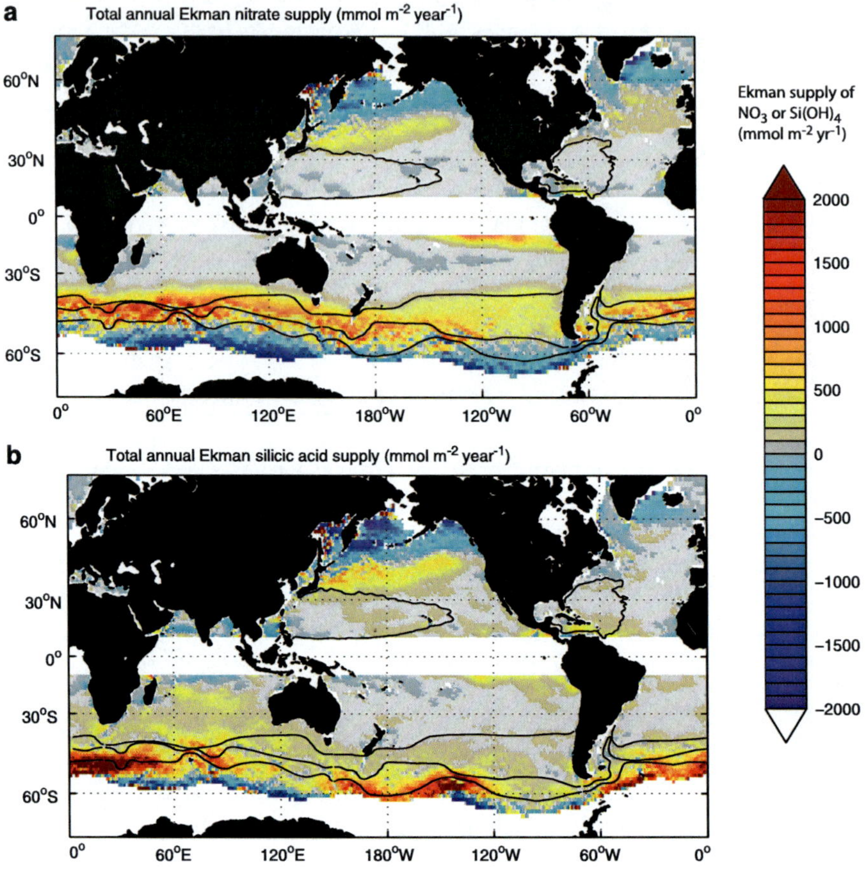

Fig. 13 Total annual Ekman supply of (**a**) nitrate and (**b**) silicic acid in mmol m^{-2} $year^{-1}$ as deduced from the monthly satellite wind climatology of Risien and Chelton [78] and the World Ocean Atlas nutrient climatology [87], according to Eqs. (1 & 2) in the text. The averaging regions used for the budget calculations presented in Fig. 13 are outlined in black. In the Southern Ocean, these averaging regions are the Subantarctic Zone (bounded to the north by the Subtropical Front and to the south by the Subantarctic Front) and the Polar Frontal Zone (bounded to the north by the Subantarctic Front and the south by the Polar Front).These Southern Ocean Fronts are defined as in Belkin and Gordon [48]. In the Northern Hemisphere the averaging region is within the largest SSH contour encircling the gyres (1.71 m in the North Atlantic, and 2.12 m in the North Pacific). This contour is related to the largest closed surface geostrophic streamline of the anticyclonic circulation for the annual mean (differing by a factor of f, the Coriolis parameter)

The convergence of the Ekman nitrate and silicic acid transport, as deduced from an 8-year climatology of satellite wind stress [78] and a monthly nutrient climatology [79], confirms our expectation regarding the Ekman supply of nutrients: nitrate is exported from the subpolar surface ocean and converges in subtropical regions (Fig. 13). In the North Atlantic, the nutrient concentrations in the surface ocean are severely depleted, the Ekman supply of nitrate is low relative to the other basins, and its influence extends over a limited region near the northern edge of the subtropical gyre. In contrast, the Pacific Ekman nitrate supply penetrates into the interior subtropical gyre. Ayers and Lozier [12] suggest that this Ekman nutrient supply governs the position of the Pacific transition zone chlorophyll front that extends into the central gyre [80]. In the Southern Ocean, a considerable Ekman supply of nitrate extends from the Antarctic Polar Front well into the subtropics. In contrast, the Ekman silicic acid supply is largely restricted to the Polar Frontal Zone, south of the SAF, where it converges in a narrower meridional band than the nitrate supply. This flux convergence is consistent with the hypothesis of silicic acid removal by diatom production and opal export north of the SAF [14, 15].

4.3 *Ekman Fluxes in the Context of Other Nutrient Supply and Demand Terms*

How does the cross-frontal Ekman transport compare with other physical and biological sources and sinks of nutrients? This question is best considered in the context of the steady state conservation equation for a nutrient (C) integrated over the annual maximum mixed layer depth (MLD):

$$\underbrace{\int_{z=-\mathrm{MLD}}^{0}(\nabla\cdot\vec{u}C}_{1}-\underbrace{\nabla\kappa_{\mathrm{h}}\nabla C)\partial z}_{2}-\underbrace{K_{\mathrm{v}}\frac{\partial C}{\partial z}\bigg|_{z=-\mathrm{MLD}}}_{3}-\underbrace{wC\bigg|_{z=-\mathrm{MLD}}}_{4}=\underbrace{\int_{z=-\mathrm{MLD}}^{0}\mathrm{SMS}\partial z}_{5} \quad (2)$$

Integrating over the annual maximum mixed layer eliminates the complication of resolving the seasonal redistribution of nutrients that are remineralized above this depth throughout the year and entrained into the mixed layer each winter (see [13] for a more detailed discussion of the limited role of local convection in maintaining euphotic zone nutrient concentrations). The terms of the equation are (1) the horizontal advective supply of the tracer above the annual maximum mixed layer depth; (2) the turbulent horizontal mixing of the tracer above the annual maximum mixed layer depth, parameterized here as a diffusive flux; (3) the mixing of the tracer across the base of the maximum mixed layer; (4) the vertical advection across the base of the maximum mixed layer; and (5) the biological source (remineralization of organic matter above the depth of the mixed layer) minus sink (export of organic matter across the base of the mixed layer). Because we integrate to the base of the annual maximum mixed layer, the only vertical fluxes that must be considered are the turbulent mixing supply (term 3), upwelling supply (term 4), and the flux of particulate nutrients across the base of that maximum mixed layer (in the source/sink term). To the degree that ocean biogeochemistry is in steady state on timescales of a year or more, a large influx of nutrients into the annually mixed layer should be balanced by the biological consumption and export of these nutrients. Equivalently, a large nutrient outflux would imply the remineralization of organic nutrients.

Our goal is to perform an observationally based, order-of-magnitude assessment of the dominant supply mechanisms in the different regions for nitrate and silicic acid. To do so, we provide an approximation for as many of these terms as possible in the subtropical North Atlantic and North Pacific and in the Polar Frontal Zone and Subantarctic Zone of the Southern Ocean. The Ekman supply is embedded in term 1 of the conservation equation, as the mixed layer velocity is the sum of its Ekman and mean geostrophic components, as well as velocity arising from turbulent motion:

$$-\int_{z=-\mathrm{MLD}}^{0} \nabla \cdot \vec{u}C\partial z = -\nabla \cdot \left(U_{\mathrm{Ek}}C_{\mathrm{Ek}} + \left\langle \overline{u}_{\mathrm{geo}}C \right\rangle H + \langle u^{*}C \rangle H\right) \quad (3)$$

Here we have decomposed the velocity into three components: U_{Ek} is the Ekman transport deduced from Eq. (1), which is vertically integrated over the Ekman layer; $\overline{u}_{\mathrm{geo}}$ is the time-mean geostrophic velocity; and u^* is the mean bolus velocity due to mesoscale motions including geostrophic turbulence [74]. In Eq. (3), H is the depth of the annual maximum mixed layer. The negative sign ensures that a nutrient supply is positive, i.e., transport convergence. The angled brackets represent the mean over the annual maximum mixed layer. C_{Ek} is a weighted average of the tracer concentration over the Ekman layer, with weights meant to mimic the velocity decay in the Ekman spiral. The weights are defined by an exponential decay with an e-folding scale of 22.1 m, as deduced from observations of velocity decaying in the Ekman spiral in the Drake Passage [81]. We note that because climatological nutrient concentrations are relatively homogenous in the upper 100 m of the ocean, our choice of e-folding depth only minimally impacts the Ekman supply term. Because the Ekman depth is generally above the annual maximum mixed layer depth, we make the reasonable assumption that the entire Ekman convergence is a supply term to this layer. The two-dimensional Ekman supply of nutrients is presented in Fig. 13, discussed in detail in Sect. 4.2. For the regions outlined in Fig. 13, we compare the annual Ekman supply to other terms in the conservation equation (Fig. 14). Because the Southern Ocean nutrient fluxes in the mixed layer are so much larger than those in the Northern subtropical gyres, they are plotted on separate scales.

The annual mean geostrophic transport at the ocean's surface is taken from Rio and Hernandez [23], which estimates the dynamic topography as the sum of sea-level anomalies averaged over the period 1992–2008 with a model of the geoid, constrained by observations. We have assumed that the surface geostrophic currents are uniform to the base of the annual maximum mixed layer, a simplification justified for our order-of-magnitude style analysis, as shallow mixed layers do not have the vertical distance for thermal wind shear to change the surface current significantly and deeper mixed layers are found in regions with weak stratification and more barotropic currents. The mixed layer depths are taken from the global 2° climatology constructed from hydrographic and Argo float profiles by de Boyer Montégut

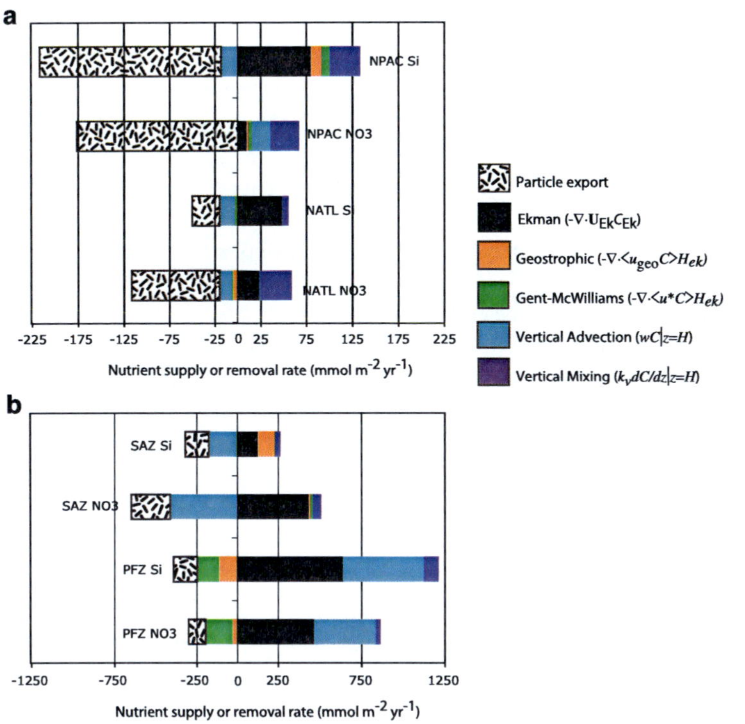

Fig. 14 The annual mean nitrate and silicic acid supply and removal rates (mmol m^2 year^{-1}) in the mixed layer of (**a**) North Atlantic and North Pacific and (**b**) the Southern Ocean Subantarctic Zone (SAZ) and Polar Front Zone (PFZ). The averaging region is outlined in Fig. 13. Note that the supply terms of nutrients in the Southern Ocean are roughly an order of magnitude greater than those in the Northern Hemisphere subtropical gyres. The calculation of each supply term is described in the text, along with an estimate of the uncertainty associated with each term, which is more than 50% of the estimated average value in many cases

et al. [82], recently updated to include profiles collected through 2008. While geostrophic currents are among the swiftest in the world ocean, these currents will only act as a supply term if the nutrient transport diverges, rather than simply recirculates around the gyres. Because the baroclinic component of the geostrophic flow is aligned with density fronts and density fronts coincide with nutrient fronts, the geostrophic flow is directed primarily along isolines of nutrients. This alignment would suggest small nutrient transport divergence for the time-mean geostrophic flow. However, our diagnostic of the divergence of geostrophic nutrient transport is not uniformly low; rather it is large and noisy. Thus, the mean geostrophic nutrient supply is very sensitive to the region over which it is averaged, and even the sign of this term is uncertain. In Fig. 14, the geostrophic term can be seen either as a supply or removal term.

The nutrient supply arising from mesoscale motions is also difficult to quantify since it depends on observations at fine spatial and temporal resolutions. As discussed briefly in Sect. 4.2, eddies give rise to two tracer transport terms: one associated with a seawater mass flux ($<u^*C>H$ in Eq. (3)) and the other associated with the turbulent mixing of the tracer with no corresponding seawater mass flux ($\int_{z=\mathrm{MLD}}^{0} \nabla\kappa_h \nabla C \partial z$ in Eq. (2)) [75]. Here, u^* represents the bolus velocity due to mesoscale eddies, flattening sloping isopycnals and therefore opposing the Ekman transport, which raises isopycnals in regions of Ekman divergence. Despite its critical role in tracer budgets [e.g., 83], the eddy bolus term is often neglected in observational studies because of chronically insufficient spatial and temporal sampling to directly observe the phenomenon.

To estimate the size of the eddy bolus nutrient supply, we make use of the Ocean Comprehensive Atlas (OCCA). OCCA is publicly available (http://www.ecco-group.org/) for the data-rich Argo period from 2004 to 2006 [84]. OCCA is an ocean climatology produced by calculating the least squares fit of the MIT general circulation model (MITgcm) to satellite and in situ data. A detailed description and validation of the global atlas, as well as applications for water mass formation studies in the Southern Ocean and North Atlantic, have demonstrated the skill and utility of this framework [84, 85]. The MITgcm uses the Gent and McWilliams (GM) formulation (with a thickness diffusion coefficient of 1,000 $m^2\ s^{-1}$) to parameterize the bolus velocities arising from mesoscale motions [86], and these velocities are included in the OCCA. We multiply these GM velocities by the lateral nutrient gradients from World Ocean Atlas [87]. A limitation of this framework is that it neglects spatial gradients in the diffusion coefficient, though these gradients can be substantial across fronts, as elucidated by the kinematic studies discussed in Sect. 4.1. Averaged over the North Pacific and North Atlantic Subtropical gyres, the GM term is much smaller than the other terms (Fig. 14). In both the Polar Frontal Zone and the Subantarctic Zone, the nutrient transport due to the GM velocities clearly opposes the Ekman nutrient transport. This opposition is expected given that Ekman transport tends to tilt isopycnals into the vertical, whereas the net effect of eddies is to lay the isopycnals flat. Our scale analysis suggests that, in the Polar Frontal Zone, the removal of nitrate by the GM term is of the same order and opposite sign as the Ekman supply term. For other all other fronts, the estimate for the GM term is an order of magnitude smaller than the Ekman term.

The last of the lateral supply terms, the turbulent mixing of tracers, is often parameterized as the Laplacian diffusion of the tracer as depicted in term 2 of Eq. (2). This term can be scaled as $\kappa_h\left(\frac{\Delta_x C}{L_x^2}+\frac{\Delta_y C}{L_y^2}\right)H$. Values of κ_h are generally 1,000 $m^2\ s^{-1}$ or lower [88], but are thought to be four times smaller across the barrier regions of fronts [5]. Recalling Fig. 2, we can see that the maximum change in nitrate and silicic acid is about 3 mmol m^{-3} and 6 mmol m^{-3}, respectively, over a length scale of 100 km at the edges of the Northern Hemisphere subtropical gyres. Integrating over an annual maximum mixed layer depth of 250 m and applying a mixing coefficient, κ_h, of 250 $m^2\ s^{-1}$, we arrive at a lateral mixing supply on the order of several hundred mmol m^{-2} $year^{-1}$ in the high-gradient region at the front. Because the high-gradient region is restricted to the gyre's edges, the diffusive supply averaged over the entire gyre is at least one to two orders of magnitude smaller than this estimate for the region near the front. For values more appropriate to the ACC of $L_y = 1{,}000$ km, $\Delta_y NO_3 = 25$ mmol m^{-3}, and $\Delta_x NO_3 = 0$ mmol m^{-3} and mixed layer of 400 m, the down-gradient diffusion is on the order of 100 mmol m^{-2} year^{-1}. Thus, the lateral mixing supply is likely a significant nutrient source for the low-nutrient subtropics near their boundaries.

Our scale analysis of the lateral mixing supply is highly sensitive to the choice of diffusion coefficients, which may vary significantly between the frontal regions where nutrient concentrations change rapidly and the homogenous region between the fronts [88]. Characterizing the entire region by one value of κ_h and one length scale ignores the fact that most of the horizontal change in nutrient concentrations happens over a narrow region near the front where horizontal mixing is inhibited. Given these caveats, this term is uncertain even in its order of magnitude, and is therefore excluded from Fig. 14. The mixing of tracers is a ripe area for future research, and eddy-resolving models are rapidly challenging how we conceptualize the influence of mesoscale motions on frontal dynamics [89], tracer budgets [73, 75], and biogeochemistry [90].

Having tackled the lateral supply terms, we turn our attention to the vertical supply. The vertical mixing term (term 3 in Eq. (2)) is calculated from the nutrient gradient from the World Ocean Atlas [87] interpolated to the depth of the annual maximum mixed layer depth from de Boyer Montégut et al. [82] and multiplied by a typical upper open-ocean value of $\kappa_v = 4 \times 10^{-5}$ m^2 s^{-1} [91–94]. Often considered the dominant supply term of nutrients to the surface of the subtropical gyres, this analysis suggests that the vertical mixing flux is no bigger than the Ekman supply in these regions (Fig. 13). However, this rough scaling of the vertical mixing is also subject to considerable uncertainty, as the vertical mixing coefficient is subject to spatial and temporal variability [95, 96].

The last physical supply term, the vertical advection of nutrients (term 4 in Eq. (2)) is estimated from the vertical velocities from the OCCA multiplied by the climatological mean nutrient concentration [79] at the base of the de Boyer Montégut [82] maximum mixed layer depth. OCCA incorporates the observed winds into a model with full physics and therefore simulates vertical velocities resulting from wind forcing along with other sources of vertical motion. As expected for the downwelling gyres of the subtropical North Atlantic and North Pacific, vertical advection generally removes nutrients from the surface mixed layer (Fig. 14). Likewise in the Subantarctic Zone, where the wind stress curl promotes downwelling conditions, nutrients are removed from the surface mixed layer by vertical advection. On the other hand, the Polar Frontal Zone, poleward of the zero wind stress curl line, receives an enormous upwelling supply of nitrate and silicic acid. The upwelling supply of nitrate dominates all other supply terms in the PFZ. In contrast, the Ekman supply of silicic acid to the PFZ is comparable to the upwelling supply, because of the rapid convergence of silicic acid in the PFZ (Fig. 13).

Finally, we examine the export of particulate organic nitrogen and silica across the base of the annual maximum mixed layer depth (term 5 in Eq. (2)) for each region. The export of nutrients from the euphotic zone (nominally 75 m) has been provided by John Dunne from his work using empirical relationships to deduce the global export of organic matter from satellite data [29]. We propagate the export of nutrients to the depth of the annual maximum mixed layer depth with the Martin formulation: $F(z) = F_{75} \times (z/75)^{-0.9}$, where $F(z)$ is the export as a function of depth and F_{75} is the export at 75 m.

It is tempting to speculate about the mismatches between nutrient supply and demand terms in Fig. 14 and search for sources or sinks to close the gap. However, these mismatches are very likely smaller than the sum of uncertainties on each term, the calculation of which is far from straightforward. Of the physical terms that are depicted in Fig. 14, the geostrophic nutrient supply may have the largest percent errors. The satellite-derived geostrophic velocities have an RMS error relative to drifters of 10–15 cm s^{-1} in swift boundary currents and 3–5 cm s^{-1} in low variability areas [23], which is between 10 and 100% of the mean velocities. The size of the geostrophic term is also very sensitive to averaging region. In addition, the annual mean nitrate concentrations in World Ocean Atlas have an average standard error of 7% of mean concentration in the top 500 m of the water column [87]. The eddy-driven advection of nutrients is also uncertain, in large part because of our choice of a constant GM diffusion coefficient of 1,000 m^2 s^{-1}. A recent statistical analysis of surface drifters and satellite altimetry suggests that average cross-ACC diffusion near the Polar Front is between 2,000 and 4,000 m^2 s^{-1}[97]. A higher diffusion coefficient would translate to a stronger eddy transport of nutrients out of the Polar Frontal Zone, helping offset the huge vertical supply. Uncertainties associated with the organic particle sinking flux are also considerable; these are derived from a multistep process accounting for known and unknown errors at each step, leading to an estimated 40% uncertainty in the export of organic carbon [29]. In addition, to convert the total export of organic carbon to nutrient export, spatially variable stoichiometric ratios are used, introducing an additional error of up to 4% [29]. Thus, the error for the estimate of nutrient export is of the order 50%.

In addition to the errors on the terms appearing in Fig. 14, there are terms that are neglected entirely in this depiction. First, our scale analysis of mixing of nutrients across the fronts was too uncertain to be included in Fig. 14. Next, the eddy heaving of isopycnals laden with nutrients into the euphotic zone, while not being part of the diapycnal mixing term, has been invoked to close the nutrient budget [98], though a quantitative estimate of this supply mechanism remains controversial [99]. Moreover, the eddy heaving hypothesis does not explain the summertime drawdown of dissolved inorganic carbon that occurs in the absence of a corresponding inorganic nutrient drawdown, observed at the Bermuda Atlantic Time Series and the Hawaii Ocean Time Series [100]. Models and observations have suggested an important role for dissolved organic nutrients in fueling subtropical productivity [101–104], which would help explain carbon fixation in the absence of inorganic nutrient drawdown. Nutrients in their dissolved organic form persist in the euphotic zone longer than their inorganic counterparts. In theory, the conversion of particulate organic matter to dissolved organic matter would allow the cross-frontal flux of nutrients to persist further from the fronts, since it effectively lengthens the timescale for the particles to sink past the base of the mixed layer. Finally, nitrogen fixation may provide a complementary biological mechanism capable of supplying fixed nitrogen to the subtropical gyres [e.g., 76, 105, 106], as long as phosphate and iron are available.

Despite large uncertainties in the scale analysis of nutrient supply and demand, Fig. 14 suggests two salient messages. First, lateral processes provide critical supply terms of nutrients in both the Northern Hemisphere subtropical gyres and the downwelling Subantarctic Zone of the Southern Hemisphere. Thus, we should abandon any expectation that the vertical sinking of particles be balanced locally by vertical transport processes. Second, the search for a missing nutrient supply in the subtropics [e.g., 94] might be more accurately reframed as a quest for smaller errors and particularly better constraints on the turbulent mixing supply across the gyre's bounding fronts. The sign of the imbalances in the Southern Ocean especially suggest this to be an area of needed improvement: the down-gradient transport of nutrients from the PFZ to the SAZ would bring our scale analysis closer to balance in both regions.

5 Conclusions and Open Questions

In our tour of the ocean's large-scale biogeochemical fronts, we first discussed the processes that maintain these sharp gradient regions and then explored the mechanisms that act against them. We have shown that, on timescales greater than a year, cross-frontal exchange plays a critical role in fueling primary productivity in the vertically stratified and laterally homogenous subtropical regions. Though diffusivity across density fronts is often reduced relative to background diffusivity, the exchange that is permitted acts on a sharp biogeochemical gradient, thus reconciling the apparently disparate views of fronts as both barriers to mixing and essential gateways of exchange.

Ekman transport proves to be a powerful exchange mechanism across the boundaries of the subtropical gyres, particularly at their poleward flanks, and is a crucial supplier of nutrients to the subtropics, as first revealed for the subtropical North Atlantic by Williams and Follows [13]. The Ekman supply is also easy to diagnose, requiring only knowledge of wind stress and nutrient concentrations. We suspect that isopycnal mixing due to mesoscale motions is also an important term in nutrient budgets. However, there are still relatively few observations or models that resolve the spatial scale at which this exchange happens (tens of kilometers) and can simultaneously quantify the impact at basin to global scales. The parameterizations of isopycnal mixing in the coarse-resolution models used to simulate climate and biogeochemistry generally do not include the impact of jets on the suppression of isopycnal mixing at the jet core and the enhancement along critical levels. Thus, the quantitative role of mixing across fronts remains an open frontier for exploration in terms of its biogeochemical impact. Moreover, this exploration is becoming timely given both that parameterizations of isopycnal mixing more accurately reflecting jet dynamics are rapidly evolving [e.g., 107] and that the use of eddy-resolving models for biogeochemical studies is now feasible.

In the Northern Hemisphere subtropics, our estimate of the nutrient supply rate is within the (large) errors of estimate of the nutrient demand. As more robust estimates of isopycnal mixing, the vertical eddy supply of nutrients, nitrogen fixation, and primary productivity fueled by dissolved organic nutrients become available, we expect these budgets to approach a more quantitative balance. In the mixed layer of the downwelling Subantarctic Zone, nutrient concentrations are much higher than in the Northern Hemisphere subtropics. Here, the Ekman flux of nutrients across the Subantarctic Front is balanced in large part by the downwelling of nutrients from the mixed layer into the pycnocline. This balance underscores the role of the ACC fronts as nutrient gateways. Between the Polar Front and Subantarctic Front, powerful westerly winds drive the upwelling of nutrients from the abyss; Ekman transport drives these nutrients across the Subantarctic Front; and north of the Subantarctic Front, these nutrients are pumped into the pycnocline. Thus, the fronts of the ACC provide a critical connection between the dense ocean interior and the pycnocline. Without this link, the nutrients that sink past the base of the pycnocline would depend entirely on diapycnal mixing to make their upward return trip, and the global distribution of productivity would be significantly altered [16].

Even as we seek a fundamental understanding of the role of cross-frontal processes in setting annual mean nutrient budgets, intriguing questions about the temporal variability of these processes are on the horizon. Variability in Ekman exchange across fronts, which is largely modulated by variability in wind stress, influences seasonality in chlorophyll concentrations in the subtropical North Pacific [12]. In addition, intra-annual variability in the cross-ACC flux of biogeochemical tracers appears to be tied to Ekman transport variability with little compensation by eddy advection [90]. However, connections between cross-frontal Ekman transport and subtropical biogeochemical variability have yet to be fully explored on interannual or interdecadal timescales. Similarly, recent observations have revealed substantial interannual variability in the shallow limb of the Meridional Overturning Circulation (MOC) [108, 109], which is an important conduit of nutrients to the North Atlantic [36, 37, 45, 110]. Yet, a quantitative assessment of impact of MOC's variability on the biogeochemistry of the North Atlantic awaits future study. Likewise, in the North Pacific, ring-mediated exchange across the Kuroshio is hypothesized to set interannual variability in the density stratification of the subtropical mode water in the North Pacific [42, 43], but the implications for nutrients and phytoplankton remain largely unexplored. Deploying our physical understanding of cross-frontal exchange to answer these unresolved questions about the living ocean offers a wealth of prospects for future research.

Acknowledgements The authors are grateful to Stephanie Schollaert-Uz, Ric Williams, and Igor Belkin for insightful reviews, which improved an early draft of this chapter. We thank John Dunne for his use of the particulate nutrient export data. We also gratefully acknowledge the many data and modeling resources used throughout this work that have been made available online: the AVISO group for compiling and making available satellite altimetry data and related products; the Estimating the Circulation and Climate of the Ocean (ECCO) group for publishing several versions of the ECCO model output, including the Ocean Circulation on Climate Atlas (OCCA) used here; de Boyer Montegut and colleagues who have calculated mixed layer depths for the global ocean and continue updating the calculations with ARGO data; Risien and Chelton who have posted their climatology of global wind stress; and the NODC for publishing the World Ocean Atlas. Support from the Canada's NSERC Discovery program, NOAA-Cooperative Institute for Climate Science Grant #NA08OAR4320752; the National Oceanic and Atmospheric Administration, US Department of Commerce award NA07OAR4310096; and the Office of Science (BER), US Department of Energy, Grant No. DE-FG02-07ER64467 are gratefully acknowledged. This material is based upon work partially supported by the National Science Foundation under Grant No. 0701252. All statements, findings, conclusions, and recommendations are those of the authors and do not necessarily reflect the views of the National Oceanic and Atmospheric Administration, the US Department of Commerce, the US Department of Energy, NSERC, or the National Science Foundation.

References

1. Hoskins BJ (1982) The mathematical theory of frontogenesis. Annu Rev Fluid Mech 14(1): 131–151. doi:10.1146/annurev.fl.14.010182.001023
2. Rhines P (1994) Jets. Chaos: an interdisciplinary. J Nonlinear Sci 4(2):313–339
3. Rudnick D, Ferrari R (1999) Compensation of horizontal temperature and salinity gradients in the ocean mixed layer. Science 283:526–529
4. Macvean MK, Woods JD (1980) Redistribution of scalars during upper ocean frontogenesis: a numerical model. QJR Meteorol Soc 106:293–311. doi:10.1002/qj.49710644805
5. Bower AS, Rossby HT, Lillibridge JL (1985) The gulf stream: barrier or blender? J Phys Oceanogr 15:24–32
6. Bower AS (1991) A simple kinematic mechanism for mixing fluid parcels across a meandering jet. J Phys Oceanogr 21:173–180
7. Bower AS, Lozier, MS (1994) A closer look at particle exchange in the gulf stream. J Phys Oceanogr 24(6):1399–1418. doi:10.1175/1520-0485(1994)024<1399:ACLAPE>2.0.CO;2
8. Williams RG, Follows MJ (2003) Physical transport of nutrients and the maintenance of biological production. In: Fasham MJR (ed) Ocean biogeochemistry. Springer-Verlag, Berlin, pp 19–51
9. Klaas C, Archer DE (2002) Association of sinking organic matter with various types of mineral ballast in the deep sea: implications for the rain ratio. Global Biogeochem Cycles 16(4): 63–61
10. Lutz M, Dunbar R, Caldeira K (2002) Regional variability in the vertical flux of particulate organic carbon in the ocean interior. Global Biogeochem Cycles 16(3):11–18
11. Martin JH, Knauer GA, Karl DM, Broenkow WW (1987) VERTEX: carbon cycling in the northeast Pacific. Deep Sea Res A Oceanogr Res Pap 34(2):267–285
12. Ayers J, Lozier M (2009) Physical controls on the seasonal migration of the North Pacific Transition Zone Chlorophyll Front. J Geophys Res. doi:10.1029/2009JC005596
13. Williams RG, Follows MJ (1998) The Ekman transfer of nutrients and maintenance of new production over the North Atlantic. Deep Sea Res I Oceanogr Res Pap 45(2–3):461–489
14. Sarmiento JL, Gruber N, Brzezinski MA, Dunne JP (2004) High-latitude controls of thermocline nutrients and low latitude biological productivity. Nature 427:56–60
15. Sarmiento JL, Simeon J, Gnanadesikan A, Gruber N, Key RM, Schlitzer R (2007) Deep ocean biogeochemistry of silicic acid and nitrate. Global Biogeochem Cycles 21(1):GB1S90
16. Gnanadesikan A, Dunne JP, Key RM, Matsumoto K, Sarmiento JL, Slater RD, Swathi PS (2004) Oceanic ventilation and biogeochemical cycling: understanding the physical mechanisms that produce realistic distributions of tracers and productivity. Global Biogeochem Cycles 18(4): 1–17
17. Palter JB, Sarmiento J, Gnanadesikan A, Simeon J, Slater R (2010) Fueling export production: nutrient return pathways from the deep ocean and their dependence on the Meridional Overturning Circulation. Biogeosciences 7(11):3549–3568. doi:10.5194/bg-7-3549-2010

18. Tomczak M, Godfrey J (2003) Regional oceanography: an introduction, 2nd edn. Pergamon, New York
19. Sharples J, Simpson J (2001) Shelf sea and shelf slope fronts. In: Steele JH, Thorpe SA, Turekian KK (eds) Encyclopedia of ocean sciences. Academic, San Diego, pp 2760–2768
20. Simpson J (1998) Tidal processes in shelf seas. In: Brink KH, Robinson AR (eds) The sea, vol 10. Wiley, New York
21. Luyten J, Pedlosky J, Stommel H (1983) The ventilated thermocline. J Phys Oceanogr 13:292–309
22. Pedlosky J (1990) The dynamics of the oceanic subtropical gyres. Science 248:316–322
23. Rio MH, Hernandez F (2004) A mean dynamic topography computed over the world ocean from altimetry, in situ measurements, and a geoid model. J Geophys Res 109(C12), C12032. doi:10.1029/2003jc002226
24. Fine RA, Maillet KA, Sullivan KF, Willey D (2001) Circulation and ventilation flux of the Pacific Ocean. J Geophys Res 106. doi:10.1029/1999jc000184
25. Sarmiento JL (1983) A tritium box model of the North Atlantic thermocline. J Phys Oceanogr 13(7):1269–1274
26. Jenkins WJ, Wallace DWR (1992) Tracer based inferences of new primary production in the sea. In: Falkowski PG, Woodhead AD (eds) Primary productivity and biogeochemical cycles in the sea, vol 37. Springer, Plenum, New York, pp 299–316
27. McCartney MS (1982) The subtropical recirculation of Mode Waters. J Mar Res 40:427–464
28. Stommel H (1979) Determination of water mass properties of water pumped down from the Ekman layer to the geostrophic flow below. Proc Natl Acad Sci 76:3051–3055
29. Dunne JP, Sarmiento JL, Gnanadesikan A (2007) A synthesis of global particle export from the surface ocean and cycling through the ocean interior and on the seafloor. Global Biogeochem Cycles 21(4):16
30. Sarmiento JL, Herbert TD, Toggweiler JR (2008) Causes of anoxia in the world ocean. Global Biogeochem Cycles 2. doi:10.1029/GB002i002p00115
31. Najjar RG, Sarmiento JL, Toggweiler JR (1992) Downward transport and fate of organic matter in the ocean: simulations with a general circulation model. Global Biogeochem Cycles 6. doi:10.1029/91gb02718
32. Palter JB, Lozier MS, Barber RT (2005) The effect of advection on the nutrient reservoir in the North Atlantic subtropical gyre. Nature 437(7059):687–692
33. Brandt P, Hormann V, Körtzinger A, Visbeck M, Krahmann G, Stramma L, Lumpkin R, Schmid C (2010) Changes in the ventilation of the oxygen minimum zone of the Tropical North Atlantic. J Phys Oceanogr 40(8):1784–1801. doi:10.1175/2010JPO4301.1
34. Deutsch C, Sigman DM, Thunell RC, Meckler AN, Haug GH (2004) Isotopic constraints on glacial/interglacial changes in the oceanic nitrogen budget. Global Biogeochem Cycles 18(4): 1–22
35. Gruber N, Sarmiento JL (1997) Global patterns of marine nitrogen fixation and denitrification. Global Biogeochem Cycles 11(2):235–266
36. Moore CM, Mills MM, Achterberg EP, Geider RJ, Laroche J, Lucas MI, McDonagh EL, Pan X, Poulton AJ, Rijkenberg MJA, Suggett DJ, Ussher SJ, Woodward EMS (2009) Large-scale distribution of Atlantic nitrogen fixation controlled by iron availability. Nat Geosci 2(12): 867–871
37. Williams RG, Roussenov V, Follows MJ (2006) Nutrient streams and their induction into the mixed layer. Global Biogeochem Cycles 20(1):GB1016. doi:10.1029/2005gb002586
38. Lewis MR, Kuring N, Yentsch C (1988) Global patterns of ocean transparency: implications for the new production of the open ocean. J Geophys Res 93(C6):6847–6856
39. Follows MJ, Dutkiewicz S (2002) Meteorological modulation of the North Atlantic spring bloom. Deep Sea Res II 49:321–344
40. Sverdrup H (1953) On conditions for the vernal blooming of phytoplankton. J Int Council Exploration Sea 18:287–295
41. Broeker WS, Peng T-H (1982) Tracers in the Sea. Lamont-Doherty Earth Observatory, Palisades

42. Qiu B, Chen S, Hacker P (2007) Effect of mesoscale eddies on subtropical mode water variability from the Kuroshio Extension System Study (KESS). J Phys Oceanogr 37(4): 982–1000
43. Qiu B, Chen S (2006) Decadal variability in the formation of the North Pacific subtropical mode water: oceanic versus atmospheric control. J Phys Oceanogr 36(7):1365–1380
44. Wiggins S (2005) The dynamical systems approach to Lagrangian transport in oceanic flows. Annu Rev Fluid Mech 37:295–328
45. Palter JB, Lozier MS (2008) On the source of Gulf Stream nutrients. J Geophys Res 113: C06018
46. Tsuchiya M (1989) Circulation of the Antarctic intermediate water in the North Atlantic Ocean. J Mar Res 47:747–755
47. Williams RG, McDonagh EL, Roussenov VM, Torres-Valdes S, King B, Sanders R, Hansell DA (2011) Nutrient streams in the North Atlantic: advective pathways of inorganic and organic nutrients. Global Biogeochem Cycles 25(GB4008). doi:10.1029/2010GB003853
48. Belkin IM, Gordon AL (1996) Southern Ocean fronts from the Greenwich meridian to Tasmania. J Geophys Res C Oceans 101(C2):3675–3696
49. Deacon GER (1937) The hydrology of the Southern Ocean. Discovery Rep 15:1–124
50. Orsi AH, Whitworth T, Nowlin WD (1995) On the meridional extent and fronts of the Antarctic Circumpolar Current. Deep Sea Res I Oceanogr Res Pap 42(5):641–673
51. Sokolov S, Rintoul SR (2009) Circumpolar structure and distribution of the Antarctic Circumpolar Current fronts: 1. Mean circumpolar paths. J Geophys Res 114(19). doi:10.1029/2008JC005108
52. Hughes CW, Ash ER (2001) Eddy forcing of the mean flow in the Southern Ocean. J Geophys Res 106(C2):2713–2722. doi:10.1029/2000jc900332
53. Naveira Garabato AC, Jullion LØ, Stevens DP, Heywood KJ, King BA (2009) Variability of subantarctic mode water and Antarctic intermediate water in the drake passage during the late-twentieth and early-twenty-first centuries. J Climate 22(13):3661–3688. doi:10.1175/2009jcli2621.1
54. Sokolov S, Rintoul SR (2007) On the relationship between fronts of the Antarctic Circumpolar Current and surface chlorophyll concentrations in the Southern Ocean. J Geophys Res 112. doi:10.1029/2006jc004072
55. Pollard R, Lucas M, Read J (2002) Physical controls on biogeochemical zonation in the Southern Ocean. Deep Sea Res II 49:3289–3305
56. Whitworth T, III, Orsi AH, Kim SJ, Nowlin WD, Jr, Locarnini RA (1998) Water masses and mixing near the Antarctic slope front. In: Ocean, ice, and atmosphere: interactions at the antarctic continental margin, vol 75. Antarctic Research Series, AGU, Washington, pp 1–27
57. Jacobs SS (1991) On the nature and significance of the Antarctic Slope Front. Mar Chem 35 (1–4):9–24. doi:10.1016/S0304-4203(09)90005-6
58. Mosby H (1934) The waters of the Atlantic Antarctic Ocean. In: Dybwad J (ed) Scientific results of The Norwegian Antarctic Expeditions 1927–1928, vol 11. Consul Lars Christensen, Oslo
59. Boyd P, LaRoche J, Gall M, Frew R, McKay RML (1999) Role of iron, light, and silicate in controlling algal biomass in subantarctic waters SE of New Zealand. J Geophys Res Oceans 104(C6):13395–13408. doi:10.1029/1999jc900009
60. Marinov I, Gnanadesikan A, Toggweiler JR, Sarmiento JL (2006) The Southern Ocean biogeochemical divide. Nature 441(7096):964–967
61. Takeda S (1998) Influence of iron availability on nutrient consumption ratio of diatoms in oceanic waters. Nature 393(6687):774–777
62. Leynaert A, Bucciarelli E, Claquin P, Dugdale RC, Martin-Jézéquel V, Pondaven P, Ragueneau O (2004) Effect of iron deficiency on diatom cell size and silicic acid uptake kinetics. Limnol Oceanogr 49(4I):1134–1143
63. Franck VM, Brzezinski MA, Coale KH, Nelson DM (2000) Iron and silicic acid concentrations regulate Si uptake north and south of the Polar Frontal Zone in the Pacific Sector of the Southern Ocean. Deep Sea Res II Topical Studies Oceanogr 47(15–16): 3315–3338

64. Matsumoto K, Sarmiento, JL, Brzezinski MA (2002) Silicic acid leakage from the Southern Ocean: a possible explanation for glacial atmospheric pCO_2. Global Biogeochem. Cycles 16. doi:10.1029/2001gb001442
65. Brzezinski MA, Pride CJ, Franck VM, Sigman DM, Sarmiento JL, Matsumoto K, Gruber, N, Rau GH, Coale KH (2002) A switch from Si(OH)4 to NO-3 depletion in the glacial Southern Ocean. Geophys Res Lett 29(12). doi:10.1029/2001gl014349
66. Ito T, Follows MJ (2005) Preformed phosphate, soft tissue pump and atmospheric CO_2. J Mar Res 63(4):813–839
67. Marinov I, Gnanadesikan A, Sarmiento JL, Toggweiler JR, Follows M, Mignone BK (2008) Impact of oceanic circulation on biological carbon storage in the ocean and atmospheric pCO_2. Global Biogeochem Cycles 22(3):GB3007
68. Marinov, I., Follows, M.J., Gnanadesikan, A., Sarmiento, J.L., Slater, R.D.: How does ocean biology affect atmospheric pCO2? Theory and models. Journal of Geophysical Research C: Oceans 113(7) (2008).
69. Archer DE, Eshel G, Winguth A, Broecker W, Pierrehumbert R, Tobis M, Jacob R (2000) Atmospheric pCO_2 sensitivity to the biological pump in the ocean. Global Biogeochem Cycles 14(4):1219–1230
70. Rogerson A, Miller P, Pratt L, Jones C (1999) Lagrangian motion and fluid exchange in a barotropic meandering jet. J Phys Oceanogr 29:2635–2655
71. Samelson RM (1992) Fluid exchange across a meandering jet. J Phys Oceanogr 22:431–440
72. Lozier M, Pratt L, Rogerson A, Miller P (1997) Exchange geometry revealed by float trajectories in the Gulf Stream. J Phys Oceanogr 27:2327–2341
73. Abernathey R, Marshall J, Mazloff M, Shuckburgh E (2010) Enhancement of mesoscale eddy stirring at steering levels in the Southern Ocean. J Phys Oceanogr 40(1):170–184. doi:10.1175/2009JPO4201.1
74. Gent PR, Willebrand J, McDougall TJ, McWilliams JC (1995) Parameterizing eddy-induced tracer transports in ocean circulation models. J Phys Oceanogr 25(4):463–474
75. Lee MM, Williams RG (2000) The role of eddies in the isopycnic transfer of nutrients and their impact on biological production. J Mar Res 58(6):895–917
76. Palter JB, Lozier MS, Sarmiento JL, Williams RG (2011) The supply of excess phosphate across the Gulf Stream and the maintenance of subtropical nitrogen fixation. Global Biogeochem Cycles 25(4), GB4007. doi:10.1029/2010gb003955
77. Gruber N, Lachkar Z, Frenzel H, Marchesiello P, Munnich M, McWilliams JC, Nagai T, Plattner G-K (2011) Eddy-induced reduction of biological production in eastern boundary upwelling systems. Nature Geosci 4(11):787–792
78. Risien CM, Chelton DB (2008) A global climatology of surface wind and wind stress fields from eight years of QuikSCAT scatterometer data. J Phys Oceanogr 38(11):2379–2413
79. Conkright ME, Locarnini RA, Garcia HE, O'Brien TD, Boyer TP, Stephens C, Antonov JI (2001) World Ocean Atlas 2001: objective analyses, data statistics, and figures. In: National Oceanographic Data Center, Silver Spring, p 17
80. Polovina JJ, Howell E, Kobayashi DR, Seki MP (2001) The transition zone chlorophyll front, a dynamic global feature defining migration and forage habitat for marine resources. Progr Oceanogr 49(1–4):469–483
81. Lenn YD, Chereskin TK (2009) Observations of Ekman currents in the Southern Ocean. J Phys Oceanogr 39(3):768–779
82. de Boyer Montégut C, Madec G, Fischer AS, Lazar A, Iudicone D (2004) Mixed layer depth over the global ocean: an examination of profile data and a profile-based climatology. J Geophys Res Oceans 109(C12), C12003. doi:10.1029/2004jc002378
83. Danabasoglu G, McWilliams JC, Gent PR (1994) The role of mesoscale tracer transports in the global ocean circulation. Science 264(5162):1123–1126
84. Forget G (2009) Mapping ocean observations in a dynamical framework: a 2004–2006 ocean atlas. J Phys Oceanogr 40(2010):1201–1221

85. Maze G, Forget G, Buckley M, Marshall J, Cerovecki I (2009) Using transformation and formation maps to study the role of air- sea heat fluxes in North Atlantic Eighteen Degree Water Formation. J Phys Oceanogr 39(8):1818–1835
86. Gent PR, McWilliams JC (1990) Isopycnal mixing in ocean circulation models. J Phys Oceanogr 20:150–155
87. Garcia HE, Locarnini RA, Boyer TP, Antonov JI, Zweng MM, Baranova OK, Johnson DR (2010) World Ocean Atlas 2009, Volume 4: Nutrients (phosphate, nitrate, silicate). In: Levitus S (ed) NOAA Atlas NESDIS 71, vol 4. U.S. Government Printing Office, Washington, p 398
88. Visbeck M, Marshall J, Haine T, Spall M (1997) Specification of eddy transfer coefficients in coarse-resolution ocean circulation models. J Phys Oceanogr 27:381–402
89. Hallberg R, Gnanadesikan A (2006) The role of eddies in determining the structure and response of the wind-driven southern hemisphere overturning: results from the modeling eddies in the southern ocean (MESO) project. J Phys Oceanogr 36(12):2232–2252
90. Ito T, Woloszyn M, Mazloff M (2010) Anthropogenic carbon dioxide transport in the Southern Ocean driven by Ekman flow. Nature 463(7277):80–83
91. Winkel DP, Gregg MC, Sanford TB (2002) Patterns of shear and turbulence across the Florida Current. J Phys Oceanogr 32(11):3269–3285
92. Ledwell JR, Watson AJ, Law CS (1993) Evidence for slow mixing across the pycnocline from an open-ocean tracer-release experiment. Nature 364:701–703
93. Gregg MC, Sanford TB (1980) Signatures of mixing from the Bermuda Slope, the Sargasso Sea and the Gulf Stream. J Phys Oceanogr 10:105–127
94. Lewis MR, Harrison WG, Oakey NS, Hebert D, Platt T (1986) Vertical nitrate fluxes in the oligotrophic ocean. Science 234:870–873
95. Naveira Garabato AC, Polzin KL, King BA, Heywood KJ, Visbeck M (2004) Widespread intense turbulent mixing in the Southern Ocean. Science 303(5655):210–213
96. Heywood KJ, Naveira Garabato AC, Stevens DP (2002) High mixing rates in the abyssal Southern Ocean. Nature 415(6875):1011–1014
97. Sallée JB, Speer K, Morrow R, Lumpkin R (2008) An estimate of Lagragian eddy statistics and diffusion in the mixed layer of the Southern Ocean. J Mar Res 66(4):441–463
98. McGillicuddy DJ, Robinson AR, Siegel DA, Jannasch HW, Johnson R, Dickey TD, McNeil J, Michaels AF, Knap AH (1998) Influence of mesoscale eddies on new production in the Sargasso Sea. Nature 419:263–266
99. Oschlies A, Garcon V (1998) Eddy-induced enhancement of primary production in a model of the North Atlantic Ocean. Nature 394:266–269
100. Karl DMEA (2003) Temporal studies of biogeochemical processes determined from ocean time-series observations during the JGOFS era. In: Fasham MJR (ed) Ocean biogeochemistry. Academic, New York, pp 239–267
101. Roussenov V, Williams RG, Mahaffey C, Wolff GA (2006) Does the transport of dissolved organic nutrients affect export production in the Atlantic Ocean? Global Biogeochem Cycles 20(3):GB3002
102. Mahaffey C, Williams RG, Wolff GA, Anderson WT (2004) Physical supply of nitrogen to phytoplankton in the Atlantic Ocean. Global Biogeochem Cycles 18, GB1034
103. Mather RL, Reynolds SE, Wolff GA, Williams RG, Torres-Valdes S, Woodward EMS, Landolfi A, Pan X, Sanders R, Achterberg EP (2008) Phosphorus cycling in the North and South Atlantic Ocean subtropical gyres. Nat Geosci 1(7):439–443
104. Torres-Valdés S, Roussenov VM, Sanders R, Reynolds S, Pan X, Mather R, Landolfi A, Wolff GA, Achterberg EP, Williams RG (2009) Distribution of dissolved organic nutrients and their effect on export production over the Atlantic Ocean. Global Biogeochem Cycles 23. doi:10.1029/2008gb003389
105. Davis CS, McGillicuddy DJ Jr (2006) Transatlantic abundance of the N2-fixing colonial cyanobacterium Trichodesmium. Science 312(5779):1517–1520. doi:10.1126/science.1123570

106. Mahaffey C, Michaels AF, Capone DG (2005) The conundrum of marine N2 fixation. Am J Sci 305(6-8 SPEC. ISS.):546–595
107. Ferrari R, Nikurashin M (2010) Suppression of eddy diffusivity across jets in the Southern Ocean. J Phys Oceanogr 40(7):1501–1519. doi:10.1175/2010JPO4278.1
108. Hakkinen S, Rhines PB (2009) Shifting surface currents in the northern North Atlantic Ocean. J Geophys Res C Oceans 114(4):C04005
109. Hátún H, Sando AB, Drange H, Hansen B, Valdimarsson H (2005) Influence of the Atlantic subpolar gyre on the thermohaline circulation. Science 309(5742):1841–1844
110. Sarmiento JL, Slater R, Barber R, Bopp L, Doney SC, Hirst AC, Kleypas J, Matear R, Mikolajewicz U, Monfray P, Soldatov V, Spall SA, Stouffer R (2004) Response of ocean ecosystems to climate warming. Global Biogeochem Cycles 18(3)
111. Campbell JW (1995) The lognormal distribution as a model for bio-optical variability in the sea. J Geophys Res Oceans 100(C7):13237–13254. doi:10.1029/95jc00458
112. Palter JB, Lozier MS (2008) On the source of Gulf Stream nutrients. J Geophys Res Oceans 113(C6), C06018. doi:10.1029/2007jc004611
113. Iudicone D, Speich S, Madec G, Blanke B (2008) The global conveyor belt from a Southern Ocean perspective. J Phys Oceanogr 38(7):1401–1425
114. Talley LD, Reid JL, Robbins PE (2003) Data-based meridional overturning streamfunctions for the global ocean. J Climate 16(19):3213–3226

The Pacific-Atlantic Front in the East Siberian Sea of the Arctic Ocean

Matthew B. Alkire, Robert Rember, and Igor Polyakov

Contents

Abstract It is the purpose of this paper to introduce the reader to the various water masses that are found in the Arctic Ocean and describe how fronts, separating these water masses, are identified through the application of physical and biogeochemical measurements. Variations in the positions/alignments of these fronts help to describe changes in circulation patterns that impact local physical, chemical, and biological

M. B. Alkire (✉)
School of Oceanography, University of Washington, Seattle, WA, USA
e-mail: alkirem@uw.edu

R. Rember and I. Polyakov
International Arctic Research Center, University of Alaska Fairbanks, Fairbanks, AK, USA

Igor M. Belkin (ed.), *Chemical Oceanography of Frontal Zones*,
Hdb Env Chem (2022) 116: 63–96, DOI 10.1007/698_2021_795,

Published online: 11 August 2021

systems. In this chapter, the distributions of various physical and geochemical parameters representing summer conditions in the Arctic during 2015, a year of particularly good spatial coverage in terms of scientific observations, will be presented and analyzed with a focus on identifying the front separating Atlantic and Pacific halocline waters in the East Siberian Sea by qualitative means.

Keywords Arctic, Fronts, Halocline, NO parameter, Tracers, Water masses

Abbreviations

AOU	Apparent oxygen utilization
AW	Atlantic water
BSB	Barents Sea Branch
ECSW	Eastern Chukchi Summer Water
FSB	Fram Strait Branch
LHW	Lower halocline water
NABOS	Nansen and Amundsen Basins Observation System
NO_3^-	Nitrate
O_2	Dissolved oxygen
PO_4^{3-}	Phosphate
PW	Pacific water
PWW	Pacific Winter Water
S	Salinity
$Si(OH)_4$	Silicic acid
SUNA	Submersible ultraviolet nitrate analyzer
TPC	Transpolar current
UHW	Upper halocline water
WCSW	Western Chukchi Summer Water
θ	Potential temperature

1 Introduction

1.1 Background

The world ocean is vertically stratified by both temperature and salinity. At low latitudes, the upper layers are primarily stratified by temperature (*alpha oceans*) whereas at high latitudes, upper layers are primarily stratified by salinity (*beta oceans*) [1]. The Arctic Ocean is very cold and vertical stratification is controlled primarily by salinity, which varies over a relatively wide range, from approximately 25 to 35. Temperature exhibits a comparatively narrow range in the Arctic Ocean, from approximately −1.9°C to +5°C. In comparison, most of the rest of the world ocean (excluding the Red and Mediterranean Seas) exhibits a salinity range of 32–36 and a temperature range of 2–30°C. The rapid increase in salinity with increasing depth is referred to as a halocline layer and the Arctic halocline is sustained by the movement and mixing of different water masses. It is the meeting of these water masses that form the fronts that will be the subject of this chapter. In order to identify these fronts and their importance to the physics and biogeochemistry of the Arctic Ocean, the different water masses must be presented. First, the basic geography of the Arctic Ocean will be introduced, followed by a brief discussion of the water masses found in the top ~500 m of the water column as well as their distinguishing chemical and physical characteristics. The general circulation of these waters will then be outlined as well as the dramatic changes that have been observed in the Arctic Ocean over the past two decades. In subsequent sections, data collected from four different expeditions during the spring and summer of 2015 will be described and analyzed to provide a snapshot of what is arguably the most important, and largest, front in the Arctic Ocean: the Pacific-Atlantic front.

1.2 Geography

The Arctic Ocean is the smallest of the five oceans (Pacific, Atlantic, Indian, Arctic, and Southern Oceans); its area represents ~5% of global ocean area and ~1% of global ocean volume [2]. In addition to its relatively small area, the Arctic Ocean is also quite shallow. While deep basins are present in this ocean, ~50% of its area is comprised of shallower continental shelves, averaging $\leq$200 m depth. Moving counterclockwise from the Fram Strait (Fig. 1), these shelf seas include the Barents (mean depth $\approx$200 m), Kara (~110 m), Laptev (~56 m), East Siberian (~48 m), Chukchi (~58 m), Beaufort (~80 m), and Lincoln (~300 m) Seas [2].

The Arctic Ocean is split approximately in half by the Lomonosov Ridge, an undersea mountain chain that is centrally located and extends roughly the length of the ocean. The two halves are referred to as the Eurasian and Amerasian Basins. Another ridge, the Mendeleyev-Alpha Ridge, further splits the Amerasian Basin into the Makarov and Canada Basins; similarly, the Gakkel Ridge separates the Eurasian Basin into the Nansen and Amundsen Basins (see Fig. 1). The shelves are separated from the deep basins by a steep continental slope that runs around the periphery of the deep basins. The slope and the system of ridges cutting across the deep basins play integral roles in the circulation of water masses in the Arctic (this will be discussed in more detail in Sect. 1.5).

1.3 Freshwater Contributions

The most saline water flowing into the Arctic is Atlantic Water (AW) and it enters via Fram Strait and the Barents Sea Opening; the former is referred to as Fram Strait

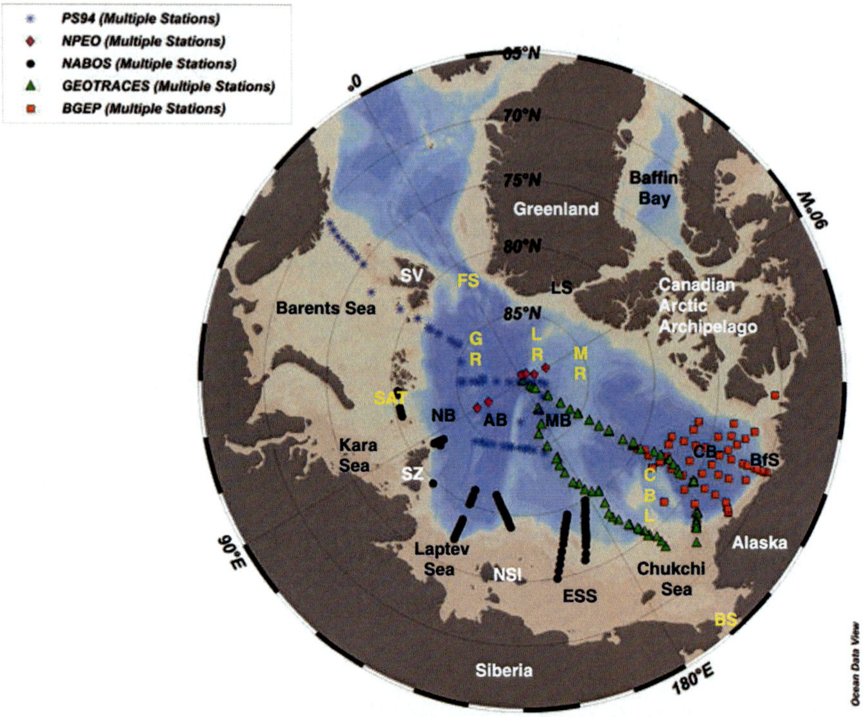

Fig. 1 Map of stations occupied during spring and summer 2015 as part of the PS94 expedition (blue stars), North Pole Environmental Observatory (purple diamonds), Nansen and Amundsen Basin Observation System (black circles), GEOTRACES (green triangles), and Beaufort Gyre Exploration Program (red squares). Relevant land masses are listed in white, marginal seas and deep basins are listed in black, and bathymetric features are listed in yellow. *BfS* Beaufort Sea, *LS* Lincoln Sea, *NB* Nansen Basin, *AB* Amundsen Basin, *MB* Makarov Basin, *CB* Canada Basin, *BS* Bering Strait, *FS* Fram Strait, *SAT* St. Anna Trough, *CBL* Chukchi Borderlands, *GR* Gakkel Ridge, *LR* Lomonosov Ridge, *MR* Mendeleyev-Alpha Ridge, *SV* Svalbard, *SZ* Severnaya Zemlya, and *NSI* New Siberian Islands. Figure created using Ocean Data View Software [68]

Branch (FSB) AW and is significantly warmer than its counterpart, Barents Sea Branch (BSB) AW, due to significant cooling of the latter branch during its transit through the Barents Sea. The FSB AW constitutes the largest source of heat to the Arctic Ocean below the mixed layer. On the opposite side of the Arctic, Pacific water (PW) enters via Bering Strait and is considered to be a source of freshwater to the Arctic as its mean salinity ($31.5 \leq S \leq 32.5$) is lower than that of AW ($34.8 \leq S \leq 35$). As such, the annual PW inflow represents the second largest source of freshwater to the Arctic Ocean, with an annual freshwater input of ~2,500–3,000 km^3 [3].

Additional freshwater is supplied by river runoff (3,000–4,000 km^3 $year^{-1}$), net precipitation over evaporation (1,800–2,200 km^3 $year^{-1}$), and smaller (but growing) contributions from glacial meltwater [4–6]. Despite its small size, the Arctic receives >10% of global annual river runoff. More than 60% of this runoff comes from six large rivers that drain the Eurasian (Ob, Yenisey, Lena, and Kolyma Rivers) and North American (Yukon and Mackenzie Rivers) continents. The remaining runoff drains from smaller rivers and streams that, quite literally, surround the Arctic Ocean. This large volume of freshwater enters the various broad and shallow peripheral seas where they reside for an average of 3.5 ± 2 years [7]. The river waters are eventually exported to the deep basins, typically moving offshore in parallel with either the Lomonosov or Mendeleyev Ridges, depending largely on local wind conditions on the shelves during summer months as well as larger scale sea-level pressure patterns that influence the general circulation regimes of the Arctic [8–13].

Sea ice meltwater contributes to the freshwater budget only seasonally – over the course of an annual cycle, more sea ice is produced in the Arctic Ocean during autumn and winter than is melted in summer. Therefore, on average, there is a net removal of freshwater from the ocean as this water is stored as solid ice. This ice can be partially or entirely melted during summer or exported from the Arctic Ocean to the North Atlantic (primarily via Fram Strait [4]). Both the melting and formation (freezing) of sea ice contributes to the stratification of the Arctic Ocean. Sea ice melt freshens and cools warm AW upon entry, forming one variety of halocline water. The formation of sea ice and the associated release of high salinity brines on the shallow shelves serves to increase the salinity (and density) of shelf waters (that have been freshened by river runoff) to high enough salinity that is sufficient to ventilate the halocline layer, resulting in the formation of another variety of halocline water.

Together, these freshwater inputs (PW, river runoff, precipitation, and sea ice melt) contribute to the ventilation of the halocline layer.

1.4 Hydrography and Water Masses

The water column of the Arctic Ocean can generally be split into four vertical layers: the surface mixed layer (SML), halocline layer, AW layer, and deeper waters (Fig. 2b). The halocline layer is comprised of numerous water sources and separates the cold and relatively fresh surface mixed layer (and overlying cover of sea ice) from the warm ($\theta > 0°C$) and saline ($S > 34.8$) AW at greater depth (~150–400 m). Below the AW layer, deeper waters are somewhat more saline and colder; thus, a typical temperature profile from the Arctic Ocean exhibits an increase in temperature from a near-freezing mixed layer to a mid-depth maximum (marking the core of the warm AW layer) and a subsequent decrease in temperature with increasing depth (Fig. 2a). A typical vertical profile of salinity exhibits a rapid increase in salinity with increasing depth through the halocline, followed by a slower increase with depth as the AW layer is approached; salinity then asymptotes to a maximum value ($S \leq 35$) and remains uniform with increasing depth (Fig. 2b). These features are also recognizable on a θ-S diagram (Fig. 2c). For example, the cold halocline layer, a layer with uniformly cold temperatures but increasing salinities with depth, plots as a nearly straight line of increasing salinity near the freezing point. Below the cold

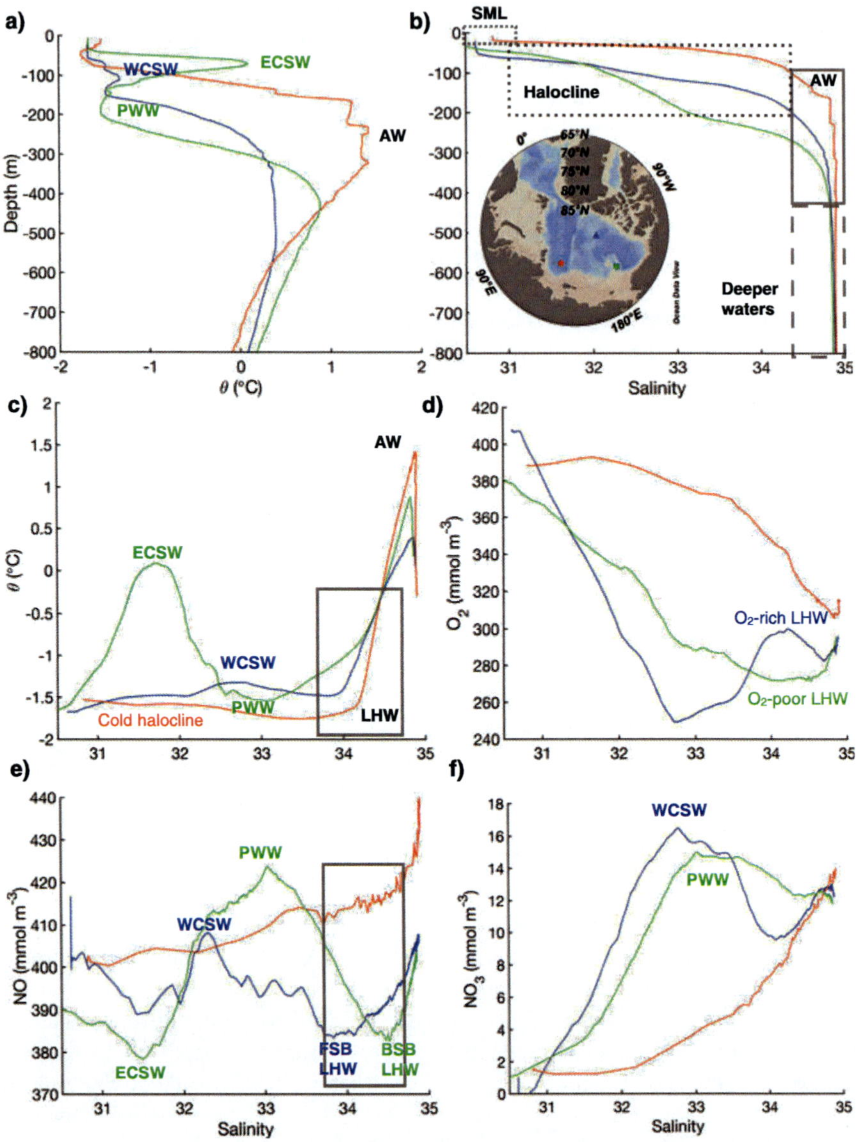

Fig. 2 Vertical profiles of (**a**) potential temperature (θ) and (**b**) salinity and the corresponding (**c**) θ-S diagram for three example stations in the Arctic Ocean (see inset map). Plots of (**d**) dissolved oxygen (O_2), (**e**) the NO parameter ($=9*[NO_3^-] + [O_2]$) and (**f**) nitrate (NO_3) versus salinity are also given for these three stations. Distinct water masses are identified in different panels; the acronyms of these water masses are color-coded to match the station where the water mass is found, or colored black when the water mass is observed at all three stations. Boxes in panel (**b**) approximately separate the surface mixed layer (SML), halocline layer, Atlantic Water (AW), and deeper waters among the three stations. Boxes in panels (**c**) and (**e**) indicate the typical salinity range and associated geochemical features of Lower Halocline Water (LHW). Additional water mass acronyms include: *ECSW* East Chukchi Summer Water, *WCSW* West Chukchi Summer Water, and *PWW* Pacific Winter Water

halocline, the lower halocline water (LHW) can be identified as a bend (or kink) on θ-S diagrams, generally found between salinities of 34 and 34.3. Below the LHW, the temperature-salinity slope increases until the potential temperature maximum marking the core of the AW layer is reached. These general features of the temperature and salinity vertical profiles become more complex with the presence of additional water masses. For example, the presence of summer and winter Pacific halocline waters can be identified via potential temperature maxima and minima, respectively, in vertical profiles of potential temperature (Fig. 2a) and θ-S diagrams (Fig. 2c). Although these features may be easily recognizable in some cases, additional tracers are needed to further distinguish the number of different water masses that contribute to the Arctic halocline.

There are multiple varieties of halocline waters in the Arctic Ocean, but they can be generally split into two types: those originating from PW versus those originating from AW. The Atlantic halocline waters are typically split into two additional groups, those that are formed in the deep basins and those that are formed on the shallower continental shelves. Basin-derived halocline waters are sometimes referred to as Fram Strait Branch lower halocline waters (FSB LHW) (e.g., [14]), or O_2-rich halocline waters [15, 16]. They are cold (near the freezing point) and are characterized by relatively high O_2 concentrations (O_2-rich LHW on Fig. 2d) and lower nutrient concentrations (Fig. 2f). Basin-derived halocline waters are thought to be formed by the initial freshening of AW entering the Arctic Ocean by sea ice meltwater, followed by multiple years of modification due to cooling and salinization in winter (via ice formation and vertical mixing), freshening in summer (via ice melt), and the utilization of nutrients via biological production. The final step in the formation of these halocline waters involves the outflow of fresher waters from the Siberian shelves; this freshwater intrusion increases the stratification and pushes the basin-derived halocline water deeper, isolating them from the surface and preventing further modification via air–ice–ocean interaction [17]. The relatively fresh outflow will be cooled and salinized during winter, forming the cold halocline layer (see Fig. 2c) that overlies the basin-derived LHW [18].

Shelf-derived halocline waters have both Atlantic and Pacific origins. The most saline (S ~ 34.5) variant of shelf-derived halocline waters is commonly referred to as Barents Sea Branch lower halocline water (BSB LHW) [14]. The higher salinity of the BSB LHW (relative to the FSB variety) is attributed to the lower temperatures of the BSB AW as cooler waters melt less sea ice. The higher salinity of the BSB LHW puts this halocline water in closer proximity to warm, FSB AW. This closer proximity results in more mixing; thus, the BSB LHW is typically warmer than the FSB LHW (see Fig. 2c). In addition to its higher salinity and warmer temperatures, the BSB LHW is typically characterized by lower O_2 concentrations (O_2-poor LHW in Fig. 2d) and somewhat higher nutrient concentrations (Fig. 2f), compared to FSB LHW, due to a presumed interaction with shelf (or slope) sediments or bottom waters that have been subjected to organic matter remineralization along its circulation pathway on (or close to) the slope. Minimum values in the NO parameter have been used to identify LHW in the western Arctic (e.g., [19, 20]). Alkire et al. [21] have also shown that, while the NO minima of FSB LHW and BSB LHW may be similar in magnitude, they occur at lower ($34 \leq S \leq 34.3$) and higher (~34.5) salinities, respectively, in accordance with their respective circulation pathways (see Fig. 2e). Therefore, the salinity of the NO minimum can be used to distinguish between FSB and BSB LHW varieties. It is also important to note that less saline, shelf-derived halocline waters of Atlantic origin are also formed on the Kara, Laptev, and/or East Siberian Sea shelves and that these waters may exhibit even lower NO values than those typically associated with the so-called NO minimum associated with LHW [21, 22].

Pacific-origin halocline waters are also shelf-derived, as they are formed/modified on the Chukchi Sea shelf. There are three main varieties of Pacific halocline waters: Eastern Chukchi Summer Water (ECSW), Western Chukchi Summer Water (WCSW), and Pacific Winter Water (PWW) [15]. These water masses may be alternatively referred to as Alaskan Coastal Water, Summer Bering Seawater, and Winter Bering Seawater, respectively [23]. The former group of names will be used in this chapter. All three Pacific halocline water varieties are characterized by low dissolved oxygen concentrations and both WCSW and PWW are additionally characterized by high nutrient concentrations (see Fig. 2d, e) that are, in part, derived from organic matter remineralization occurring in the sediments of the Bering Sea (particularly the Gulf of Anadyr) and the Chukchi Sea [19, 24, 25].

The summer Pacific halocline waters are characterized by localized maximum values of potential temperature in the salinity ranges of $31 < S < 32$ (ECSW) and $32 < S < 33$ (WCSW) (Fig. 2a, c). In addition to the lower salinity, ECSW is also characterized by lower nutrient concentrations and somewhat higher dissolved oxygen concentrations (Fig. 2d, f), relative to the WCSW. Winter Pacific Water is characterized by a potential temperature minimum near the freezing point and high nutrient concentrations centered around a salinity of ~33 ($32.8 \leq S \leq 33.2$).

1.5 General Circulation Patterns

A simplification of the surface circulation of the Arctic Ocean can be generally described as the meeting of two gyres. The Beaufort Gyre is an anticyclonic circulation that is typically centered in the Beaufort Sea/southern Canada Basin and dominates the circulation in the Amerasian Basin. The surface circulation in the Eurasian Basin, in contrast, is generally cyclonic (though there are regions of the Eurasian Basin that exhibit more complex and even anticyclonic flows). These two gyres meet in the central Arctic and form a density front. The Transpolar Current (TPC) is a geostrophic ocean surface current resulting from the density contrast across this front and is the liquid ocean companion of the Transpolar Drift of sea ice. It is a major circulation feature of the Arctic Ocean that moves both Pacific- and Atlantic-derived waters from an origin roughly located around the New Siberian Islands, through the central Arctic and out via the western side of Fram Strait (see Fig. 3).

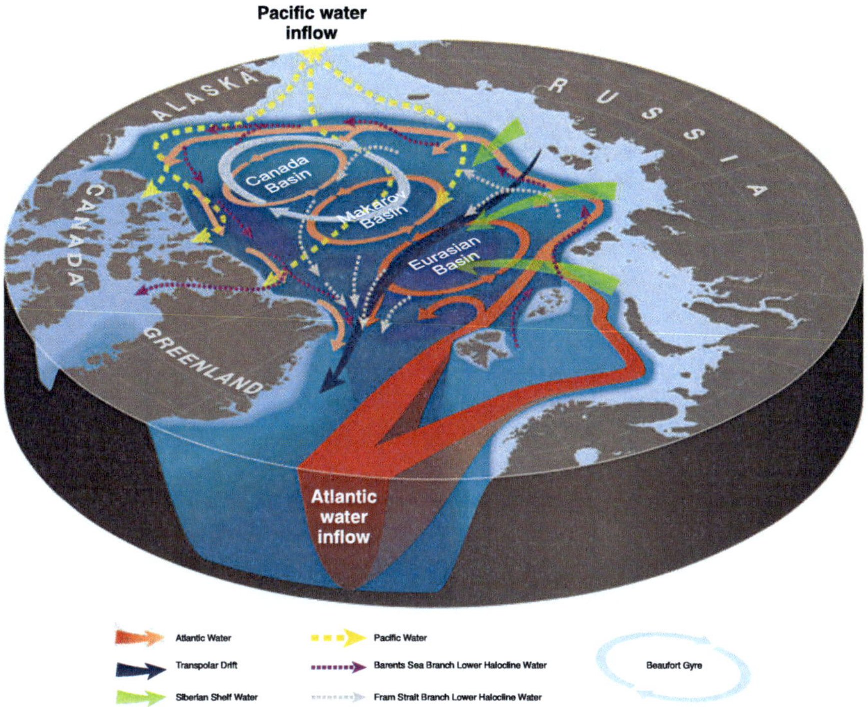

Fig. 3 General circulation features of various water masses of interest in the Arctic Ocean. The Transpolar Drift and Beaufort Gyre are shown as dark and light blue arrows, respectively. The Atlantic water inflow through Fram Strait and the Barents Sea Opening and its cyclonic circulation around the Arctic Ocean is shown as large red arrows. The inflow of Pacific water and its separation into three branches is shown as yellow, dashed arrows. Siberian shelf water flows into the deep basins are shown as green arrows. The circulations of the Fram Strait and Barents Sea Branch Lower Halocline Waters are shown as purple and gray dashed arrows, respectively. Figure adapted from Carmack et al. [69]

The AW layers generally circulate cyclonically around the periphery of the Arctic Ocean (see red arrows in Fig. 3) along the continental slope via the Arctic Boundary Current [26]. This current penetrates the deep basins via movement along the ridges; thus, the Arctic Boundary Current is said to be "topographically steered" as it follows along steep topographic features. Atlantic-derived halocline waters are thought to roughly follow the circulation pattern of the deeper Atlantic core waters [27]. The BSB LHW is thought to leave the shelf via Saint Anna Trough and circulate around the periphery of the Arctic Ocean on the continental slope. The BSB displaces the FSB LHW further off the slope to ventilate the Amundsen and Makarov Basins, as well as the northern Canada Basin. The BSB LHW ventilates the halocline of the southern Makarov and Canada Basins [14].

Pacific waters transit the Chukchi Sea and enter the deep basins predominately via three pathways [28]. The eastern pathway transports waters along the Alaskan continental slope and enters Canada Basin via Barrow Canyon whereas the central and western pathways transport waters northward via the Central Channel and Herald Canyon, respectively (see yellow arrows in Fig. 3). The spreading of the different Pacific halocline water varieties is split among these three pathways. For example, ECSW is generally advected along the eastern pathway via the Alaskan Coastal Current. Shimada et al. [29] also suggest that ECSW spreads northward over the central and eastern flank of the Chukchi Plateau. WCSW is generally found west of the Chukchi Cap and spreads northward into the southern Makarov Basin along the flank of the Mendeleyev Ridge. PWW is thought to spread into the Canada Basin primarily east of the Chukchi Cap; however, PWW has been found west of the Chukchi Cap in the western section of the East Siberian Sea and southern Makarov Basin [30, 31]. Pacific halocline waters are generally restricted to the Amerasian Basin and primarily exit the Arctic Ocean via the passages of the Canadian Arctic Archipelago (CAA); however, these waters also cross the Lomonosov Ridge in the region north of the Lincoln Sea and exit via Fram Strait during periods of predominately anticyclonic circulation [32–34]. PW has been mostly (or entirely) absent from the Fram Strait during extended periods (e.g., 2004–2010) of cyclonic and neutral (neither strongly cyclonic nor anticyclonic) circulation regimes [35, 36] as the spatial extent of Pacific waters varies considerably with the expansion and contraction of the Beaufort Gyre. For example, the extent of WCSW can stretch into the Makarov Basin as far as the Lomonosov Ridge/North Pole during periods characterized by anticyclonic circulation and an expanded Beaufort Gyre; thus, WCSW is associated with the front separating Pacific and Atlantic halocline waters [29]. The spatial extent of ECSW has also been linked to the expansion and contraction of the Beaufort Gyre and has been found to accompany WCSW in the expanded, northern limb of the gyre during periods of predominately anticyclonic circulation [23].

1.6 Variations in the Pacific-Atlantic Front

The primary difference in the water column structure of the Amerasian versus Eurasian Basins of the Arctic Ocean is the presence or absence of Pacific halocline waters. Though the boundary separating Pacific and Atlantic waters has been typically found to be aligned roughly along the Lomonosov Ridge (in the center of the Arctic Ocean), this front has shifted back and forth with the intrusion and retreat of Atlantic waters into and out of the Makarov Basin. These realignments of the Pacific-Atlantic front are associated with similar adjustments of the Transpolar Current and the expansion/contraction of the Beaufort Gyre. Therefore, the precise alignment of the front is strongly linked to variations in the large-scale circulation regime of the Arctic Ocean. Tracking the change in the position of this front over time aids in the understanding of the response of the ocean circulation to changes in atmospheric forcing and other variables (e.g., sea ice extent and thickness) as well as the impacts of these circulation changes on biogeochemical cycles, such as the availability of nutrients for photosynthesis [30, 31, 37, 38] and susceptibility to ocean acidification [39–41].

The alignment of the front has been a topic of many studies, with the focus of observations concentrated in the central Arctic [8, 13, 21, 34, 42–46] and in the Lincoln Sea [23, 47, 48]. However, the East Siberian Sea is also known to be a region of interaction between Atlantic waters progressing eastward and Pacific waters moving northward and westward [49]. In fact, the origin of the Transpolar Current is generally thought to be located in the vicinity of the New Siberian Islands, between the northeastern Laptev Sea and northwestern East Siberian Sea (e.g., [50–51]). However, the separation of Atlantic versus Pacific influences in the East Siberian Sea is difficult due to active sedimentary respiration in both the East Siberian Sea (Atlantic origin) and Chukchi Sea (Pacific origin) that produce similar NO_3^-:PO_4^{3-} relationships [52, 53] and other geochemical characteristics, such as relatively high NO_3^- and $Si(OH)_4$ concentrations and low O_2 concentrations. Such difficulties have led prior studies focusing on the quantitative separation of water masses to use a Mixed East Siberian and Chukchi Shelf water endmember, rather than attempt to separate contributions from these two shelf seas [54, 55].

2 Data and Methods

2.1 *Data Sets*

This paper takes advantage of multiple oceanographic cruises that were conducted during the late spring and summer of 2015 to provide a fairly comprehensive and quasi-synoptic picture of the hydrographic structure of the Arctic Ocean. These data sets can be accessed online via the websites listed in Table 1. Particular focus will be placed upon data collected during the NABOS program; however, the other data sets will be used to place these observations into greater spatial context with respect to Arctic water mass distributions and circulation patterns.

2.2 *Quality Assurance/Quality Control*

The combined data set includes measurements of pressure, temperature, salinity, and dissolved oxygen collected from sensors deployed at each station and matched to specific depths at which water samples were collected. Concentrations of nitrate, phosphate, and silicic acid as well as values of $\delta^{18}O$ included in the data set were derived from chemical analyses on the collected water samples, with the exception of the nitrate concentrations reported from the NABOS expedition. The majority of the nitrate concentrations reported from the NABOS expedition were collected using a Deep SUNA (Submersible Ultraviolet Nitrate Analyzer). At stations where the SUNA was not deployed, reported nitrate concentrations were derived from

Table 1 Projects/expeditions that provided data collected from the Arctic Ocean during spring and summer 2015

Project	Dates	Study region	Vessel	Website/data access
North pole environmental observatory (NPEO)	April 13–20	85–90°N along 90°E & 90°W longitudes	Twin otter aircraft, Kenn Borek, ltd.	http://psc.apl.washington.edu/northpole/Data.html https://arcticdata.io/catalog/view/doi%3A10.18739%2FA25T3FZ8X
Beaufort gyre exploration program (BGEP)	September 21–October 14	70–80°N, 123–160°W	*CCGS Louis S. St-Laurent*	http://www.whoi.edu/website/beaufortgyre/data
Nansen & Amundsen Basin Observation System (NABOS)	August 28–September 26	75–82.5°N, 64.5–178°E	*R/V Akademik Tryoshnikov*	http://research.iarc.uaf.edu/NABOS2/ https://arcticdata.io/catalog/view/doi%3A10.18739%2FA2VM42X8F https://arcticdata.io/catalog/view/doi%3A10.18739%2FA2QV3C437
U.S. GEOTRACES	August 12–October 7	60–90°N along ~180° and 150°W	*USCGC Healy*	https://www.ncei.noaa.gov/data/oceans/ncei/ocads/data/0156924/
TransArc II (PS94)	August 9–October 9	71–90°N, 0–180°E	*R/V Polarstern*	https://www.ncei.noaa.gov/data/oceans/ncei/ocads/data/0170256/

seawater samples. We note that the nutrient concentrations measured on the NABOS samples were not of the highest quality, due to the partial thawing of frozen samples during shipment from the field to laboratory facilities in Fairbanks, AK. The quality of the phosphate data from these samples was determined to be particularly poor and the results are not reported. The nitrate and silicic acid data were determined to be acceptable, but questionable; these data should therefore be interpreted with caution. A subset of higher quality nitrate concentrations was used to help calibrate the SUNA instrument.

The overall quality and intercomparability of the data collected as part of the different expeditions was determined via the comparison of mean concentrations of nitrate, dissolved oxygen, and $\delta^{18}O$ in deeper waters (≥990 m) where concentrations vary much less compared to shallower waters. Stations with concentrations deviating from the mean by more than two standard deviations (2σ) were removed from the final data set. For example, the mean (±1σ) concentrations of NO_3^- and O_2 determined from all available observations was 14.3 ± 0.8 mmol m^{-3} and 299 ± 13 mmol m^{-3}, respectively. Concentrations were also compared at the ~500 m depth horizon ($495 \leq z \leq 505$ m) as the SUNA did not collect data below ~1,000 m due to pressure restrictions on the battery pack needed to provide power to the instrument. The mean (± 1σ) concentrations of NO_3^- and O_2 determined from all observations available at this depth level were 12.8 ± 0.7 mmol m^{-3} and 305 ± 8 mmol m^{-3}, respectively.

Mean $\delta^{18}O$ values from depths ≥300 m were 0.22 ± 0.05‰ (NABOS), 0.20 ± 0.06‰ (GEOTRACES), and 0.20 ± 0.02‰ (NPEO). No $\delta^{18}O$ data were publicly available for the BGEP and PS94 expeditions at the time of this writing.

2.3 *Definition of Geochemical Parameters*

High biological productivity on the broad and shallow Arctic shelves creates organic matter that is deposited in the sediments and eventually remineralized, releasing inorganic nutrients back to the water column. Bacterial respiration in the sediments may be aerobic and/or anaerobic; in fact, dissolved oxygen concentrations can approach zero within a few centimeters of the sediment–seawater interface. Once the oxygen has been depleted, nitrate may be utilized as an electron acceptor to facilitate organic matter reduction; this process is called denitrification and results in the removal of nitrogen from the water column via conversion of biologically available sources of nitrogen (e.g., NH_4^+, NO_3^-, and NO_2^-) to inert, nitrogen gas (N_2). Active sedimentary denitrification in the Bering and Chukchi Seas [56] preferentially removes nitrogen relative to phosphorus, resulting in a lower N:P ratio (and excess phosphate) compared to that of Atlantic waters entering the Arctic Ocean via Fram Strait and the Barents Sea Opening [57]. Within the domains of the Arctic Ocean, N:P ratios are expected to vary according to Redfield dynamics; thus, Pacific-origin waters should retain their relatively low N (high P) character compared to Atlantic-origin waters. This nitrogen deficit/phosphate excess has been exploited as a means to separate Atlantic and Pacific water contributions to the Arctic halocline, since Atlantic waters do not encounter denitrification upon entry to the Arctic Ocean [57–59]. However, evidence of denitrification has been found on the Laptev and East Siberian Sea shelves [52] calling into question the ability of these nutrient relationships to reliably separate Atlantic from Pacific waters [46, 53].

Though the N:P method may be rendered unreliable for the quantitative separation of Atlantic and Pacific waters in the Arctic Ocean in the near future, the geochemical signatures of higher nutrient concentrations and lower dissolved oxygen concentrations remain defining characteristics to qualitatively identify shelf water influences. For example, dissolved oxygen is utilized in the aerobic respiration of organic matter present in the water column and the sediments. In regions that are sufficiently deep or strongly stratified such that these waters are not ventilated via direct exposure to the atmosphere or mixing with recently ventilated surface waters, the dissolved oxygen concentration is depleted relative to its saturation value (the concentration expected for waters in equilibrium with the atmosphere). This deviation from the saturation value is called Apparent Oxygen Utilization (AOU). Positive values of AOU indicate that O_2 has been consumed whereas negative values indicate oversaturation of dissolved oxygen due to the net production of oxygen over respiration and/or physical mechanisms, such as bubble injection, that may increase the dissolved oxygen concentration in excess of the saturation concentration.

The NO_3^- concentration measured in the water column can be split into two fractions: remineralized and preformed nitrate. Remineralized nitrate is that produced during the biological respiration of organic matter and can be calculated via multiplying the AOU by an assumed Redfield ratio (in this paper, an NO_3^-:O_2 remineralization ratio of 16:175 is applied). The preformed nitrate is equal to the nitrate present in the water at the surface just prior to subduction below the mixed layer/euphotic zone. In other words, it is the nitrate that was not utilized by photosynthesis at the surface. Preformed nitrate is computed as the observed nitrate concentration minus the remineralized nitrate concentration. In general, waters that have spent significant time on the shelves carry a relatively high remineralized nutrient concentration; in fact, nutrient concentrations in these waters may be mostly, if not entirely, derived from remineralization. In contrast, both Atlantic and Pacific waters entering the Arctic Ocean are associated with significant preformed nutrient concentrations before being subducted below the less dense polar mixed layer. Note that, since the calculation of remineralized and preformed nutrient concentrations are dependent upon the AOU, waters containing a significant remineralized nutrient component that are upwelled to the surface and ventilated/re-oxygenated will be subsequently characterized as containing a significant quantity of preformed nutrient concentrations. Deep Pacific waters upwelled onto the East Bering and Anadyr shelves prior to their northward transport into the Arctic Ocean via Bering Strait are two such examples of this phenomenon.

The NO parameter is a quasi-conservative tracer that combines concentrations of nitrate and oxygen to correct for biological production and remineralization according to Redfield dynamics. NO is defined as: $9*[NO_3^-] + [O_2]$ [60]. NO can be generally considered as conservative in waters below the surface mixed layer. At such depths, where air-sea exchange does not impact the dissolved oxygen concentration, NO can be used to assess mixing among water masses as it corrects for decreases in O_2 and increases in NO_3^- resulting from biological respiration. As such, NO can be thought of as a sort of preformed dissolved oxygen concentration, since it is corrected for apparent oxygen utilization. However, non-Redfield processes such as denitrification, anammox, and nitrogen fixation influence the NO parameter in a non-conservative manner, limiting its quantitative application. NO has been successfully used to qualitatively separate Atlantic versus Pacific sources to the halocline layer in the Amerasian Basin, as Atlantic-derived lower halocline waters are characterized by an NO minimum whereas Pacific halocline waters are generally associated with NO maxima [19] (also see Fig. 2e). The NO parameter has an advantage over preformed NO_3^- concentrations as a water mass tracer since it does not rely on the assumption that waters are 100% saturated with respect to O_2 prior to subduction below the mixed layer.

The N* parameter is another useful tracer of waters influenced by shelf processes. N* is defined as: $0.87*([NO_3^-] - 16*[PO_4^{3-}] + 2.9)$ [61]. N* values are positive in waters impacted by net nitrification and negative in those impacted be net denitrification. The N* parameter has been used in various Arctic studies (e.g., [31, 62]) to identify halocline waters in the Makarov and Canada Basins that have been impacted by net denitrification on the surrounding shelves prior to their advection offshore.

3 Results

The NABOS 2015 expedition deployed sensors and collected water samples at 94 stations arranged along five transects running from the continental slope toward the deep basins, approximately along the 95°E, 126°E, 145°E, 165°E, and 175°E longitude lines (Fig. 4, inset map), as well as an additional transect across the

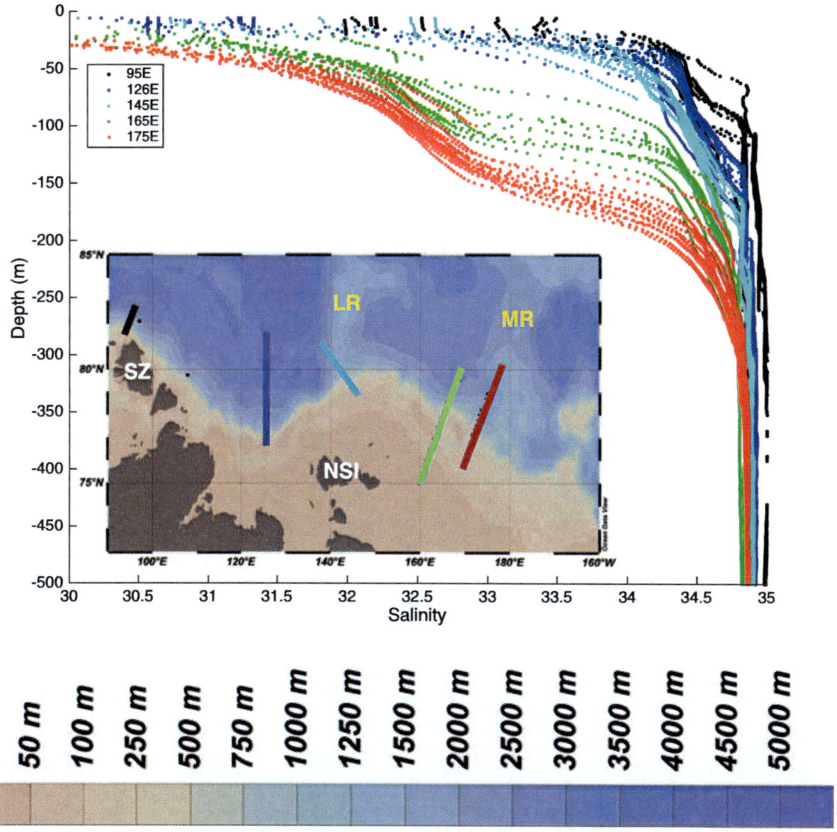

Fig. 4 Vertical profiles of salinity in the top 0–500 m of the water column at stations occupied during the 2015 NABOS expedition. The colors of the profiles correspond to the five transects occupied approximately along the 95°E (black), 126°E (blue), 145°E (cyan), 165°E (green), and 175°E (red) longitude lines (see inset map for locations). *SZ* Severnaya Zemlya, *NSI* New Siberian Islands, *LR* Lomonosov Ridge, *MR* Mendeleyev Ridge. The inset map was created using Ocean Data View software [68]. The colorbar corresponds to the bathymetry of the inset map (depth in meters)

St. Anna Trough (not shown). The following discussion will focus on vertical profiles of temperature, salinity, O_2, and NO_3^- collected via sensors deployed on the rosette at stations located east of the St. Anna Trough.

3.1 Distinct Hydrographic Structures to the West and East of the Lomonosov Ridge

The surface salinity was highest and the halocline thinnest along the 95°E line (black dots in Fig. 4). Moving east toward the Lomonosov Ridge, the surface salinity decreased and the halocline became progressively thicker (blue and cyan dots in Fig. 4). Crossing the Lomonosov Ridge and moving toward the western flank of the Mendeleyev Ridge, the surface salinity decreased markedly and the halocline thickened considerably.

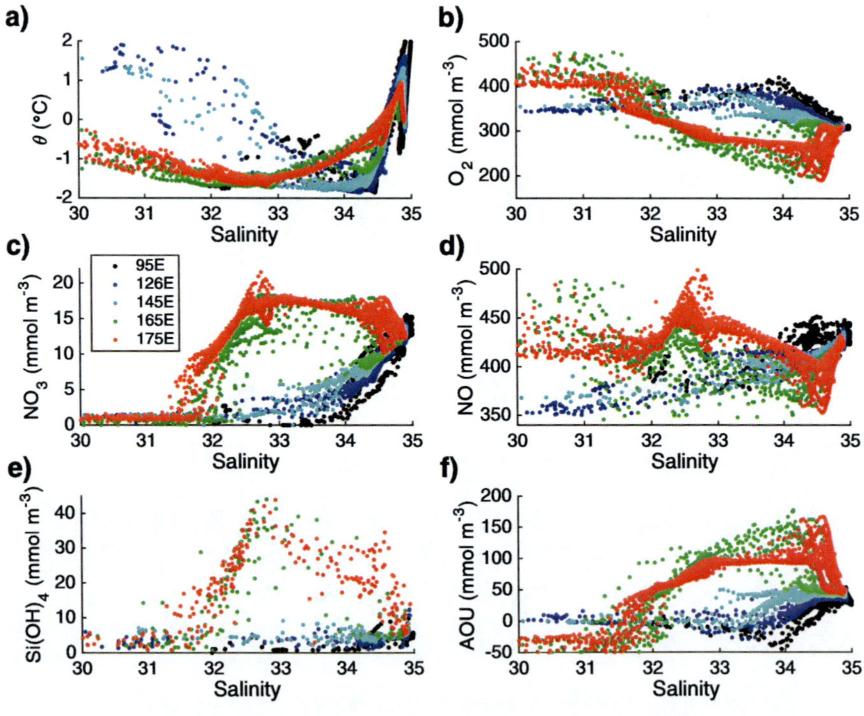

Fig. 5 Plots of salinity versus **(a)** potential temperature (θ), **(b)** dissolved oxygen (O_2), **(c)** nitrate (NO_3), **(d)** NO (=9*[NO_3^-] + [O_2]), **(e)** silicic acid ($Si(OH)_4$), and **(f)** apparent oxygen utilization (AOU) from data collected from transects aligned approximately along the 90°E (black), 126°E (blue), 145°E (cyan), 165°E (green), and 175°E (red) longitude lines during the 2015 NABOS expedition. Note that three shelf stations from the 126°E line (10–12) were omitted from the figure for ease of interpretation

Salinity was not the only variable that expressed considerable change moving from west to east across the study area. Plots of potential temperature (θ), O_2, NO_3^-, $Si(OH)_4$, and NO also demonstrate considerable change in the hydrographic structure across the Lomonosov Ridge (Fig. 5). Surface layers at all stations may exhibit relatively warm ($\theta > 0°C$) or cold temperatures, depending on the presence or absence of sea ice (or recent ice melt). Both western and eastern stations exhibited relatively warm temperatures at salinities <33 as these waters were close to the surface and absorbed solar radiation when the ice cover was absent. Below the mixed layer, the differences in hydrography on either side of the ridge were more robust.

The largest differences between stations located west (95°E, 126°E, and 145°E) and east (165°E and 175°E) of the Lomonosov Ridge were observed in the halocline ($32 \leq S \leq 34.5$) and Atlantic ($34.8 \leq S \leq 34.9$) layers. It has been well established from previous work that the halocline thickens and warms whereas the Atlantic layer cools moving from west to east across the Siberian continental slope (e.g., [8, 17, 63]). For example, the θ_{max} associated with the core of the Atlantic layer was colder at the eastern stations ($\theta_{max} < 1°C$) compared to those observed at the western stations ($\theta_{max} > 1.2°C$). Some part of the heat lost from the Atlantic layer is transferred upward to the overlying halocline layer, resulting in the west-to-east increase in potential temperature observed in the halocline. While the vertical transfer of heat from the Atlantic layer to the halocline layer is an important contributor to the continued and accelerating loss of Arctic sea ice [63], changes to the Atlantic layer are not discussed further as this chapter is focused primarily on the physical and biogeochemical fronts associated with the various contributions to the upper and lower halocline layers.

Within the salinity range of the upper halocline ($32 < S < 34$), the eastern transects exhibited potential temperature minima (θ_{min}) centered around a salinity of ~32.8 that are associated with high NO_3^- ($\geq$14 mmol m^{-3}) and $Si(OH)_4$ (>30 mmol m^{-3}) but low O_2 ($\leq$300 mmol m^{-3}) concentrations (Fig. 5b, c, e). Transects west of the Lomonosov Ridge did not exhibit a θ_{min} at this salinity; instead, temperatures typically remained near the freezing point from a salinity of ~33 toward higher salinities that are most typically associated with the lower halocline ($34 \leq S \leq 34.5$). An abrupt transition was then observed, as temperatures rapidly increased with increasing salinity (and depth) toward the Atlantic water core. This sharp transition, or kink, in the θ-S diagram is characteristic of basin-derived lower halocline water [14, 17, 18]. Over the extent of the halocline ($32 < S < 34.5$), NO_3^- ($\leq$8 mmol m^{-3}) and $Si(OH_4)$ ($\leq$8 mmol m^{-3}) concentrations were lower and O_2 concentrations higher (>320 mmol m^{-3}) at the western transects compared to the eastern transects.

Derived parameters, NO (Fig. 5d) and AOU (Fig. 5f), also indicated significant differences between the western and eastern transects. For example, the western transects exhibited weakly positive values of AOU (<40 mmol m^{-3}) in the halocline, indicative of some influence from the net respiration of organic matter; however, the AOU values observed along the 165°E and 175°E transects mostly exceeded 75 mmol m^{-3}, suggesting a much more substantial impact from respiration compared to the western transects. The differences in the NO parameter observed between the western and eastern transects were somewhat more complex. Although concentrations of O_2 and NO_3^- exhibited strong negative correlations, the relative positions of maxima and minima in these variables did not always strictly coincide; these slight differences resulted in an NO parameter that may more strongly resemble either NO_3^- or O_2 at different salinities. For example, the eastern transects exhibited maximum NO values at salinities of 32.5–32.8, coincident with the NO_3^- maxima; however, minimum NO values were observed between salinities of 34.2 and 34.5, corresponding with minimum O_2 concentrations, though relatively high NO_3^- concentrations. Although the NO values exhibited more complex structure at individual stations along the western transects, they generally increased with increasing salinity through the halocline toward maximum values associated with the Atlantic layer.

3.2 *Variations Within the East Siberian Sea*

A closer examination of the data collected along the eastern transects indicated significant differences to the halocline layer between 165°E and 175°E. For example, NO values between salinities of 32.5 and 34.3 were higher along the 175°E transect than those observed along 165°E (Fig. 5). These differences primarily resulted from higher NO_3^- concentrations along the 175°E line at salinities $\leq$32.8 and lower O_2 concentrations (and higher AOU) at salinities ranging between 32.5 and 34.5. Consequently, the O_2 and NO minima observed at 165°E generally occurred at $34 \leq S \leq 34.2$ whereas those along 175°E occurred at $S \sim 34.5$ (Fig. 5b, d). In addition to these general differences, there were also a number of stations along the 165°E transect that exhibited colder temperatures, higher O_2 concentrations, and lower NO_3^- concentrations at salinities >34.2 that generally resembled stations occupied west of the Lomonosov Ridge.

These specific differences can be viewed more easily by contrasting individual stations at similar isobaths along each transect (Fig. 6). At the ~200 m isobath, the physical and geochemical features were rather similar at the two transects; however, O_2, NO_3^- (not shown), and NO were all somewhat higher at 175°E. Features diverged more strongly moving further offshore. At the ~500 m and ~1,000 m isobaths, O_2 and NO were considerably higher at 175°E (compared to 165°E) in the salinity range $32.5 \leq S \leq 34$; however, at higher salinities ($34 \leq S \leq 34.8$), potential temperatures were warmer and both O_2 and NO lower at 175°E. At stations farthest offshore (~2,500 m water depth), the two transects mostly converged to show similar features throughout the water column, with the exception of the lower halocline layer ($34 < S < 34.5$), where potential temperatures were colder and both O_2 and NO higher at 165°E (versus 175°E). Thus, there were two significant distinctions between the two transects north of the ~200 m isobath. The first was the persistently higher O_2 and NO along 175°E (relative to 165°E) at salinities <34. The second difference occurred in the lower halocline ($34 \leq S \leq 34.5$) and was manifest as a cold and O_2-rich feature along 165°E compared to a warmer and

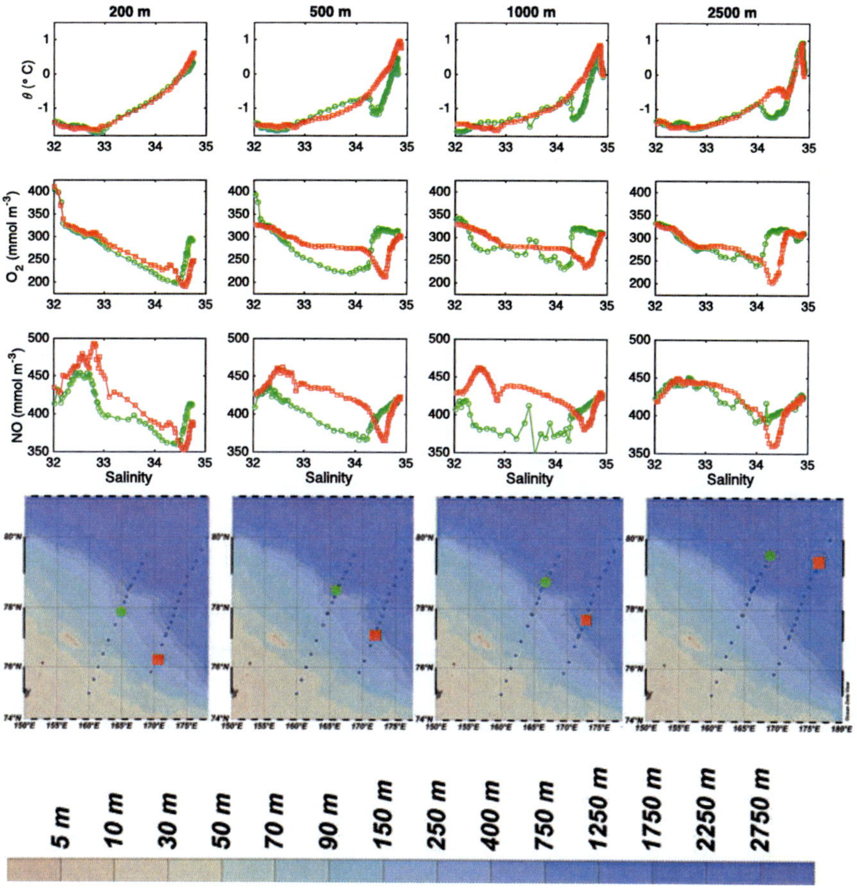

Fig. 6 Direct comparisons of potential temperature (θ), dissolved oxygen (O_2), and NO at stations along the 165°E (green) and 175°E (red) transects with similar bottom depths, approximately at the ~200 m, ~500 m, ~1,000 m, and ~ 2,500 m isobaths. The specific stations compared include: 45 and 70 (~212 m), 48 and 67 (450–500 m), 50 and 65 (1,000 m), and 54 and 59 (2,300–2,600 m). Station locations indicated on corresponding inset maps created using Ocean Data View software [68]. The colorbar corresponds to the bathymetry of the inset maps (depth in meters)

O_2-depleted feature along 175°E. In the following section, it will be argued that these differences resulted from distinct sources ventilating the upper and lower halocline layers along these two transect lines.

4 Discussion

The eastward increase in the thickness of the halocline layer shown in Fig. 4 corresponded to the northward movement of Siberian shelf waters, freshened by river runoff, from the shelves to the continental slope and deep basins between the Lomonosov and Mendeleyev Ridges (e.g., [10]) as well as the westward intrusion of Pacific-origin waters from the Chukchi Sea (e.g., [31]). The Siberian shelf and Chukchi Sea waters converged in the East Siberian Sea, forming physical and biogeochemical fronts that separated Atlantic- and Pacific-origin waters. Such fronts have been previously observed in both the water column and the sediment record of the East Siberian Sea; furthermore, these fronts have also been found to shift with anticyclonic and cyclonic regimes of ocean circulation [49], similar to that of the Transpolar Drift [34, 42, 43]. In fact, as the region north of the Laptev and East Siberian Seas has been implicated as the origin of the Transpolar Drift, we argue that the fronts observed in the East Siberian Sea extended northward to describe the alignment of the Pacific-Atlantic front through the central Arctic Ocean.

4.1 Tracing the Chukchi-Siberian (Pacific-Atlantic) Front Throughout the Arctic Ocean

An analysis of the physical and geochemical concentrations along specific isohaline surfaces of interest indicates the position of fronts separating the contributions of different water types to the Arctic halocline. By combining available data collected during the various expeditions conducted across the Arctic in 2015 (see Table 1), Figs. 7, 8, 9, and 10 provide a quasi-synoptic picture of the distribution of useful geochemical and physical variables. In this section, these observations are utilized to qualitatively assess the distribution of Siberian/Atlantic versus Chukchi/Pacific shelf waters in the Arctic halocline during summer 2015; these distributions are then inferred to provide a basic picture of the general circulation during this specific time period.

The values of θ, O_2, $Si(OH)_4$, NO, and N* on salinity surfaces of 32.8, 33.5, 34.0, and 34.5 are illustrated in Figs. 7, 8, 10, and 11, respectively. These specific isohalines were chosen as they generally correspond to specific water types of interest. The $S = 32.8$ isohaline identifies the potential temperature minimum water identified by *Nishino* et al. [30, 31] as WCSW that has been modified in the East Siberian Sea. The major fronts/features identified on the $S = 32.8$ surface were very similar to those observed on the $S = 32.5$ and $S = 33$ surfaces (not shown) that are typically associated with Pacific summer ($S \sim 32.5$) and winter ($S \sim 33$) halocline waters [23]. The $S = 33.5$ isohaline was previously associated with NO minima marking Siberian shelf water influences in the Makarov Basin [21]. The $S = 34.0$ and $S = 34.5$ isohalines are typically associated with basin- and shelf-derived lower halocline waters of Atlantic origin, respectively (e.g., [14, 18]).

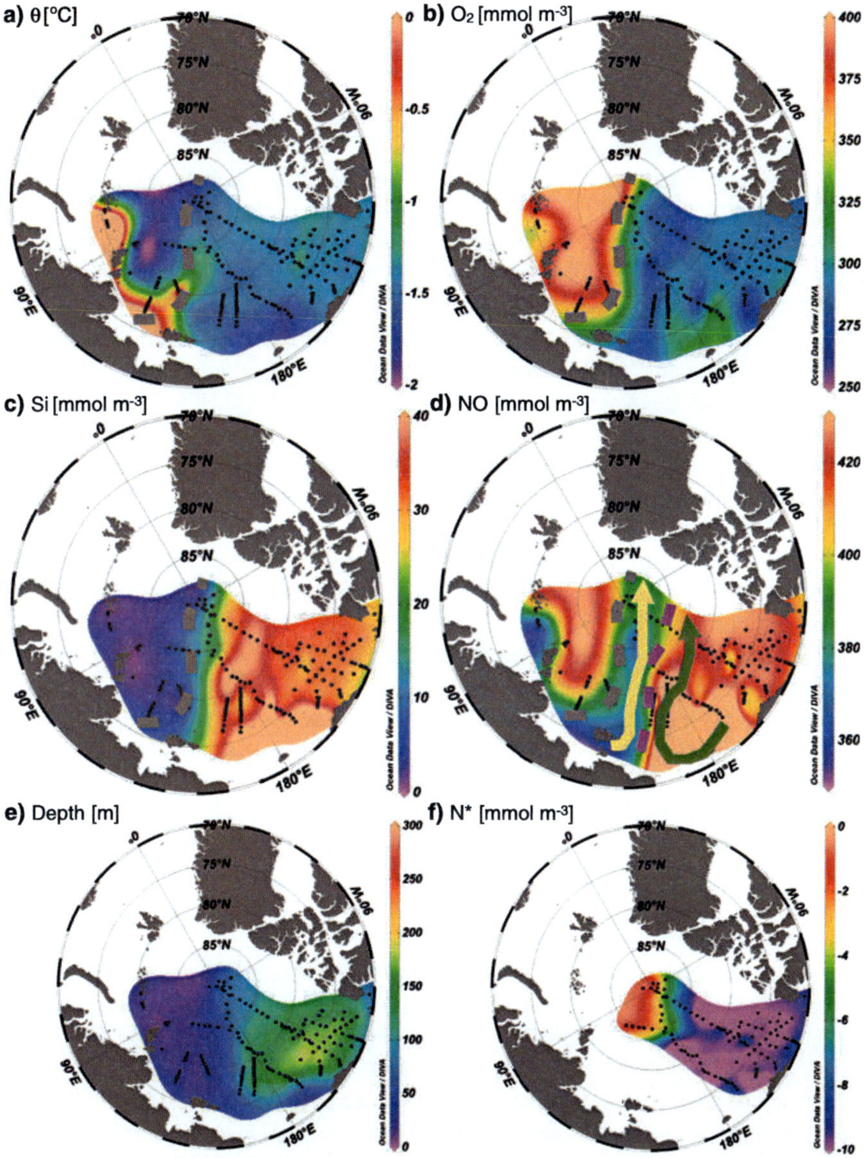

Fig. 7 (**a**) Potential temperature, (**b**) dissolved oxygen, (**c**) silicic acid, (**d**) NO, (**e**) depth, and (**f**) N* along the S = 32.8 isohaline. Colorbars denote the values of the variable presented in each panel, where cooler colors represent lower/more negative values and warmer colors indicate higher/less negative values. Figure created using Ocean Data View Software [68]. Colors are contoured using the DIVA Gridding setting in Ocean Data View; interpolations resulting from the gridding algorithm must be interpreted with caution, particularly in regions with low (or absent) data coverage. Gray and magenta dashed lines indicate the positions of hydrographic fronts. The gray line was drawn along the front in dissolved oxygen and then copied over to the other panels; while the frontal positions are generally similar among the different variables, some differences in alignment exist, indicating that not all variables delineate the water masses similarly. The magenta

In the upper halocline layer, along isohalines of S = 32.8 (Fig. 7) and S = 33.5 (Fig. 8), O_2 concentrations were relatively high and $Si(OH)_4$ concentrations relatively low in regions west of the Lomonosov Ridge. Values of N* were also generally high in the Nansen and Amundsen Basins. Potential temperatures were primarily cold throughout the Arctic Ocean, but warmer temperatures were occasionally observed at locations closer to shore on the continental slope where sea ice was generally absent and these isohalines were close to the surface; the depths of these two isohalines were <30 m west of the Lomonosov Ridge and therefore susceptible to warming via solar insolation. These characteristics generally reflect near-surface waters of Atlantic origin that have not been modified by interaction with shelf sediments. In contrast, O_2 concentrations and values of N* were lower and Si $(OH)_4$ concentrations higher east of the Lomonosov Ridge; these abrupt changes in the geochemical characteristics were not restricted to the shallow shelf seas but extended into the Makarov and Canada Basins. This transition was associated with an increase in the isohaline depths to >100 m at ~165°E and >200 m east of 180°E. These shifts are indicative of a hydrographic front (illustrated by gray, dashed lines in Figs. 7 and 8) that highlights the offshore spread of Siberian shelf waters east of the Lomonosov Ridge. These waters have been influenced by shelf processes; thus, they are characterized by negative N* as well as lower O_2 and higher $Si(OH)_4$ concentrations.

A second front was visible in the NO distribution (Figs. 7d and 8d) that is not as clearly apparent in the distributions of θ, O_2, $Si(OH)_4$, or N*. This front was aligned approximately along the ~165°E section and separated lower NO waters to the west from higher NO waters to the east. These higher NO waters reflect Pacific-origin waters that have expanded westward into the East Siberian Sea and then progressed northward via the Canadian Branch of the Transpolar Drift [34]. These interpretations suggest that Pacific halocline waters were generally restricted to the northern Canada Basin/Mendeleyev Ridge whereas halocline waters from the Laptev and/or East Siberian Seas spread northward into the Makarov Basin as far as the North Pole.

Using similar, high-resolution observations of NO_3^- and O_2 collected in the central Arctic Ocean during spring 2007 and 2008, Alkire et al. [21] argued that Siberian shelf waters characterized by low NO values ventilated the upper halocline of the Makarov Basin. More recent work using Ra tracers collected as part of the 2015 GEOTRACES expedition has also suggested the East Siberian Sea as the origin of surface waters in the central Arctic Ocean [64]. The observations presented here support the conclusions made by both Alkire et al. [21] and Kipp et al. [64] that the East Siberian Sea serves as an important source of upper halocline waters to the deep basins of the Arctic Ocean. Furthermore, the NO distributions depicted in Figs. 7 and 8 indicate that Pacific halocline waters were restricted to the Canadian Basin side of the Mendeleev-Alpha Ridge (at least as far east as 130°W).

Fig. 7 (continued) line was drawn along the front in NO. The yellow and green arrows denote the presumed spreading pathways of Siberian shelf water and Pacific water, respectively

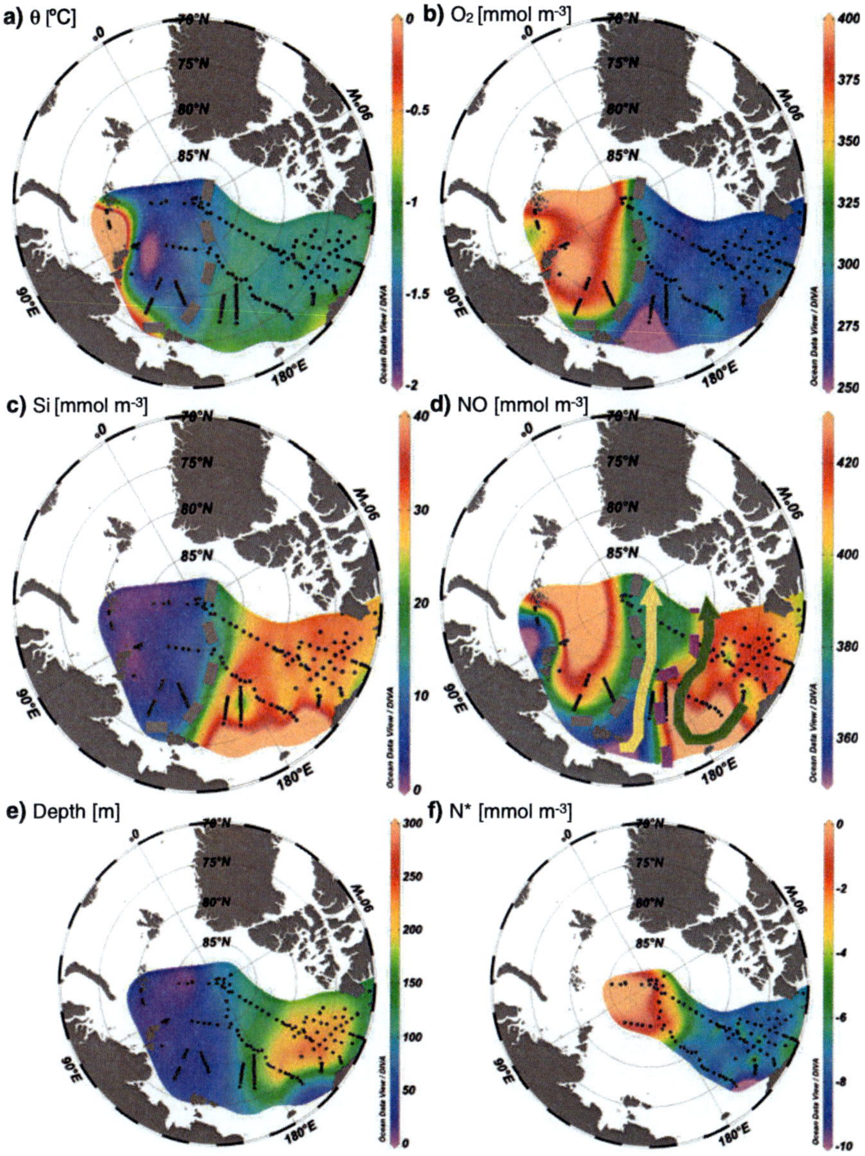

Fig. 8 Same as for Fig. 7, but values are plotted along the S = 33.5 isohaline

The interpretation of the available data is further supported by plots of dynamic heights relative to 250 m depth (Fig. 9). Each of the panels in Fig. 9 indicates a maximum center of dynamic height in the Canada Basin that marks the position and extent of the Beaufort Gyre. The gradient between this maximum height and lower values that are observed in the Eurasian Basin is illustrated by contours that are

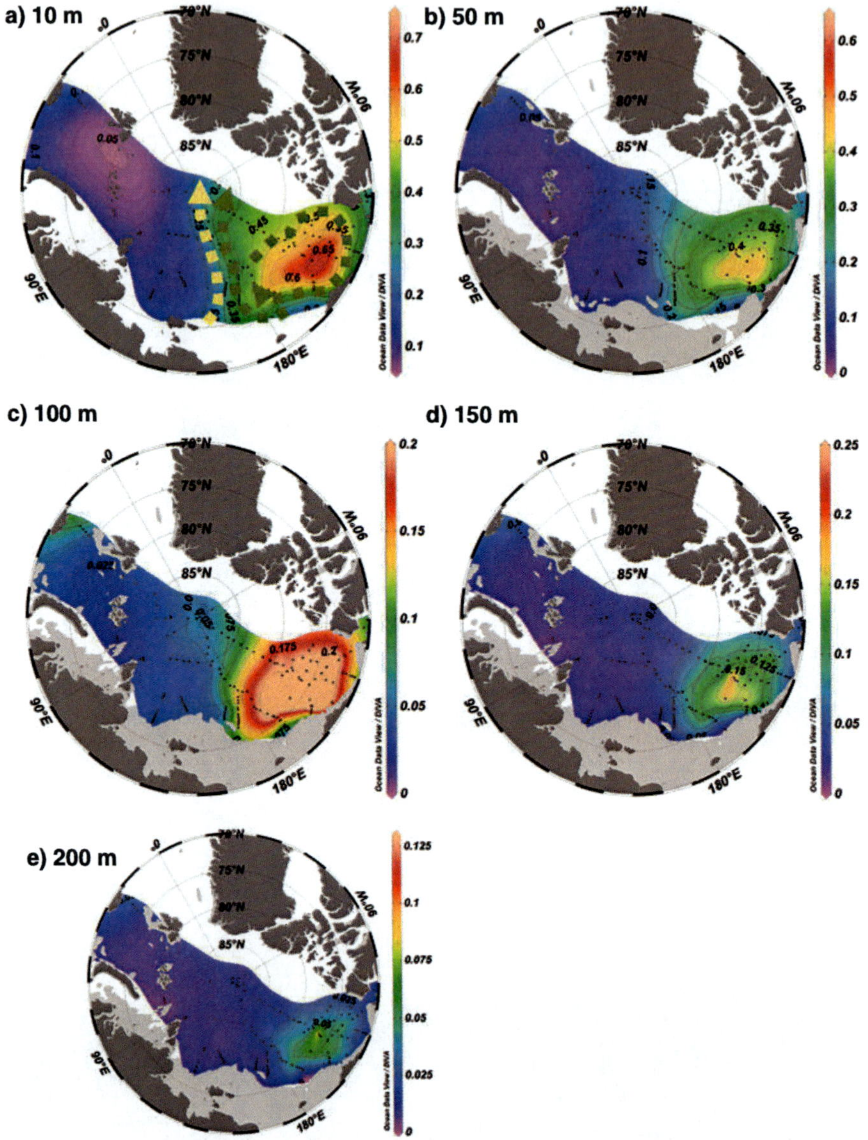

Fig. 9 Dynamic heights calculated relative to an assigned level of no motion at 250 m. Heights plotted for depths of (**a**) 10 m (near-surface), (**b**) 50 m, (**c**) 100 m, (**d**) 150 m, and (**e**) 200 m. The depth horizons 50, 100, and 150 m generally coincide with the depths of the isohalines 32.8, 33.5, and 34.0, respectively, near the hydrographic front along ~165°E. Arrows in panel (**a**) show the implied northward flow of Siberian shelf (yellow) and Pacific (green) waters along the Transpolar Current as well as the anticyclonic (clockwise) circulation of the Beaufort Gyre. Note that the ranges of the colorbars vary with each panel, as the dynamic heights decrease with increasing depth (toward the presumed level of no motion). Figure created using Ocean Data View Software [68], with DIVA Gridding (automatic scale lengths). Contour intervals at 0.05 for panels (**a**, **b**) and 0.025 for panels (**c**–**e**)

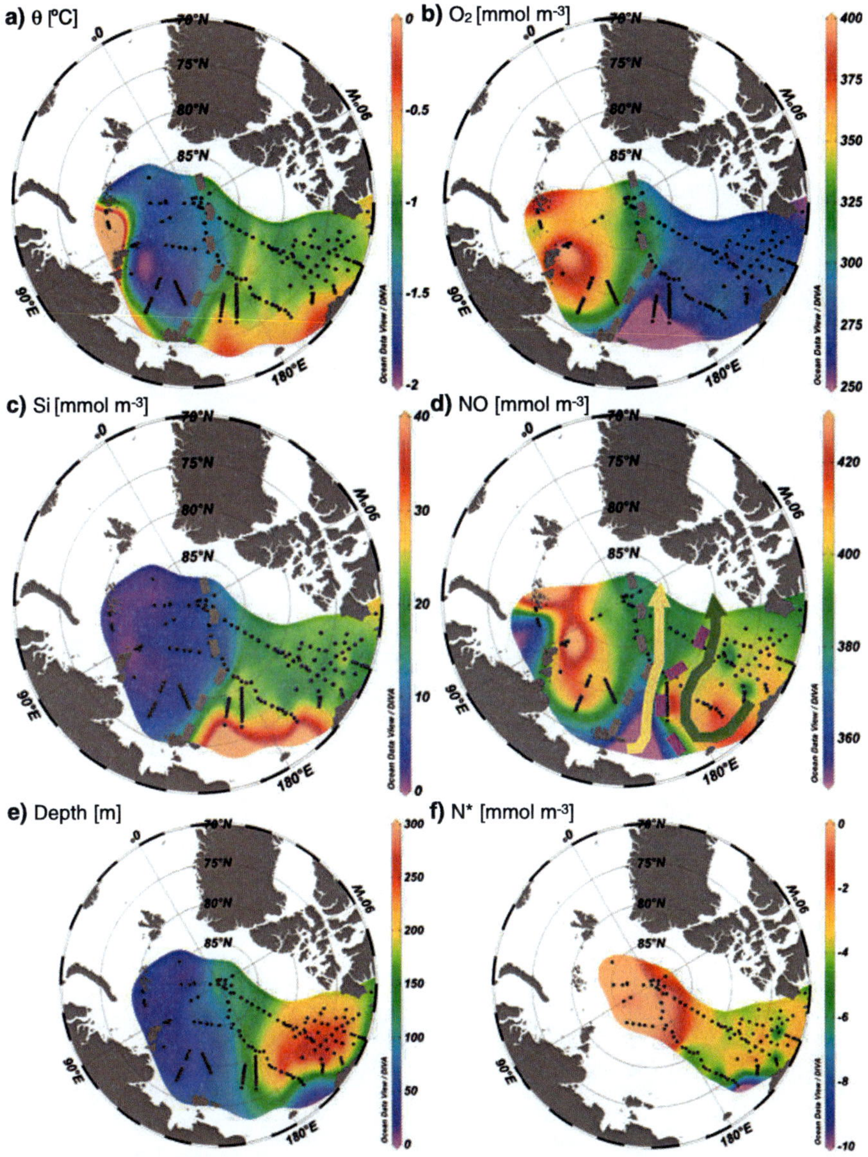

Fig. 10 Same as for Fig. 7, but plotting values along the S = 34.0 isohaline

roughly aligned in a meridional direction and extend between the Lomonosov and Mendeleyev Ridges. In the northern hemisphere, contours of dynamic height indicate the direction of relative geostrophic currents with areas of higher dynamic heights to the right of the flow direction. Therefore, these panels show the anticyclonic circulation around the Beaufort Gyre as well as the flow of shelf waters

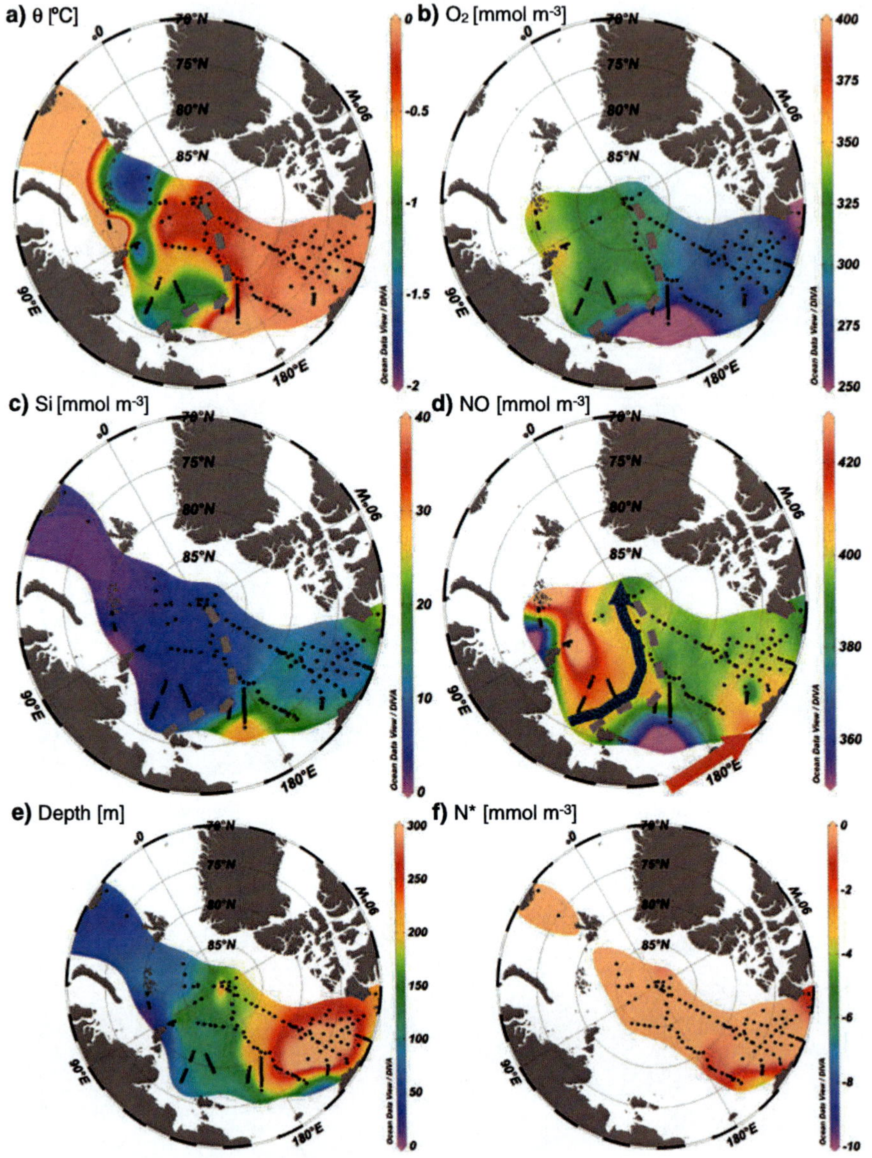

Fig. 11 Same as for Fig. 7, but plotting values along the S = 34.5 isohaline. In addition, the blue arrow indicates the inferred circulation pathway of Fram Strait Branch (FSB) lower halocline water (i.e., along the NO front) and the red arrow illustrates a general/presumed pathway for Barents Sea Branch (BSB) lower halocline water

northward from the East Siberian Sea toward the North Pole along the route of the Transpolar Current (Fig. 9a). In fact, the largest differences in dynamic heights are aligned very closely with the front separating lower and higher NO values between the 165°E and 175°E transects (Figs. 7, 8, and 10). Note that the dynamic heights decrease with increasing depth (approaching the assigned level of no motion at 250 m); however, the differences between the central maximum in the Beaufort Gyre and the lower heights of the Eurasian Basin are persistent at least to a depth of 150 m (Fig. 9d). Below this depth, the extent of the Beaufort Gyre is still visible, but the strong gradient marking the alignment of the Transpolar Current is absent. For context, we briefly note that the strength of the Beaufort Gyre in 2015, quantified in terms of the mean annual Ekman vertical velocity, was characterized as average relative to the 2003–2018 time series [65].

The geochemical distinctions marking the positions of the two fronts are relatively consistent down to the S = 34 isohaline (Fig. 10), indicating Siberian shelf and Chukchi shelf water influences/modifications to waters more saline than 33.5. Woodgate et al. [66] previously suggested upwelling of warmer Atlantic waters onto the Chukchi slope and diapycnal mixing with Pacific-origin waters to explain relatively warm temperatures, high concentrations of $Si(OH)_4$, and low concentrations of O_2 observed at salinities of ~34 in the southern Makarov Basin and Chukchi Cap, compared to characteristics observed farther west. The persistence of the NO front to salinities of ~34, in combination with the low O_2, high $Si(OH)_4$, and relatively warm temperatures observed on both sides of the front suggests that similar upwelling and/or diapycnal mixing are occurring on the East Siberian Sea shelf, in accordance with hypotheses raised in previous studies (e.g., [62]). Nishino et al. [30, 31] and Ardyna et al. [67] have previously discussed the increased capacity of upwelling along the Siberian continental slope and southern Makarov Basin as the sea ice cover in this region has declined and the period of open water has extended into the fall and early winter. In addition to the consequences of this larger potential for upwelling on biological production, there is also an enhanced opportunity for halocline water modification. Despite this potential increase in shelf modification of lower halocline waters, the horizontal and vertical extent of this influence is still limited. For example, the NO front separating geochemical characteristics observed along the 165°E and 175°E lines is much less recognizable at S = 34.2 (data not shown), indicating that such mixing, if present, was generally restricted to salinities <34.2. In addition, the higher temperatures and $Si(OH)_4$ concentrations (and lower O_2 and N*) were generally restricted to the region south of ~82°N; thus, shelf-influenced lower halocline waters seem to have been prevalent only on the Canadian side of the Mendeleyev Ridge.

4.2 *Identifying the Split Between Fram Strait and Barents Sea Branches of Lower Halocline Water*

Along the S = 34.5 isohaline (Fig. 11), the front between the 165°E and 175°E transects was no longer prominent and the primary front separating conditions west and east of the Lomonosov Ridge was shifted somewhat toward the east and south.

In fact, the front extended across the northernmost stations of the 165°E line (specifically between the ~200 and ~500 m isobaths) such that the northernmost stations along this line were characterized by colder temperatures, lower $Si(OH)_4$ concentrations, and higher O_2 and NO compared to stations occupied further south on the transect; these conditions are reflective of basin-derived LHW (or FSB LHW) that dominated the halocline west of the Lomonosov Ridge. In contrast, stations further south along the 165°E line, as well as all stations farther east, exhibited warmer temperatures, lower O_2 and NO, and higher $Si(OH)_4$ concentrations, indicating significant shelf modification. These attributes are associated with shelf-derived LHW (or BSB LHW).

Similar differences in O_2 and NO have been observed in the southern Canada Basin by Alkire et al. [21]; these differences roughly corresponded with the circulation patterns described by the basin-derived/FSB (O_2-rich) LHW and shelf-derived/BSB (O_2-poor) LHW discussed by Rudels et al. [14]. That is, the O_2-rich LHW waters were observed in the northern and eastern regions of the southern Canada Basin whereas O_2-poor LHW were observed closer to the continental slope, in the western and southern regions. The similar differences observed in O_2 and NO along the 165°E transect line indicate that this longitude was roughly the position where the basin-derived (or FSB) LHW broke away from the near-slope region (and may progress northward to ventilate the Makarov and northern Canada Basins) while the BSB LHW was more prominent along the slope and may eventually ventilate the halocline in the southern Canada Basin.

In addition, the southernmost (shallowest) stations on both of these transects exhibited waters at $S = 34.5$ characterized by very low O_2 and NO, but higher $Si(OH)_4$ and θ. Although N* values could not be calculated for these stations (see Methods section for details), stations occupied during the GEOTRACES expedition exhibited very low N* values farther to the east. Thus, it is apparent that waters of high salinity, relatively warm temperatures, and a strong influence from shelf processes were circulating on the inner slope/shelf break in the East Siberian and Chukchi Seas during summer 2015; perhaps, this is BSB LHW that has been more heavily modified by shelf processes or an entirely distinct water mass that has been formed directly on the shelves. Additional observations are needed to further investigate the origin(s), potential modification, and spreading pathway(s) of this water mass.

5 Summary and Conclusions

- We take advantage of the unusually good spatial coverage afforded by multiple expeditions operating in the Arctic Ocean during spring and summer of 2015. We identify and describe multiple fronts separating different water masses using physical (temperature and salinity) and chemical (nitrate, dissolved oxygen, phosphate, silicic acid) data, with a particular emphasis on sensor-based measurements (as these offer the highest vertical resolution).

- Distributions of the salinity, potential temperature, dissolved oxygen, nitrate, and limited observations of silicic acid and phosphate were combined to determine that Pacific-origin (i.e., Chukchi shelf) waters characterized by high values of the NO parameter met Atlantic-origin (i.e., East Siberian shelf) waters characterized by lower NO values at a front located at ~165°E in the central East Siberian Sea. The NO parameter proved particularly valuable in identifying this front as potential temperature, dissolved oxygen, and silicic acid concentrations were similar in these two distinct water types.
- Waters from the East Siberian Sea shelf were found to ventilate the upper halocline of the Makarov Basin as lower NO values appeared to spread northward from the shelf break/continental slope into the central Arctic. Pacific waters, in contrast, appeared to have been restricted to the Canada Basin and Mendeleyev Ridge (i.e., east of the Pacific-Atlantic front). This front generally coincides with the origins of the Transpolar Current in the vicinity of the New Siberian Islands and marks the contribution from both Canadian/Pacific and Siberian/Atlantic branches to its confluence.
- In the lower halocline, basin-derived waters from the Eurasian sector were also advected offshore in the vicinity of the Pacific-Atlantic front and spread northwards into the Makarov and northern Canada Basins. Shelf-derived lower halocline waters, likely originating in the Barents Sea [14], in contrast, continued to spread eastward along the slope and into the Chukchi shelf region. These Barents Sea Branch lower halocline waters were presumably heavily modified by interaction with shelf sediments and/or bottom waters, as they were characterized by very low O_2 concentrations (<290 mmol m^{-3}) and elevated NO_3^- (>12 mmol m^{-3}) and $Si(OH)_4$ (>10 mmol m^{-3}) concentrations. Potential temperatures along the $S = 34.5$ isohaline associated with these lower halocline waters also increased from west to east, implying substantial warming during transit. This warming must have resulted from mixing with warmer waters of the thermocline layer directly underlying the lower halocline. The available data suggest that the upwelling of thermocline waters onto the Siberian continental slope may have facilitated this mixing; however, additional analyses are needed to confirm these assertions.
- The divergence of multiple water masses ventilating both the upper and lower halocline of the Arctic Ocean in the East Siberian Sea is an interesting observation, and may not be coincidental. The offshore advection of East Siberian shelf/slope waters as well as the split of cold, O_2-rich (FSB) LHW and comparatively warmer and O_2-deficient (BSB) LHW at the Pacific-Atlantic front may be a consequence of the convergence of Atlantic waters moving eastward and Pacific waters spreading westward at the initiation of the Transpolar Current.
- The results of this study generally confirm circulation patterns and spreading pathways of different water masses that have already been proposed by previous studies. However, this study clearly shows the utility of the NO parameter as a high-resolution, qualitative tracer of different water masses contributing to the

Arctic halocline as well as an effective means to separate waters originating (or modified on) the East Siberian and Chukchi Sea shelves. Therefore, NO may be applied more widely as a qualitative check on more quantitative water type analyses that attempt to compute Pacific versus Atlantic water contributions using other tracers, such as N:P relationships and the PO* parameter.

References

1. Carmack EC (2007) The alpha/beta ocean distinction: a perspective on freshwater fluxes, convection, nutrients and productivity in high-latitude seas. Deep-Sea Res II 54:2578–2598
2. Jakobsson M, Grantz A et al (2004) Bathymetry and physiography of the Arctic Ocean and its constituent seas. In: Stein R, Macdonald RW (eds) The Arctic Ocean organic carbon cycle: present and past. Springer, Berlin, pp 1–6
3. Woodgate RA, Weingartner TJ et al (2012) Observed increases in Bering Strait oceanic fluxes from the Pacific to the Arctic from 2001 to 2011 and their impacts on the Arctic Ocean water column. Geophys Res Lett 39(L24603). https://doi.org/10.1029/GL054092
4. Serreze MC, Barrett AP et al (2006) The large-scale freshwater cycle of the Arctic. J Geophys Res 111. https://doi.org/10.1029/2005JC003424
5. Haine TWN, Curry B et al (2015) Arctic freshwater export: status, mechanisms, and prospects. Global Planet Change 125:13–35
6. Alkire MB, Morison J et al (2017) A meteoric water budget for the Arctic Ocean. J Geophys Res 122. https://doi.org/10.1002/2017JC012807
7. Schlosser P, Bauch D et al (1994) Arctic river-runoff: mean residence time on the shelves and in the halocline. Deep-Sea Res I 41(7):1053–1068
8. Steele M, Boyd T (1998) Retreat of the cold halocline layer in the Arctic Ocean. J Geophys Res 103:10419–10435
9. Bauch D, Schlosser P et al (1995) Freshwater balance and the sources of deep and bottom waters in the Arctic Ocean inferred from the distribution of $H_2{}^{18}O$. Prog Oceanogr 35:53–80
10. Bauch D, Dmitrenko IA et al (2009) Exchange of Laptev Sea and Arctic Ocean halocline waters in response to atmospheric forcing. J Geophys Res 114(C05008). https://doi.org/10.1029/2008JC005062
11. Guay CK, Falkner KK et al (2001) Wind-driven transport pathways for Eurasian Arctic river discharge. J Geophys Res 106:11469–11480
12. Dmitrenko I, Kirillov S et al (2005) Wind-driven summer surface hydrography of the eastern Siberian shelf. Geophys Res Lett 32. https://doi.org/10.1029/2005GL023022
13. Morison J, Kwok R et al (2012) Changes in Arctic Ocean circulation and freshwater measure with ICESat altimetry, GRACE gravimetry, and in situ observations. Nature 481. https://doi.org/10.1038/nature10705
14. Rudels B, Jones EP et al (2004) Atlantic sources of the Arctic Ocean surface and halocline waters. Polar Res 23(2):181–208
15. Shimada K, Itoh M et al (2005) Halocline structure in the Canada Basin of the Arctic Ocean. Geophys Res Lett 32. https://doi.org/10.1029/2004GL021358
16. Itoh M, Carmack E et al (2007) Formation and spreading of Eurasian source oxygen-rich halocline water into the Canadian Basin in the Arctic Ocean. Geophys Res Lett 34. https://doi.org/10.1029/2007GL029482
17. Rudels B, Anderson LG et al (1996) Formation and evolution of the surface mixed layer and halocline of the Arctic Ocean. J Geophys Res 101:8807–8821
18. Alkire MB, Polyakov I et al (2017) Lower halocline water formation and modification, a comparison of physical and geochemical methods. Ocean Sci 13:983–995. https://doi.org/10.5194/os-13-983-2017

19. Jones EP, Anderson LG (1986) On the origin of the chemical properties of the Arctic Ocean halocline. J Geophys Res 91:10759–10767
20. Salmon DK, McRoy CP (1994) Nutrient-based tracers in the Western Arctic: a new lower halocline water defined. In: Johannessen OM, Muench RD, Overland JE (eds) The polar oceans and their role in shaping the global environment. AGU, Washington, pp 47–61
21. Alkire MB, Falkner KK et al (2010) Sensor-based profiles of the NO parameter in the central Arctic and southern Canada Basin: new insights regarding the cold halocline. Deep-Sea Res I 57:1432–1443
22. Alkire MB, Rember R et al (2019) Discrepancy in the identification of the Atlantic/Pacific front in the central Arctic Ocean: NO versus nutrient relationships. Geophys Res Lett 46. https://doi.org/10.1029/2018GL081837
23. Steele M, Morison J et al (2004) Circulation of summer Pacific halocline water in the Arctic Ocean. J Geophys Res 109. https://doi.org/10.1029/2003JC002009
24. Cooper LW, Whitledge TE et al (1997) The nutrient, salinity, and stable oxygen isotope composition of Bering and Chukchi Seas waters in and near the Bering Strait. J Geophys Res 102:12563–12573
25. Cooper LW, Cota GF et al (1999) Modification of NO, PO, and NO/PO during flow across the Bering and Chukchi shelves: implications for use as Arctic water mass tracers. J Geophys Res 104(C4):7827–7836
26. Aksenov Y, Ivanov VV et al (2011) The Arctic circumpolar boundary current. J Geophys Res 116(C09017). https://doi.org/10.1029/2010JC006637
27. Pnyushkov A, Polyakov I et al (2015) Structure and variability of the boundary current in the Eurasian Basin of the Arctic Ocean. Deep-Sea Res I 101(7):80–97. https://doi.org/10.1016/j.dsr.2015.03.001
28. Weingartner T, Aagaard K et al (2005) Circulation on the north central Chukchi Sea shelf. Deep-Sea Res II 52:3150–3174
29. Shimada K, Carmack EC et al (2001) Varieties of shallow temperature maximum waters in the western Canadian Basin of the Arctic Ocean. Geophys Res Lett 28(18):3441–3444
30. Nishino S, Shimada K et al (2008) East-west differences in water mass, nutrient, and chlorophyll a distribution in the sea ice reduction region of the western Arctic Ocean. J Geophys Res 113(C00A01). https://doi.org/10.1029/2007JC004666
31. Nishino S, Itoh M et al (2013) Shoaling of the nutricline with an increase in near-freezing temperature water in the Makarov Basin. J Geophys Res 118(1–15). https://doi.org/10.1029/2012JC008234
32. Taylor JR, Falkner KK et al (2003) Quantitative considerations of dissolved barium as a tracer in the Arctic Ocean. J Geophys Res:108. https://doi.org/10.1029/2002JC001635
33. Falck E, Kattner G et al (2005) Disappearance of Pacific water in the northwestern Fram Strait. Geophys Res Lett 32(L14619). https://doi.org/10.1029/2005GL023400
34. Alkire MB, Falkner KK et al (2007) The return of Pacific waters to the upper layers of the central Arctic Ocean. Deep-Sea Res I 54:1509–1529
35. Dodd PA, Rabe B et al (2012) The freshwater composition of the Fram Strait outflow derived from a decade of tracer measurements. J Geophys Res 117(C1105). https://doi.org/10.1029/2012JC008011
36. Rabe B, Dodd PA et al (2013) Liquid export of Arctic freshwater components through the Fram Strait 1998-2011. Ocean Sci 9:91–109
37. Torres-Valdes S, Tsubouchi T et al (2013) Export of nutrients from the Arctic Ocean. J Geophys Res 118:1625–1644
38. McLaughlin FA, Carmack EC (2010) Nutricline deepening in the Canada Basin, 2003-2009. Geophys Res Lett 37(L24602). https://doi.org/10.1029/2010GL045459
39. Yamamoto-Kawai M, McLaughlin FA et al (2009) Aragonite undersaturation in the Arctic Ocean: effects of ocean acidification and sea ice melt. Science 326. https://doi.org/10.1126/science.1174190

40. Azetsu-Scott K, Clarke A et al (2010) Calcium carbonate saturation states in the waters of the Canadian Arctic Archipelago and the Labrador Sea. J Geophys Res 115. https://doi.org/10.1029/2009JC005917
41. Fransson A, Chierici M et al (2013) Impact of sea-ice processes on the carbonate system and ocean acidification at the ice-water interface of the Amundsen Gulf. Arctic Ocean J Geophys Res 118. https://doi.org/10.1002/2013JC009164
42. McLaughlin FA, Carmack E et al (1996) Physical and geochemical properties across the Atlantic/Pacific water mass front in the southern Canadian Basin. J Geophys Res 101:1183–1197
43. Morison J, Steele M et al (1998) Hydrography of the upper Arctic Ocean measured from the nuclear submarine U.S.S. Pargo. Deep-Sea Res I 45:15–38
44. Morison JH, Aagaard K et al (2000) Recent environmental changes in the Arctic: a review. Arctic 53:4
45. Morison J, Steele M et al (2006) The relaxation of central Arctic Ocean hydrography to pre-1990s climatology. Geophys Res Lett 33. https://doi.org/10.1029/2006GL026826
46. Alkire MB, Morison J et al (2015) Variability and trends in the meteoric water, sea-ice melt, and Pacific water contributions to the central Arctic Ocean, 2000-2013. J Geophys Res 120:1573–1598
47. de Steur L, Steele M et al (2013) Hydrographic changes in the Lincoln Sea in the Arctic Ocean with focus on an upper ocean freshwater anomaly between 2007 and 2010. J Geophys Res 118:4699–4715. https://doi.org/10.1002/jgrc.20341
48. Jackson JM, Lique C et al (2014) On the waters upstream of Nares Strait, Arctic Ocean, from 1991 to 2012. Cont Shelf Res 73:83–96
49. Semiletov I, Dudarev O et al (2005) The east Siberian Sea as a transition zone between Pacific-derived waters and Arctic shelf waters. Geophys Res Lett 32(L10614). https://doi.org/10.1029/GL022490
50. Rutgers van der Loeff MM, Key RM et al (1995) ^{228}Ra as a tracer for shelf water in the Arctic Ocean. Deep-Sea Res II 42(6):1533–1553
51. Rigor IG, Wallace JM et al (2002) Response of sea ice to the Arctic oscillation. J Climate 15:2648–2663
52. Nitishinsky M, Anderson LG et al (2007) Inorganic carbon and nutrient fluxes on the Arctic Shelf. Cont Shelf Res 27:1584–1599
53. Bauch D, Rutgers van der Loeff MM et al (2011) Origin of freshwater and polynya water in the Arctic Ocean halocline in summer 2007. Prog Oceanogr 91(4):482–495. https://doi.org/10.1016/j.pocean.2011.07.017
54. Ekwurzel B, Schlosser P et al (2001) River runoff, sea ice meltwater, and Pacific water distribution and mean residence times in the Arctic Ocean. J Geophys Res 106:9075–9092
55. Newton R, Schlosser P et al (2013) Canadian Basin freshwater sources and changes: results from the 2005 Arctic Ocean section. J Geophys Res 118(2133–2154). https://doi.org/10.1002/jgrc.20101
56. Devol AH, Codispoti LA et al (1997) Summer and winter denitrification rates in western Arctic shelf sediments. Cont Shelf Res 17:1029–1050
57. Jones EP, Anderson LG et al (1998) Distribution of Atlantic and Pacific waters in the Arctic Ocean: implications for circulation. Geophys Res Lett 25(6):765–768
58. Yamamoto-Kawai M, Carmack E et al (2006) Nitrogen balance and Arctic throughflow. Nature 443. https://doi.org/10.1038/443043a
59. Yamamoto-Kawai M, McLaughlin FA et al (2008) Freshwater budgets of the Canada Basin, Arctic Ocean, from salinity, $\delta^{18}O$, and nutrients. J Geophys Res 113. https://doi.org/10.1029/2006JC003858
60. Broecker WS (1974) "NO", a conservative water mass tracer. Earth Planet Sci Lett 23:100–107
61. Gruber N, Sarmiento JL (1997) Global patterns of marine nitrogen fixation and denitrification. Global Biogeochem Cycles 11(2):235–266

62. Anderson LG, Andersson PS et al (2013) Source and formation of the upper halocline of the Arctic Ocean. J Geophys Res 118(1–12). https://doi.org/10.1029/2012JC008291
63. Polyakov IV, Timokhov LA et al (2010) Arctic Ocean warming contributes to reduced polar ice cap. J Phys Oceanogr 40:2743–2756. https://doi.org/10.1175/2010/JPO4339.1
64. Kipp LE, Charette MA et al (2018) Increased fluxes of shelf-derived materials to the central Arctic Ocean. Sci Adv 4:eaao1302
65. Proshutinsky A, Krishfield R et al (2019) Analysis of the Beaufort Gyre freshwater content in 2003–2018. J Geophys Res 124. https://doi.org/10.1029/2019JC015281
66. Woodgate RA, Aagaard K et al (2005) Pacific ventilation of the Arctic Ocean's lower halocline by upwelling and diapycnal mixing over the continental margin. Geophys Res Lett 32. https://doi.org/10.1029/2005GL023999
67. Ardyna M, Babin M et al (2014) Recent Arctic Ocean sea ice loss triggers novel fall phytoplankton blooms. Geophys Res Lett 41:6207–6212. https://doi.org/10.1002/2014GL061047
68. Schlitzer R (2006) Ocean data view. http://odv.awi.de
69. Carmack EC, Polyakov IV et al (2015) Toward quantifying the increasing role of oceanic heat in sea ice loss in the new Arctic. Bull Am Meteorol Soc 96(12):2079–2105. https://doi.org/10.1175/BAMS-D-13-00177.1

Major Nutrient Fronts in the Northeastern Atlantic: From the Subpolar Gyre to Adjacent Shelves

Hjálmar Hátún, Karin Margretha H. Larsen, Sólvá Káradóttir Eliasen, and Moritz Mathis

Contents

Abstract Diatom-dominated spring blooms in the northeastern Atlantic, both in the open ocean and on adjacent shelves, become silicate-limited every spring/summer. We here review the fertilizing silicate fluxes from the large subpolar gyre source, across the major oceanic Subarctic Front and further across shelf edge and tidal mixing fronts and onto adjacent shelves. As a case study, we illustrate potential linkages between the open ocean dynamics and the primary production, fish larvae

H. Hátún (✉), K. M. H. Larsen, and S. K. Eliasen
Faroe Marine Research Institute, Tórshavn, Faroe Islands
e-mail: hjalmarh@hav.fo; karinl@hav.fo; solvae@hav.fo

M. Mathis
Max Planck Institute for Meteorology, Hamburg, Germany
e-mail: moritz.mathis@hzg.de

Igor M. Belkin (ed.), *Chemical Oceanography of Frontal Zones*,
Hdb Env Chem (2022) 116: 97–142, DOI 10.1007/698_2021_794,

Published online: 13 August 2021

abundances and seabird breeding success within the Faroe shelf ecosystem. The "boosting effect" of vigorous winter convection occurring every 5–8 years is illustrated, and we also discuss the pre-bloom silicate decline, which has taken place throughout the entire subpolar North Atlantic since the early 1990s. Reduced winter convection due to global warming is projected by most climate models, and this is expected to have severe impact on the North Atlantic Ocean primary production.

Keywords Lateral exchange, Mixed layer depth, Productivity, Silicate, Subpolar gyre

1 Introduction

The open ocean boreal waters in the northeastern Atlantic host some of the largest pelagic fish stocks in the world [1], and the adjacent continental shelves support rich marine ecosystems – including commercially important fish stocks [2] and large seabird colonies [3, 4]. There is a growing recognition of how climate variability and large-scale oceanography can regulate both the biomass and regional distribution of these important marine resources. However, establishing possible climate-ecosystem linkages has been generally limited to the utilization of sea temperatures as an environmental proxy [5, 6], without much consideration for nutrient dynamics.

Nutrient limitation of primary production at lower latitudes is "textbook knowledge" [7], but the oceanographic community has often overlooked the likely importance of this fundamental ecosystem driver in boreal/subarctic parts of the North Atlantic Ocean. Only very few studies have investigated possible linkages between ocean dynamics and nutrient concentrations there. The subpolar North Atlantic is characterized by strong spring blooms [8, 9]. These generally ensue after positive air-sea heat fluxes stratify the near-surface layer around April–May, although several other hypotheses for the bloom initiation have been proposed ([10] – and references therein). These blooms consist primarily of diatoms, which are fast-growing algae that, in addition to phosphate and nitrate, require silicate to sustain their growth [11, 12]. Macronutrients are quickly drawn out of the shallow summer mixed layer, and in the North Atlantic, silicate typically becomes the limiting nutrient for diatom growth [11, 13] – occasionally confounded by seasonal iron limitation [14]. After this, new production becomes restricted to upwelling regions, e.g., along the large-scale nutrient fronts and along the continental slopes [11]. Diatoms are an important food source for secondary producers, and in particular for calanoid copepods such as *Calanus finmarchicus*, which itself is a key prey item linking phytoplankton to higher trophic levels in subpolar ecosystems [15]. Predators like pelagic fish and seabirds will therefore congregate at such upwelling feeding hotspots during summer [16–18].

However, the processes through which the larger basin scale could alter the productivity along fronts are not well understood. If rectified supply of nutrients is the dominant process and involves vertical transfer near fronts [11], then this raises the question of how the underlying nutrients in the thermocline are maintained. Without a continuous supply to the thermocline, any near-front rectified supply will only provide an initial enhancement in productivity. This will gradually decrease in time until an equilibrium state is reached at lower productivity. One task is therefore to trace the supply of nutrients in the deeper waters.

The northeastern (NE) Atlantic receives a mix of nutrient-poor subtropical waters from the Bay of Biscay and the Gulf Stream, and a contrasting nutrient-rich subarctic water contribution from the subpolar gyre (SPG, a list of acronyms is provided in Table 2) in the west [19] (Fig. 1). Strong nutrient gradients are co-located with the Subarctic Front (SAF), which divides the subtropical and subarctic water masses – and which thus outlines the main periphery of the SPG.

The NE Atlantic encompasses the confluence of these contrasting water masses. Consequently the nutrient dynamics within this region are highly sensitive to the strength and size of the subpolar gyre [22, 24, 26] and lateral shifts of the SAF (Fig. 1b) – the large-scale *nutrient front*. However, in order to fertilize the poleward flowing Atlantic waters, subarctic waters of the SPG must be transported across the SAF. The SAF is a three-dimensional structure (Fig. 1e), therefore the cross-frontal transport also has a vertical dimension. Strong nutrient gradients are located at the base of the winter mixed layer, separating the nutrient-poor Atlantic waters from nutrient-rich waters below. The nutrient concentrations off the south Iceland and European continental slopes are therefore regulated by the SAF dynamics and the winter convection – both associated with the dynamics of the SPG [26, 27] – as well as the summer mixed layer dynamics [28].

Further north, in the southern Norwegian Sea, the northward flowing Atlantic water meets southeastward flowing subarctic waters from the Greenland and Iceland Seas. This establishes the Iceland-Faroe Front (IFF) (Fig. 1b) – another large-scale nutrient front [11]. The Atlantic waters surrounding the Faroe shelf are influenced by the IFF dynamics, in addition to the upstream SPG regulation and the strength of the summer mixed layer.

The ecosystems of the adjacent European, south Iceland and Faroe shelves are critically dependent on nutrient fluxes from the open ocean [29, 30], since riverine nutrient input only enriches the very near-coast band along these seaboards. This ocean-to-shelf *onwelling* of nutrients [31] requires cross-frontal exchanges (shelf edge fronts along the margins and tidal mixing fronts in shallower waters, Fig. 1b) in order to bring nutrients from the open ocean onto the shelves [28, 32].

In this chapter, we illustrate the nutrient pathway from the North Atlantic SPG, across open-ocean nutrient fronts and slope fronts, to adjacent shelves. Emphasis is put on silicates and the Faroe shelf ecosystem. This should not be regarded as a complete review, but rather a story which builds on some selected and relevant publications, complemented by more general reviews of important sub-topics. In Sect. 2, we describe the large-scale open-ocean nutrient fronts and other basic SPG-related oceanographic features, which could be important for the pre-bloom

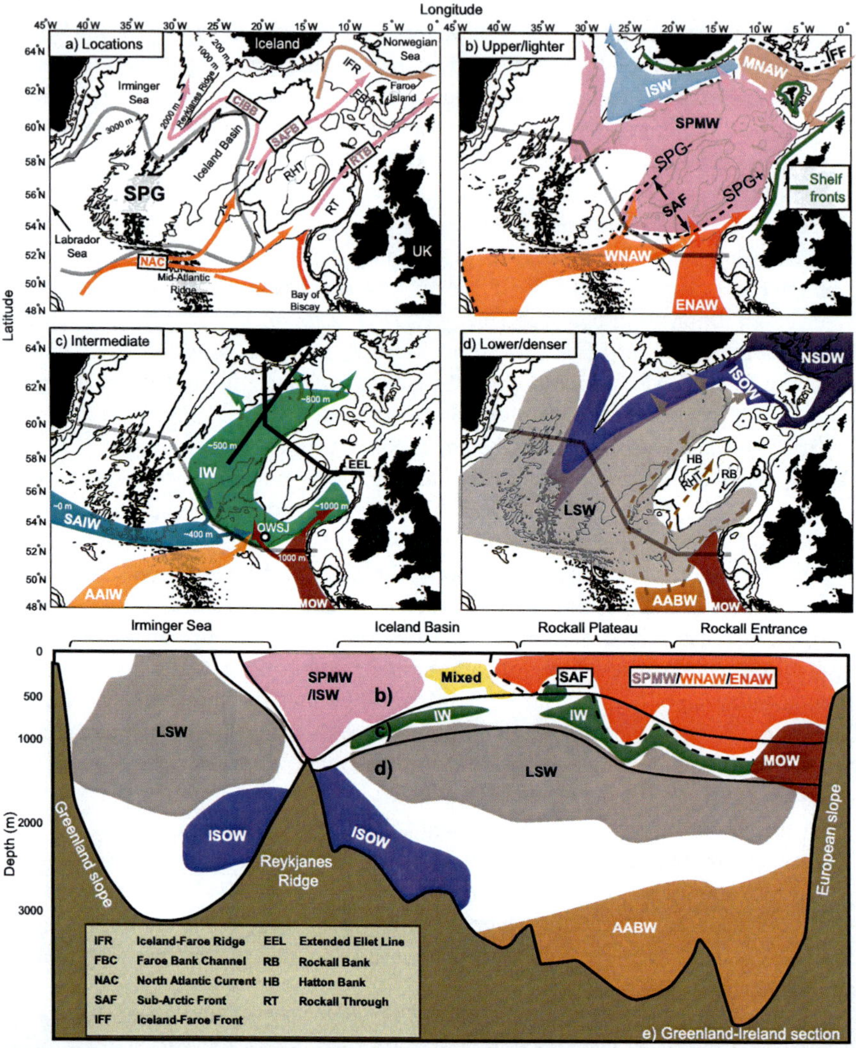

Fig. 1 Overview. (**a**) Bathymetry, toponymes, the approximate outline of the subpolar gyre (SPG, gray) and the main Atlantic inflows current branches (reddish arrows). A vertical view of main water masses along the A1E WOCE section from Greenland to Ireland (gray line in panels **b**, **c** and **d**) is shown in (**e**). This is based on the occupation of this section in 1991, presented in [20]. The rough horizontal distribution of the water masses is illustrated by vertically dividing them into an upper/lighter layer (**b**), an intermediate layer aligned with the permanent pycnocline (**c**), and a deeper/denser layer (**d**). The approximate position of the oceanic Sub-Arctic Front (SAF) is illustrated with black dashed lines (**b** and **e**) and differentiated between a strong subpolar gyre state (SPG+) and a weak state (SPG-, respectively in panel (**b**). Near-shelf fronts are also sketched (green in **b**). The black lines in (**c**) show the sections presented in Figs. 5, 7, and 8. The water mass acronyms are summarized in Table 1, and acronyms for the current branches are: The North Atlantic Current (NAC), the Rockall Trough Branch (RTB), the Sub-Arctic Front Branch (SAFB), and the Central Iceland Basin Branch (CIBB) (following [25])

nutrient concentrations. Key aspects of open ocean silicate limitation during summer and its influence on zooplankton and pelagic fish are also outlined in Sect. 2. Evidence for oceanic regulation of shelf ecosystems – across the shelf edge and tidal mixing fronts – is provided in Sect. 3. A summary is given in Sect. 4.

2 Nutrient Fronts in the Open Ocean

2.1 The Main Thermocline: Currents and Water Masses

Large parts of the World oceans are characterized by a layer of oxygen-poor, nutrient-rich intermediate water [33]. From a climatological spatial perspective, there is a general poleward shallowing of the isopycnals, bringing this biogeochemically distinct thermocline layer toward the surface in the north [34]. In the North Atlantic, the isopycnals steepen markedly between the major subtropical and subpolar gyres, establishing large-scale, persistent nutrient fronts near the surface – aligned with the North Atlantic Current (NAC) [35]. As the NAC – the boundary between the subtropical and subpolar gyre – passes east of the Mid-Atlantic Ridge, it encounters the Rockall-Hatton Plateau and subsequently the European Continental slope (Fig. 1). These topographical features split the NAC into three branches: (1) a branch that is deflected south toward the Bay of Biscay, (2) a branch that is channeled into the Rockall Trough, and (3) a branch that exists as a partial retroflection into the Iceland Basin (IB) (Fig. 1a). The IB is surrounded by the Rockall-Hatton Plateau, south Iceland slope and the Reykjanes Ridge. In this region, the NAC further splits into the SAF branch that hugs the western slope of the Rockall-Hatton Plateau (Figs. 1a, b) and the Central Iceland Basin branch that veers west and marks the periphery of the SPG in the IB [25].

Numerous studies have presented zonal cross-sections between the European and the Greenland slope [20, 36]. Main characteristics are the relatively homogenous and saline Eastern North Atlantic Water mode water in the east (ENAW, Figs. 1b, e, Table 1), bounded by the stratified "floor" of the westward surfacing intermediate water (IW, Figs. 1c, e, Table 1) associated with the main thermocline. The IW is strongly influenced by the low-saline and nutrient-rich Subarctic Intermediate Water (SAIW, Figs. 1c, e, Table 1) that is formed near the surface in the central SPG/Labrador Current region and flows along isopycnals to greater depths while being transported eastward [35, 37] (Fig. 1c). The subtropical water carried eastward by the NAC, as it is influenced by the admixture SAIW, is referred to as Western North Atlantic Water (WNAW, [19, 38]), and this outlines the "*south-eastern*" periphery of the SPG (Fig. 1b, Table 1). The thermocline/IW layer envelopes the SPG and constitutes a major deep nutrient front in the NE Atlantic. The confluence and mixing of ENAW and WNAW produces Subpolar Mode Water (SPMW) – the dominant upper layer water mass in the study region [39] (Fig. 1b, e). The works by Brambilla and colleagues [25, 40] furthermore divided the SPMW into several density classes, separated horizontally by highly stratified water masses not

Table 1 List of water masses and their acronyms

Water masses		
Acronym	**Color**	**Name**
AAIW		Antarctic Intermediate Water
ENAW		Eastern North Atlantic Water
IW		Intermediate Water
ISW		Iceland Slope Water
ISOW		Iceland-Scotland Overflow Water
LSW		Labrador Sea Water
MNAW		Modified North Atlantic Water
MOW		Mediterranean Overflow Water
NADW		North Atlantic Deep Water
SAIW		Subarctic Intermediate Water
SPMW		Subpolar Mode Water
WNAW		Western North Atlantic Water
AABW		Antarctic Bottom Water

belonging to any SPMW class. These are carried polewards by the abovementioned intense current branches (Fig. 1a). Previous studies have treated the various SPMW classes as individual water masses (e.g., Modified North Atlantic Water, Iceland Slope Water, introduced later), and the popular term *Atlantic water* is a general reference to this water mass complex. Underlying these waters, and under the *IW* envelope, the low-salinity and relatively nutrient-rich Labrador Sea Water (LSW) comprises the main body of the SPG [41] (Figs. 1d, e).

South of the Rockall-Hatton complex, the IW is composed of oxygen-poor and silicate-rich Antarctic Intermediate Water (AAIW), transported northward under the Gulf Stream, and salty Mediterranean Overflow Water (MOW) ([42], Fig. 1c, Table 1). At larger depths east of the Mid-Atlantic Ridge, silicate-rich Antarctic Bottom Water (AABW) is also transported northwards extending into the Rockall Through and IB and onto the Rockall-Hatton Plateau (Fig. 1d, Table 1).

2.2 *From the Thermocline to the Photic Zone*

In order to contribute to the rich primary production in the NE Atlantic Ocean, nutrients must somehow be transported from and/or through the IW layer toward the photic zone. The gyre-scale circulation transfers fluid between the thermocline

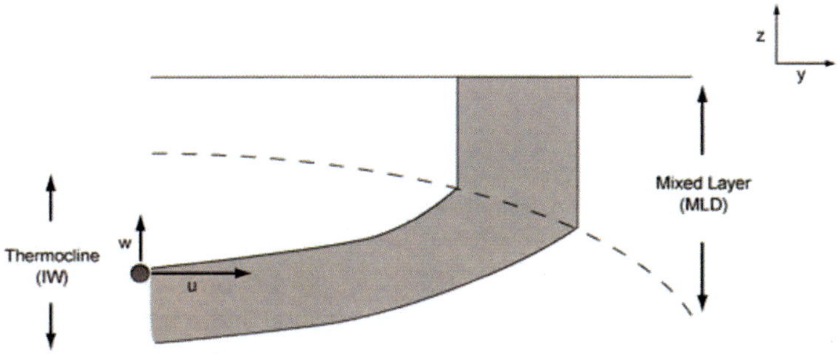

Fig. 2 A schematic figure of the Induction flux into the mixed layer. Fluid is being transferred from the stratified thermocline and into the mixed layer (gray region). A fluid particle is advected by the vertical velocity w and the lateral velocity u. (Redrawn from [33]; their Fig. 2.11a)

(IW layer) and the mixed layer/seasonal boundary layer [33]. This exchange is due to the time-mean circulation and consists of vertical and horizontal contributions:

$$\text{Exchange} = w_b + u_b \cdot \nabla \text{MLD} \tag{1}$$

where w_b and u_b are the vertical and horizontal velocity vectors at the base of the seasonal boundary, the depth of this layer is represented by the mixed layer depth (MLD), and ∇ is the lateral gradient operator (Fig. 2, from [33]). In the IB region, u_b is generally directed polewards, and the MLD also increases northwards ($\frac{\partial(\text{MLD})}{\partial y} > 0$) (see Fig. 6). This results in net transport into the seasonal boundary layer – a so-called *induction flux* [33], or *obduction* in the terminology of [40]. Over the subpolar gyre in general, fluid is transferred from the thermocline into the seasonal boundary layer by this *induction flux*. This process is essentially the reverse of the classical subduction process that produces the mode water on the warm side of the Gulf Stream [43]. The largely lateral induction flux in the SPG region totals 300 m yr^{-1}, compared to a vertical Ekman flux of only about 50 m yr^{-1} (where the volume fluxes are expressed per unit horizontal area, [33]). Hence, the surface waters around the periphery of the subpolar gyre are likely sustained through the horizontal and advective influx from the nutrient-rich thermocline/IW layer.

In contrast to the work of [33, 40] reported a net subduction in the northern NE Atlantic – not in the classical sense, but due to entrainment of thermocline waters into the overflows (Fig. 1d). The thermocline-to-boundary layer exchange estimates by these previous works were, however, solely based on air–sea interaction data and therefore lack details of the realistic flows and their interaction with the topography. If the near-thermocline flows are rectified by bathymetry (e.g., by the Rockall-Hatton Plateau), this could force the nutrient-rich IW into the seasonal boundary layer and thus strongly increase the induction flux (obduction) locally.

The region south of the Iceland-Faroe Ridge has previously been identified as a key water mass transformation area in the subpolar North Atlantic [44]. Through the variable entrainment of SPMW/LSW/IW, northeastward extension of the SPG-limb is reported to regulate the properties of the Iceland-Scotland Overflow Water (ISOW, Fig. 1d) and thus the North Atlantic Deep Water downstream [45]. However, since the regional seasonal boundary layer can deepen to the depth range of the overflow and SPG water complex during winter (see Fig. 6), the energetic overflow has the potential to amplify an induction flux from the IW to the photic zone – an important process which has not received much attention in the literature.

By reviewing how the SPG shifts the frontal systems laterally in relations to the complex bathymetry in the NE (especially on and around the Rockall-Hatton Plateau), we discuss the induction fluxes and suggest locations for possible nutrient fertilizing "hotspots."

2.3 *The Subpolar Gyre*

Linkages between the energetic atmospheric and oceanic dynamics and hydrography in the northeastern Atlantic are thoroughly discussed in the literature [19, 24]. The regional state of the atmosphere is typically represented by the North Atlantic Oscillation (NAO) index [46], while the state of the subpolar Atlantic marine climate has been proxied by the so-called gyre index [22, 24, 47] (Fig. 3) in addition to the Atlantic Multidecadal Oscillation (AMO) [49]. It is now well established that periods with a high NAO index (NAO+ states) are associated with a strong and expanded SPG [44], (SPG+ state, Fig. 1b, represented by a high gyre index, Fig. 3). This involves a northeastward expansion of the frontal complex and increased contribution of WNAW to the SPMW, at the expense of the ENAW. On the other hand, a weak NAO-/SPG- state is associated with generally southwestward frontal shifts (Fig. 1b) and an increased relative contribution of saline and nutrient-poor ENAW to the SPMW. The major salinification and warming of the NE Atlantic after the early 1990s was linked to an abrupt drop in the NAO index and a subsequent weakening and westward retraction of the SPG [24]. This is, admittedly, a simplification of this complex system. The relative importance of horizontal gyre changes and vertical mixing processes for influencing regional hydrography is still widely debated [40, 50].

Few studies have examined regional nutrient dynamics in the context of large-scale oceanographic features. Rey [51] linked a persistent decline in the upper ocean, pre-bloom (late winter) silicate concentrations along the Norwegian slope after the early 1990s to the weakening SPG. This was based on correlations between the silicate data and the gyre index. Similarly, in the Rockall Trough, Johnson et al. [38] linked a decline in the upper ocean nitrate and phosphate, but not silicate, to a weakening SPG and the associated dominance of nutrient-poor ENAW. More recently, Hátún et al. [52] revealed a steady silicate decline throughout a broad area of the subpolar Atlantic, including the Labrador Sea, Irminger Sea (Fig. 3), IB and the northern Nordic Seas since the early 1990s. In agreement with [51], this broad, pre-bloom silicate decline was primarily linked to the decline of the SPG,

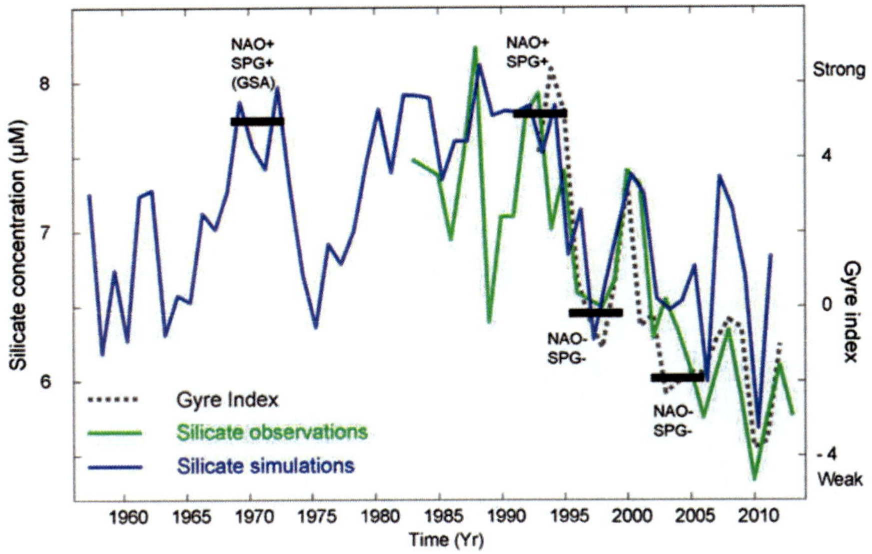

Fig. 3 Temporal evolution of silicate and the gyre index. The observed (green) and simulated (blue) pre-bloom upper ocean silicate concentrations (0–200 m, March) in the northern Irminger Sea (Ir in Fig. 4) are compared to the gyre index (dashed gray). The dimensionless gyre index is associated with the leading North Atlantic sea-surface height mode obtained from altimetry observations [48]. The samples have been taken in the pre-bloom homogeneous winter mixed layer several hundred meters thick. Periods with a strong/weak atmospheric forcing and subpolar gyre (NAO+/SPG+ and NAO-/SPG-, respectively) are emphasized. The 1970s Great Salinity Anomaly (GSA) period is also noted

with a possible contribution from a reduction in trans-Arctic silicate. In agreement with [38, 52] did not report any marked silicate decline in the Rockall Trough or south of this. By extrapolating station-based silicate records with a realistic ocean carbon cycle model HAMOCC [53], it is evident that the signal of silicate decline originates at the frontal zone at the southern tip of the Rockall-Hatton Plateau (Fig. 4).

The degree of silicate decline intensifies across the IB with especially strong imprints in the Reykjanes Ridge frontal region and along the IFF (Fig. 4). This spatial analysis indicates that oceanographic processes responsible for the observed silicate decline are likely to occur adjacent to and downstream of the IB. However, previous studies [38, 51, 52] did not attempt to quantify the relative importance of the vertical and horizontal SPG-related processes for explaining the silicate decline.

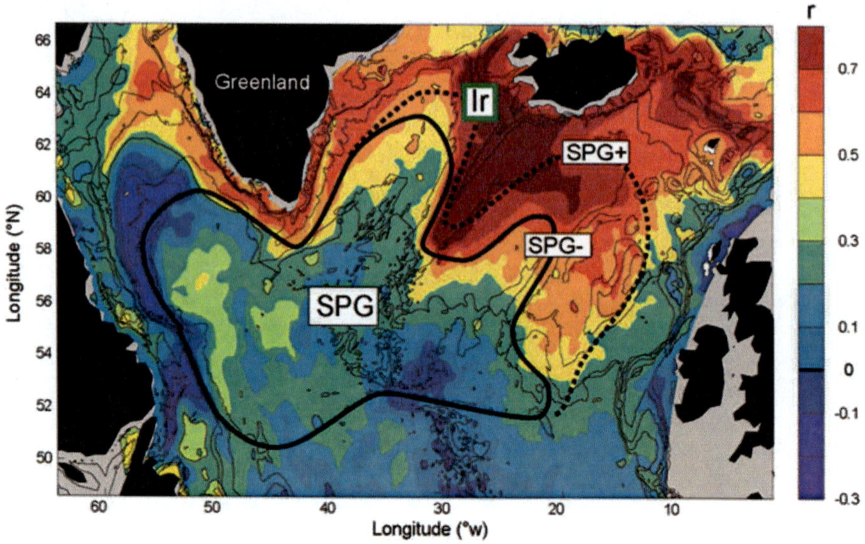

Fig. 4 Analysis of the simulated (HAMOCC) near-surface (0–150 m) silicate during March. The map shows the correlation between the simulated time series in the northern Irminger Sea (Ir, green box) and the time series at each individual model grid cell (1958–2011). The black lines outline the approximate boundary of the subpolar gyre during a weak state (SPG-, dashed) and a strong state (SPG+)

2.4 On and Around Rockall

2.4.1 South of Rockall

The SPG dynamics might impact the IW layer, even south of the Rockall Region. A long-term decrease in the oxygen concentration of the IW, with the largest changes occurring from 1993 to 2002 has been observed along a zonal repeat section at 48°N [54]. These changes were associated with the SPG contraction and thus reduced penetration of oxygen-rich IW of subpolar origin into the region. This allowed an increased northward transport of IW of subtropical origin, which is lower in oxygen. A decrease of preformed phosphate during the same period, especially in the density range of IW and MOW (along the 48°N section), supports the conclusion of a reduced influence from the SPG after the early 1990s ([54], see Fig. 3).

2.4.2 The Rockall Trough

The WNAW and ENAW meet in the Rockall Trough [19, 38] (Fig. 1b). In the deeper intermediate layer below, the confluence of contrasting SAIW/IW and MOW forms sharp lateral physical and biogeochemical gradients [55] (Fig. 1c, e). A strong, eastward extended SPG increases the influx of SAIW into the Rockall Trough (Fig. 1b). It is thoroughly documented that such events lead to a marked freshening of the region and of the poleward flowing Atlantic water [24, 38, 56]. Strong SPG+ states and freshening periods – also referred to as Great Salinity Anomalies (GSAs, Fig. 3) [57, 58] – took place during the early 1970s and the early 1990s. Frequent sampling at the strategically located Ocean Weather Ship J at the southern tip of the Rockall-Hatton Plateau (52°30'N, 20°00'W, Fig. 1c) showed that SAIW comprised more than 35% of the 400–700 m water column during 1972–1975 [37]. This fractional contribution was suddenly reduced to below 5% during 1976 and remained very low during the rest of this record (was terminated in 1980). The influx of SPG-related subarctic water into the Rockall Trough increased by more than 2 Sv (1 Sv = 10^6 m^3 s^{-1}) during the 1972–1975 and decreased again by the same amount from 1976 to 1980 ([24], their Fig. 4).

There has been a recent salinity decline in the NE Atlantic [59], which consisted of two marked drops – one between 2010 and 2012, and a very abrupt drop after 2015 (Fig. 5). Hydrographic data from the standard, repeat Extended Ellet Line section across the northern part of the Rockall Trough (Fig. 1c) revealed major salinity drops through the water column over the IW layer (0–1,000 m) along the eastern side of the Rockall-Hatton Plateau during both events (Figs. 5c, d). After remaining within the ENAW salinity range (>35.30) for about two decades (~1996–2016), the salinity in this region dropped below 35.2 during 2016. This is a clear indication of increased presence of SAIW. Both the salinity values and the temporal trend in the Rockall Trough Branch are very similar to the properties in the Atlantic water inflow to the Nordic Seas observed in the Faroe Bank Channel (Fig. 5b). There has also been a marked salinity decrease in the SAF Branch in the IB since the early 2000s (Figs. 1a and 5b). However the salinity of this flow is much lower than the salinity observed in the Faroe Bank Channel, and the 2010–2012 and 2015–2016 drops are less evident.

This all indicates that the Rockall Trough Branch is a more direct source to the waters west of the Faroe Plateau than is the SAF Branch. In the same way that the increased influx of low-salinity SAIW leads to freshening events, this nutrient-rich water mass also leads to increased nutrient concentrations – especially along the western side of the Rockall Trough entrance [56]. The major salinity drop around 2015–2016 was accompanied by a marked silicate increase of about 1 μM off the Faroe shelf, in the northern IB and in the northern Irminger Sea (see Fig. 15 herein, adapted from [60]). This disrupted the persistent silicate decline that has characterized the entire subpolar Atlantic since the early 1990s [52] (Fig. 3).

In summary, (1) a strong SPG shifts the SAF from the IB to east of the Rockall-Hatton Plateau (Fig. 1b), which leads to strong near-pycnocline flows into the deep seasonal boundary layer of the Rockall Trough, and (2) this intensified and topographically rectified Rockall Trough Branch will markedly increase the induction flux of nutrient-rich subarctic water to the Atlantic inflow waters. We propose this area (referred to as Region 1 in Fig. 6b) as a key *fertilization hotspot*.

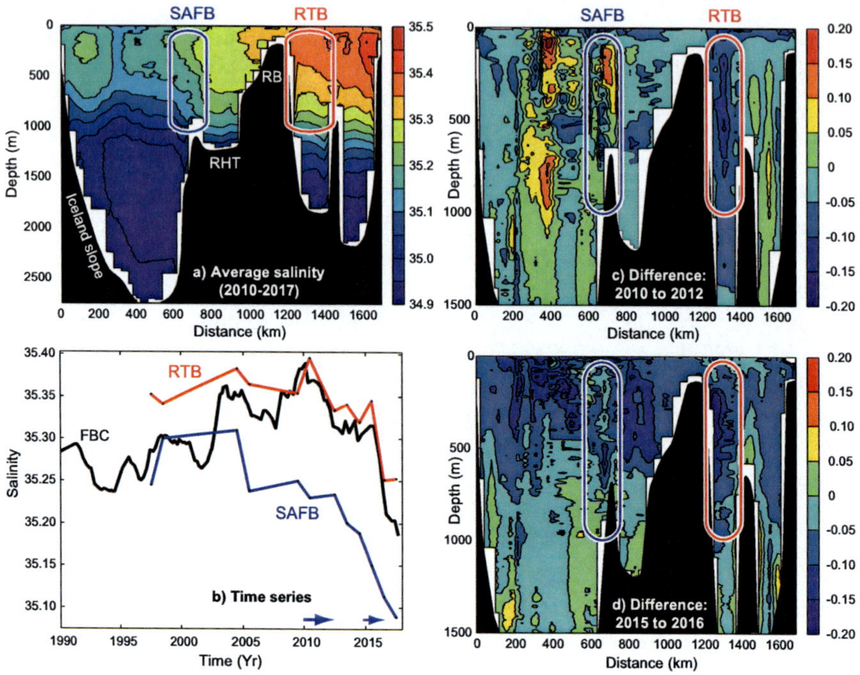

Fig. 5 Salinity along the Extended Ellet Line (EEL, Fig. 1a). (**a**) Average over the period 2010–2017 based on yearly CTD sections (no section available for 2011). (**b**) Time series, based on averages across the Sub-Arctic Front Branch (SAFB), the Rockall Trough Branch (RTB) – blue and red rectangles – and the salinity in the upper Atlantic water layer in the downstream Faroe Bank Channel (FBC, black). Sectional views of the salinity difference over the two major declining steps evident in the RTB and FBC series (blue arrows in **b**) are shown in (**c**) 2010-to-2012 and (**d**) 2015-to-2016. Bathymetric abbreviations: Rockall-Hatton Trough (RHT) and Rockall Bank (RB)

2.4.3 On the Rockall-Hatton Plateau

A near-bottom flow below the oxygen minima brings silicate-rich water (Si > 15 μM), enriched by Antarctic Bottom Water (AABW), northwards through the Rockall-Hatton Trough, which is located between the Rockall and Hatton Banks (Fig. 1d, Table 1) [42]. A modeling study [23] identified especially large inter-annual variability in winter MLDs in the very same region (Region 2 in Fig. 6b), which likely is enhanced by topographic effects of the Rockall-Hatton Plateau. Intensified surface heat losses and wind stress curls prior to 1995 (NAO+ state, Fig. 3) resulted in deep MLDs, with a potential to induce large upwelling of silicate. The resulting vertically homogenous mode water advects northward between the Faroese banks and joins the SPMW complex south of the Iceland-Faroe Ridge. The SPMW complex is a reservoir for the Atlantic water flowing westward toward Greenland and northeastward between Iceland and the Faroe Islands, also referred to as Modified North Atlantic Water (MNAW, Fig. 1b, Table 1) [61]. After 1995, the atmospheric forcing weakened considerably and the MLDs on the Rockall-Hatton Plateau became

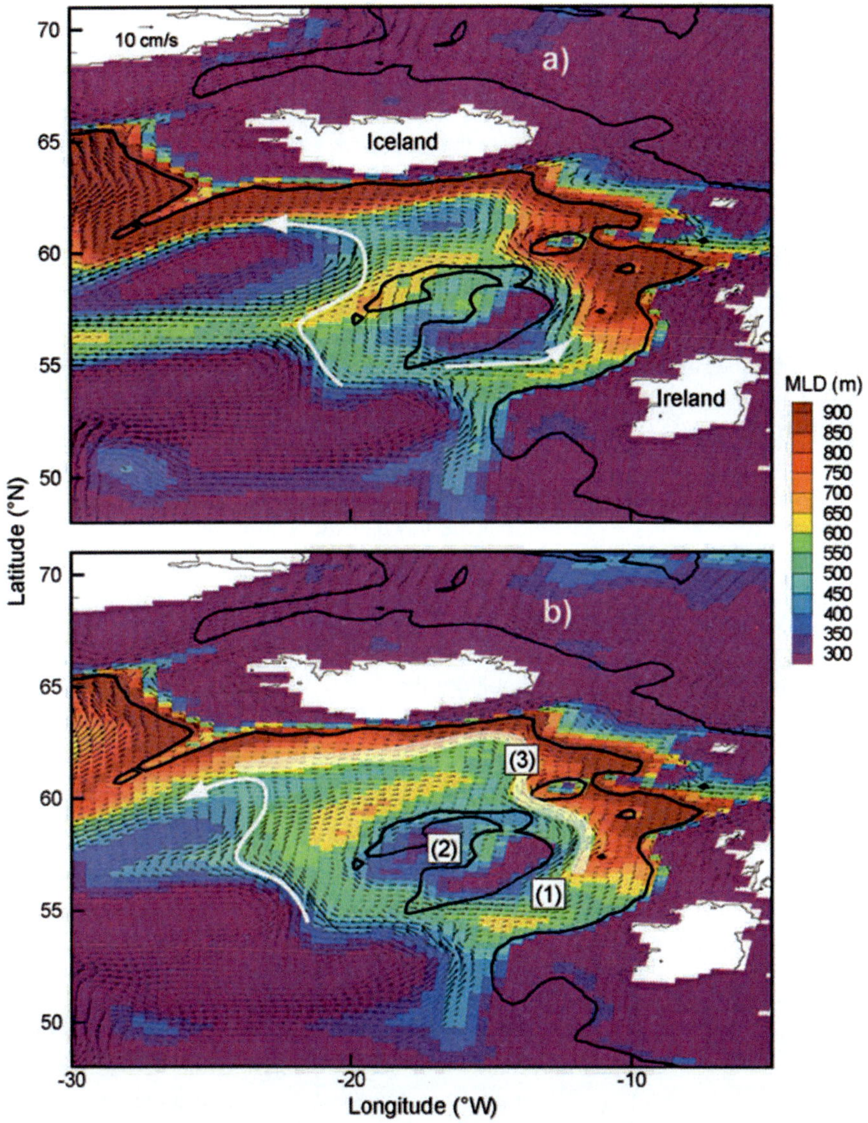

Fig. 6 Simulated mixed layer depths (colors) and velocities (small arrows) in March during (**a**) an intense NAO+/SPG+ period (1990–1995) and (**b**) a more relaxed NAO-/SPG- period (1996–2001). The figure is adapted from the study by [23], which was based on the Miami Isopycnal Ocean Model (MICOM) – the same model system as used by [24] to calculate the first gyre index. The black contour line corresponds to 1,000 m depth. Potential fertilization hotspots locations are shown with the numbers in (**b**), where the gray band outlines hotspot (3). Large white arrows emphasize flow anomalies in the Rockall Trough Branch (**a**), and in the Central Iceland Basin Branch (**a, b**) (see Fig. 1a), respectively

gradually shallower (NAO- state, Figs. 3 and 6b). The re-intensified atmospheric forcing (NAO+/SPG+) state after 2014 resulted in a marked densification of the seasonal boundary layer and deeper MLDs again in this region (see Fig. 8, below). Although the linkage to frontal dynamics is less direct here, we propose the Rockall-Hatton Trough (Region 2, Fig. 6b) as the *second nutrient fertilization hot spot*.

2.5 The Iceland Basin

When integrated across the Ellett Line in the northern Rockall Trough (Fig. 1a), the SPG-related fertilization is evident in nitrate and phosphate but not in silicate [38]. However, an analysis of individual oceanographic stations has revealed a negative trend (1990 to present), which becomes increasingly evident as one moves west from the European Continental Slope across the Rockall-Hatton Plateau into the central IB and toward the south Iceland slope ([52], their Fig. S1). These data support our previous conclusion that the mechanism behind the silicate decline is likely to be found in the vicinity of the IB, as demonstrated with the model result in Fig. 4.

The boundary between the relatively well-stratified SPG water masses and the deep seasonal boundary layers manifests itself as a band of steep isopycnals all the way from the northern Rockall Trough to the eastern flank of the Reykjanes Ridge (Fig. 6). According to Eq. (1), the strong flows in this region (large u_b) and the steep isopycnals (large ∇ MLD) establish a potential for large induction fluxes through this broad, deep front.

The real flows are more complex than those outlined by [25] (Fig. 1a). Interannual variability in the currents' strength and position is pronounced (Fig. 6) [24]. The SAF-associated current branches east and west of the Rockall-Hatton Plateau are locked to the topography (Figs. 1a, b and 6), and the relative strength of these branches varies in relation to the strength and size of the SPG. So horizontal frontal shifts in this region are "discretized" in a similar manner as the NAC is channeled through the fracture zones farther west along the Mid-Atlantic Ridge [55]. The SAF Branch in the IB is always present, but it remains uncertain whether it continues directly through the Iceland-Faroe gap [25, 62], or if the Rockall Trough Branch folds west on it, and a confluenced flow veers westward in the northern IB (Fig. 6). The results in Fig. 5 (Sect. 2.4.2) indicate that the SAF Branch cannot be the sole source for the water west of the Faroe Islands, while the Rockall Trough Branch is a more viable, direct candidate.

The location and strength of the Central Iceland Basin Branch is shifting continuously along a northeast-southwest axis, intimately related to the SPG dynamics (see below). The schematic representation from [21] of a cross-section downstream of the Faroe Bank Channel overflow plume (here extend southwestward to the central IB, Fig. 7) gives a vertical perspective of the frontal shifts in the IB, and associated impacts on the induction fluxes and water mass composition. The SPG-limb in the IB consists of an oxygen-poor (Figs. 7 and 8), nutrient-rich IW layer over a thicker

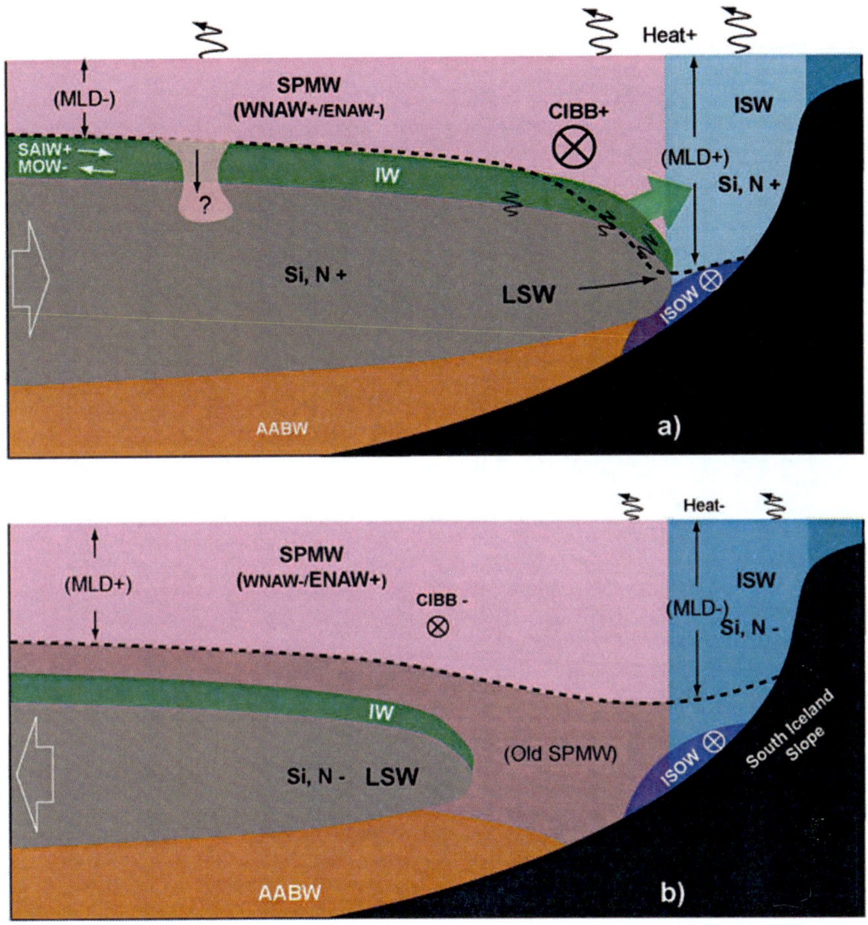

Fig. 7 A schematic representation of a cross-section downstream of the Faroe Bank Channel overflow plume. It extends from the central Iceland Basin (left) to the South Iceland slope (right) (see Fig. 1c) during a (**a**) NAO+/SPG+ state and (**b**) NAO-/SPG- state. Abbreviations: Silicate (Si), nitrate (N), the Central Iceland Basin Branch (CIBB), and the mixed layer depth (MLD). The black dashed line illustrates the depth of this mixed layer. Water masses are color coded in correspondence to Fig. 1, and Table 1, and the water mass acronyms are provided in this table. Inspired by [21] (their Fig. 13). Note that the details around whether or not vertical winter mixing interacts with the LSW are tentative

body of LSW. The IW layer continues north through the IB, where it has a doming structure over the central basin, in addition to its general westward shoaling toward the Reykjanes Ridge [63] (Figs. 1e and 8a). It terminates at about 2000 m depth along the south Iceland slope (at 20°W, [64]), but might reach as far northeast as the Faroe Bank Channel overflow plume [21, 65] (Fig. 1c).

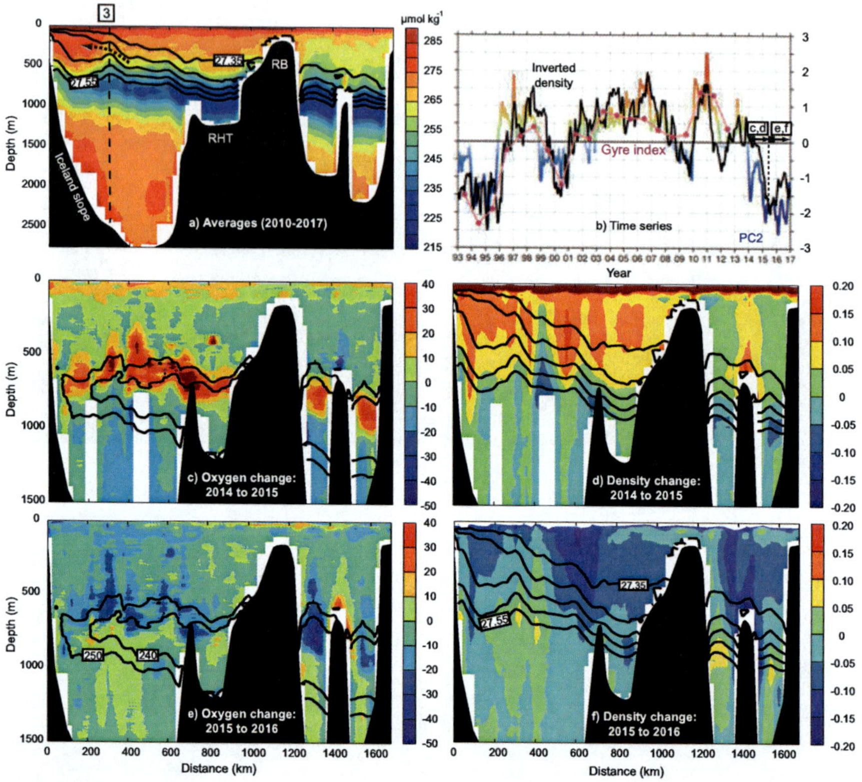

Fig. 8 Oxygen and potential density (σ) along the Extended Ellet Line (EEL, Fig. 1c). (**a**) Averages over the period 2010–2017 showing oxygen with colors and σ with black contours, based on yearly CTD sections (no section available for 2011). (**b**) Time series of the gyre index (pink), the density over the Iceland Basin, Reykjanes Ridge and Irminger Sea region (black) and the second principal component (PC2) from an updated analysis of the re-worked satellite altimetry data (colored, from [22]). The oxygen changes from 2014 to 2015 (the strongest recent convection event, see **b**) and from 2015 to 2016 (weaker atmospheric forcing) are shown in panels (**c**) and (**e**), respectively. The averaged oxygen from panel (**a**) is, in (**c**) and (**d**), contoured at 240 and 250 μmol kg^{-1}, for comparison (black lines). The corresponding changes in σ are shown in (**d**) and (**f**), respectively. The black dashed arrow in (**a**) illustrates the potentially large induction flux; the suggested nutrient fertilization Region 3 (see Fig. 6) is highlighted. Bathymetric abbreviations: Rockall-Hatton Trough (RHT) and Rockall Bank (RB)

Over the SPG-limb, we have a slab of SPMW. The slope-hugging, oxygen-rich ISOW (Figs. 1d, 7 and 8a, Table 1) plume underlies the Iceland Slope Water (introduced below) near Iceland.

SPG dynamics strongly regulate the water mass contribution into the IW layer in the IB [63]. A strong gyre leads to more nutrient-rich SAIW and less MOW [56, 66] (Figs. 1c and 7) and thus higher nutrient concentrations in the IW. Whether or not the gyre influences the AAIW contribution to the IW is an open question. The silicate concentration in the LSW also increases with an intensified SPG through the associated deeper convection in the Labrador-Irminger Seas [52]. This signal should reach the IB (Fig. 7) after a 1–2 years advective lag [67].

2.5.1 The Central Iceland Basin Branch

Northward expansions of the SPG-limb and an associated northward shift of the Central Iceland Basin Branch in the IB during NAO+/SPG+ states (Figs. 1b, 6a and 7a) are caused by the combined impact of increased volumes of LSW and intensified Ekman pumping [33]. According to model simulations ([23], Fig. 6), the main IW-layer rises in the central IB during NAO+ states, resulting in reduced winter MLDs there (Figs. 6a and 7a). In contrast, winter MLDs deepen in the Rockall-Hatton Trough and along the south Iceland slope and the eastern side of the Reykjanes Ridge (Fig. 6). This results in markedly increased pycnocline gradients (∇ MLD) around the central IB (Figs. 6a and 7a). At the same time, both the meridional SAF Branch west of the Rockall-Hatton Plateau and the Central Iceland Basin Branch strengthen (stronger u_b). The increased ∇ MLD, u_b, in addition to the increased volume of – and nutrient concentrations in – the SPG-limb water, all likely contribute to stronger induction fluxes of nutrients to the northern IB. We propose that the front associated with the steep isopycnals along the northern rim of the SPG in the IB is a broad nutrient fertilization region (Region 3, Figs. 6b and 8a).

2.5.2 Interaction with the Overflows

The SPG-limb in the IB can bring IW/LSW all the way north to the slopes of the Iceland Plateau and the Iceland-Faroe Ridge [21, 65]. Here IW/LSW can interact directly with SPMW above and ISOW below (Figs. 1 and 7a). While the importance of entrainment, or subduction, of lighter water masses into the overflows has been thoroughly discussed in the literature [21, 44], the possibility that overflows could also induce induction fluxes, or obduction, has not received much attention.

The secondary transverse circulation around the overflow plume from the Faroe Bank Channel transports ISOW "downhill" in a thin Ekman layer and draws the nutrient-rich/oxygen-poor IW northwards over the plume [68] (Figs. 1c and 7). The energetic overflow plume induces mixing, most strongly near the seafloor, but also in the ambient overlying waters due to e.g. breaking internal waves [68]. Increased activity of topographic Rossby waves are also observed along the southern slope of the Iceland-Faroe Ridge, all the way from the Faroe Bank Channel plume to the Icelandic continental rise [69, 70]. Seaglider data furthermore reveal that vertical dissipation is elevated along the Iceland-Faroe Ridge, although much intensified near the overflow plume and in the fast deep flows along steep topography near Iceland [71]. Bottom-up mixing effects of topography and the overflow plume likely invigorate regional winter convection. This could thus partly account for the very deep seasonal boundary layer along the Faroe-Bank Channel-to-Reykjanes Ridge swath, slope-ward of the fertilization band marked (3) in Figs. 6b and 8a.

Mixing along the Iceland-Faroe Ridge produces the Iceland Slope Water (ISW), residing within the 2000 m isobaths south of Iceland and along the eastern side of the Reykjanes Ridge (Figs. 1c and 7). It has previously been suggested that the ISW is a mixture primarily of SPMW and ISOW, with some additional LSW in varying mixing ratios [64]. Recent studies near the overflow plume, however, suggest that the IW will contribute to the ISW as well [21]. We therefore find it likely that the overflow plume amplifies the induction flux of nutrient-rich LSW and IW to the SPMW and ISW (Region 3, Figs. 6 and 8).

This effect is probably stronger during SPG+ states, when the northeastward extended SPG-limb in the IB brings the subarctic waters closer to the topography/overflow “blender” [21]. On the other hand, the transport of ISOW becomes weaker during years with strong wind stress curl (NAO+/SPG+ state) over the Nordic Seas, which concurs with an SPG+ state [72–74]. At present, we can only speculate whether this potentially weaker “blender” will have any appreciable effect on the mixing efficiency along the south Iceland-Faroe Ridge slope.

2.5.3 Potential Convection into the IW Layer

An important remaining question is to what degree winter convection reaches into the thermocline in the central IB. According to simulations, the rise of the thermocline during NAO+/SPG+ states reduces the winter MLDs in the central IB (Fig. 6), which gives an impression of reduced vertical mixing of deep water. However, this also brings the nutrient-rich IW-layer closer to the convection zone, which together with the intense surface heat losses could increase the upwelling of nutrients from the IW-layer.

Johnson et al. [75] reported much higher oxygen concentrations in the IW during the strong NAO+/SPG+ state in 1993 compared to the weak NAO-/SPG- state in 2003 (see Fig. 3). They interpreted this as evidence for a stronger ventilation of the IW during 1993. The strong atmospheric forcing during the winter 2014–2015 resulted in revived deep convection in the Labrador-Irminger Sea [76, 77] and generally denser seawater in the Irminger Sea, Reykjanes Ridge and IB region – that is an intensified SPG (Fig. 8b, from [22]). New data from the Extended Ellet Line reveal a marked densification throughout the upper layer waters during 2015 (Fig. 8d), and a major positive oxygen anomaly is evident along the top of the IW-layer (500–800 m depths, Fig. 8c) – an indication of deeper convection and ventilation of the IW-layer. This is in agreement with [75]. The subsequent winter (2015–2016), with again weaker atmospheric forcing, resulted in a negative upper layer density anomaly (Fig. 8f) and a negative oxygen anomaly in the upper part of the IW-layer (Fig. 8e).

The strong oxygen-nutrient correlation in the IB [42] implies that vigorous ventilation events like in 1993 and 2015 would also increase the flux of nutrient-rich IW toward the surface. So vertical contribution to the induction flux (w_b) can potentially be substantial in the central IB during years with strong air-sea forcing. That is, the gray region in the seasonal boundary layer in Fig. 2 would be broad, higher in the water column, and the nutrient fertilization would be more efficient throughout the IB.

Despite the identifiable impact of convection in the Rockall Trough, recent studies conclude that frontal shifts and its impact on lateral advection are much more important for the water mass properties in the Rockall Trough than vertical convection [78].

2.6 An Integrated Perspective: In the Context of Sea-Surface Height, MLD, and the Gyre Index

Sea-surface height (SSH) variability over the open ocean subpolar Atlantic is primarily determined by steric effects – that is changes in the integrated density over the entire water column [79, 80]. We discuss here the proposed induction flux hotspots against the backdrop of the simulated winter MLD and SSH fields in the North Atlantic combined with satellite altimetry.

In regions with strong winter convection, surface heat losses intensely cool the entire seasonal boundary layer, and this densification determines the depth of the boundary layer (Fig. 8). In such *convective domains*, a winter with strong heat losses (NAO+ state) results in both large MLDs and depressed SSHs induced by the steric contraction of the anomalously deep and dense boundary layer. That is, correlations between the MLD (positive with increasing depth) and SSH are negative in a *convective domain.*

There are also regions where the depth of the permanent thermo/pycnocline is determined by other processes than convection, e.g. Ekman pumping or lateral pressure gradients in the deep sub-thermocline layer. In such *advective domains*, a rise of the thermocline fills a larger part of the water column with cold and dense deep water, and the SSH drops – again due to steric contraction. That is, correlations between MLD and SSH are positive in an *advective domain.*

Utilizing the model system presented in [81] – the Max Planck Institute Ocean Model (MPIOM) – we identify the following areas as *convective domains* (negative SSH-MLD correlations, blue in Fig. 9): most of the Labrador Sea, the entire Irminger Sea, the Reykjanes Ridge, and the near-slope zone all around the NE Atlantic. In contrast, the central IB and waters south of the Rockall-Hatton Plateau and into the Rockall Trough entrance clearly outline an *advective domain* (positive SSH-MLD correlations, reddish in Fig. 9). As a digression, the biologically productive frontal zone in the NE Labrador Sea [82], associated with the strong west Greenland Current and the shedding of large Irminger Rings (eddies) [83], is also an advection-dominated region (Fig. 9).

During an intense winter (NAO+ state), the SSH is depressed throughout the subpolar North Atlantic, and the MLD in the convective domains deepens (blue in Fig. 9). The western mode waters (mostly LSW, [81]), produced in the Labrador and Irminger Seas, slide eastwards in and under the main thermocline in the advective

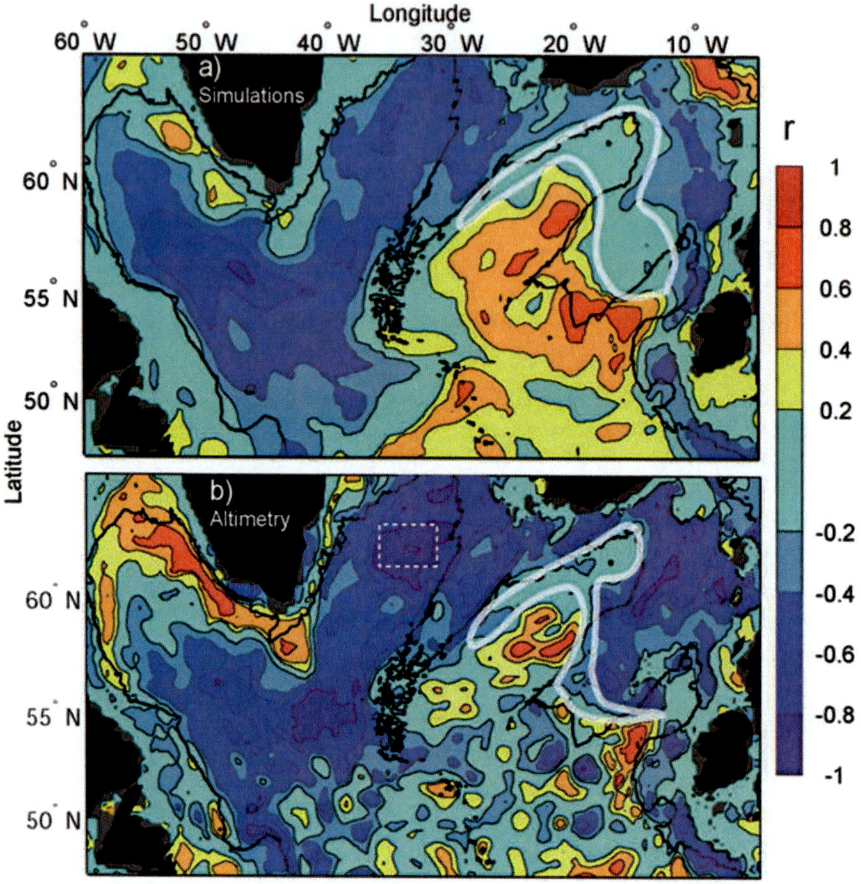

Fig. 9 Convective and advective domains. Correlation coefficients (*r*) between simulated March MLD and annually averaged (**a**) simulated SSH and (**b**) satellite altimetry. The simulations are obtained from the MPIOM model system presented in [81]. Negative values (blue) show *convective domains*, where deep winter mixing regulates the SSH, and positive values show *advective domains* (reddish), where the SSH variability is mainly determined by remotely forced undulations of the permanent thermo/pycnocline. The boundary between these regions in the NE, where large lateral induction fluxes are expected, is emphasized with the thick white curve. The 2000 m isobath is shown for reference, and the white dashed rectangle shows the region represented by the MLD time series in Fig. 10

domain. This, in turn, lifts the IW-layer in the central IB and Rockall Trough region and the winter MLDs there become shallower. That is, the eastern SPG-limb becomes inflated, and the gyre expands eastwards – partly driven by the curl of the wind stress field (e.g., [36]). During such NAO+/SPG+ states, the boundary between the convective and advective domains in the NE Atlantic (Fig. 9) must be characterized by steep isopycnals (large ∇ MLD) and thus high induction fluxes (Eq. 1). The high induction flux frontal zone from the Rockall Trough, crossing the

Rockall-Hatton Plateau and roughly following the 2000 m isobaths around the IB (Fig. 9) roughly coincides with the previously proposed nutrient fertilization hotspots.

The gyre index, which is calculated from the SSH field over the North Atlantic [24], is intrinsically linked to the above-discussed morphology of the seasonal boundary layer (Fig. 9). The temporal development of these processes since 1950 is summarized in Fig. 10, using the model-based gyre index [24], the altimetry-based gyre index [48], and the simulated SSH and MLD in the northern Irminger Sea, previously presented by [81].

2.7 *After Established Summer Stratification*

As soon as air-sea heat exchanges during spring/early summer become positive into the ocean, a relatively shallow mixed layer forms (~50 m). Around this period – potentially advance or delayed by other eddies and other turbulence inducing processes [10] – the spring bloom is initiated [8]. The silicate content in the summer mixed layer within the central IB is exhausted by the fast-growing diatoms within a week or two after the onset of a major bloom [12]. In the absence of an upwelling agent after bloom initiation, the growth of diatoms terminates and these relatively large and heavy algae quickly subside in a major export event out of the photic zone [84]. In the southeastern part of the IB, silicate depletion (<2 μM) takes place already in early May, earlier than in the Irminger Sea where it typically occurs in July–August [12]. In general, the Irminger Sea is a biologically much more productive region compared to the IB [85]. At first, this appears as a conundrum, considering that the IB region appears to determine pre-bloom upper ocean Si concentrations in the Atlantic inflow toward the Arctic and the Labrador Sea [52]. Although the altimetry-derived eddy kinetic energy – which expectedly should enhance vertical mixing – is very high in the IB [86], eddy slumping and/or wind-driven Ekman transports might only initiate stratification and blooms in the IB a bit earlier than expected from explicit consideration of air-sea heat exchanges [9]. The turbulent mixing of nutrients up through the seasonal thermocline is still lower in the central IB (*advective domain*, Fig. 9) compared to the adjacent Irminger Sea (*convective domain*) [85]. Consequently, the IB becomes biologically poor during summer, with low nutrient, phytoplankton/chlorophyll, zooplankton, and fish abundances/concentrations [85, 87]. According to logbooks, little fishery activity has historically taken place in the central IB (G. J. Óskarsson, pers. comm.) and pelagic fish surveys (redfish, mackerel, blue whiting, and herring) all end at the border of the IB [88].

2.7.1 A Bottom-Up Mackerel Case Study

In order to illustrate potential impacts of nutrient limitation on higher pelagic trophic levels, we summarize a recent study involving phytoplankton, zooplankton, and

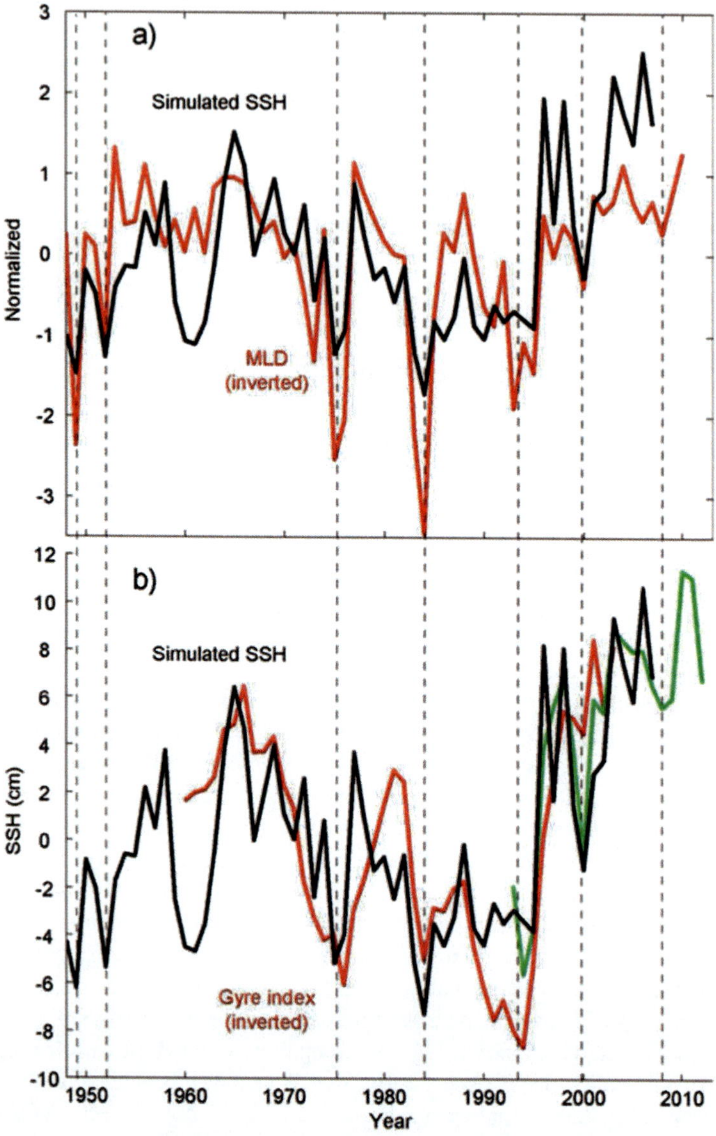

Fig. 10 Temporal development of key metrics for the thickness of the seasonal boundary layer (Fig. 9). (**a**) Simulated SSH (black) and MLD anomalies (red, inverted) within the Northern Irminger Sea box (white dashed rectangle in Fig. 9), normalized by the standard deviation. (**b**) The same simulated SSH time series (black) compared to two previously published gyre indices (inverted), based on satellite altimetry (green) [48] and simulations (red) [24]. The vertical dashed lines emphasize NAO+/SPG+ states, with strong air-sea forcing, deeper winter convection and an intensified SPG circulation

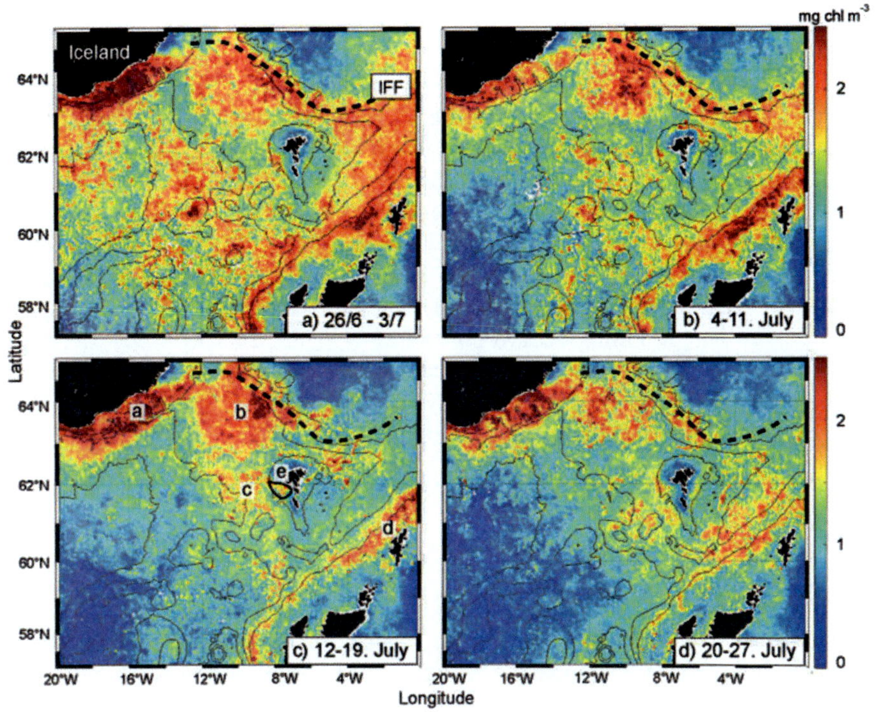

Fig. 11 Average near-surface chlorophyll distribution (mg Chla m^{-3}). Climatological averages (1998–2015) over four eight-day time segments from late June to late July (specific dates are shown in the part labels) are shown (from [17]). The dashed black line outlines the Iceland-Faroe Front (IFF). The letters in panel **c**) refer to the regions mentioned in the text (Sect. 2.7.1). These gridded satellite data were downloaded from the GlobColour Project, ACRI-ST (http://www.globcolour.ifo). Eight-day temporal averages of Level 3, merged (SeaWIFS, MODIS, MERIS, VIIRS), GSM-gridded [89] chlorophyll (CHL1) data were used on a 4-km horizontal grid for the period 1998–2015

Atlantic mackerel (*Scomber scombrus*) [17]. After having spawned along the European continental slope near Ireland during spring, starved mackerel are in critical need of food. Very low abundance of their preferred food item – the copepod *C. finmarchicus* [90] – in the spawning region forces mackerel to migrate north to more productive waters.

Accumulation of phytoplankton, and therefore likely of calanoid copepods (mackerel food), is confined to the near-topography or near-front regions upwelling hotspots through the summer (Fig. 11). Through July, the IB becomes virtually devoid of surface chlorophyll, while elevated concentrations persist in the following regions: (a) on and immediately off the south Iceland shelf, (b) shallow parts of the Iceland-Faroe Ridge and along the IFF, (c) the continuously mixing Faroe Bank Channel plume, (d) along the European Continental slope frontal zone, and (e) in a small indentation on the western side of the Faroe shelf (Western Region, see Sect. 3.2.1).

Nutrient limitation has particularly strong impact on mackerel migration along the northern rim of the IB, where the mackerel is not near any thermal threshold. Data from a meridional repeat section along 20°W (Extended Ellet Line and WOCE A20, Fig. 1c) shows that the upper 50 m in the central IB becomes Si-exhausted in June, while there is a "corridor" near the south Iceland shelf with persistently non-limiting Si concentrations (Fig. 12, north of 63°N). Furthermore, data from international mackerel surveys during July show that both the zooplankton and the mackerel abundance are very low in the central IB. The biomasses are much higher in the "corridor" in Fig. 12 during both SPG+ and SPG- states. This "corridor" overlaps with the frontal zone along the ISW band (Fig. 1b) and the shelf [17], and the increased biological production is likely associated with topographic/overflow plume mixing processes and lateral induction fluxes from the rim of the SPG.

2.7.2 Gyre Impact on the Summer Conditions

NAO+/SPG+ states will set higher pre-bloom nutrient concentrations in the seasonal mixed layer and thus increase the new production. However, if not replenished by eddy stirring or summer storms, the initial mixed layer nutrient content is quickly exhausted [84]. Elevated winter concentrations may therefore only have a moderate impact on higher trophic levels, which are in continuous demand of food throughout the summer. However, in locations that experience recurrent mixing events throughout the summer, such as frontal zones, biological production can be sustained. The dynamics associated with the SPG can contribute to this process by increasing nutrient inventories in the sub-surface waters near frontal and topographic upwelling regions, and thus the replenishment of surface nutrient stocks in these production hotspots. In the case of mackerel, the Norwegian Sea and especially the IFF region have historically been the preferred feeding areas. A marked shift occurred around 2006, when mackerel started to expand west toward Greenlandic waters. Pacariz et al. [17] discussed this major mackerel expansion in the context of the SPG-related nutrient decline (Fig. 3), and the strong zonal Si gradient with climatologically low concentrations west of the British Isles, and high concentrations in the Irminger Sea.

Although the NAO primarily documents variability in winter weather conditions, there are also indications that NAO positive periods are associated with a higher frequency of summer storms [91]. This could drive enhanced vertical mixing events and, through associated nutrient replenishment, increase new production in summer months in the central IB. In contrast, [92] find that the establishment of a summer mixed layer is probably independent of the preceding winter conditions. Understanding the possible impacts of NAO variability on summer nutrient replenishment therefore warrants future research.

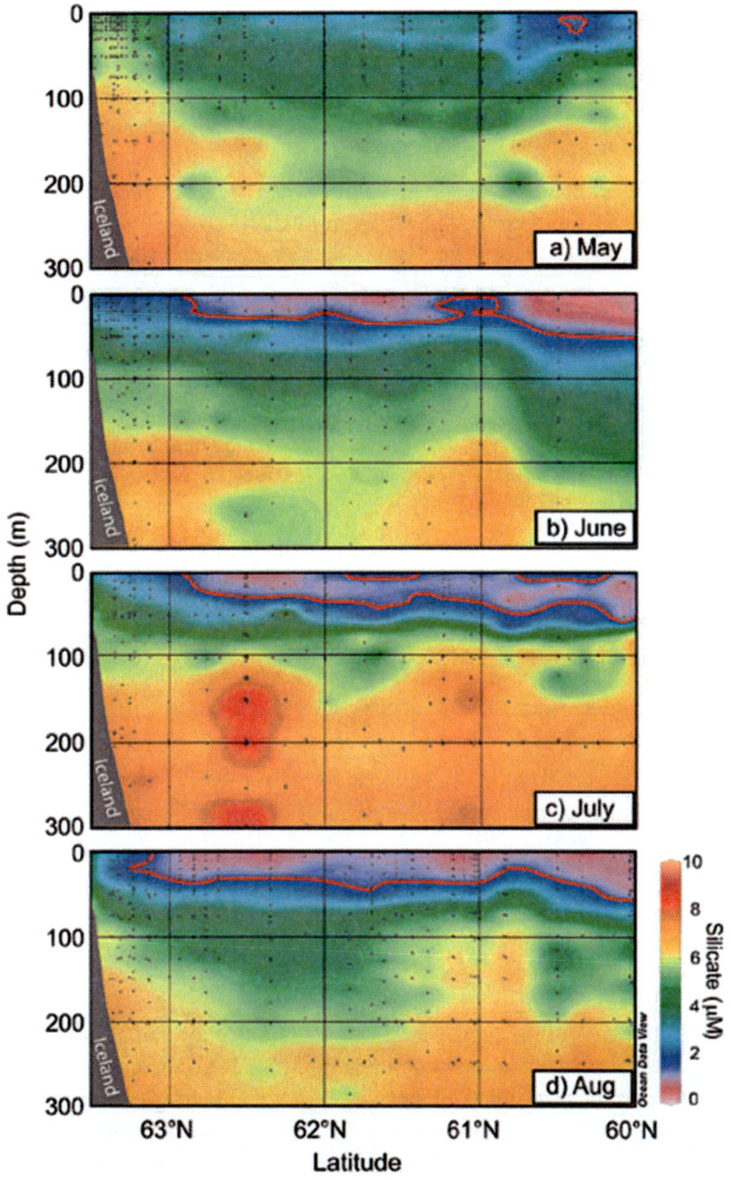

Fig. 12 Climatological silicate concentrations (µM) along the meridional WOCE/EEL section at 20°W (Fig. 1c) in (**a**) May, (**b**) June, (**c**) July, and (**d**) August. The red contour line emphasizes the limiting silicate concentration of 1.5 µM (from [17])

3 Onwelling to Adjacent Shelves

Broad-scale nutrient dynamics in the NE Atlantic likely exert a major impact on adjacent shelf ecosystems (i.e., the Northwestern European slope (NWES), North Sea, Faroe shelf, and the south Iceland shelf). In this section, we review recent literature on NE Atlantic-to-North Sea linkages. These considerations are based primarily on model simulations, but also aim to take into account emerging observations of shelf ecosystems. The Faroe Shelf area is described in detail since observations from this system, which is spatially confined and thus "surveyable," could provide a useful framework for enhancing our general understanding of ocean–shelf interactions.

3.1 *From the Northeast Atlantic to the North Sea*

There is now a broad consensus that nutrient budgets in the open shelf areas of the NWES, and in the deeper parts of the North Sea, are dominated by nutrient fluxes from the NE Atlantic [27–29, 31, 93–97] – also referred to as "nutrient onwelling" [31]. Nutrient fluxes depend both on water mass exchange rates and nutrient gradients between the open ocean and the shelf seas. Some previous studies have reported increased nutrient fluxes during NAO+ phases, when intensified winds likely increase the transport of oceanic water onto the shelf [29, 96]. Recent studies, on the other hand, conclude that it is the open ocean nutrient concentration, more so than the influx volume, that primarily controls nutrient onwelling to the North Sea [27, 31, 94, 95]. Many of the initial NE Atlantic-North Sea studies, however, considered climate change scenarios, based on model simulations.

3.1.1 Future Projections

A general shallowing of the winter MLD during the twenty-first century, caused by increasing air temperature and accompanying weakened air-sea heat fluxes, is a robust feature in most climate models used in the Intergovernmental Panel on Climate Change (IPCC) reports [98]. These climate models project a particularly strong decline in the NE Atlantic MLD and pre-bloom nutrient inventories, which in turn could lead to a marked reduction in primary production within the present century [95]. The simulated MLD projections rely primarily on one-dimensional (vertical) mixing schemes and are thus heavily influenced by reduced heat losses from the oceans to the atmosphere [27]. It has been acknowledged that large-scale circulation changes may also play a role, but up until recently only vague statements about shifting currents have been made [95].

Recent studies have identified the importance of variable horizontal gyre circulation and its potential to regulate water mass balance in the NE Atlantic (Sect. 2) –

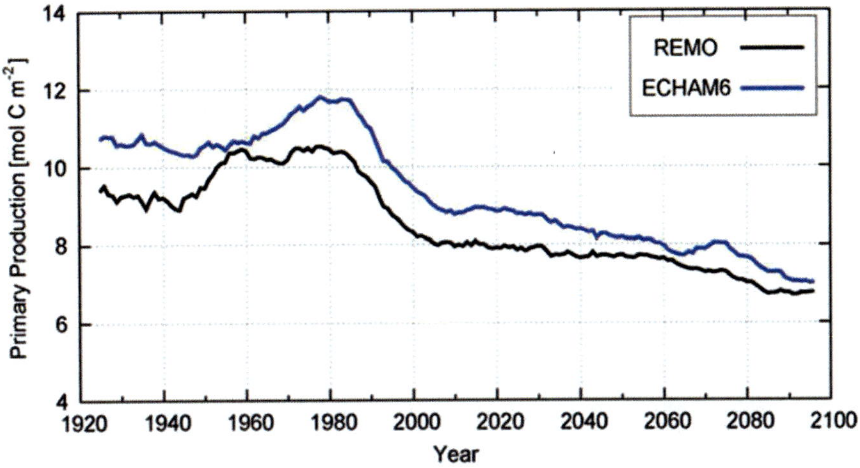

Fig. 13 Time series of annual accumulated depth-integrated primary production in the North Sea (10-year running mean, simplified from [27]). Shown are outputs from the regional climate model REMO and from the general circulation model, ECHAM, developed by the Max Planck Institute for Meteorology

and subsequently to the North Sea [27]. Global models have large deficiencies in the SPG domain, and elaborate bias corrections are required to introduce realistic hydrographic gradients, such as encountered in frontal zones [27]. The predicted nutrient decline in the SPG and in the NE Atlantic [98] is especially evident in regionalized models that are characterized by a higher spatial resolution. The first study to explicitly discuss the potential connections between the SPG and the North Sea was based on a global version of HAMOCC [27]. The same model system was used by [52] to contextualize the observed silicate decline in the subpolar North Atlantic both in time and space (Figs. 3 and 4). The projection run by [27] shows a reduction in primary production throughout the twenty-first century, although at slower rates than during the 1980–2000 period (Fig. 13).

The projected properties in the North Sea are inherited from the open ocean, but climate models have deficiencies in water mass exchanges between the NE Atlantic and the NWES [27, 99] that is along the slope/shelf frontal zone (Fig. 1b). The open ocean and the deeper parts of the NWES become strongly stratified during summer. More focused climate change model runs, with increased resolution in the North Sea region, find that increased summer stratification will augment the reduction in onwelling. These processes have a potential to reduce biological production on the NWES by ~30% at the end of the twenty-first century [97]. On the other hand, a recent study [100] suggests that vertical mixing processes at the shelf break will maintain a connection to nutrient-enriched sub-pycnocline water masses in a future climate change scenario. This decoupling from the upper ocean conditions in the North Atlantic moderates the decline in new production projected by [97].

On the shallow part of the NWES, strong tidal currents persistently erode stratification and keep inner shelf waters permanently well-mixed. A tidal mixing front separates these well-mixed inner shelf waters from the seasonally stratified outer shelf waters [101]. Regional modeling of the NWES suggests that the cross-front exchanges are characterized by net on-shelf transports in the upper water layer and a net off-shelf transport to the lower layer of the outer shelf. This results in a net downwelling along the shelf [28]. Consequently, the projected increase and/or prolonged duration of summer stratification will reduce the degree of nutrient *onwelling* through nutrient depletion of upper layer outer shelf waters and thus amplify nutrient limitation on the near-shore well mixed water [28, 97].

3.1.2 Observations

Increasing recognition of the importance of the NE Atlantic for the North Sea ecosystem has motivated dedicated observational campaigns focused on the ocean–shelf interface, e.g., Fastnet and the UK Shelf Sea Biogeochemistry (SSB) Programme. Previous observations have shown that nutrient concentrations in the North Sea declined during the 1990s [96] primarily due to reduced oceanic *onwelling*, but also due to a reduction in the riverine nutrient loads to the southern domain. Also, the observed increase in northern North Sea nutrient concentrations during NAO+ phases has been explained by a combination of increased ocean-shelf water mass exchanges [29], decreased near-shelf stratification [94], and increased off-shelf winter MLD [95] (e.g., Sect. 2). Most previous nutrient observations have focused on nitrate and phosphate [93], although results from more recent surveys demonstrate that only silicate reaches limiting concentrations on the Hebridean shelf [102]. A mass balance approach suggests that 73% of the shelf water silicate must be resupplied from the open ocean in order to maintain shelf-wide estimates of productivity. This is considerably larger than similar estimates for nitrate (34%) and phosphate (28%) [102]. Although general aspects of the *onwelling* process apply to all major nutrients, we suggest that future research should pay more attention to silicate dynamics.

The stratification in the northern/deeper part of the North Sea intensified markedly between 2001 and 2002 and remained very strong during the following warm 2003–2005 period [103]. Apparent impacts of the intensified stratification on the phytoplankton concentration are here illustrated by seasonally-averaged near-surface chlorophyll (Chla) from ocean color satellites for the contrasting years, 2001 and 2002, respectively (Fig. 14). During 2001, Chla was highest on the shallower part of the NWES (<100 m depths, Fig. 14a) and much lower in deeper water. In contrast, during 2002, the near-shore Chla concentrations declined, while the Chla concentrations on the outer shelf/slope band from west of Ireland to north of Shetland increased significantly (Fig. 14b). This increase is particularly evident in the 100–200 m depth range, although also evident in the broader oceanic domain, e.g. the Faroe-Shetland Channel and the Norwegian Trench. It is plausible that the high Chla content in 2002 reflects a stratified and nutrient-depleted upper layer

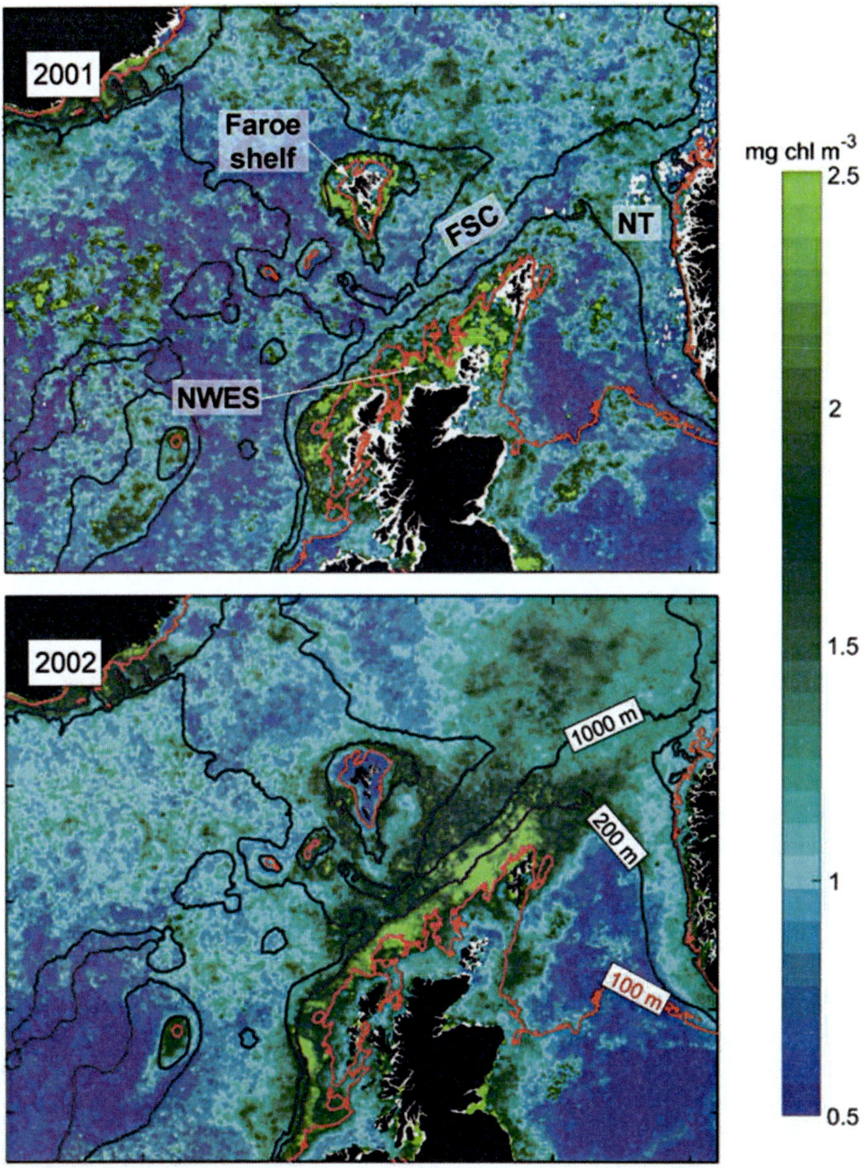

Fig. 14 Shift in phytoplankton concentrations from a year with weak stratification (2001, upper panel) to a year with strong stratification (2002, lower). The images are based on gridded near-surface satellite chlorophyll data, downloaded from the GlobColour Project, ACRI-ST (http://www.globcolour.info). Annual mean for the years 2001 and 2002 were computed from monthly averages of Level 3, merged (SeaWIFS, MODIS, MERIS), GSM-gridded [89] chlorophyll (CHL1) data on a 4-km horizontal grid from the months April–September. Abbreviations: Faroe-Shetland Channel (FSC), the Norwegian Trench (NT), and the NorthWestern European Slope (NWES)

containing a rich community of both large and small phytoplankton species, fueled by both new and regenerated production. Although the seasonally-averaged Chla concentration is not a robust proxy for primary production, the low shallow-shelf (<100 m) concentrations during 2002 do probably reflect reduced production. This could be caused by the low nutrient concentration in the onwelled outer shelf water, as discussed below (Sect. 3.2.3, Fig. 18c).

A significant reduction in landings of the key prey fish species – sand eel (*Ammodytes tobianus*) in 2003 [105] – also suggests that higher trophic level productivity in the northern North Sea might have decreased after 2002. The average landings of the period 1994–2002 was 880,000 tons whereas the average landings of the period 2003–2009 was only 288,000 tons. This period also coincided with a broad-scale breeding failure of the black-legged kittiwakes (*Rissa tridactyla*) in Britain [106].

The current level of understanding regarding impacts of *onwelling* to productivity of the North Sea is inherently limited by (a) the lack of detailed long-term observations and (b) the long stretch and openness of this system. Monitoring all key ocean-shelf exchange hotspots with adequate temporal resolution over decades might be an unfeasible task. Furthermore, fish and seabirds are relatively free to roam this large shelf region, rendering it difficult to trace how potential bottom-up processes propagate through the food web.

3.2 The Faroe Shelf

The upper layer water masses that flow northeastwards along the NWES also immerse the Faroe shelf, which is located centrally on the Iceland-Scotland Ridge (Fig. 1). In contrast to the NWES, the Faroe shelf is spatially confined in a manner that facilitates detailed observations of major ecosystem components. Comprehensive survey data at high spatial resolution and a large number of valuable time series are available from this system. And these span several decades and cover key aspects of phyto- and zooplankton communities, major fish stocks and seabird species [107–110]. The Faroe Shelf system therefore represents a good observational framework to examine impacts of *onwelling* on all major shelf ecosystem components – from nutrient-driven plankton dynamics to upper trophic levels. This system may thus serve as a useful model for understanding oceanic influence on boreal shelf systems in general.

The abovementioned 2001-to-2002 shift in the North Sea, from a productive inner shelf to a productive outer shelf/slope (and low inner shelf Chla concentrations) was more accentuated on the Faroe Shelf than on the NWES (Fig. 14). A previously defined Primary Production Index (PPI) for Faroese shelf waters [110] (Fig. 15) showed a major drop between 2001 and 2002. The PPI remained generally low until 2008, and these biologically poor conditions extended throughout all trophic levels on the Faroe shelf [108]. This period was also poor for seabirds, e.g., no fledged kittiwake chicks were observed at the Faroese colonies during this period [26].

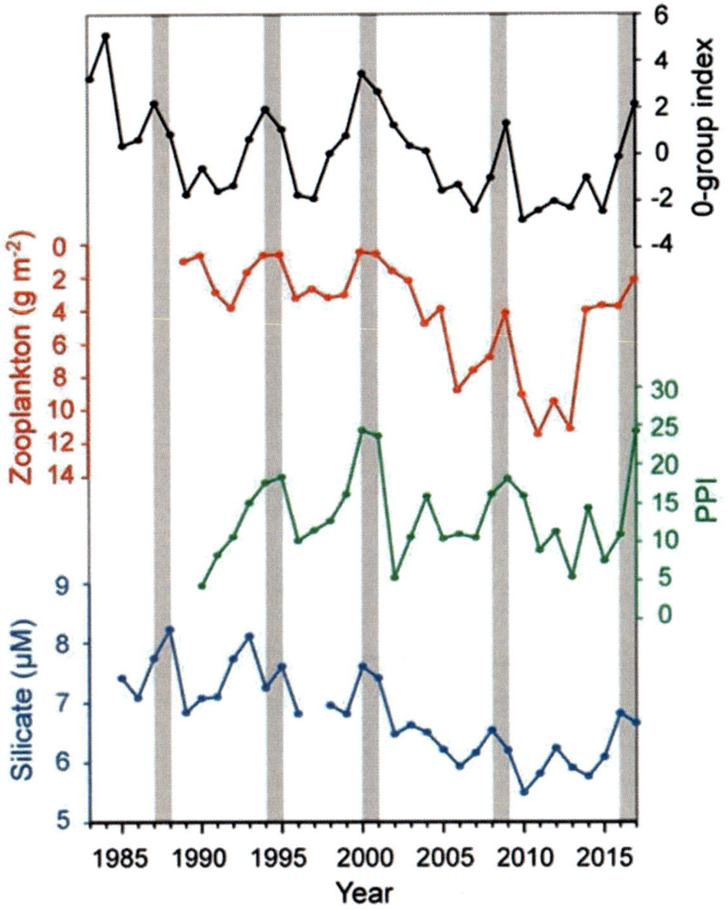

Fig. 15 Key ecological indicators for the Faroe shelf. Temporal development in the 0-group index estimated from four main fish stocks (black), zooplankton biomass in late June (red) (note the reversed axis), primary production index (PPI) from spring to late June (green), and open-ocean pre-bloom silicate concentrations (blue). The gray bars highlight convective (NAO+/SPG+) periods of simultaneous high 0-group index, low zooplankton biomass, high PPI, and high silicate concentration. From [60]

3.2.1 Primary Production Variability: Intensity and Phenology

Much effort has been dedicated to constrain the underlying mechanisms that govern the highly variable primary production on the Faroe shelf [108, 111, 112]. Here, cross-shelf front exchanges have been a central theme. Opposed to the view from the NWES, it has been hypothesized that increased exchanges may impede the primary production, since this might "flush" algae and spores off the shelf [111, 113]. These

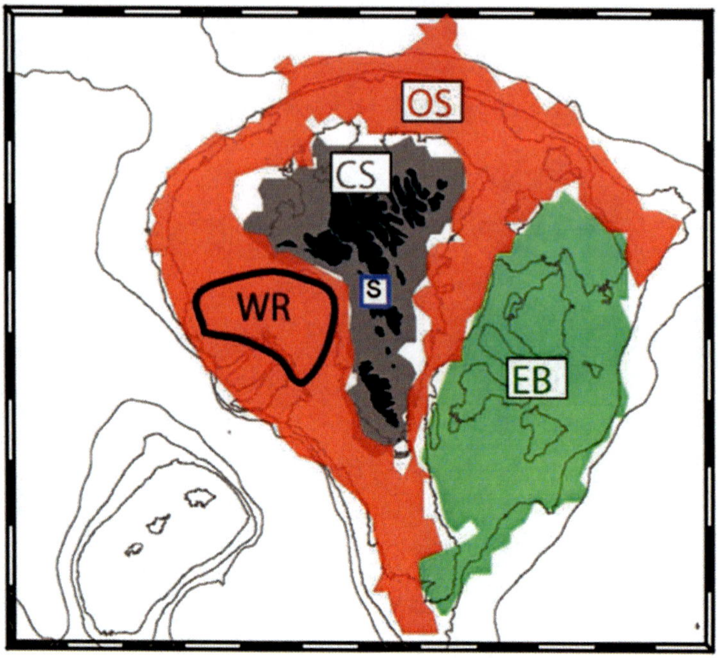

Fig. 16 The Faroe shelf bio-geographical zones: the Central Shelf (CS, gray), the Outer Shelf (OS, red) and the especially productive Western Region (WR), and the Eastern Banks (EB, green). The blue square shows station S. The 100 m, 200 m, 300 m, and 500 m isobaths are shown (from [104])

studies have, however, not explicitly considered variable stratification or the fact that the outer shelf/slope region also hosts a rich phytoplankton community.

The permanently well-mixed water on the central part of the shelf has been termed Faroe Shelf Water [114]. Eliasen et al. [32] showed that biomass observations at a fixed coastal Station (Station S, Fig. 16) provide a valid representation of the near-surface satellite-based Chla throughout the extent of the Central Shelf (CS) area. This is a region which roughly overlaps with the summer expanse of the Faroe Shelf Water domain (gray in Fig. 16). Phytoplankton biomass at Station S generally starts increasing during spring, however interrupted by halts in the Chla concentrations, apparent as local minima in the Chla time series [104] (Fig. 17).

Based on the phytoplankton concentration and phenology, observed from ocean color satellites, [104] have further divided the Faroe Plateau into the following three main bio-geographical zones: the Central Shelf (CS), the Eastern Banks (EB), and the Outer Shelf (OS) (Fig. 16). These non-overlapping regions cover the entire shelf within the 300 m isobath. Since the oceans, and in particular oceanic fronts, are dynamic, the borders of the zones should not be seen as rigid but merely as approximate positions with emphasis on summer conditions. Region CS blooms early, but Chla concentrations diminish quickly after the peak bloom, which

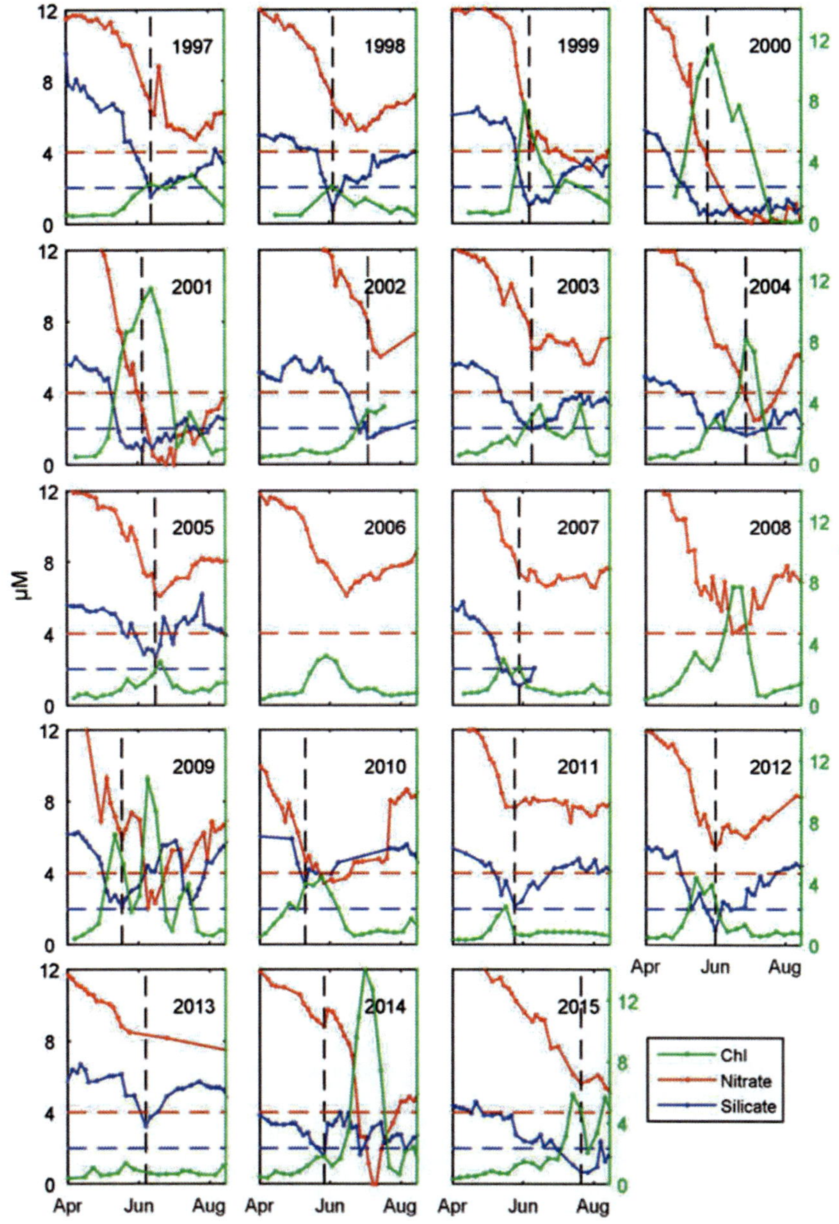

Fig. 17 Phytoplankton biomass (Chla) and nutrients from the coastal station S (see Fig. 16) on the Faroe shelf. The 2 μM and 4 μM lines are indicated as well as the date of the lowest observed silicate concentration (from [32]). The observations are made in water which is pumped from 18 m depth. This is located in a vertically and horizontally well-mixed Inner shelf, and the observations thus represent these shelf waters

typically takes place around late May–early June (Fig. 17). Chla concentrations remain relatively low during June and July, and in some years a short fall bloom occurs. The seasonally-averaged Chla concentration is lower in the CS compared to the surface layers in the surrounding waters during warm years (e.g., 2002, Fig. 14), when the stratification is generally intensified [103]. The OS blooms later than the CS, but the elevated Chla concentrations are sustained for a longer period. This region is therefore more visible in temporally averaged near-surface Chla images like in Figs. 11 and 14. Both studies by Eliasen et al. [104] and [32] highlighted the seasonally stratified western side of the OS – the Western Region (WR) (Fig. 16) – as a region which blooms relatively late, but which has the highest surface Chla concentrations on the Faroe Plateau during June–August.

A high-resolution (~1 km) version of the Hybrid Coordinate Ocean Model (HYCOM) [115] has been setup for the Faroese waters, and the output of key physical parameters compare favorably with in situ observations [116]. This model provides the first spatio-temporally comprehensive view of the stratification dynamics of the Faroe shelf. Similar to the NWES [117], a net *downwelling* is identified with net on-shelf transports from the upper OS-layer, and a net off-shelf transport from the CS to the lower OS-layer. Rasmussen et al. [116] presented the first discussion of the inter-annual primary production variability in the context of stratification. They showed that the growth-intensive years, 2000 and 2001, were also characterized by an early and sudden onset of stratification in the OS. Complementing the same HYCOM simulations with in situ oceanographic, atmospheric, and sunlight data, [118] explained the bloom timing by a *Critical Volume Hypothesis*. The onset of persistent OS stratification establishes a distinct tidal mixing front between the CS and the OS, which shifts shoreward during spring as the insolation and heat fluxes into the ocean strengthen. According to the *Critical Volume Hypothesis*, spring bloom initiates when the total primary production within the CS (which is shrinking at this point of the year) equals the net respiration of the same volume of water. The position of the tidal mixing front plays, in the horizontal *Critical Volume Hypothesis*, a role equivalent to the mixed layer depth in Sverdrup's classical (vertical) *Critical Depth Hypothesis* [118].

3.2.2 Silicate Limitation and Interplay Between the Outer and Central Shelf

Winter (pre-bloom) concentrations of nitrate and silicate on the Faroe Shelf are ~12 μM and ~5–7 μM, respectively (i.e., a Si:N ratio of ~1:2) [119, 120]. These concentrations have been declining since the early 1990s, in a similar manner as in the surrounding open ocean (Fig. 3) [52]. As mentioned, nutrient-related studies generally focus on nitrate, but since this nutrient only reaches limiting concentrations (< 4 μM) on the CS during the most productive years (e.g., 1994–1995, 2000–2001, 2004, 2009 and 2014 Figs. 15 and 17), nutrient dynamics have generally been overlooked in the early studies that examined bottom-up controls on inter-annual Faroe shelf bloom variability.

The average diatom ratio Si:N of 1.12 ± 0.33 [119] implies that silicate and nitrate should be taken up at approximately equal rates. Observations from the Faroe Shelf, together with those from the NWES [102], suggest that silicate limitation is equally, if not more, important for bloom dynamics in these shelf environments.

Maxima in the Chla concentrations for the CS typically occur after May and are associated with minima in silicate concentrations in 16 out of the 17-year time series (1997–2015; no silicate data for 2006 and 2008). In most cases (12 out of 17), silicate concentrations fall below 2 μM (Fig. 17) whereas it is, as mentioned, only the most productive years that nitrate concentrations are driven to limiting concentrations. From the end of May and throughout June, average silicate and nitrate concentrations, associated with Chla peaks, were in the range 1.9–2.2 μM and 5.0–7.1 μM, respectively.

A main *onwelling* conduit to the CS originates from the upper layer WR [121], where the silicate concentrations are driven to even lower values than those on the CS [32]. After the first major CS Chla peak, when the environment has reached silicate limitation (Fig. 17), successive events between erosion/strengthening of the WR stratification and peaks/decreases in the CS Chla are observed [32]. This is idealistically sketched for a single de-stratification/re-stratification event in Fig. 18. The process is interpreted as follows: strong phytoplankton growth will quickly deplete the nutrients in the CS and in the upper WR layer, and nutrient supply will therefore be essential for the ongoing growth (Fig. 18c). The erosion of the established stratification leads to new nutrients from the bottom layer in the WR (green arrows in Fig. 18b) being supplied to the upper layer. If this nutrient input is not exhausted by growth in the outer shelf, it may help to maintain nutrient *onwelling* to the CS. Local and intermittent convection events will probably also admix nutrient-rich water from the lower layer directly onto the CS. The CS growth therefore alternates between nutrient enrichment by erosion of the pycnocline (Fig. 18a, b) and re-stratification and nutrient limitation (Fig. 18a, c). Phytoplankton dynamics on the species level have only been examined in 2004, 2005, and 2010 on the Faroe shelf. In these years a consistent shift from a community dominated by large diatoms to small/diatoms and flagellates has been observed when silicate concentrations reach <2 μM [120, 122].

3.2.3 Higher Trophic Levels

Inter-annual variability in the size of fish larvae appears to be correlated with the primary production on the Faroe Shelf [60]. These authors have developed an *O-group index*, which is estimated from data on the four main fish stocks, cod (*Gadus morhua*), haddock (*Melanogrammus aeglefinus*), sand eel, and Norway pout (*Trisopterus esmarkii*), that together account for >90% of all the fish larvae on the shelf. The *O-group index* displays accentuated peaks every 7–10 years, which coincide with peaks in both the PPI and open ocean pre-bloom Si concentrations (Fig. 15). Jacobsen et al. [60] used the previously discussed silicate record from the northern Irminger Sea (Fig. 3) as a proxy for the open ocean surrounding the Faroe

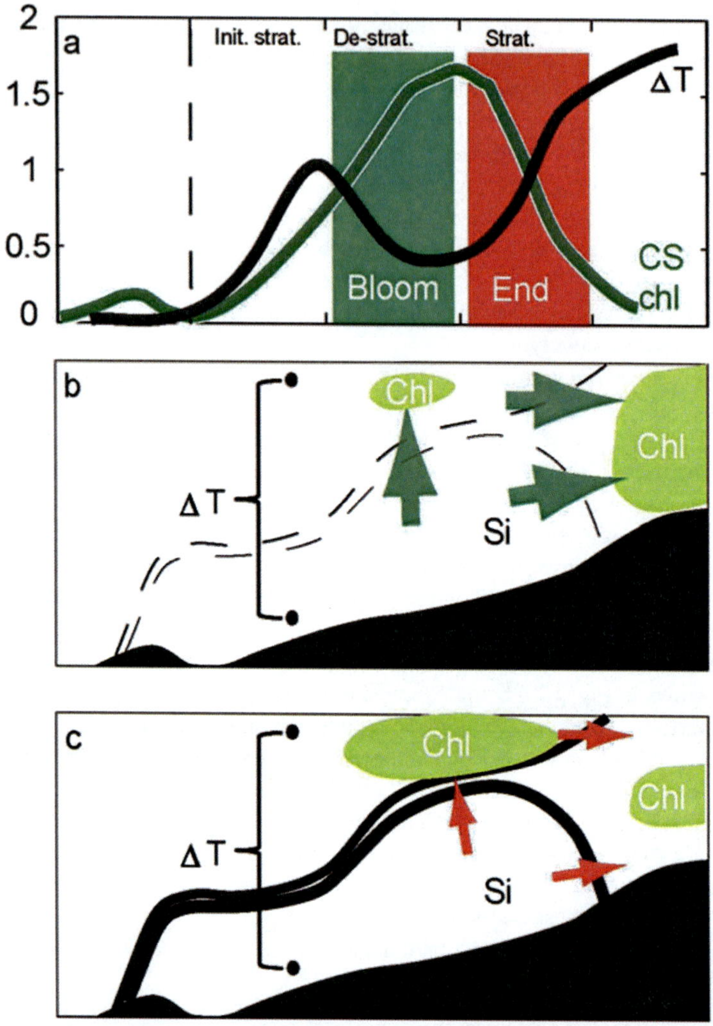

Fig. 18 An idealized sketch, describing the Faroe shelf bloom. (**a**) Time series of the temperature stratification (black) and Chla concentrations (green) on the western part of the outer shelf (the Western Region, see Fig. 16) and the Central Shelf (CS) and two idealized transects across the shelf during the green (**b**) and red phases (**c**), respectively. The red and green arrows symbolize small and large nutrient fluxes in the red and green phase (panel **a**), respectively, from the nutrient-rich "cold pool" (from [32])

shelf (Fig. 15). The relatively long record from the Irminger Sea is closely correlated to shorter records of pre-bloom silicate concentrations in the northern IB and north of the Faroes (not shown here).

The peaks associated with NAO+/SPG+ phases (Fig. 3) also concur with improved breeding success of the Faroe kittiwakes (*R. tridactyla*) [26] and improved recruitment and average weight gain of the main commercial fish stocks. Among many possible underlying mechanisms, we propose the following two as the most plausible:

1. NAO+/SPG+ dynamics are linked to oceanic Si peaks that fuel Faroe shelf primary production (PPI), which supports the growth of local zooplankton [123], key seabird fish-prey species (sand eel and Norway pout) and thus kittiwake chicks.
2. Increased volumes of zooplankton-rich subarctic water masses could increase the concentration of the copepod *C. finmarchicus* around the Faroe shelf. Thus, the on-shelf transport of eggs, nauplii, and adults of this key food item can increase, which in turn more directly supplies the fish and birds.

The impact on the kittiwakes could originate from the fact that these birds spend their non-breeding period, which is fall, winter, and early spring – and thus most of their lives – in the SPG region where they congregate near major oceanic frontal systems [16]. It has been shown that a strong gyre is associated with increased abundance of *C. finmarchicus* in the overwintering region in the Labrador and Irminger Sea [81]. It is therefore plausible that the improved non-breeding feeding conditions could translate to better breeding success the subsequent summer through a so-called carry-over effect [18, 26].

Finally, a similar SPG-shelf link, although focused on zooplankton, has been demonstrated for the south Iceland shelf. NAO+/SPG+ phases, associated with larger MLD in the northern Irminger Sea (Fig. 10), and northeastward-shifted SPG all coincide with markedly increased total zooplankton abundance on the south Iceland shelf during May (dominated by *C. finmarchicus,* [81]).

4 Summary

The fact that the North Atlantic subpolar gyre regulates temperature and salinity variability in the poleward flowing Atlantic water is now thoroughly described in the literature. It was, however, not until recently that awareness has been raised about the gyre's potential to also impact nutrient concentrations in the NE Atlantic. Subtropical waters are generally low in nutrients, but on their journey northwards they are being mixed with nutrient-rich subpolar waters – mainly along the frontal zones between the Atlantic and subarctic water masses. In years when the subpolar gyre circulation is intensified, more nutrients are upwelled from the nutrient-rich Intermediate Water level which forms the permanent pycnocline in the NE Atlantic. In contrast, when the gyre is weak, these *induction fluxes*, or *obduction*, are weak and nutrient-poor subtropical water masses become more prevalent in this region.

Silicate concentrations in the upper layer Atlantic waters have been decreasing since the early 1990s, likely linked to the declining subpolar gyre. This, in turn, was associated with weaker winter overturning in the Labrador and Irminger Seas. The declining trend becomes identifiable in the vicinity of the Rockall-Hatton Plateau and extends downstream to the Nordic Seas (in north) and the Labrador Sea (in west). Here, we have suggested potential fertilization hotspots, located along the northeastern periphery of the subpolar gyre, and potential links to primary and secondary production, as well as the feeding migration of pelagic fish.

The gyre-regulated nutrient concentrations along the adjacent continental slopes will furthermore influence the ocean-to-shelf nutrient fluxes across shelf and tidal mixing fronts – the so-called nutrient *onwelling*. We review how this *onwelling* likely impacts the Northwestern European shelf seas and the North Sea and illustrate how variations in the nutrient *onwelling* propagate up through the food chain – using the Faroe shelf as a case study. We hypothesize that ecosystems on the south Iceland shelf, Faroe shelf, Northwestern European shelf seas, and potentially the Norwegian shelf are to some degree influenced by pre-bloom nutrient concentrations in the Atlantic-derived water masses that connect these regions. Our results from the spatially confined Faroe shelf, which are based on an extensive amount of observational data collected over more than two decades, could be important to the understanding of other larger shelf ecosystems as well. It is our recommendation that these observations are continued and complemented with additional observations of essential ocean variables both in the Faroes and at key locations on neighboring shelves (Table 2).

Table 2 List of acronyms, arranged as introduced in the text

Acronyms	Name
SPG	Subpolar gyre
SAF	Subarctic Front
NAC	North Atlantic Current
RTB	Rockall Trough Branch
SAFB	Sub-Arctic Front Branch
CIBB	Central Iceland Basin Branch
IFF	Iceland-Faroe Front
MLD	Mixed Layer Depth
AMO	Atlantic Multidecadal Oscillation
NAO	North Atlantic Oscillation
GSA	Great Salinity Anomaly
IB	Iceland Basin
SSH	Sea-surface height
Si	Silicate
NWES	Northwestern European slope
PPI	Primary Production Index
CS	Central Shelf
EB	Eastern Banks
OS	Outer Shelf
WR	Western Region

References

1. Trenkel VM, Huse G, MacKenzie BR, Alvarez P, Arrizabalaga H, Castonguay M, Goñi N, Grégoire F, Hátún H, Jansen T, Jacobsen JA, Lehodey P, Lutcavage M, Mariani P, Melvin GD, Neilson JD, Nøttestad L, Óskarsson GJ, Payne MR, Richardson DE, Senina I, Speirs DC (2014) Comparative ecology of widely distributed pelagic fish species in the North Atlantic: implications for modelling climate and fisheries impacts. Prog Oceanogr 129:219–243. https://doi.org/10.1016/j.pocean.2014.04.030
2. Sundby S (2000) Recruitment of Atlantic cod stocks in relation to temperature and advection of copepod populations. Sarsia 85:277–298. https://doi.org/10.1080/00364827.2000.10414580
3. Frederiksen M, Moe B, Daunt F, Phillips RA, Barrett RT, Bogdanova MI, Boulinier T, Chardine JW, Chastel O, Chivers LS, Christensen-Dalsgaard S, Clément-Chastel C, Colhoun K, Freeman R, Gaston AJ, Gonzáléz-Solís J, Goutte A, Grémillet D, Guilford T, Jensen GH, Krasnov Y, Lorentsen SH, Mallory ML, Newell M, Olsen B, Shaw D, Steen H, Strøm H, Systad GH, Thórarinsson TL, Anker-Nilssen T (2011) Multicolony tracking reveals the winter distribution of a pelagic seabird on an ocean basin scale. Divers Distrib 18:530–542. https://doi.org/10.1111/j.1472-4642.2011.00864.x
4. Wanless S, Frederiksen M, Daunt F, Scott BE, Harris MP (2007) Black-legged kittiwakes as indicators of environmental change in the North Sea: evidence from long-term studies. Prog Oceanogr 72:30–38. https://doi.org/10.1016/j.pocean.2006.07.007
5. Jansen T, Campbell A, Kelly C, Hátún H, Payne MR (2012) Migration and fisheries of north east Atlantic mackerel (*Scomber scombrus*) in autumn and winter. PLoS One 7(12):e51541. https://doi.org/10.1371/journal.pone.0051541
6. Mackenzie BR, Payne MR, Boje J, Høyer JL, Siegstad H (2014) A cascade of warming impacts brings bluefin tuna to Greenland waters. Glob Chang Biol 20:2484–2491. https://doi.org/10.1111/gcb.12597
7. Lalli CM, Parsons TR (1997) Biological oceanography: an introduction.2nd edn. Elsevier, Oxford, pp 1–309
8. Longhurst A (2006) Ecological geography of the sea.2nd edn. Academic Press, London, pp 1–541
9. Mahadevan A, D'Asaro E, Lee C, Perry MJ (2012) Eddy-driven stratification initiates North Atlantic spring phytoplankton blooms. Science 337:54–58. https://doi.org/10.1126/science.1218740
10. Chiswell SM, Calil PHR, Boyd PW (2015) Spring blooms and annual cycles of phytoplankton: a unified perspective. J Plankton Res 37:500–508. https://doi.org/10.1093/plankt/fbv021
11. Allen JT, Brown L, Sanders R, Moore CM, Mustard A, Fielding S, Lucas M, Rixen M, Savidge G, Henson S, Mayor D (2005) Diatom carbon export enhanced by silicate upwelling in the northeast Atlantic. Nature 437:728–732. https://doi.org/10.1038/nature03948
12. Henson SA, Robinson I, Allen JT, Waniek JJ (2006) Effect of meteorological conditions on interannual variability in timing and magnitude of the spring bloom in the Irminger Basin, North Atlantic. Deep-Sea Res Part I 53:1601–1615. https://doi.org/10.1016/j.dsr.2006.07.009
13. Egge JK, Aksnes DL (1992) Silicate as regulating nutrient in phytoplankton competition. Mar Ecol Prog Ser 83:281–289. https://doi.org/10.3354/meps083281
14. Nielsdóttir MC, Moore CM, Sanders R, Hinz DJ, Achterberg EP (2009) Iron limitation of the postbloom phytoplankton communities in the Iceland Basin. Global Biogeochem Cycles 23: GB3001. https://doi.org/10.1029/2008GB003410
15. Meyer-Harms B, Irigoien X, Head R, Harris R (1999) Selective feeding on natural phytoplankton by *Calanus finmarchicus* before, during, and after the 1997 spring bloom in the Norwegian Sea. Limnol Oceanogr 44:154–165. https://doi.org/10.4319/lo.1999.44.1.0154
16. Pacariz SV, Hátún H, Jacobsen JA, Johnson C, Eliasen S, Rey F (2016) Nutrient-driven poleward expansion of the Northeast Atlantic mackerel (*Scomber scombrus*) stock: a new hypothesis. Elem Sci Anthr 4:000105. https://doi.org/10.12952/journal.elementa.000105

17. Edwards EWJ, Quinn LR, Wakefield ED, Miller PI, Thompson PM (2013) Tracking a northern fulmar from a Scottish nesting site to the Charlie-Gibbs fracture zone: evidence of linkage between coastal breeding seabirds and Mid-Atlantic Ridge feeding sites. Deep-Sea Res Part II Top Stud Oceanogr 98:438–444. https://doi.org/10.1016/j.dsr2.2013.04.011
18. Bogdanova MI, Daunt F, Newell M, Phillips RA, Harris MP, Wanless S (2011) Seasonal interactions in the black-legged kittiwake, *Rissa tridactyla*: links between breeding performance and winter distribution. Proc R Soc B Biol Sci 278:2412–2418. https://doi.org/10.1098/rspb.2010.2601
19. Holliday NP (2003) Air-sea interaction and circulation changes in the northeast Atlantic. J Geophys Res 108(C8):3259. https://doi.org/10.1029/2002JC001344
20. Bersch M, Meincke J, Sy A (1999) Interannual thermohaline changes in the northern North Atlantic 1991-1996. Deep-Sea Res Part II Top Stud Oceanogr 46:55–75. https://doi.org/10.1016/S0967-0645(98)00114-3
21. Ullgren JE, Fer I, Darelius E, Beaird N (2014) Interaction of the Faroe Bank Channel overflow with Iceland Basin intermediate waters. J Geophys Res Oceans 119:228–240. https://doi.org/10.1002/2013JC009437
22. Hátún H, Chafik L (2018) On the recent ambiguity of the North Atlantic subpolar gyre index. J Geophys Res Oceans 123:5072–5076. https://doi.org/10.1029/2018JC014101
23. Sandø AB, Furevik T (2008) Relation between the wind stress curl in the North Atlantic and the Atlantic inflow to the Nordic seas. J Geophys Res 113:C06028. https://doi.org/10.1029/2007JC004236
24. Hátún H, Sandø AB, Drange H, Hansen B, Valdimarsson H (2005) Influence of the Atlantic subpolar gyre on the thermohaline circulation. Science 309:1841–1844. https://doi.org/10.1126/science.1114777
25. Brambilla E, Talley LD (2008) Subpolar mode water in the northeastern Atlantic: 1. Averaged properties and mean circulation. J Geophys Res Ocean 113:C04025. https://doi.org/10.1029/2006JC004062
26. Hátún H, Olsen B, Pacariz SV (2017) The dynamics of the North Atlantic subpolar gyre introduces predictability to the breeding success of kittiwakes. Front Mar Sci 4:123. https://doi.org/10.3389/fmars.2017.00123
27. Mathis M, Elizalde A, Mikolajewicz U (2018) Which complexity of regional climate system models is essential for downscaling anthropogenic climate change in the Northwest European Shelf? Clim Dyn 50:2637–2659. https://doi.org/10.1007/s00382-017-3761-3
28. Holt J, Butenschön M, Wakelin SL, Artioli Y, Allen JI (2012) Oceanic controls on the primary production of the northwest European continental shelf: model experiments under recent past conditions and a potential future scenario. Biogeosciences 9:97–117. https://doi.org/10.5194/bg-9-97-2012
29. Emeis K-C, Beusekom JV, Callies U, Ebinghaus R, Kannen A, Kraus G, Kröncke I, Lenhart H, Lorkowski I, Matthias V, Möllmann C, Pätsch J, Scharfe M, Thomas H, Weisse R, Zorita E (2015) The North Sea — a shelf sea in the Anthropocene. J Mar Syst 141:18–33. https://doi.org/10.1016/j.jmarsys.2014.03.012
30. Gröger M, Maier-Reimer E, Mikolajewicz U, Moll A, Sein D (2013) NW European shelf under climate warming: implications for open ocean – shelf exchange, primary production, and carbon absorption. Biogeosciences 10:3767–3792. https://doi.org/10.5194/bg-10-3767-2013
31. Pätsch J, Kühn W (2008) Nitrogen and carbon cycling in the North Sea and exchange with the North Atlantic—a model study. Part I. nitrogen budget and fluxes. Cont Shelf Res 28:767–787. https://doi.org/10.1016/j.csr.2007.12.013
32. Eliasen SK, Hátún H, Larsen KMH, Jacobsen S (2017) Faroe shelf bloom phenology – the importance of ocean-to-shelf silicate fluxes. Cont Shelf Res 143:43–53. https://doi.org/10.1016/j.csr.2017.06.004

33. Williams RG, Follows MJ (2003) Physical transport of nutrients and the maintenance of biological production. In: Ocean biogeochemistry. Springer, Berlin, pp 19–51. https://doi.org/10.1007/978-3-642-55844-3_3
34. McCartney MS (1992) Recirculating components to the deep boundary current of the northern North Atlantic. Prog Oceanogr 29:283–383. https://doi.org/10.1016/0079-6611(92)90006-L
35. Arhan M (1990) The North-Atlantic current and SubArctic intermediate water. J Mar Res 48:109–144. https://doi.org/10.1357/002224090784984605
36. Bersch M (2002) North Atlantic oscillation–induced changes of the upper layer circulation in the northern North Atlantic Ocean. J Geophys Res 107(C10):3156. https://doi.org/10.1029/2001JC000901
37. Wade IP, Ellett DJ, Heywood KJ (1997) The influence of intermediate waters on the stability of the eastern North Atlantic. Deep-Sea Res Part I Oceanogr Res Pap 44:1405–1426. https://doi.org/10.1016/S0967-0637(97)00023-X
38. Johnson C, Inall M, Häkkinen S (2013) Declining nutrient concentrations in the northeast Atlantic as a result of a weakening subpolar gyre. Deep-Sea Res Part I Oceanogr Res Pap 82:95–107. https://doi.org/10.1016/j.dsr.2013.08.007
39. McCartney MS, Talley LD (1982) The sub-polar mode water of the North-Atlantic Ocean. J Phys Oceanogr 12:1169–1188. https://doi.org/10.1175/1520-0485(1982)012<1169:TSMWOT>2.0.CO;2
40. Brambilla E, Talley LD, Robbins PE (2008) Subpolar mode water in the northeastern Atlantic: 2. Origin and transformation. J Geophys Res Oceans 113:C04026. https://doi.org/10.1029/2006JC004063
41. Talley LD, McCartney MS (1982) Distribution and circulation of Labrador Sea water. J Phys Oceanogr 12:1189–1205. https://doi.org/10.1175/1520-0485(1982)012<1189:DACOLS>2.0.CO;2
42. Van Aken HM, Becker G (1996) Hydrography and through-flow in the north-eastern North Atlantic Ocean: the NANSEN project. Prog Oceanogr 38:297–346. https://doi.org/10.1016/S0079-6611(97)00005-0
43. Thomas LN, Joyce TM (2010) Subduction on the northern and southern flanks of the Gulf stream. J Phys Oceanogr 40:429–438. https://doi.org/10.1175/2009JPO4187.1
44. Sarafanov A (2009) On the effect of the North Atlantic oscillation on temperature and salinity of the subpolar North Atlantic intermediate and deep waters. ICES J Mar Sci 66:1448–1454. https://doi.org/10.1093/icesjms/fsp094
45. Sarafanov A, Mercier H, Falina A, Sokov A (2010) Cessation and partial reversal of deep water freshening in the northern North Atlantic: observation-based estimates and attribution. Tellus A Dynamic Meteorol Oceanogr 62:80–90. https://doi.org/10.1111/j.1600-0870.2009.00418.x
46. Hurrell JW (1995) Decadal trends in the North-Atlantic oscillation: regional temperatures and precipitation. Science 269:676–679. https://doi.org/10.1126/science.269.5224.676
47. Häkkinen S, Rhines PB (2004) Decline of subpolar North Atlantic circulation during the 1990s. Science 304:555–559. https://doi.org/10.1126/science.1094917
48. Larsen KMH, Hátún H, Hansen B, Kristiansen R (2012) Atlantic water in the Faroe area: sources and variability. ICES J Mar Sci 69:802–808. https://doi.org/10.1093/icesjms/fss028
49. Goldenberg SB, Landsea CW, Mestas-Nunez AM, Gray WM (2001) The recent increase in Atlantic hurricane activity: causes and implications. Science 293:474–479. https://doi.org/10.1126/science.1060040
50. de Boisséson E, Thierry V, Mercier H, Caniaux G, Desbruyères D (2012) Origin, formation and variability of the subpolar mode water located over the Reykjanes Ridge. J Geophys Res 117:C12005. https://doi.org/10.1029/2011JC007519
51. Rey F (2012) Declining silicate concentrations in the Norwegian and Barents Seas. ICES J Mar Sci 69:208–212. https://doi.org/10.1093/icesjms/fss007
52. Hátún H, Azetsu-Scott K, Somavilla R, Rey F, Johnson C, Mathis M, Mikolajewicz U, Coupel P, Tremblay J-É, Hartman S, Pacariz SV, Salter I, Ólafsson J (2017) The subpolar

gyre regulates silicate concentrations in the North Atlantic. Sci Rep 7:14576. https://doi.org/10.1038/s41598-017-14837-4
53. Ilyina T, Six KD, Segschneider J, Maier-Reimer E, Li H, Núñez-Riboni I (2013) Global Ocean biogeochemistry model HAMOCC: model architecture and performance as component of the MPI-Earth system model in different CMIP5 experimental realizations. J Adv Model Earth Syst 5:287–315. https://doi.org/10.1029/2012MS000178
54. Stendardo I, Kieke D, Rhein M, Gruber N, Steinfeldt R (2015) Interannual to decadal oxygen variability in the mid-depth water masses of the eastern North Atlantic. Deep-Sea Res Part I Oceanogr Res Pap 95:85–98. https://doi.org/10.1016/j.dsr.2014.10.009
55. Bower AS, Le Cann B, Rossby T, Zenk W, Gould J, Speer K, Richardson PL, Prater MD, Zhang H-M (2002) Directly measured mid-depth circulation in the northeastern North Atlantic Ocean. Nature 419:603–607. https://doi.org/10.1038/nature01078
56. McGrath T, Nolan G, McGovern E (2012) Chemical characteristics of water masses in the Rockall Trough. Deep Sea Res Part I Oceanogr Res Pap 61:57–73. https://doi.org/10.1016/j.dsr.2011.11.007
57. Dickson RR, Meincke J, Malmberg SA, Lee AJ (1988) The great salinity anomaly in the northern North-Atlantic 1968-1982. Prog Oceanogr 20:103–151. https://doi.org/10.1016/0079-6611(88)90049-3
58. Belkin IM (2004) Propagation of the "great salinity anomaly" of the 1990s around the northern North Atlantic. Geophys Res Lett 31:L08306. https://doi.org/10.1029/2003GL019334
59. Holliday NP, Bersch M, Berx B, Chafik L, Cunningham S, Florindo-López C, Hátún H, Johns W, Josey SA, Larsen KMH, Mulet S, Oltmanns M, Reverdin G, Rossby T, Thierry V, Valdimarsson H, Yashayaev I (2020) Ocean circulation causes the largest freshening event for 120 years in eastern subpolar North Atlantic. Nat Commun 11:585. https://doi.org/10.1038/s41467-020-14474-y
60. Jacobsen S, Gaard E, Hátún H, Steingrund P, Larsen KMH, Reinert J, Ólafsdóttir SR, Poulsen M, Vang HBM (2019) Environmentally driven ecological fluctuations on the Faroe shelf revealed by fish juvenile surveys. Front Mar Sci 6:559. https://doi.org/10.3389/fmars.2019.00559
61. Hansen B, Østerhus S (2000) North Atlantic–Nordic Seas exchanges. Prog Oceanogr 45:109–208. https://doi.org/10.1016/S0079-6611(99)00052-X
62. Childers KH, Flagg CN, Rossby T, Schrum C (2015) Directly measured currents and estimated transport pathways of Atlantic water between 59.5°N and the Iceland-Faroes-Scotland Ridge. Tellus A Dyn Meteorol Oceanogr 67:28067. https://doi.org/10.3402/tellusa.v67.28067
63. Sarafanov A, Falina A, Sokov A, Demidov A (2008) Intense warming and salinification of intermediate waters of southern origin in the eastern subpolar North Atlantic in the 1990s to mid-2000s. J Geophys Res 113:C12022. https://doi.org/10.1029/2008JC004975
64. Van Aken HM, De Boer CJ (1995) On the synoptic hydrography of intermediate and deep-water masses in the Iceland Basin. Deep-Sea Res Part I Oceanogr Res Pap 42:165–189. https://doi.org/10.1016/0967-0637(94)00042-Q
65. Beaird NL, Rhines PB, Eriksen CC (2016) Observations of seasonal subduction at the Iceland-Faroe front. J Geophys Res Oceans 121:4026–4040. https://doi.org/10.1002/2015JC011501
66. Pollard RT, Read JF, Holliday NP, Leach H (2004) Water masses and circulation pathways through the Iceland Basin during Vivaldi 1996. J Geophys Res 109:C04004. https://doi.org/10.1029/2003JC002067
67. Bersch M, Yashayaev I, Koltermann KP (2007) Recent changes of the thermohaline circulation in the subpolar North Atlantic. Ocean Dyn 57:223–235. https://doi.org/10.1007/s10236-007-0104-7
68. Seim KS, Fer I (2011) Mixing in the stratified interface of the Faroe Bank Channel overflow: the role of transverse circulation and internal waves. J Geophys Res Oceans 116:C07022. https://doi.org/10.1029/2010JC006805
69. Darelius E, Fer I, Quadfasel D (2011) Faroe Bank Channel overflow: mesoscale variability. J Phys Oceanogr 41:2137–2154. https://doi.org/10.1175/JPO-D-11-035.1

70. Darelius E, Ullgren JE, Fer I (2013) Observations of Barotropic oscillations and their influence on mixing in the Faroe Bank Channel overflow region. J Phys Oceanogr 43:1525–1532. https://doi.org/10.1175/JPO-D-13-059.1
71. Beaird NL, Fer I, Rhines P, Eriksen C (2012) Dissipation of turbulent kinetic energy inferred from seagliders: an application to the eastern Nordic Seas overflows. J Phys Oceanogr 42:2268–2282. https://doi.org/10.1175/JPO-D-12-094.1
72. Bringedal C, Eldevik T, Skagseth Ø, Spall MA, Østerhus S (2018) Structure and forcing of observed exchanges across the Greenland–Scotland ridge. J Clim 31:9881–9901. https://doi.org/10.1175/JCLI-D-17-0889.1
73. Yang J, Pratt LJ (2013) On the effective capacity of the dense-water reservoir for the Nordic seas overflow: some effects of topography and wind stress. J Phys Oceanogr 43:418–431. https://doi.org/10.1175/JPO-D-12-087.1
74. Chafik L, Hátún H, Kjellsson J, Larsen KMH, Rossby T, Berx B (2020) Discovery of an unrecognized pathway carrying overflow waters toward the Faroe Bank Channel. Nat Commun 11:3721. https://doi.org/10.1038/s41467-020-17426-8
75. Johnson GC, Gruber N (2007) Decadal water mass variations along 20°W in the northeastern Atlantic Ocean. Prog Oceanogr 73:277–295. https://doi.org/10.1016/j.pocean.2006.03.022
76. de Jong MF, de Steur L (2016) Strong winter cooling over the Irminger Sea in winter 2014-2015, exceptional deep convection, and the emergence of anomalously low SST. Geophys Res Lett 43:7106–7113. https://doi.org/10.1002/2016GL069596
77. Yashayaev I, Loder JW (2016) Recurrent replenishment of Labrador Sea water and associated decadal-scale variability. J Geophys Res Oceans 121:8095–8114. https://doi.org/10.1002/2016JC012046
78. Crocket KC, Hill E, Abell RE, Johnson C, Gary SF, Brand T, Hathorne EC (2018) Rare earth element distribution in the NE Atlantic: evidence for benthic sources, longevity of the seawater signal, and biogeochemical cycling. Front Mar Sci 5:147. https://doi.org/10.3389/fmars.2018.00147
79. Richter K, Maus S (2011) Interannual variability in the hydrography of the Norwegian Atlantic current: frontal versus advective response to atmospheric forcing. J Geophys Res 116:C12031. https://doi.org/10.1029/2011JC007311
80. Siegismund F, Johannessen J, Drange H, Mork KA, Korablev A (2007) Steric height variability in the Nordic seas. J Geophys Res Oceans 112:C12010. https://doi.org/10.1029/2007JC004221
81. Hátún H, Lohmann K, Matei D, Jungclaus JH, Pacariz S, Bersch M, Gislason A, Ólafsson J, Reid PC (2016) An inflated subpolar gyre blows life toward the northeastern Atlantic. Prog Oceanogr 147:49–66. https://doi.org/10.1016/j.pocean.2016.07.009
82. Frajka-Williams E, Rhines PB, Eriksen CC (2009) Physical controls and mesoscale variability in the Labrador Sea spring phytoplankton bloom observed by Seaglider. Deep-Sea Res I 56:2144–2161. https://doi.org/10.1016/j.dsr.2009.07.008
83. Hátún H, Eriksen CC, Rhines PB (2007) Buoyant eddies entering the Labrador Sea observed with gliders and altimetry. J Phys Oceanogr 37:2838–2854. https://doi.org/10.1175/2007JPO3567.1
84. Alkire MB, D'Asaro E, Lee C, Perry MJ, Gray A, Cetinić I, Briggs N, Rehm E, Kallin E, Kaiser J, González-Posada A (2012) Estimates of net community production and export using high-resolution, Lagrangian measurements of O_2, NO_3^-, and POC through the evolution of a spring diatom bloom in the North Atlantic. Deep Sea Res Part I Oceanogr Res Pap 64:157–174. https://doi.org/10.1016/j.dsr.2012.01.012
85. Painter SC, Henson SA, Forryan A, Steigenberger S, Klar J, Stinchcombe MC, Rogan N, Baker AR, Achterberg EP, Moore CM (2014) An assessment of the vertical diffusive flux of iron and other nutrients to the surface waters of the subpolar North Atlantic Ocean. Biogeosciences 11:2113–2130. https://doi.org/10.5194/bg-11-2113-2014
86. Lankhorst M, Zenk W (2006) Lagrangian observations of the middepth and deep velocity fields of the northeastern Atlantic Ocean. J Phys Oceanogr 36:43–63. https://doi.org/10.1175/JPO2869.1

87. Gudfinnsson H, Debes H, Falkenhaug T, Gaard E, Gislason A, Petursdottir H, Sigurdsson T, Stupnikova A, Valdimarsson H (2008) Abundance and productivity of the pelagic ecosystem across the northern Mid-Atlantic Ridge in June 2003. ICES CM2008/C:12
88. ICES (2020) Working group on widely distributed stocks (WGWIDE). ICES Sci Rep 2 (82):1019. https://doi.org/10.17895/ices.pub.7475
89. Maritorena S, d'Andon OHF, Mangin A, Siegel DA (2010) Merged satellite ocean color data products using a bio-optical model: characteristics, benefits and issues. Remote Sens Environ 114:1791–1804. https://doi.org/10.1016/j.rse.2010.04.002
90. Langøy H, Nøttestad L, Skaret G, Broms C, Fernö A (2012) Overlap in distribution and diets of Atlantic mackerel (*Scomber scombrus*), Norwegian spring-spawning herring (*Clupea harengus*) and blue whiting (*Micromesistius poutassou*) in the Norwegian Sea during late summer. Mar Biol Res 8:442–460. https://doi.org/10.1080/17451000.2011.642803
91. Ogi M, Yamazaki K, Tachibana Y (2004) The summertime annular mode in the northern hemisphere and its linkage to the winter mode. J Geophys Res 109:D20114. https://doi.org/10.1029/2004JD004514
92. Waniek JJ, Holliday NP (2006) Large-scale physical controls on phytoplankton growth in the Irminger Sea, part II: model study of the physical and meteorological preconditioning. J Mar Syst 59:219–237. https://doi.org/10.1016/j.jmarsys.2005.10.005
93. Galloway JN, Howarth RW, Michaels AF, Nixon SW, Prospero JM, Dentener FJ (1996) Nitrogen and phosphorus budgets of the North Atlantic Ocean and its watershed. Biogeochemistry 35:3–25. https://doi.org/10.1007/BF02179823
94. Holt J, Schrum C, Cannaby H, Daewel U, Allen I, Artioli Y, Bopp L, Butenschon M, Fach BA, Harle J, Pushpadas D, Salihoglu B, Wakelin S (2016) Potential impacts of climate change on the primary production of regional seas: a comparative analysis of five European seas. Prog Oceanogr 140:91–115. https://doi.org/10.1016/j.pocean.2015.11.004
95. Steinacher M, Joos F, Frölicher TL, Bopp L, Cadule P, Cocco V, Doney SC, Gehlen M, Lindsay K, Moore JK, Schneider B, Segschneider J (2010) Projected 21st century decrease in marine productivity: a multi-model analysis. Biogeosciences 7:979–1005. https://doi.org/10.5194/bg-7-979-2010
96. Vermaat JE, McQuatters-gollop A, Eleveld MA, Gilbert AJ (2008) Past, present and future nutrient loads of the North Sea: causes and consequences. Estuar Coast Shelf Sci 80:53–59. https://doi.org/10.1016/j.ecss.2008.07.005
97. Gröger M, Maier-Reimer E, Mikolajewicz U, Moll A, Sein D (2013) NW European shelf under climate warming: implications for open ocean - shelf exchange, primary production, and carbon absorption. Biogeosciences 10:3767–3792. https://doi.org/10.5194/bg-10-3767-2013
98. IPCC (2019) IPCC special report on the ocean and cryosphere in a changing climate. https://www.ipcc.ch/srocc/cite-report/
99. Holt J, Allen JI, Anderson TR, Brewin R, Butenschön M, Harle J, Huse G, Lehodey P, Lindemann C, Memery L, Salihoglu B, Senina I, Yool A (2014) Challenges in integrative approaches to modelling the marine ecosystems of the North Atlantic: physics to fish and coasts to ocean. Prog Oceanogr 129:285–313. https://doi.org/10.1016/j.pocean.2014.04.024
100. Mathis M, Elizalde A, Mikolajewicz U (2019) The future regime of Atlantic nutrient supply to the Northwest European Shelf. J Mar Syst 189:98–115. https://doi.org/10.1016/j.jmarsys.2018.10.002
101. Sharples J, Holt J, Dye SR (2013) Impacts of climate change on shelf sea stratification. MCCIP Sci Rev:67–70. https://doi.org/10.14465/2013.arc08.067-070
102. Painter SC, Hartman SE, Kivimäe C, Salt LA, Clargo NM, Daniels CJ, Bozec Y, Daniels L, Allen S, Hemsley VS, Moschonas G, Davidson K (2017) The elemental stoichiometry (C, Si, N, P) of the Hebrides Shelf and its role in carbon export. Prog Oceanogr 159:154–177. https://doi.org/10.1016/j.pocean.2017.10.001
103. Pätsch J, Burchard H, Dieterich C, Gräwe U, Gröger M, Mathis M, Kapitza H, Bersch M, Moll A, Pohlmann T, Su J, Ho-hagemann HTM, Schulz A, Elizalde A, Eden C (2017) An evaluation of the North Sea circulation in global and regional models relevant for ecosystem simulations. Ocean Model 116:70–95. https://doi.org/10.1016/j.ocemod.2017.06.005

104. Eliasen SK, Hátún H, Larsen KMH, Hansen B, Rasmussen TAS (2017) Phenologically distinct phytoplankton regions on the Faroe Shelf - identified by satellite data, in-situ observations and model. J Mar Syst 169:99–110. https://doi.org/10.1016/j.jmarsys.2017.01.015
105. ICES (2019) Report of the Herring Assessment Working Group for the Area South of 62°N (HAWG). ICES Sci Rep 1(2):971. www.ices.dk
106. Coulson JC (2011) The Kittiwake. T & AD Poyser, London, pp 1–299
107. Eliasen K, Reinert J, Gaard E, Hansen B, Jacobsen JA, Grønkjær P, Christensen JT (2011) Sandeel as a link between primary production and higher trophic levels on the Faroe Shelf. Mar Ecol Ser 438:185–194. https://doi.org/10.3354/meps09301
108. Hansen B, Eliasen SK, Gaard E, Larsen KMH (2005) Climatic effects on plankton and productivity on the Faroe Shelf. ICES J Mar Sci 62:1224–1232. https://doi.org/10.1016/j.icesjms.2005.04.014
109. Steingrund P, Gaard E (2005) Relationship between phytoplankton production and cod production on the Faroe Shelf. ICES J Mar Sci 62:163–176. https://doi.org/10.1016/j.icesjms.2004.08.019
110. Gaard E, Hansen B, Heinesen SP (1998) Phytoplankton variability on the Faroe Shelf. ICES J Mar Sci 55:688–696. https://doi.org/10.1006/jmsc.1998.0373
111. Eliasen SK, Gaard E, Hansen B, Larsen KMH (2005) A 'horizontal Sverdrup mechanism' may control the spring bloom around small oceanic islands and over banks. J Mar Syst 56:352–362. https://doi.org/10.1016/j.jmarsys.2005.03.005
112. Gaard E (2003) Plankton variability on the Faroe Shelf during the 1990s. ICES Mar Sci Symp 219:182–189
113. Eliasen SK, Hansen B, Larsen KMH, Hátún H (2016) The exchange of water between the Faroe Shelf and the surrounding waters and its effect on the primary production. J Mar Syst 153:1–9. https://doi.org/10.1016/j.jmarsys.2015.08.004
114. Larsen KMH, Hansen B, Svendsen H (2008) Faroe shelf water. Cont Shelf Res 28:1754–1768. https://doi.org/10.1016/j.csr.2008.04.006
115. Bleck R (2002) An oceanic general circulation model framed in hybrid isopycnic-Cartesian coordinates. Ocean Model 4:55–88. https://doi.org/10.1016/S1463-5003(01)00012-9
116. Rasmussen TAS, Olsen SM, Hansen B, Hátún H, Larsen KMH (2014) The Faroe shelf circulation and its potential impact on the primary production. Cont Shelf Res 88:171–184. https://doi.org/10.1016/j.csr.2014.07.014
117. Holt J, Wakelin S, Huthnance J (2009) Down-welling circulation of the northwest European continental shelf: a driving mechanism for the continental shelf carbon pump. Geophys Res Lett 36:L14602. https://doi.org/10.1029/2009GL038997
118. Eliasen SK, Hátún H, Larsen KMH, Vang HBM, Rasmussen TAS (2019) The Faroe shelf spring bloom onset explained by a 'critical volume hypothesis'. J Mar Syst 194:91–101. https://doi.org/10.1016/j.jmarsys.2019.02.005
119. Brzezinski MA (1985) The Si:C:N ration of marine diatoms: interspecific variability and the effect of some environmental variables. J Phycol 21:347–357. https://doi.org/10.1111/j.0022-3646.1985.00347.x
120. Djurhuus A, Jørgensen J, Hátún H, Debes HH, Christiansen DH (2015) Seasonal progression of microbial communities on the Faroe shelf. Mar Biol Res 11:895–908. https://doi.org/10.1080/17451000.2015.1041532
121. Hátún H, Larsen KMH, Debes H (2013) The western region. Technical report No 13-07, Faroe Marine Research Institute, 52p. http://www.hav.fo/PDF/Ritgerdir/2013/TecRep1307.pdf
122. Debes H, Gaard E, Hansen B (2008) Primary production on the Faroe shelf: temporal variability and environmental influences. J Mar Syst 74:686–697. https://doi.org/10.1016/j.jmarsys.2008.07.004
123. Jacobsen S, Gaard E, Larsen KMH, Eliasen SK, Hátún H (2018) Temporal and spatial variability of zooplankton on the Faroe shelf in spring 1997–2016. J Mar Syst 177:28–38. https://doi.org/10.1016/j.jmarsys.2017.08.004

Fronts in the Baltic Sea: A Review with a Focus on Its North-Eastern Part

Ülo Suursaar, Jüri Elken, and Igor M. Belkin

Contents

Abstract Oceanic fronts have a crucial impact on the physical, chemical, and biological processes of the Baltic Sea. Due to its complex morphometry and numerous freshwater inflows, the Baltic Sea hosts a multitude of fronts that separate water masses with different properties. Intense thermohaline and nutrient fronts occur as high-gradient zones in the Baltic Sea proper, Gulf of Finland, Bothnian Bay, and Gulf of Riga. These sub-basins feature freshwater and nutrient inflows from numerous rivers. Coastal upwelling fronts are formed under favourable wind conditions. Due to the complex shape of the Baltic Sea, wind from any direction blows parallel to some section of its coastline and hence can cause coastal upwelling at virtually any section. In the Gulf of Finland, upwelling may extend up to 400 km

Ü. Suursaar (✉)
Estonian Marine Institute, University of Tartu, Tallinn, Estonia
e-mail: ulo.suursaar@ut.ee

J. Elken
Department of Marine Systems, Tallinn University of Technology, Tallinn, Estonia

I. M. Belkin
College of Marine Science and Technology, Zhejiang Ocean University, Zhoushan, China

Igor M. Belkin (ed.), *Chemical Oceanography of Frontal Zones*,
Hdb Env Chem (2022) 116: 143–182, DOI 10.1007/698_2021_813,

Published online: 19 October 2021

along either the northern or southern coast, depending on the wind direction. River plumes have the largest extent – up to several tens of km – during high discharge and low winds. Some of the water-mass fronts are detected in ocean colour imagery where these fronts delineate river plumes, wind- and dredging-induced turbidity patches, phytoplankton blooms, and pollution discharges.

Keywords Eutrophication, Nutrients, River plumes, Thermohaline fronts, Turbidity plumes, Upwelling

1 Introduction

Due to the diversity of the physical nature of oceanic fronts and their widely different spatiotemporal scales [1, 2], a universal front definition does not exist. Oceanic fronts are typically defined as narrow zones of enhanced gradients separating different water masses [3]. According to K.N. Fedorov, a front is "a zone in which the spatial gradients of the main thermodynamic characteristics are very high in comparison with the mean uniform distribution between the stable climatic or other extremes" [1, 4]. The broad importance of fronts on ecosystems stems from their considerable impacts on the physical, chemical, and biological environments of the sea [5]. Oceanic fronts are associated with strong mixing and stirring, current jets, and bioproductivity maxima [6]. Along-front length scales vary from 100 m to 10,000 km, while cross-front length scales vary from 1 m to 100 km, and vertical scales vary from 1 m to a few kilometres. Cross-frontal temperature and salinity differences can be as large as 10–15°C and 2–3 psu, respectively, or even more [4]. Large-scale fronts are associated with elements of the general circulation such as subtropical convergence, equatorial upwelling, western boundary currents, shelf-slope fronts, etc. [5]. As a rule, large-scale fronts are in geostrophic equilibrium according to the Margules formula; they are also affected by Ekman transport and buoyancy fluxes. Meso- and small-scale fronts are characterized by additional ageostrophic motion components, causing secondary circulation and mixing [2, 4]. Such fronts are associated with river plumes, tidal mixing in bays and estuaries, currents–topography interaction, and coastal wind-induced upwelling. The process of horizontal gradient enhancement by external forcing is called fronto-genesis [7], while the opposite process of gradient relaxation is called front-olysis. Continuity and sharpness of quasi-permanent fronts is maintained by a rough balance of generation and decay processes over seasonal and longer time scales [8].

The Baltic Sea features numerous fronts of diverse physical nature due to salinity gradients maintained by the Atlantic inflow in the southwest, river inflows mostly in the east and north, water-mass convergences, and wind-driven coastal upwelling. Both temperature and salinity play important roles in forming the Baltic Sea fronts. This review concentrates on regional- and local-scale fronts in the Baltic Sea (Fig. 1), particularly in its north-eastern part (Fig. 2). According to the Large Marine

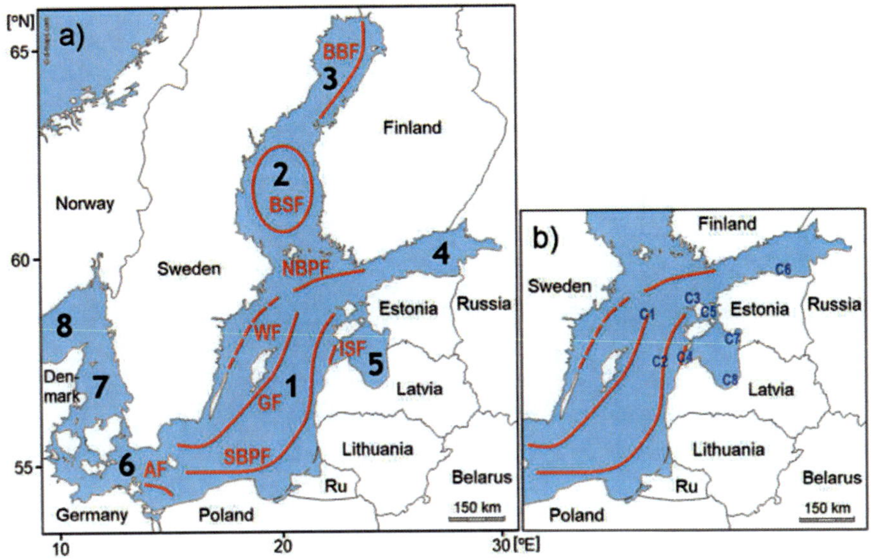

Fig. 1 Main thermal fronts of the Baltic Sea from satellite SST data [9] (**a**): AF, Arkona Front; BBF, Bothnian Bay Front; BSF, Bothnian Sea Front; GF, Gotland Front; ISF, Irbe Strait Front; NBPF, North Baltic Proper Front; SBPF, South Baltic Proper Front; WF, Western Front. Shown are also the Baltic subregions: 1 – Baltic Proper, 2 – Bothnian Sea, 3 – Bothnian Bay, 4 – Gulf of Finland, 5 – Gulf of Riga, 6 – Arkona Sea (part of Baltic Proper), 7 – Kattegat, and 8 – Skagerrak (part of the North Sea). Locations of some case studies (C1–C8) reviewed in this article (**b**)

Ecosystems (LMEs) concept pioneered by Kenneth Sherman (NOAA), there are 66 LMEs worldwide. The Baltic Sea is one of them. Using the Cayula-Cornillon edge detection algorithm [10], which is a robust, automated tool for detecting thermal fronts, a global survey of sea surface temperature (SST) fronts was conducted by Belkin and co-authors [5]. The fronts were detected from the NOAA/NASA AVHRR SST fields over the period 1985–1996, with two 9-km resolution fields per day. Based on that study, Fig. 1 reproduces the original map of the Baltic Sea thermal fronts presented by Belkin and Cornillon [9]. The SST fronts in the Baltic Sea are mostly topographically controlled, either due to differential cooling/warming in shallow/deep marine areas or due to wind-induced upwelling [9, 11]. To identify thermohaline and dynamical characteristics of fronts, both temperature and salinity are required. Previous satellite-based studies of the Baltic Sea fronts used either SST or chlorophyll (CHL) to identify fronts [5]. The advent of space-borne sensors that allow sea surface salinity (SSS) to be estimated (in such satellite missions as Aquarius, SMOS, and SMAP) opens a new era of mapping SSS fronts. This new development is especially promising regarding the Baltic Sea fronts. According to the front classification developed by Belkin, Cornillon, and Sherman [5], the Baltic Sea features two major types of fronts – type F (due to freshwater inflow) and type U (upwelling).

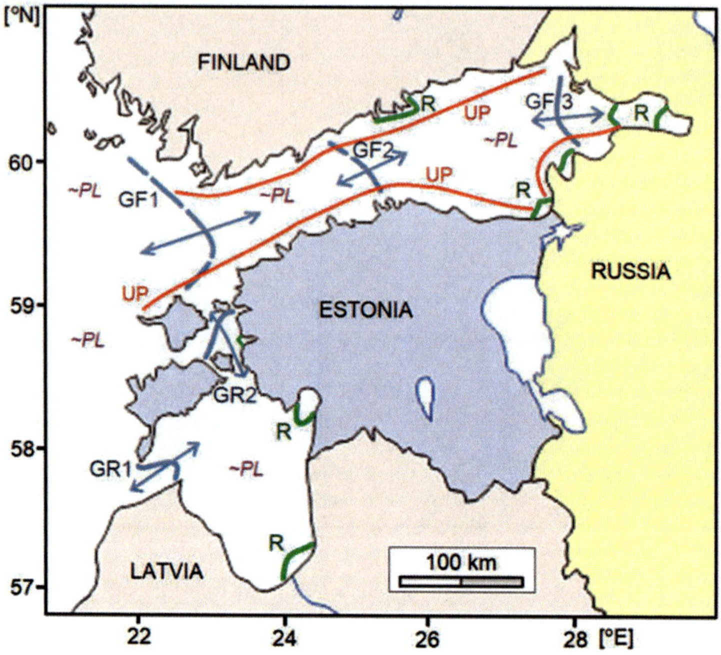

Fig. 2 Fronts in the north-eastern Baltic (schematic). Arrows indicate variations in front location. Shown are quasi-permanent thermohaline fronts (blue) in the Gulf of Finland (GF1, GF2, and GF3, the) and Gulf of Riga (GR1, the Irbe Strait Front, and GR2, the Väinameri Sea Front), upwelling fronts in the Gulf of Finland (UP; red), river plume fronts (R; green), and occasional summertime plankton (~PL) bloom patterns all over the sea

The Baltic Sea has substantial terrestrial influence and anthropogenic pressure, whereas there are crucially important along-basin and transverse salinity gradients [12]. There is significant horizontally variable vertical stratification as well. The interplay of waters originating from different sub-basins creates contrasts of biological and optical parameters that quite often match thermohaline contrasts. As an example of an early interdisciplinary study approach, a thorough catalogue of patterns of eddies, fronts and upwelling was compiled in 1983 by combining visible ocean colour and infrared temperature data [13]. To capture eddying meanders and filaments (squirts) of the fronts, the grid step of the data field should be 1 km or less, i.e., several times smaller than the baroclinic Rossby deformation radius [12]. Hence, early satellite data [14, 15] often lack meso- and submesoscale features on the sea surface.

Basin-wide hydrographic characteristics of water masses, as well as small- and mesoscale hydrophysical features, such as fronts, eddies, gyres, and "patches" of various types, have been extensively studied in the Baltic Sea since the 1980s [16–21]. To date, a vast amount of data on the thermohaline structure (T, S, density) and biochemical functioning of the sea have been accumulated [12, 22–25]. Observational knowledge was first obtained from surveys performed on research vessels, including vertical and towed CTD casts along transects or in polygons, and from time series measurements on mooring stations. Many of them were carried out through bilateral cooperation [20, 26–29] or international programmes, such as the Baltic Sea Patchiness Experiment PEX-86 [30, 31], or by contributing to Baltic-wide oceanographic activities such as BOOS [32] and BALTEX [23, 33]. In recent decades, several new in situ study methods and approaches have been applied, such as FerryBox systems [34, 35], profiling autonomous stations [36], profiling ADCP and RDCP current meters [37–40], ARGO floats [41–43], and gliders [43]. In addition, there has been significant development in regional remote sensing techniques [44–50] and hydrodynamic and ecological modelling [51–57]. The main contemporary ocean data service portals, useful for frontal studies, are the Copernicus Marine Environment Monitoring Service (CMEMS, marine.copernicus.eu/), ICES Data Centre (ices.dk/data/), and European Marine Observation and Data Network (EMODNET, emodnet.eu/en). Many process-oriented observations are stored in SeaDataNet (www.seadatanet.org/).

In this article, an overview of fronts in the Baltic Sea is given, with a focus on high-gradient areas in the north-eastern part of the sea. While Fig. 1 depicts Baltic-wide frontal pattern [9], Fig. 2 presents fronts in the northeast. Following the presentation of the hydrographic and environmental settings of the area, frontal characteristics of the large but low-gradient Baltic Proper are given. Subsequently, fronts located on the quasi-permanent salinity gradient, running from the Baltic Proper into the Gulf of Finland and Gulf of Riga, are outlined. Then, coastal upwelling-related thermohaline fronts are highlighted, and finally, local-scale river plumes, turbidity, phytoplankton, and pollution "clouds" are described. In fact, the basic features of these local-scale fronts are not unique, and examples of fronts from other Baltic Sea areas are also considered. At the end of the article, we discuss the prospects of further studies on frontal dynamics using a combination of the recently collated high-resolution data sets from FerryBox and other modern in situ techniques, from new remote sensing products, and from model products.

2 Hydrographic and Environmental Setting

The Baltic Sea is a large semi-enclosed sea that has an elongated shape and complex multi-basin morphometric peculiarities (see overviews in textbooks [12, 58]). While along-basin length scales can be estimated to be approximately 1,500 km (from the southwestern Baltic to the farthest parts of Bothnian Bay or the Gulf of Finland), cross-basin lengths are shorter, from 50 km (Gulf of Finland) to 250 km (Baltic Proper and Bothnian Sea). The surface area of the Baltic Sea, including Kattegat, is $417{\cdot}10^3\ km^2$, and the sea volume is $21.9{\cdot}10^3\ km^3$ [59], which produces a mean depth of only 52.5 m. The exchange of saline and brackish waters between the Arkona Sea and Kattegat (Fig. 1) is limited by narrow and shallow Danish Straits. Another water budget component – freshwater influx – is dominated by river discharge, which during 1950–2014 comprised $454\ km^3\ year^{-1}$ [60] or approximately 2% of the sea

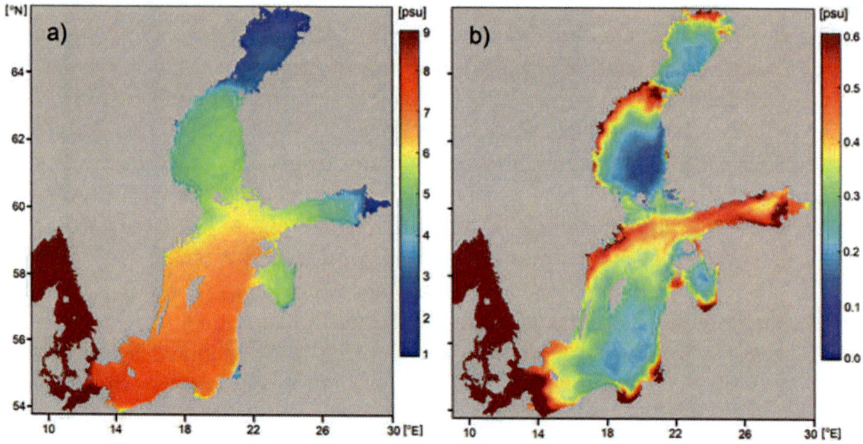

Fig. 3 Long-term annual mean salinity (**a**) and its temporal standard deviation for the period 1993–2019 (**b**), based on the CMEMS daily reanalysis data (www.copernicus.eu/en/access-data/copernicus-services-catalogue/baltic-sea-physics-reanalysis)

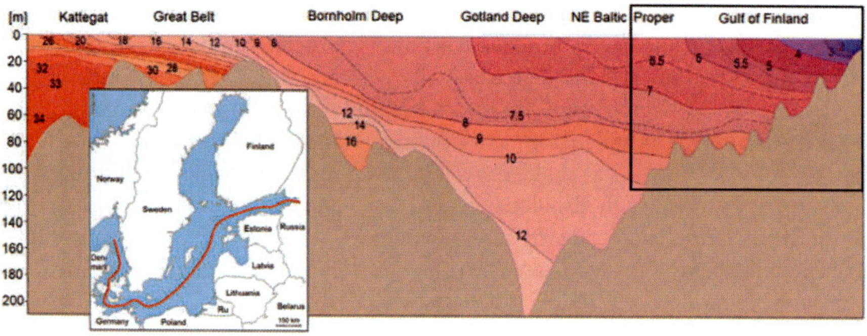

Fig. 4 Salinity distribution along a thalweg from Kattegat to the Neva River estuary. Location of the transect is shown in the inset. Modified from [62, 63]

volume per year. About 74% of river discharge is located far from the Danish Straits [58, 61]: Bothnian Sea and Bothnian Bay – 40%, Gulf of Finland – 26%, and Gulf of Riga – 8%.

The Baltic Sea hosts a multitude of physical and chemical gradients [12] that vary in space and occasionally form fronts that separate water masses with different properties (Figs. 3 and 4). High-gradient zones occurring below the surface form vertical stratification (thermo- or haloclines); when the same feature curves up to the sea surface, it forms a front on a horizontal plane (Fig. 4). In addition to salinity, water temperature differences create stratification and fronts. A seasonal thermocline usually forms in May; it develops at a depth of 10–25 m during the summer and erodes in August–September [22, 25]. The halocline and thermocline usually do not coincide in time and space, and both create their own density gradients (pycnoclines), which in turn govern the distribution of other substances [12]. Rivers carry substantial amounts of nutrients and polluting substances from the catchment area, making environmental imprints in the water masses [60, 61]. The Baltic Sea is dominated by westerlies and affected by alternating maritime and continental weather patterns, with precipitation exceeding evaporation [64].

The estuarine salinity gradients are the strongest (i.e., spatially "packed" into fronts) in the Danish Straits, where salinity drops from ca. 30 to 8–10 psu [51] (this salinity range is not shown in Fig. 3a). To the east from these straits, there are constrictions in the sea topography that define certain sub-basins of the sea and create frontal salinity differences from 1 to 3 psu between Bothnian Bay and the Bothnian Sea and from the Baltic Proper to the Bothnian Sea and the Gulf of Riga (Fig. 3a). Therefore, the intra-basin salinity contrasts are much smaller than the inter-basin drops, justifying the use of box models of connected basins [65, 66].

Because the positions of fronts vary, the frontal zones are characterized by higher temporal variance than the intra-basin areas. Indeed, high values of temporal standard deviation of salinity, based on reanalysis in 1993–2019 (Fig. 3b), reveal several structures:

1. Quark Front between the Bothnian Sea and Bothnian Bay,
2. Irbe Strait Front between the Gulf of Riga and Baltic Proper,
3. North Baltic Proper Front between freshened waters of the Gulf of Finland and Bothnian Sea and saltier waters of the Baltic Proper,
4. fronts and high-variance zones in the Gulf of Finland,
5. river influence areas near the coasts, and
6. highly variable Danish Straits and Kattegat.

Surface distributions of temperature and salinity are controlled by circulation and mixing processes that occur in the cascade of stratified sub-basins. The convergence and mixing of freshwater received at the surface and of saline water received in the deep layers result in tilted isohalines, whereas their deepening matches the surface salinity decrease (Fig. 4).

The Gulf of Finland lies in the north-eastern part of the Baltic Sea (Fig. 1). The estuary-like, nearly 400-km long embayment is relatively narrow (width varies from 50 to 130 km) and shallow (mean depth 37.3 m). The Gulf has a surface area of 29,600 km^2 and a water volume of 1,103 km^3. Greater depths (up to 123 m [61]) occur in the transition area to the Baltic Proper in the west, which enables considerable vertical stratification (Fig. 4; [12]). Hydrographically, the Gulf can be regarded as a stratified ("salt wedge") estuary in its western part and a partially mixed estuary in its eastern end (also a gulf-type ROFI – Region of Freshwater Influence [67]). Freshwater is mostly (~70%) discharged into the eastern part of the Gulf by the Neva River (Fig. 5) with the mean discharge of 79 km^3 $year^{-1}$ in 1981–2010 [60]. The other large rivers are the Narva (12.6 km^3 $year^{-1}$) and Kymijoki (8.9 km $year^{-1}$) Rivers. The surface layer salinity increases horizontally from about 1 psu near the Neva River to 6–7 psu at the Gulf's entrance area [71]. In

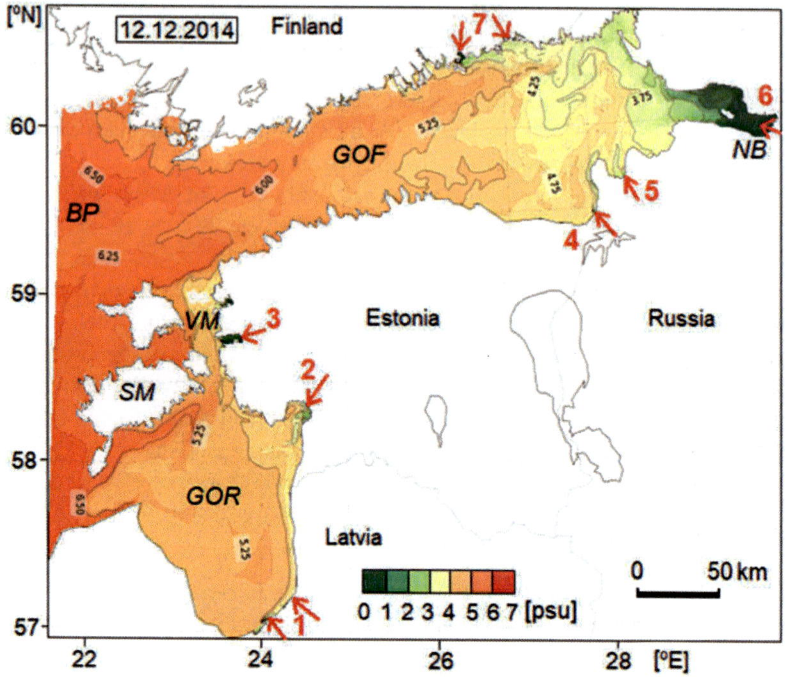

Fig. 5 A sample surface salinity map (forecast) in the north-eastern Baltic Sea retrieved from the Estonian Weather Service (EWS) website (www.ilmateenistus.ee) for 12 December 2014. The forecast is made using the HBM-EST model [68–70]. Acronyms: BP, Baltic Proper; GOF, Gulf of Finland; NB, Neva Bay; GR, Gulf of Riga; VM, Väinameri Sea; SM, Saaremaa Island. Red arrows denote river inflows (1-Daugava, Lielupe, and Gauja; 2-Pärnu, 3-Kasari, 4-Narva, 5-Luga, 6-Neva, 7-Kymijoki)

deeper areas, salinity may reach up to 9–10 psu below the halocline (Fig. 4), which is usually located at a depth of 60–70 m [12, 72–74].

Water exchange processes in the Gulf of Finland (Fig. 5) are usually described as estuarine circulation [75], where freshwater, which also imports nutrients and pollutants, spreads along the surface, while saline water from the Baltic Proper is forced to enter the Gulf in the bottom layer (Fig. 4) [76, 77]. During persistent westerlies (e.g., in autumn or winter), the reversal of estuarine circulation occurs [38, 75], whereby the Baltic Proper surface water flows into the Gulf, while the deep-layer water flows out, and stratification weakens. Because of this partial collapse of stratification, the near-bottom oxygen conditions may temporarily improve, but at the same time, the deep-layer phosphorus will be mixed upward in the water column, thus likely intensifying eutrophication in summer [38, 78].

Substantial differences in nutrients (mainly in P and N compounds) occur both vertically and horizontally in the Gulf of Finland (Table 1, Fig. 6). The trophic status of the Baltic Sea sub-basins can be best described by the wintertime concentrations of nutrients, such as inorganic phosphorus compounds (dissolved inorganic

Table 1 Typical winter and summertime values of salinity (Sal), P_{tot}, DIP, N_{tot} and DIN (all in μM) in the western and eastern Gulf of Finland (WGOF, EGOF), Neva Bay (NB), and Gulf of Riga (GOR) [61, 79–85]. The WGOF-EGOF-NB borders are at approximately 25–26°E and 29.5°E, respectively

Season	Summer				Winter			
Region	WGOF	EGOF	NB	GOR	WGOF	EGOF	NB	GOR
Sal	4–6	2–4	0–0.5	3–6	4–7	2–5	0–0.5	4–6
P_{tot}	0.3–0.7	0.4–1	0.8–1.6	0.5–0.8	1–1.4	1.5–2	2–3	1–1.8
DIP	0–0.1	0.1–0.2	0.2–0.4	0.1–0.3	0.4–0.7	0.5–0.8	0.6–1.2	0.6–1.1
N_{tot}	18–24	25–35	40–50	20–30	22–28	25–40	50–70	35–50
DIN	0–3	1–5	10–20	2–6	5–8	8–12	20–40	10–20

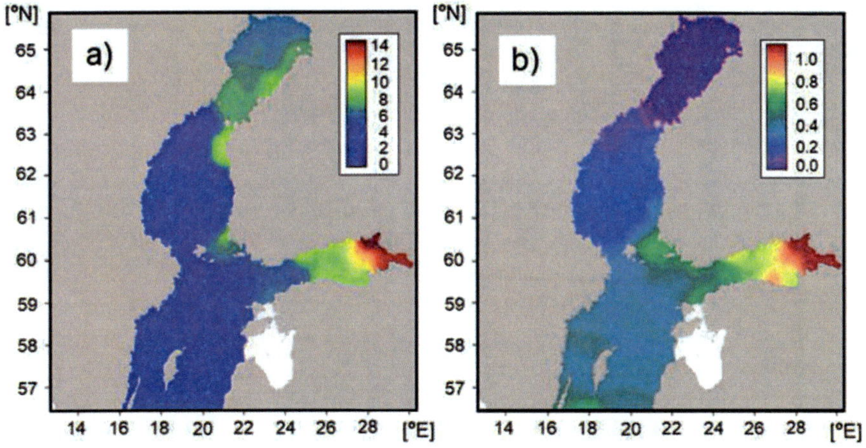

Fig. 6 An example of spatial variations in the $N\text{-}NO_3$ (**a**) and $P\text{-}PO_4$ (**b**) concentrations (μM l^{-1}) in the surface layer of the northern Baltic Sea – Gulf of Finland in January 2008 (modified from [83])

phosphorus (DIP), PO_4) and inorganic nitrogen (dissolved inorganic nitrogen (DIN), NO_3), as well as their total amounts (P_{tot} and N_{tot}, respectively; Table 1). The horizontal variations in the surface layer reveal a negative correlation between salinity and nutrients, representing the nutrient spread from the riverine source (Neva River) towards the open sea (Table 1, Fig. 6). The vertical distribution, especially in the deeper part, is maintained by a combination of the advective inflow of nutrient-rich Baltic deep waters [76, 86], the uptake of nutrients by primary producers during the warm season and the subsequent release of N and P compounds in the deep layers [74, 79]. In the eastern Gulf of Finland and especially in the Neva Estuary, a sharp (up to twofold) decline in concentrations occurs [80–83], whereas a relatively smaller decrease is spread along the full length of the Gulf of Finland (Table 1, Fig. 6).

The Neva Estuary (the easternmost narrow part of the Gulf of Finland) acts as an efficient buffer zone, where nutrients spread, dilute, and transform [79]. In the Neva Bay (the easternmost part of the Neva Estuary east of Kotlin Island and east of the flood-protection dam; Fig. 5), the nutrient concentrations are well over the detection limits even in summer [81]. According to the latest HELCOM pollution load compilation (PLC-6), the flow-normalized total N_{tot} load to the Gulf of Finland has fluctuated between ca. 107,000 and 150,000 ton per year in 1995–2015 with a very slight overall decreasing trend. The annual P_{tot} load has decreased from 10,245 ton in 1995 down to 2,853 ton in 2015 [60]. Nevertheless, the corresponding (expected) decrease in concentrations is much less obvious, partly due to the rerelease from the bottom sediments [78, 79]. For instance, increasing P (and decreasing N) trends have been reported for the Gulf of Finland since the 1970s up to the 2000s [61]. Moreover, the increase in the deep-water P concentration is probably connected to the strengthened salinity stratification in the Gulf, the declining oxygen concentrations, and the various biochemical processes at the sediment–water interface [82]. The decrease in N concentrations in the water is largely due to the decrease in N input [61].

Compared to the Gulf of Finland, the Gulf of Riga is a much shallower and morphometrically better separated sub-basin of the Baltic Sea (Figs. 2c and 3). The Gulf Proper (i.e., excluding the Väinameri sub-basin) has a surface area of 16,330 km^2, a volume of 424 km^3, an average depth of 26 m, and a maximum depth of approximately 62 m [61, 87]. Due to its shallowness, the Gulf of Riga lacks a permanent halocline, whereas the thermocline typically evolves in May, reaches a depth of 25 m in August, and disintegrates in September–October [88]. The average salinity of the Gulf of Riga varies from 1–3 psu near river inflows to 6–7 psu in the Irbe Strait (Figs. 3 and 5). In the central part of the Gulf, the average surface salinity is approximately 5.6 psu and up to ca. 7 psu in a thin near-bottom layer [87, 88]. The annual river discharge varies between 18 and 56 km^3 (on average 34 km^3) [87, 89] delivered by the Daugava (21 km^3 $year^{-1}$), Lielupe (2 km^3 $year^{-1}$), and Pärnu (2 km^3 $year^{-1}$) rivers (Fig. 5).

The Gulf of Riga is ecologically phosphorus-limited and acts as a transition zone between the nutrient-rich estuarine waters and Baltic Proper waters [89, 90]. Considering the sub-basins average salinity of 5.6 psu and the typical salinity in the Baltic Proper outside the Irbe Strait of 7.2 psu, the simple salt conservation law gives idealistic mean water exchange estimates of 113 km^3 for inflows and 147 km^3 for outflows in a year, hence giving a water residence time of 2.9 years, with a considerable seasonal and interannual variability [87].

The annual inputs of nitrogen (N_{tot}) and phosphorus (P_{tot}) are approximately 90,000–140,000 ton and 2,400–3,000 ton, respectively [60, 84]. The nutrient concentrations are set by the occasional inflows of saline and nutrient-rich deep water from the Gotland Basin via the Irbe Strait and by the year-to-year variations in river inflows [91]. Higher concentrations of nutrients are found adjacent to the river mouths (e.g., the Daugava and Pärnu Rivers; Fig. 5). High concentrations of nutrients are often observed in deep water in May–October, accessible to the surface layers through occasional mixing events [92]. Both dissolved inorganic nitrogen and phosphate pools of the upper mixed layer are exhausted by mid-May, except at river mouths. Starting from October–November, a gradual increase starts to occur again. According to the nutrient balance estimates, most of the P input is finally exported via the straits, while as much as 60–80% of N input is retained (deposited) or transformed within the Gulf's ecosystem [90, 91, 93].

3 Fronts in the Baltic Proper

Revolutionary sensor-based oceanic observation techniques, introduced in the 1970s, were also used in Baltic Sea studies [94]. One of the most notable international efforts was the Baltic Open Sea Experiment in 1977 (BOSEX-77) in the southern part of the Eastern Gotland Basin [95]. A series of high-resolution CTD surveys with station-to-station spacing of 10 km or less, which were conducted from 1978 to 1985 in the deep channel south of the Gotland Deep (>20 surveys), revealed a distinct salinity front [96]. The front was generally oriented along the slope between the western shallow and eastern deeper parts of the approximately 70-km-wide deep channel (GF in Fig. 1a). This front had a typical cross-front salinity increase of 0.2 psu per 10 km from the western shallow area to the eastern deeper area. The presence of the front was caused by SSW currents with a speed of 5–10 cm s^{-1}, and these currents carried slightly less saline water from the northern regions along the isobaths. The front occasionally made excursions from the slope area towards the deeper part of the channel, where it was found to extend below the thermocline down to a depth of 30 m [19]. In the periods with opposite current directions, fronts were not observed in the slope and deeper channel areas. The multi-ship two-week Patchiness Experiment (PEX-86), conducted in the spring of 1986 in the same area [30], revealed the same front during repeated surveys. The experiment also showed the transport of less saline water by cyclonic eddies migrating in the deeper part of the area, favouring an earlier establishment of phytoplankton spring blooms [31, 97, 98].

Across the southern part of the Gotland Front (GF in Fig. 1a) extending to the Bornholm Deep (Fig. 4), the north-south salinity gradient may exceed 0.5 psu per 5 km as observed by Kahru et al. [99], who also noted elevated nitrate and chlorophyll concentrations and phytoplankton abundance associated with this front.

An interesting study of frontal dynamics was conducted in the North Baltic Proper in September 1981 [16]. A bow-shaped front (GF, Fig. 1a) was followed for at least 50 km, and it was directed from the southwest to the northeast (Fig. 7). Frontal meanders had radii of 5–7 km, matching the local Rossby deformation radius [100]. The width of the front was typically 1–2 km, which was much less than the distance between CTD casts. The cross-frontal CTD casts from a drifting ship allowed us to reduce the distance between adjacent CTD downcasts to ~100 m. The high-resolution vertical section of temperature (Fig. 7a) exhibited a very sharp front in the surface layer, with a temperature step of 0.7°C per 100 m. North of the warmer water on the surface, there was an extremely low-salinity water mass extending over the colder surface water, with a salinity of 6.3 psu compared to the background of 7.2 psu. Therefore, the hydrostatic stability of the upper 15 m was due to the low salinity of the colder water. Vertical current profiling revealed along-front

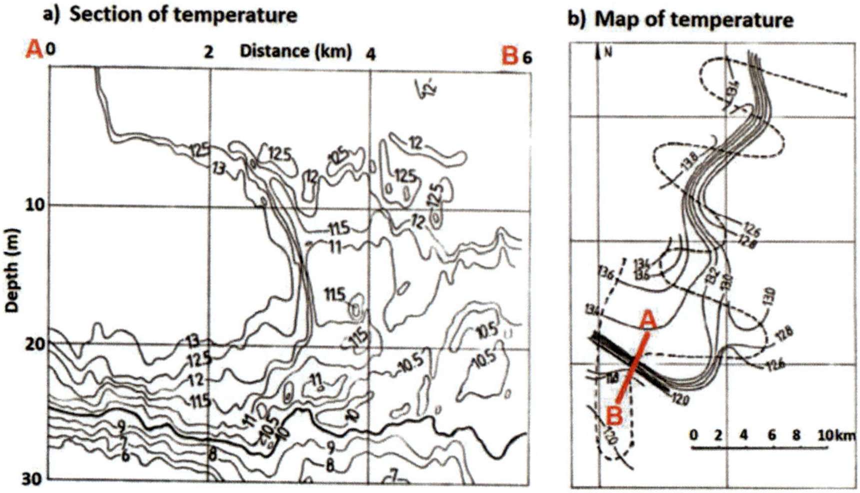

Fig. 7 Vertical cross-frontal distribution of temperature from closely spaced CTD casts (**a**) and sea surface temperature (SST) along the same front from towed thermistor data obtained by the same ship moving along and across the front in a zigzag pattern shown with dashed line (**b**). The CTD section was made along the AB line shown in the SST map (b). The CTD section (a) and SST map (**b**) are not synchronous; also note the different temperature contour intervals. Observations were made on 10–14 September 1981 in the Baltic Proper at approximately 58.9°N, 20.3°E (C1 in Fig. 1b). Modified from [16]

upper layer currents with a speed up to 10 cm s^{-1}. On both sides of the front, convergence towards the front was observed. In Fig. 7a, the density front with a cross-frontal step of 0.5 $g{\cdot}l^{-1}$ was dominated by a salinity change of 0.9 psu; in the northern part, the salinity contrasts were much smaller, and the cross-frontal density step of 0.15 $g{\cdot}l^{-1}$ was mainly caused by a temperature step of 1°C.

Observations from ships moving at full speed were made using towed undulating CTD [17, 20] or by flow-through analysis of water pumped into the ship's lab [21]. The latter technique has been developed into well-known FerryBox systems that work automatically onboard regular ferries. Such long sections are well suited for the identification of surface water masses and their frontal boundaries, although there are some difficulties in separating frontal meanders from smaller patches of water masses [101]. Although there are abundant FerryBox data available at the CMEMS portal, recent publications on fronts were not found.

As briefly mentioned in the introductory chapter, the analysis of satellite imagery can be a powerful tool for detecting thermal fronts over large marine areas. In the Baltic Sea, satellite-based SST images have helped to study basin-wide upwelling distribution patterns since the 1980s [14, 15]. Occasionally, a thermal front (an analogue of the lacustrine thermal bar) can be observed during springtime heating and autumn cooling in the Baltic Sea [102]. In recent years, various image processing techniques, such as edge detection or automated feature extraction, have been introduced worldwide [5, 10, 103–106] and subsequently used in the Baltic Sea

[9, 11, 107]. However, as noted before, SST-based studies alone cannot evaluate the role of salinity variations or hydrodynamics in frontal processes and should be better accompanied by in situ measurements or modelling.

Basin-wide dynamics of salinity fronts can be studied using numerical modelling. The numerical experiments on root-mean-squared gradients and the frontogenesis term, conducted several decades ago [108], showed that deep basins have, on average, low spatial gradients, and in the high-gradient frontal regions, the frontogenesis terms (time rate of change of squared salinity gradient due to the horizontally convergent currents) dominate over frontolysis. The identified frontal areas agreed with the generalizations given in Figs. 1 and 2. It is interesting to note that the freshwater pathway along the Swedish coast in the northern part of the Baltic Proper is sensitive to wind forcing: persistent southerly winds force alongshore northward flow, and the southward moving freshwater detaches temporarily from the coast. Similar effects appear in the Gulf of Finland, where westerly winds may detach the westward moving freshwater from the northern coast towards the axis of the Gulf. The spreading of freshwater has a clear seasonal pattern due to the high river discharges during the spring [109]. The springtime freshwater influx comes to the Baltic Proper from the Bothnian Sea and the Gulf of Finland and spreads along the northern and western coasts of the Baltic Proper from April–May to June but detaches from the coast in July–September and moves to the south along the western flanks of the Eastern Gotland Basin [110], carried by the general circulation of this region [111–113].

Numerical models have been refined in recent years to submesoscale resolution, with a horizontal grid step of 500 m or less [53, 114–117]. Submesoscale fronts, filaments, and eddies are resolved by these models only at sufficiently small grid steps, which allow resolving the flow regimes with high Rossby numbers: $Ro = (\partial v/\partial x - \partial u/\partial y)/f \approx 1$, where u and v are the horizontal velocity components and f is the Coriolis parameter. A combined observational and submesoscale modelling study conducted near the eastern slope of the Eastern Gotland Basin (front SBPF in Fig. 1) in October 2017 [117] revealed a distinct front between the colder deep area and the warmer shallower area (Fig. 8a). The temperature step of approximately 4°C was due to (1) advection of warmer waters from the south along the eastern coast by cyclonic circulation and (2) cooling the surface of deeper areas by upwelling near Gotland Island and advection of colder water from the north. Differential cooling in the well-mixed shallow waters and stratified deeper waters contributed to the formation of the front as well. In terms of density variations, the front was dominated by the temperature change. The front persisted for more than a month, confirmed both by satellite observations and by modelling, and it was resistant to moderate storm events that did not destroy the front. The near-frontal area was rich in submesoscale eddies and filaments (Fig. 8b), whereas positive cyclonic vorticity dominated, highlighting the importance of ageostrophic flow components. This study also considered frontal sharpness, defined as squared buoyancy gradient, and modulation of mixed layer depth and restratification after storm events.

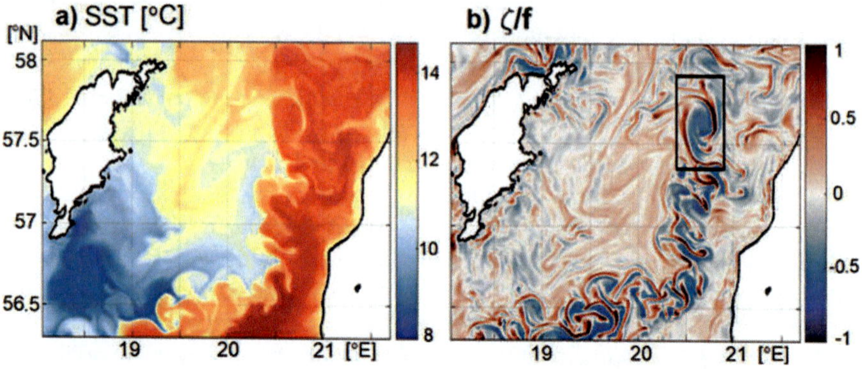

Fig. 8 Frontal quantities in the Eastern Gotland Basin (C2 in Fig. 1b) indicative of the submesoscale regime. (**a**) Sea surface temperature (SST), (**b**) Rossby number, i.e., relative vorticity $\zeta = \partial v/\partial x - \partial u/\partial y$ normalized by the Coriolis frequency f. The simulated SST (verified by satellite observations) was at 14:00 UTC on 19 October 2017. Modified from [117]

4 Estuarine Fronts in the Gulf of Finland

The occurrence of thermohaline fronts and their coupling with biological processes have been studied in the entrance section of the Gulf of Finland since the 1980s [17, 26–28, 108]. According to these studies, three main (quasi-permanent) zones of frontogenesis can be identified. First, there is a quasi-permanent density front at the entrance to the Gulf of Finland (GF1, Fig. 2). However, it is rather dynamic and cannot be considered a true basin-defining front. This front is frequently extended west-to-east [67, 108]. Regarding the depth range of 80–120 m, there is no sill in the Gulf's bottom topography, and the Gulf also widens up towards the Baltic Proper (Fig. 4). The front is formed and maintained by the topographically controlled convergence of southern inflowing saltier waters and northern outflowing freshened waters, but it is modified by wind-induced Ekman drift and upwelling, differential mixing in coastal and offshore areas, and the time-varying general circulation of the Gulf. An example of GF1 dynamics is given in Fig. 9 based on [27, 28]. By the beginning of the study in July 1993 (Fig. 9a), the front extended from southwest to northeast, which is typical for the region, whereas the salinity gradients were approximately 0.5 psu per 3 km of the width of the front. Subsequently (Fig. 9b, c), apparently due to wind forcing, the front formed a meander and was moved southward (by the Ekman drift of westerly winds) towards the shallow coastal area; after 5 days (Fig. 9d), the fronts moved out of the study region. By weakening the winds for the situation in Fig. 9e, the original position of the front was restored, but the width of the front decreased, while higher salinity contrasts were due to the arrival of more saline waters (compared to the first survey) from the Baltic Proper and less saline waters from the Gulf of Finland. The wind forcing of differential advection in the layer affected by the front was found to agree with the rate of change of the potential energy anomaly [28].

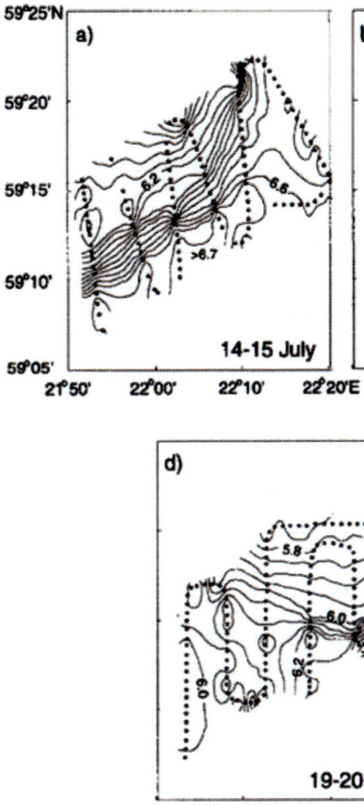

Fig. 9 (**a–e**) Salinity at 5 m at the entrance to the Gulf of Finland in July 1993 (C3 on Fig. 1b). The contour interval is 0.05 psu. Stars indicate positions of towed undulated CTD downcasts. Modified from [28]

In the central part of the Gulf of Finland, fronts occur mainly around the 25°E latitude (GF2, Fig. 2), where the basin narrows, hence different water masses interact more closely. The area is subject to variable circulation regimes from monthly to yearly time scales [67, 69, 118–120]. Rotation in the circulation cells and transport of less and more saline waters may vary (cyclonic or anticyclonic) in both the areas, west and east of 25°E. For example, repeated along-gulf transects in August 1990 [121] showed that a west-east decrease in surface salinity was interrupted by a less saline water mass between 23°E and 24°E, which was followed by a more saline water mass until 25°E. As a result of frontal meandering, three fronts were identified on a fixed transect, with salinity gradients of up to 0.1 psu·km^{-1}. The total cross-frontal salinity step could reach 1.2 psu [22]. An inspection of many unpublished along-basin observational data sets [17, 122] revealed that quasi-permanent fronts occur at the western boundary of the western and eastern sub-basins of the Gulf of Finland (GF1 and GF2 in Fig. 2). From 13 frontal crossings, the average cross-frontal salinity step was 0.5 psu. A much higher number of observations

(77) revealed randomly located fronts, which showed a mean cross-frontal salinity step of 0.3 psu. Both types of fronts had a width of 2–5 km, resulting in an along-front geostrophic current speed of 15–20 cm·s^{-1} [17, 122].

In the easternmost narrow part of the Gulf of Finland, in the 20–30 km wide and 70 km-long area called the Neva Estuary, a more prominent front (GF3, Fig. 2) is found due to the Neva River discharge. Across this front, salinity ranges from 3–4 psu in the west to 1 psu in the east within about 50 km [80, 123]. There are topographically controlled alongshore thermal fronts in the Gulf of Finland due to uneven cooling/warming in the spring/autumn and upwelling-related thermohaline fronts (Fig. 2) that are discussed in the following chapters.

5 Strait Fronts in the Gulf of Riga

The Gulf of Riga has two connections with the Baltic Proper, both containing fronts dividing water masses with different properties (Fig. 2, Table 1). The salinity differences between the Gulf of Riga and Baltic Proper are up to 1–2 psu, while nutrient concentrations may differ by twofold [84, 85], whereas turbidity, yellow substance, and some other characteristics differ even more [93, 124–127]. The main connection with the Baltic Proper is through the Irbe Strait, which has a width of 27 km and a sill depth of 21 m (Fig. 10).

The topographically-steered S-shaped Irbe Strait Front (ISF in Fig. 1a, GR1 in Fig. 2) fluctuates between the extreme inward and outward positions in a self-restoring manner [108, 128–130]. Intensive observations were carried out in 1993–1995 within the international Gulf of Riga Project. Compared to the local Rossby deformation radius (approximately 3 km), the Irbe Strait is wide enough that baroclinic inflows and outflows can occur simultaneously (Fig. 10). The S-shape appearance of the front is caused by the inflow along the southern coast and the

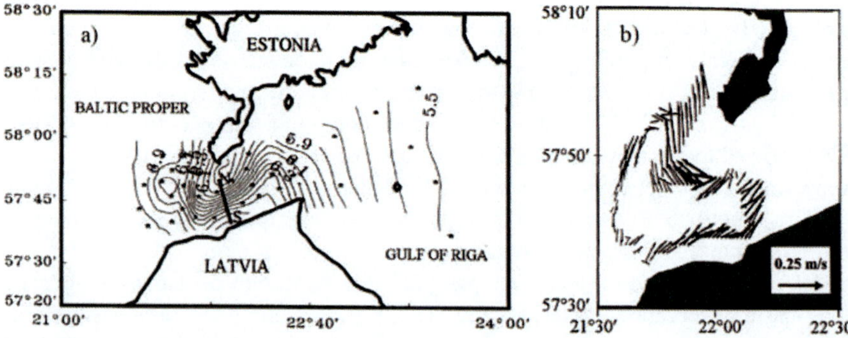

Fig. 10 Map of salinity at 15 m depth in the Irbe Strait area (C4 on Fig. 1b) from 6–8 June 1995 (**a**) and map of current velocity in the 7–9 m depth layer during an ADCP survey on 12 June 1995 (**b**). Modified from [128]

outflow along the northern coast because of the geostrophic adjustment of flows in the wide strait. The baroclinic forcing is due to the density differences between the Gulf of Riga and the Baltic Proper, whereas barotropic forcing is due to the sea level differences between these basins. Direct wind stress drives the water exchange [128]. In response to wind conditions, the front can shift by 80–100 km eastward or westward. Cross-front salinity gradients typically vary from 0.05 to 0.1 psu km^{-1} [128]. The Irbe Strait Front is well reproduced by model studies [69, 70, 130].

The second connection is to the north via the 4 km wide Suur Strait (Fig. 11) and further on, via a small sub-basin of the Väinameri Sea (also West Estonian Archipelago Sea; Figs. 5 and 11), which has an area of 2,243 km^2 and a volume of 10.6 km^3. In the shallow Väinameri Sea with a mean depth of 4.7 m, flow is usually unidirectional (northwards or southwards) at each given moment and horizontal flow separation does not occur. For that reason, in the Väinameri Sea Front (GR2, Fig. 2), separating the Gulf of Riga waters from the Gulf of Finland and Northern Baltic Proper, saltier waters make much longer excursions than near the Irbe Strait Front [132]. The behaviour of the Väinameri Sea Front in autumn 1996 is shown in Fig. 11 [131, 132]. When the wind turned from the northerly to southerly direction, the front moved northward, carrying warmer and less saline waters to the observation site, causing abrupt changes in the time series recorded. By the arrival of the Gulf of Riga waters, nitrate and phosphate concentrations increased considerably, while transparency declined.

All the straits of the Gulf of Riga respond to the external forcing to a great extent as a joint oscillatory system [133]. The Väinameri Sea Front fluctuates northward or southward (Figs. 10 and 11), and the local conditions are frequently overrun by such variations (Fig. 3b) forced by the two large nearby basins [131]. Over longer periods, diurnal or synoptic-scale to-and-fro movements cancel each other out, whereas seasonal variability becomes apparent in cumulative water exchange curves [134]. Northward fluxes usually prevail in autumn and winter, whereas south-westward fluxes usually prevail in summer [93, 134]. According to satellite images and in situ measurements [99, 131, 132], the less saline and more turbid Gulf of Riga water may reach north of the Väinameri Sea beyond the Hari Strait (Fig. 11). However, during long-lasting north-easterly winds, the front can be pushed to the northern part of the Gulf of Riga, and the entire Väinameri Sea is filled with exceptionally cold and saline waters (Fig. 11), which originate from upwelling events in the southern part of the mouth section of the Gulf of Finland (see the next subchapter).

6 Upwelling Fronts

Upwelling is a rather widespread oceanographic phenomenon that involves wind-driven offshore Ekman transport in surface layers and compensating upward motion of denser (in the Baltic Sea, it means colder during summer) and nutrient-rich water towards the sea surface, and this process replaces the surface water. As upwelled

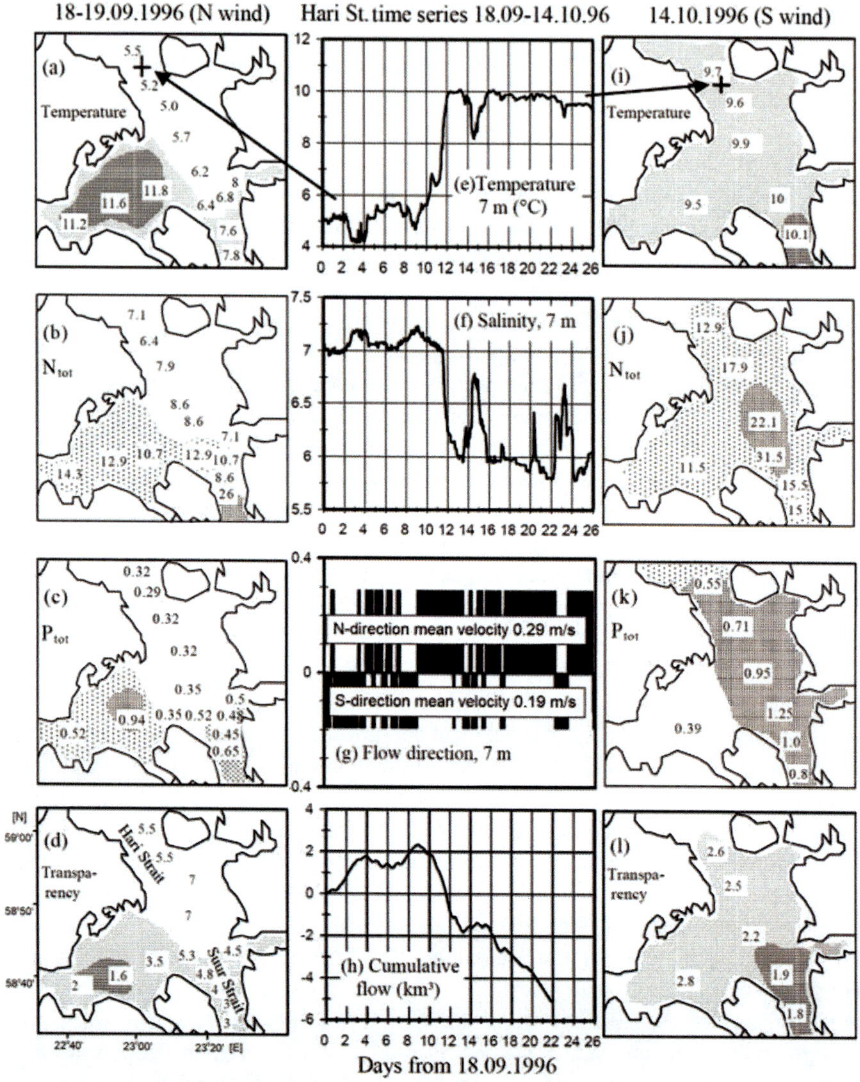

Fig. 11 Frontal excursions and their impact in the Väinameri Sea (C5 on Fig. 1b) in autumn 1996. Time series measured at the Hari Strait (**e–h**) at a location shown by cross (**a**). The maps of temperature, N_{tot}, P_{tot} and transparency are given for the inflow situation from north through the Hari Strait from the Baltic Proper (**a–d**) and from south through the Suur Strait (**d**) from the Gulf of Riga a few weeks later (**i–l**). The arrows (between **e**/**a** and **e**/**i**) connect the corresponding values. Nutrient concentrations in μM. Modified from [131]

water stimulates the growth of phytoplankton due to the nutrient supply, it holds enormous ecological importance in the World Ocean [4–6]. In the Baltic Sea (in the Northern Hemisphere), coastal upwelling is a mesoscale, spatially and temporarily

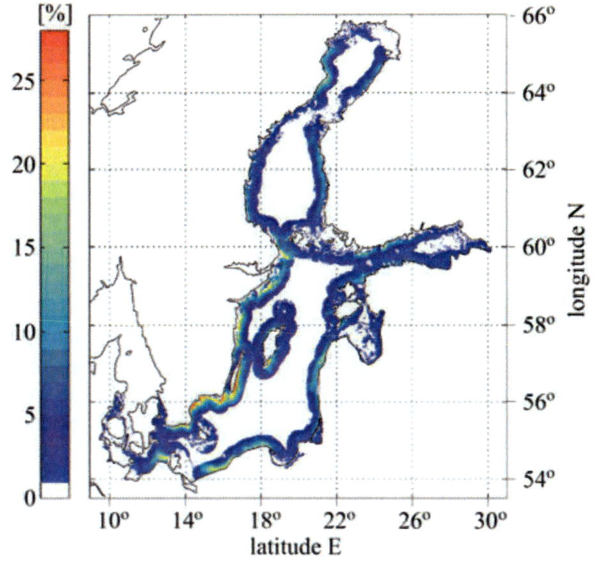

Fig. 12 Upwelling frequencies (%) obtained by automatic detection with a 2°C temperature difference threshold based on SST maps (NOAA/AVHRR satellite data for May–September 1990–2009; modified from [140])

variable process that occurs when wind blows persistently (at least for 2–3 days) parallel to a coastline on its left [12, 22, 135].

Coastal upwelling has been extensively studied in the Baltic Sea since the 1980s [15, 136–140], including in the Gulf of Finland. Due to the complex shape of the Baltic Sea, wind from virtually any direction blows parallel to some section of the coast and hence causes coastal upwelling along that section. The coastal upwelling frequency peaks at 25–40% along the south-eastern coast of Sweden [137, 140]. In the Gulf of Finland, the coastal upwelling frequency is approximately 20% along the northern coast and slightly less along the southern coast [35] because westerlies are prevalent vs. easterlies in the Baltic Sea. Typically, 15% of the Gulf of Finland area is covered by upwelled water at any given time [107]. The cross-shore extension of coastal upwelling is normally up to ca. 25 km [35], excluding filaments. The meandering filaments and vortices (Fig. 13) reveal meso- and submesoscale hydrodynamic activity that occurs in relation to the coastal topography and shape of the basin [107, 135, 139]. Upwelling-related, occasional thermal and thermohaline fronts also manifest in long-term statistical summaries for the Gulf of Finland [69].

The effects associated with upwelling include the rise of pycnoclines and nutriclines [136, 141] and a relatively fast (up to 60 cm s^{-1}) alongshore current, i.e., a coastal jet [39, 139, 142]. In the Baltic Sea, coastal upwelling can create strong physical and chemical fronts, with cross-frontal salinity steps of up to 2–3 psu and temperature steps of up to 10–20°C [107, 143]. When pycnoclines rise or sink, the spatial variations also manifest as temporal variations at a spot (Fig. 14). For instance, temperature variations up to 20°C (including a 24 h decline of 15.7°C) have been recorded near the Letipea Peninsula in the Gulf of Finland [144]. In upwelling-prone coastal regions, most of the variability in water temperature and

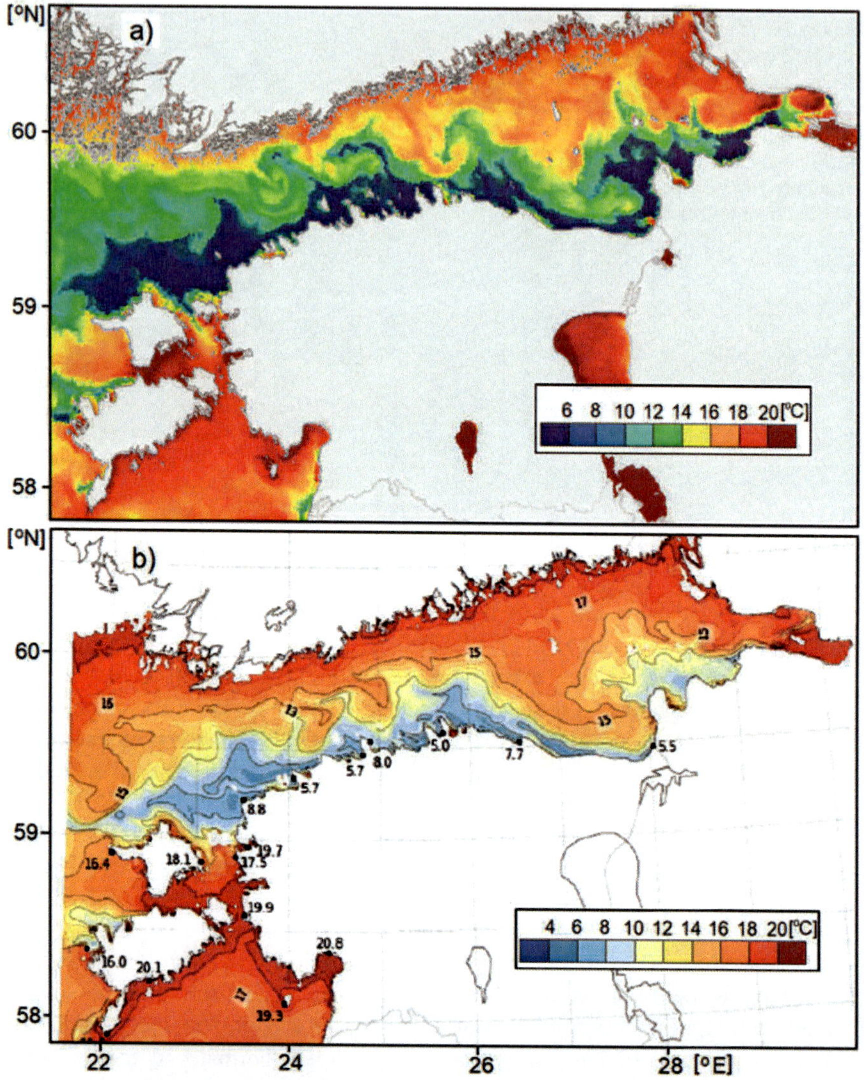

Fig. 13 Upwelling front: a comparison of satellite-based (ESA Sentinel-3 SLSTR) SST product (**a**); modified from TARKKA web map service by Finnish Environment Institute/SYKE (syke.fi.tarkka/en) with HBM-EST model-based SST forecast (**b**) from the Estonian Weather Service (EWS) website (www.ilmateenistus.ee) for 21–22 June 2020 (persistent easterly winds 5–12 m s^{-1}). Water temperature varied between 5 and 21°C across the area. The longitudinal extent of the upwelling-affected area during this rather extreme event was 400–500 km

salinity originates from upwelling-related fluctuations, aside from the smoothly varying seasonal changes [40, 144].

As a rule, a prominent front would manifest in all physical and biochemical properties (e.g., Fig. 13) [71, 138, 145]. The upwelling-related rise of nitrogen- and

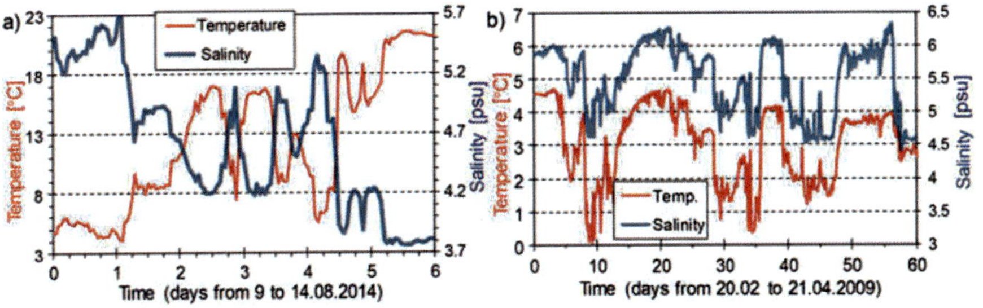

Fig. 14 Upwelling-related negatively correlated variations in water temperature and salinity on the Letipea Peninsula (C6 in Fig. 1b) (10 m depth) in summer (**a**) and positively correlated variations in winter (**b**). Modified from [40]

phosphorus-rich water during the summertime nutrient concentrations minimum may cause cyanobacterial growth and prolong blooms of other algal groups [27, 138]. Due to a steeper bottom slope and greater depth near the southern coast of the Gulf of Finland, the upwelling along this coast produced by easterly winds brings more nutrients to the sea surface than does the upwelling along the northern coast produced by westerly winds of identical strength [141]. Subsequently, nutrient-rich waters gradually spread offshore in the form of filaments, eventually contributing to elevated productivity [138, 145].

Fronts during individual coastal upwelling events can be short-lived, lasting just a few days. However, depending on the combination of wind persistence and bottom topography, upwelling events can be a recurring or quasi-stationary process [40, 72]. Hence, under a changing wind climate, current patterns may also shift, and the typical parameters of upwelling can change [140, 146]. Under a warming climate and more frequent marine heat waves, thermal contrasts during upwelling may first increase in the Gulf of Finland [144] because the climate warming of the Baltic Sea water masses progresses from the sea surface down, lagging in the deeper layers [22, 147].

Wind-driven coastal upwelling is not merely a summertime process. In situ measurements have shown that under upwelling-favourable conditions, "warm" upwelling may occur in winter at the same locations [39, 40]. The impact of SST on such wintertime upwelling has been analysed in the southern Baltic Sea using satellite imagery [148]. The temperature variations were much smaller (e.g., up to 3°C in Fig. 14), whereas the salinity variations were of the same magnitude, than those in summer. During summer, variations in temperature and salinity in upwelling zones are negatively correlated (Fig. 14a), whereas during winter, this correlation is positive (Fig. 14b). Normally, in winter, the surface water is fresher, cooler, and lighter than the deep water below, so that during upwelling the saltier and warmer deep-water outcrops. Its possible ecological impact in the Baltic Sea has not yet been studied. Nevertheless, it was shown that in the Gulf of Finland, the winter upwelling frequency along the North Estonian coast was 21–28% in 2010–2021, and it kept the Estonian coast ice-free longer and the surface water warmer than inshore waters off the Finnish coast [40].

7 River Plume Fronts

Riverine inflow brings freshwater and some endemic terrigenous substances to the sea that normally are not encountered in seawater [149]. Every river plume forms a front whose physics is governed by the buoyant freshened waters intruding into the ambient sea waters. Some of the Baltic Sea river plumes, especially those caused by large rivers such as the Oder, Vistula, and Daugava Rivers, are well studied [150–152]. The Neva River plume was discussed in an earlier subchapter. Plumes also occur near smaller rivers (Fig. 15) such as the Pärnu River (Fig. 5). Such plumes can be detected using a combination of various techniques, including in situ sampling, satellite imagery analysis, and modelling. River plumes can be traced on satellite images of suspended matter, as shown in Fig. 15 for the north-eastern Baltic Sea. River waters usually have higher turbidity than seawater. An exception can be found

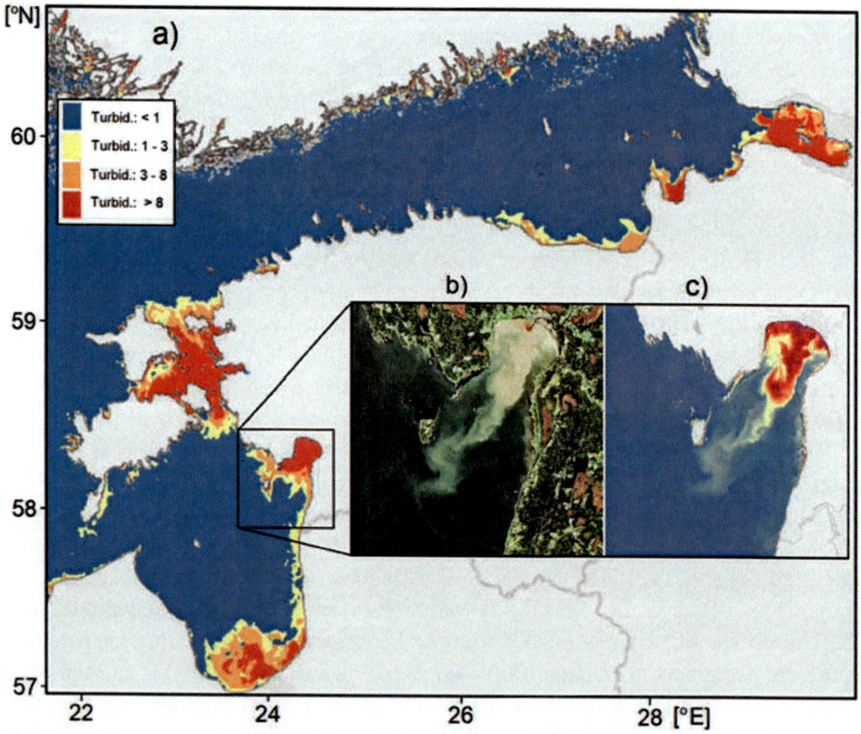

Fig. 15 A composite image (**a**; "Riverine Impact Areas" product by SYKE based on ESA Envisat MERIS data, weeks 12–22, year 2009) showing riverine turbidity plumes as well as shallow sea wind-generated turbidity (e.g., in the Väinameri Sea; Fig. 11). The Pärnu River impact area (C7 on Fig. 1b) according to the true colour image on 26 July 2020 (**b**; ESA-Copernicus Sentinel-2 data) and the corresponding turbidity product by SYKE [149] (**c**). Images were retrieved and modified from the TARKKA website (syke.fi/tarkka/en)

in large shallow areas such as the Väinameri Sea, where wind-generated sediment resuspension can be strong [131].

River plumes usually spread as a coastal buoyant jet that is deflected in the Northern Hemisphere to the right from the river mouth. Under low mixing conditions (in the tideless Baltic Sea, during low wind speeds), freshened waters tend to form a lens-like river bulge [153] that features anticyclonic circulation since the sea level is elevated and the pycnocline is lowered in the centre of the bulge. Plume dynamics are controlled mainly by the river discharge Q, reduced gravity in the plume $g' = g\ \Delta\rho/\rho$ (here, g is the gravity and $\Delta\rho/\rho$ is the relative vertical density change), Coriolis parameter f, and mixing. After the bulge formation, the bulge width increases over time relative to the Rossby radius. The spatial scale of the river bulge can be estimated according to [154, 155] by the formula $L_b = \left(2Qg'/f^3\right)^{\frac{1}{4}}$, which is derived by inserting the geostrophic depth of the buoyant flow $h_b = (2Qf/g')^{\frac{1}{2}}$, induced by the discharge, into the definition of the Rossby radius $L_b = \left(g'h_b/f^2\right)^{\frac{1}{2}}$. In the major Baltic rivers, the mean bulge scale is from 1 to 2 km but may increase by a factor of 2–3 during the spring discharge maximum.

A prominent river bulge was detected off the Daugava River mouth in March 2014 from the remote sensing data of total suspended matter (Fig. 16) [152]. Satellite observations were accompanied by high-resolution modelling and in situ measurements. In the sequence of four maps shown in Fig. 16, the first situation was after weak winds; therefore, small plumes of the Daugava River and two smaller neighbouring rivers are visible. Between the first and the second snapshot, a brief but strong wind burst destroyed the initial plumes. The river discharge increased by two thirds from the initial value. As a result, a distinct nearly circular anticyclonically rotating river bulge was formed. Its diameter increased from 8 to 20 km, whereas the bulge scale increased from 2.7 to 3.1 km. At the peak of the bulge evolution, along-front currents amounted to 50 cm s^{-1}, while off-front velocities were approximately 5 cm s^{-1}. The bulge existed for 10 days; subsequently, it was eliminated by wind mixing when wind speed increased up to 10 m s^{-1}. The wind dependence of freshwater pathways has been studied in a series of numerical experiments [36].

River plumes have been extensively studied in the Pomeranian Bight, which includes the Oder River outflow area [150, 151]. Under high discharge conditions (e.g., during the summer flood of 1997), the plume can be detected up to 70 km offshore and/or 300 km alongshore. Wind direction has an important effect on the plume spreading path: downwelling-favourable winds (westerly winds in the Pomeranian Bight) focus the plume onshore and cause downwind alongshore transport over large distances, enhancing the natural right-hand plume path. Conversely, upwelling-favourable winds transport the plume water offshore.

A river plume also exists off the Vistula River mouth [156]. The largest river of the Baltic Sea – the Neva River – drains into the Neva Bay, which is often considered to be the Neva River Estuary and acts as a buffer zone [81, 82]. The Neman River enters the Curonian Lagoon, forming a plume. A river plume has recently been studied in a small elongated semi-enclosed bay in southwestern Finland

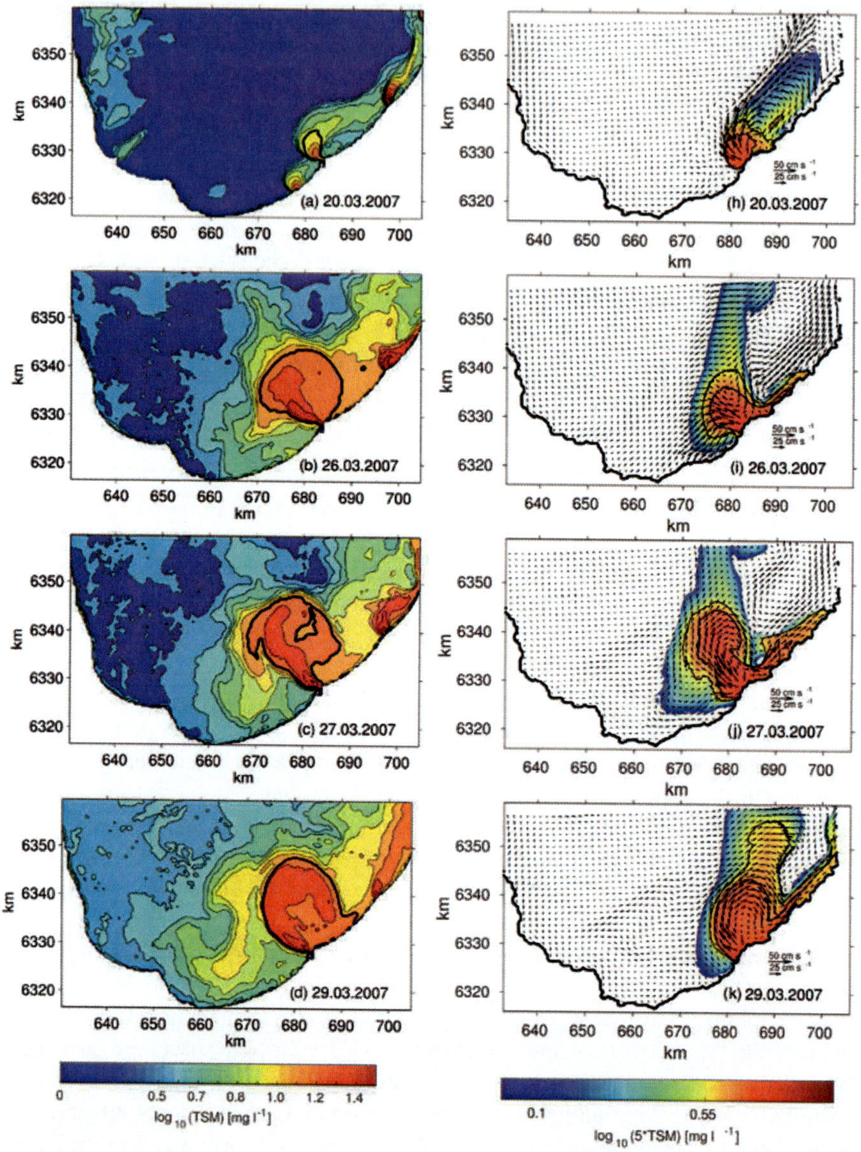

Fig. 16 Growing phase of the river bulge near the Daugava River (C8 on Fig. 1b). Maps of total suspended matter (TSM) in the southern part of the Gulf of Riga from satellite images (left column) and from numerical simulation (right column). Surface current maps from numerical simulation are added. The bold contours on the satellite images show the edge of the Daugava River bulge. Black contours on the numerical model simulation maps represent TSM concentrations of $\log_{10}(\mathrm{TSM}) = -0.15$ and -0.05. The former is used to delineate the Daugava River bulge. The coordinate system is on the UTM-34v projection. Modified from [152]

[157]. Plumes can be formed by cooling water discharges from nuclear power plants [158]. In the western Baltic Sea, near-bottom salinity fronts are formed by intrusions of high-salinity waters from the North Sea [159]. These near-bottom fronts have been simulated by high-resolution modelling [160].

8 Fronts in Satellite Ocean Colour Imagery

Space-borne and airborne instruments are widely used to provide information on optically active substances (OAS) in the sea [44, 161], such as coloured dissolved organic matter (CDOM, also called yellow substance), suspended matter (most frequently, total suspended matter or TSM), and phytoplankton pigments (chlorophyll; Chl-*a*). Optically, Chl-*a* has two absorption bands (ca. 400–440 nm and 640–690 nm); the absorption and scattering of suspended matter depends on the size, shape, and optical properties of the particles, and CDOM absorbs short wavelengths in the UV to visible range [161]. All these substances can form patterns that are clearly visibly in satellite imagery (Fig. 17).

Original sources of the satellite imagery are large governmental agencies (e.g., NASA, NOAA, ESA, JAXA) that operate satellites and sensors (e.g., MODIS, MERIS, SeaWiFS, HYPERION, ALI, OLI, and others). In Europe, the central gateway for ocean-related satellite products is CMEMS. In addition, there are many websites of regional importance from which satellite imagery and derived products can be retrieved. Among them, the TARKKA website of the Finnish Environment Institute (www.syke.fi) can be mentioned. In the Polish web-based SatBałtyk system (satbaltyk.pl/en), the gaps in SST satellite imagery are operationally filled, and a sequence of maps of various parameters is created using a coupled ecosystem model [148].

To estimate concentrations of various substances, many algorithms have been developed to calculate TSM, CDOM, Chl-*a*, and other variables in the Baltic Sea [43, 45, 46, 162–167].

For the Baltic Sea, specific remote sensing algorithms are often required due to the large riverine inflow with substantial anthropogenic contributions. Baltic Sea waters are usually referred to as case II waters, which complicates the development of remote sensing algorithms [44]. However, case II merely refers to waters with more than one optically active substance (phytoplankton pigments, CDOM and inorganic suspended matter). Anthropogenic contributions further complicate the analysis [46, 149, 162, 167]. Usually, the algorithms apply combinations of different spectral bands or channels; they should be validated (or calibrated) either against in situ measurements or modelling [46, 47]. A frequent problem is the nonuniversal nature of such algorithms that may only be valid for certain regions, seasons, or surveys because of spatially and temporally varying combinations of different water masses [149]. Gaps in satellite imagery can be filled using eco-hydrodynamic models or operational data [168]. In recent years, machine-learning procedures have also been incorporated into such studies [169].

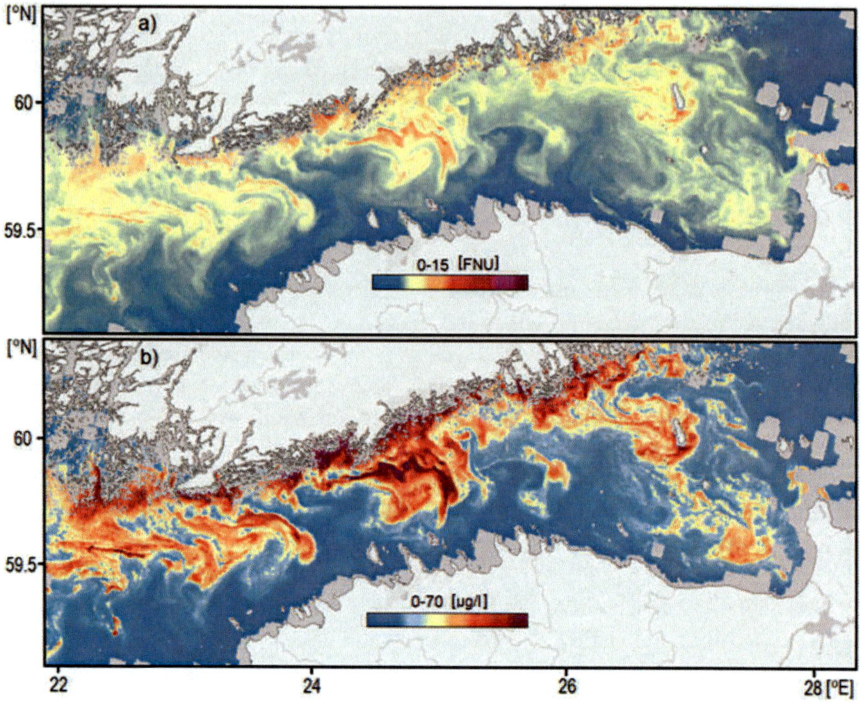

Fig. 17 Turbidity (**a**) and Chl-a (**b**; both modified from the TARKKA website: syke.fi/tarkka/en) in the Gulf of Finland on 8 July 2011. Under persisting fair weather conditions (air temperature of 20–28°C and weak winds), the turbidity in the open part of the sea was predominantly contributed by Chl-*a*. Turbidity and Chl-*a* – ESA Envisat MERIS data, SYKE products [149, 162]

Turbidity-related composites are among the most useful remote sensing products. An example of a composite image called "riverine impact areas" in TARKKA shows local-scale turbidity plumes of various origins (Fig. 15). These include riverine inflow impact areas (e.g., from the Neva, Narva, Pärnu or Daugava Rivers) and hydrodynamically induced shallow sea turbidity in the areas without direct riverine inflow (e.g., the Väinameri Sea; Fig. 15), which varies depending on depth and bottom sediment properties [131]. Short-lived resuspension-generated TSM clouds can also include a limited number of other substances (e.g., P_{tot}) [132], which alongside the TSM settle back to the sea bottom after the event. Sometimes a turbid plume may include many substances (e.g., TSM, CDOM, nutrients, pollutants), but sometimes it is just one substance (e.g., a cloud of a specific pollutant) without other freshwater-borne substances.

Dredging of the sea bottom during the construction of ports or seaways is frequently monitored using satellite imagery [166, 170, 171]. However, quite similar TSM plumes may appear, e.g., due to river inflows or storm-generated resuspension of bottom sediments. In some cases, the total amount of TSM matter resuspended by wave actions is far larger than that brought up by dredging or river input

[171, 172]. Thus, it is important to know the natural variability of suspended matter concentration before estimating the impact of dredging on aquatic environments.

Sipelgas et al. [173] assessed water quality near all important Estonian harbours using a full MERIS archive (2002–2011) and data from vessel traffic, river inflow, and wind conditions. It appeared that while the TSM could originate from port operation, wind-generated resuspension, riverine inflow, or algal blooms, the CDOM plumes were mainly associated with river mouths. While river inflows gradually spread over a larger marine area, forming fronts far from the source (Fig. 5), the concentrations of nonconservative substances (e.g., nutrients) may decrease faster because of settling or biochemical transformation; therefore, such fronts can be found at closer distances from the river mouths (Figs. 6 and 15). Nevertheless, depending on hydrodynamic conditions, the CDOM and turbidity can still spread over relatively large areas from the source [173].

The mapping of harmful algal blooms is presently one of the most important applications of remote sensing in marine optics in the Baltic Sea [72]. Summertime algal blooms (e.g., in toxic cyanobacteria) can be estimated in Chl-*a* products [45, 162, 164] and from turbidity products if applied together with supplementary (e.g., in situ) information. In such images, the time-varying patterns include fibroid filaments (Fig. 17), which are quite different from quasi-permanent thermohaline fronts. They nevertheless testify to the existence of complex, synoptic, and meso-scale hydrodynamic features on the sea surface, such as cyclonic and anticyclonic eddies, gyres and jets [11, 19, 77], which constantly move and change shape.

9 Discussion

Baltic oceanography has a long tradition of research on the features and dynamics of fronts, dating back to the 1960s. In the 1970s, it became clear that stationary geostrophic regimes governed by the Margules formula were disturbed by quasi-geostrophic and ageostrophic motions, where the baroclinic Rossby radius played a key role in scaling the cross-frontal dimensions and mesoscale frontal meanders and eddies. The Baltic Sea fronts in most aspects repeat the regularities of oceanic fronts at smaller scales since the deformation radius is 10–20 times smaller than that in the ocean. The nondimensional characteristics of the Baltic and oceanic fronts have the same range of values. While oceanic fronts are characterized by the subduction of surface waters, this process is not well documented in the Baltic. Regarding the generation mechanisms of fronts, tidal mixing is not relevant in the Baltic Sea. However, variable wind due to transient weather patterns plays a key role in the creation of unsteady circulation and coupled upwelling/downwelling; nonperiodic sea level variations and volume changes in particular sea areas that partly mimic the effect of tides; and wind waves, the variations of which cause variability in mixing. All this variability occurs on the large-scale estuarine salinity gradients and shallow-deep thermal gradients.

There is a good history of mesoscale case studies. The reviewed results on frontal studies include several detailed case studies dedicated to estuarine and strait fronts [16–21, 26–31, 36, 39, 97–102, 128–132, 150–152, 156–158], including river plume fronts. Several recent studies have been devoted to upwelling fronts [35, 40, 107, 135–145], whereas some recent studies have combined the three main high-resolution horizontal data sources – in situ, remote sensing and submesoscale modelling [52, 53, 67, 114–117, 141, 148] – with excellent interpretations. Statistical evaluation of Baltic Sea fronts is mainly based on the data and knowledge obtained 15–5 years ago using remote sensing data. The dynamics of fronts cannot be understood based on SST alone; salinity is needed to evaluate density and pressure gradients. There is vast new observational material suitable for frontal studies (e.g., FerryBox, profiles/gliders/floats, remote sensing, model reanalysis) that is currently available from CMEMS, EMODNET, and other initiatives. Unfortunately, these recent data have not yet been properly analysed regarding the gradient zones and fronts. Therefore, there is an obvious knowledge gap in understanding the role and impact of fronts in the meso- and submesoscale dynamics of the Baltic Sea, covering the physical and biogeochemical processes.

New high-resolution data sets are too large to perform usual visual detection and analysis of different frontal characteristics. Therefore, these data sets should be processed using new robust unsupervised feature detection methods that could allow the identification of full life cycles of elongated sections of fronts, frontal meanders, squirts, and small-scale filaments – from their generation until decay. Such approaches already exist for the analysis of oceanic eddies [174], but the analysis of more complex frontal patterns is still a challenge. Site-specific questions of the analysis could address the formation frequency and lifetime of fronts, the regularities in the formation of filaments and their detachment process, and the length scales, periods, and amplitudes of frontal meanders.

There could be local variations in front definitions based on features such as gradients, edge detection, or water-mass analysis. For example, the Gulf of Riga has a distinct water mass, which differs in salinity from the Baltic Proper surface water mass. In the Gulf of Finland, there is a more gradual transition from fresh to brackish waters, and local definitions could be more favourable for identifying the fronts of more dynamic origins. Oceanic feature detection is of utmost importance for concept building and dependence identification between oceanic subsystems to find intelligible explanations of the system response to climate and environmental forcing. Because the dynamics of oceanic fronts have a crucial impact on the physical, chemical, and biological environments of the sea, further investigation is therefore of vital importance in the current epoch of fast environmental change.

10 Conclusions

In the semi-enclosed Baltic Sea, a heterogeneous water body of complex shape with substantial anthropogenic pressure, a multitude of fronts exist. Holding enormous importance, e.g., to mixing processes, eutrophication, and fisheries, fronts can be viewed as agents of various mass and energy exchange processes in the sea, as well as the products of the same processes. In this article, various types of fronts have been reviewed and their typical locations are summarized (Figs. 1 and 2).

- In the nearly 400 km-long ROFI-type Gulf of Finland, the mean east-west salinity gradient with a sea surface salinity difference of 1–7 psu is split into several fronts, and these fronts are frequently found in three frontal zones located in the western, central, and eastern parts of the Gulf. While the fronts at the entrance and central part are rather variable, the front in the easternmost part features distinct cross-front differences in salinity, nutrients, turbidity, and very likely various polluting substances due to the vicinity of the Neva River, which is largest river that discharges into the Baltic Sea.
- Narrow connections of the Gulf of Riga to the Baltic Proper – the Irbe Strait and the multi-strait Väinameri Sea area – host well-defined fronts that enclose the less saline and more nutrient-rich waters of the Gulf. The fronts are rather persistent, although they exhibit back and forth excursions of several tens of km depending on meteorological conditions.
- Fronts are also evident in the Baltic Proper. Although their salinity and density differences are small compared to those of the strait and river plume fronts, meso- and submesoscale features of frontal dynamics are evident.
- Driven by a divergent Ekman drift, coastal upwelling is a relatively frequent phenomenon in the Baltic Sea. In the Gulf of Finland, it occurs 15–20% of the time, especially under persistent westerly winds along the northern coast and under easterly winds along the southern coast. The corresponding thermohaline fronts are most recognizable on SST images, and their imprints also manifest in long-term statistical summaries. The upwelling-related SST changes may reach up to 20°C, and the effect is more profound along the southern coast and in summer. Winter (warm) upwelling can occur along the Gulf's southern coast and prolong the ice-free conditions there.
- Large rivers such as the Neva, Daugava, Oder, and Vistula Rivers create river plumes that expand during high discharge and low winds, with along-front current speed up to 0.5 m s^{-1}.
- Subregional and smaller water masses may differ from adjacent water masses so sharply that the corresponding fronts and patterns can be traced from satellite imagery (mostly using SST, TSM, CDOM, and Chl-*a* products). Turbidity-related satellite products can indicate riverine impact areas, shallow sea wind-induced resuspension or dredging plumes, and cyanobacterial bloom areas. Differentiation between these pattern sources can be made based on in situ measurements (carried out routinely or under specific operational tasks) or background information obtained, e.g., from modelling.

Acknowledgements The study was supported by the Estonian Research Council grant PUT1439 and institutional funding of the Tallinn University of Technology. We thank the Finnish Environment Institute (SYKE) for the remote sensing images and the Estonian Weather Service (EWS) for hydrometeorological data. We are thankful to three anonymous reviewers and Dr. Kai Myrberg for their valuable suggestions to improve the manuscript.

References

1. Fedorov KN (1986) The physical nature and structure of oceanic fronts. Springer, New York, 333 pp (Originally published in Russian in 1983 as: Fizicheskaya priroda i struktura okeanicheskikh frontov, Gidrometeoizdat, Leningrad, 296 pp). https://www.springer.com/gp/book/9780387964454
2. McWilliams JC (2021) Oceanic frontogenesis. Annu Rev Mar Sci 13:227–253. https://doi.org/10.1146/annurev-marine-032320-120725
3. Cromwell T, Reid Jr JL (1956) A study of oceanic fronts. Tellus 8(1):94–101. https://doi.org/10.1111/j.2153-3490.1956.tb01198.x
4. Ginzburg AI, Kostianoy AG (2020) Oceanography. Fronts and mixing processes. Encyclopedia of Life Support Systems (EOLSS). https://www.eolss.net/Sample-Chapters/C01/E6-18-02-06.pdf
5. Belkin IM, Cornillon PC, Sherman K (2009) Fronts in large marine ecosystems. Prog Oceanogr 81:223–236. https://doi.org/10.1016/j.pocean.2009.04.015
6. Daly K, Smith W (1993) Physical-biological interactions influencing marine plankton production. Annu Rev Ecol Syst 24:555–585. http://www.jstor.org/stable/2097190
7. MacVean MK, Woods JD (1980) Redistribution of scalars during upper ocean frontogenesis: a numerical model. Q J Royal Meteorol Soc 106(448):293–311
8. Roden GI (1977) Oceanic subarctic fronts of the central Pacific: structure of and response to atmospheric forcing. J Phys Oceanogr 7(6):761–778. https://doi.org/10.1175/1520-0485(1977)007%3C0761:OSFOTC%3E2.0.CO;2
9. Belkin IM, Cornillon P (2007) Baltic Sea thermal fronts from satellite data. ICES CM 2007/C:25
10. Cayula J-F, Cornillon P (1992) Edge detection algorithm for SST images. J Atmos Ocean Technol 9(1):67–80. https://doi.org/10.1175/1520-0426(1992)009<0067:EDAFSI>2.0.CO;2
11. Kahru M, Håkansson B, Rud O (1995) Distributions of the sea-surface temperature fronts in the Baltic Sea as derived from satellite imagery. Cont Shelf Res 15(6):663–679. https://doi.org/10.1016/0278-4343(94)E0030-P
12. Leppäranta M, Myrberg K (2009) Physical oceanography of the Baltic Sea. Springer, Berlin, 378 pp. https://doi.org/10.1007/978-3-540-79703-6
13. Horstmann U (1983) Distribution patterns of temperature and water colour in the Baltic Sea as recorded in satellite images: indicators of plankton growth. Berichte Institut für Meereskunde Universität Kiel 106, 147 pp
14. Bychkova IA, Viktorov SV (1987) Use of satellite data for identification and classification of upwelling in the Baltic Sea. Oceanology 27:158–162
15. Gidhagen L (1987) Coastal upwelling in the Baltic Sea – satellite and in situ measurements of sea-surface temperatures indicating coastal upwelling. Estuar Coast Shelf Sci 24(4):449–462. https://doi.org/10.1016/0272-7714(87)90127-2
16. Aitsam A, Pavelson J, Laanemets J, Lilover M-J (1982) Preliminary analysis of front in the Baltic Sea. In: Proceedings of the 13th conference of the Baltic oceanographers. Valtion Painatuskeskus, Helsinki, pp 70–86
17. Pavelson J (1988) Nature and some characteristics of thermohaline fronts in the Baltic proper. In: Proc of the 16th conference of the Baltic oceanographers, Kiel, vol 2, pp 796–805

18. Kahru M, Aitsam A, Elken J (1982) Spatio-temporal dynamics of chlorophyll in the open Baltic Sea. J Plankton Res 4(4):779–790. https://doi.org/10.1093/plankt/4.4.779
19. Kahru M, Elken J, Kotta I, Simm M, Vilbaste K (1984) Plankton distributions and processes across a front in the open Baltic Sea. Mar Ecol Prog Ser 20:101–111. https://doi.org/10.3354/MEPS020101
20. Aitsam A, Hansen HP, Elken J, Kahru M, Laanemets J, Pajuste M, Pavelson J, Talpsepp L (1984) Physical and chemical variability of the Baltic Sea: a joint experiment in the Gotland Basin. Cont Shelf Res 3(3):291–310. https://doi.org/10.1016/0278-4343(84)90013-X
21. Kahru M, Nõmmann S (1990) The phytoplankton spring bloom in the Baltic Sea in 1985, 1986: multitude of spatio-temporal scales. Cont Shelf Res 10(4):329–354. https://doi.org/10.1016/0278-4343(90)90055-Q
22. Alenius P, Myrberg K, Nekrasov A (1998) The physical oceanography of the Gulf of Finland: a review. Boreal Environ Res 3(2):97–125. http://www.borenv.net/BER/archive/pdfs/ber3/ber3-097-125.pdf
23. Omstedt A, Elken J, Lehmann A, Piechura J (2004) Knowledge of the Baltic Sea physics gained during the BALTEX and related programmes. Prog Oceanogr 63:1–28. https://doi.org/10.1016/j.pocean.2004.09.001
24. Omstedt A, Elken J, Lehmann A, Leppäranta M, Meier HEM, Myrberg K, Rutgersson A (2014) Progress in physical oceanography of the Baltic Sea during the 2003–2014 period. Prog Oceanogr 128:139–171. https://doi.org/10.1016/j.pocean.2014.08.010
25. Elken J, Lehmann A, Myrberg K (2015) Recent change – marine circulation and stratification. In: The BACC II Author Team (ed) Second assessment of climate change for the Baltic Sea basin. Regional climate studies. Springer, Cham, pp 131–144. https://doi.org/10.1007/978-3-319-16006-1_7
26. Mälkki P, Talpsepp L (1988) On the joint Soviet-Finnish experiment in the Gulf of Finland in May 1987: the results of hydrophysical measurements. In: Proc 16th CBO, Kiel, vol 2, pp 687–695
27. Kononen K, Kuparinen J, Mäkelä K, Laanemets J, Pavelson J, Nõmmann S (1996) Initiation of cyanobacterial blooms in a frontal region in the entrance to the Gulf of Finland. Limnol Oceanogr 41:98–112. https://doi.org/10.4319/lo.1996.41.1.0098
28. Pavelson J, Laanemets J, Kononen K, Nõmmann S (1996) Quasi-permanent density front at the entrance of the Gulf of Finland: response to wind forcing. Cont Shelf Res 17(3):253–265. https://doi.org/10.1016/S0278-4343(96)00028-3
29. Elken J, Mälkki P, Alenius P, Stipa T (2006) Large halocline variations in the northern Baltic proper and associated meso-and basin-scale processes. Oceanologia 48(S):91–117. http://www.iopan.gda.pl/oceanologia/48Selken.pdf
30. Dybern B, Hansen HP (1989) Baltic Sea patchiness experiment PEX'86. ICES Coop Res Rep 163(1):1–100
31. Kahru M, Leppänen JM, Nõmmann S, Passow U, Postel L, Schulz S (1990) Spatio-temporal mosaic of the phytoplankton spring bloom in the open Baltic Sea in 1986. Mar Ecol Prog Ser 66:301–309. https://www.jstor.org/stable/24844566
32. She J, Meier M, Darecki M, Gorringe P, Huess V, Kõuts T, Reissmann JH, Tuomi L (2020) Baltic Sea operational oceanography-a stimulant for regional earth system research. Front Earth Sci 8:7. https://doi.org/10.3389/feart.2020.00007
33. BACC (2015) Second assessment of climate change for the Baltic Sea basin. Springer, 501 pp
34. Karlson B, Andersson LS, Kaitala S, Kronsell J, Mohlin M, Seppälä J, Wranne AW (2016) A comparison of Ferrybox data vs. monitoring data from research vessels for near surface waters of the Baltic Sea and the Kattegat. J Mar Syst 162:98–111. https://doi.org/10.1016/j.jmarsys.2016.05.002
35. Kikas V, Lips U (2016) Upwelling characteristics in the Gulf of Finland (Baltic Sea) as revealed by Ferrybox measurements in 2007-2013. Ocean Sci 12:843–859. https://doi.org/10.5194/os-12-843-2016

36. Lips U, Kikas V, Liblik T, Lips I (2016) Multi-sensor in situ observations to resolve the sub-mesoscale features in the stratified Gulf of Finland, Baltic Sea. Ocean Sci 12(3):715–732. https://doi.org/10.5194/os-12-715-2016
37. Lilover MJ, Elken J, Suhhova I, Liblik T (2017) Observed flow variability along the thalweg, and on the coastal slopes of the Gulf of Finland, Baltic Sea. Estuar Coast Shelf Sci 195:23–33. https://doi.org/10.1016/j.ecss.2016.11.002
38. Lips U, Laanemets J, Lips I, Liblik T, Suhhova I, Suursaar Ü (2017) Wind-driven residual circulation and related oxygen and nutrient dynamics in the Gulf of Finland (Baltic Sea) in winter. Estuar Coast Shelf Sci 195:4–15. https://doi.org/10.1016/j.ecss.2016.10.006
39. Suursaar Ü (2010) Waves, currents and sea level variations along the Letipea – Sillamäe coastal section of the southern Gulf of Finland. Oceanologia 52(3):391–416. https://www.iopan.pl/oceanologia/523suurs.pdf
40. Suursaar Ü (2021) Winter upwelling in the Gulf of Finland, Baltic Sea. Oceanologia 63 (3):356–369. https://doi.org/10.1016/j.oceano.2021.04.001
41. Walczowski W, Merchel M, Rak D, Wieczorek W, Goszczko I (2020) Argo floats in the southern Baltic Sea. Oceanologia 62(4, Part A):478–488. https://doi.org/10.1016/j.oceano.2020.07.001
42. Siiriä S, Roiha P, Tuomi L, Purokoski T, Haavisto N, Alenius P (2019) Applying area-locked, shallow water Argo floats in Baltic Sea monitoring. J Oper Oceanogr 12(1):58–72. https://doi.org/10.1080/1755876X.2018.1544783
43. Mack L, Attila J, Aylagas E, Beermann A, Borja A, Hering D, Kahlert M, Leese F, Lenz R, Lehtiniemi M, Liess A (2020) A synthesis of marine monitoring methods with the potential to enhance the status assessment of the Baltic Sea. Front Mar Sci 7:823. https://doi.org/10.3389/fmars.2020.552047
44. Arst H (2003) Optical properties and remote sensing of multicomponental water bodies. Springer-Praxis, Chichester, 231 pp
45. Reinart A, Kutser T (2006) Comparison of different satellite sensors in detecting cyanobacterial bloom events in the Baltic Sea. Remote Sens Environ 102(1–2):74–85. https://doi.org/10.1016/j.rse.2006.02.013
46. Woźniak B, Krężel A, Darecki M, Woźniak SB, Majchrowski R, Ostrowska M, Kozłowski Ł, Ficek D, Olszewski J, Dera J (2008) Algorithm for the remote sensing of the Baltic ecosystem (DESAMBEM). Part 1: Mathematical apparatus. Oceanologia 50(4):451–508. https://www.iopan.pl/oceanologia/504desa1.pdf
47. Darecki M, Ficek D, Krężel A, Ostrowska M, Majchrowski R, Woźniak SB, Bradtke K, Dera J, Woźniak B (2008) Algorithm for the remote sensing of the Baltic ecosystem (DESAMBEM). Part 2: empirical validation. Oceanologia 50(4):509–538. https://www.iopan.pl/oceanologia/504desa2.pdf
48. Rikka S, Pleskachevsky A, Jacobsen S, Alari V, Uiboupin R (2018) Meteo-marine parameters from Sentinel-1 SAR imagery: towards near real-time services for the Baltic Sea. Remote Sens 10(5):757. https://doi.org/10.3390/rs10050757
49. Kozlov IE, Krek EV, Kostianoy AG, Dailidienė I (2020) Remote sensing of ice conditions in the southeastern Baltic Sea and in the Curonian lagoon and validation of SAR-based ice thickness products. Remote Sens 12(22):3754. https://doi.org/10.3390/rs12223754
50. Aavaste A, Sipelgas L, Uiboupin R, Uudeberg K (2021) Impact of thermohaline conditions on vertical variability of optical properties in the Gulf of Finland (Baltic Sea): implications for water quality remote sensing. Front Mar Sci 8:537. https://doi.org/10.3389/fmars.2021.674065
51. Mälkki P, Tamsalu R (1985) Physical features of the Baltic Sea. Finnish Mar Res 252:110 pp
52. Gräwe U, Holtermann P, Klingbeil K, Burchard H (2015) Advantages of vertically adaptive coordinates in numerical models of stratified shelf seas. Ocean Model 92:56–68. https://doi.org/10.1016/j.ocemod.2015.05.008

53. Väli G, Zhurbas V, Lips U, Laanemets J (2017) Submesoscale structures related to upwelling events in the Gulf of Finland, Baltic Sea (numerical experiments). J Mar Syst 171:31–42. https://doi.org/10.1016/j.jmarsys.2016.06.010
54. Björkqvist JV, Lukas I, Alari V, van Vledder GP, Hulst S, Pettersson H, Behrens A, Männik A (2018) Comparing a 41-year model hindcast with decades of wave measurements from the Baltic Sea. Ocean Eng 152:57–71. https://doi.org/10.1016/j.oceaneng.2018.01.048
55. Burchard H, Bolding K, Feistel R, Gräwe U, Klingbeil K, MacCready P, Mohrholz V, Umlauf L, van der Lee EM (2018) The Knudsen theorem and the total exchange flow analysis framework applied to the Baltic Sea. Prog Oceanogr 165:268–286. https://doi.org/10.1016/j.pocean.2018.04.004
56. Meier HE, Edman MK, Eilola KJ, Placke M, Neumann T, Andersson HC, Brunnabend SE, Dieterich C, Frauen C, Friedland R, Gröger M (2018) Assessment of eutrophication abatement scenarios for the Baltic Sea by multi-model ensemble simulations. Front Mar Sci 5:440. https://doi.org/10.3389/fmars.2018.00440
57. Hordoir R, Axell L, Höglund A, Dieterich C, Fransner F, Gröger M, Liu Y, Pemberton P, Schimanke S, Andersson H, Ljungemyr P (2019) Nemo-Nordic 1.0: a NEMO-Based Ocean model for the Baltic and North seas–research and operational applications. Geosci Model Dev 12(1):363–386. https://doi.org/10.5194/gmd-12-363-2019
58. Voipio A (1981) The Baltic Sea. Elsevier Oceanography Series No 30. Elsevier Science, 418 pp
59. Jakobsson M, Stranne C, O'Regan M, Greenwood SL, Gustafsson B, Humborg C, Weidner E (2019) Bathymetric properties of the Baltic Sea. Ocean Sci 15(4):905–924. https://doi.org/10.5194/os-15-905-2019
60. HELCOM (2018) The sixth pollution load compilation (PLC-6). Baltic Marine Environment Protection Commission – Helsinki Commission. https://helcom.fi/media/publications/PLC-6-Executive-Summary.pdf
61. HELCOM (2009) Eutrophication in the Baltic Sea – An integrated thematic assessment of the effects of nutrient enrichment and eutrophication in the Baltic Sea region. In: Baltic Sea environment proceedings 115B, Helsinki, 148 pp. https://helcom.fi/wp-content/uploads/2019/08/BSEP115B.pdf
62. Bock KH (1971) Monatskarten des Salzgehaltes der Ostsee dargestellt für verschiedene Tiefenhorizonte. Dt. hydrogr Z, Ergänzungsheft Reihe B 12, 147 pp
63. Furman E, Pihlajamäki, M, Välipakka P, Myrberg K (2014) The Baltic Sea. Environment and ecology. https://www.syke.fi/en-US/Publications/Brochures/The_Baltic_Sea_Environment_and_Ecology(29197)
64. Kont A, Endjärv E, Jaagus J, Lode E, Orviku K, Ratas U, Rivis R, Suursaar Ü, Tõnisson H (2007) Impact of climate change on Estonian coastal and inland wetlands — a summary with new results. Boreal Environ Res 12(6):653–671. http://www.borenv.net/BER/archive/pdfs/ber12/ber12-653.pdf
65. Omstedt A (1990) Modelling the Baltic Sea as thirteen sub-basins with vertical resolution. Tellus A 4:286–301
66. Gustafsson E, Savchuk OP, Gustafsson BG, Müller-Karulis B (2017) Key processes in the coupled carbon, nitrogen, and phosphorus cycling of the Baltic Sea. Biogeochemistry 134 (3):301–317. https://doi.org/10.1007/s10533-017-0361-6
67. Maljutenko I, Raudsepp U (2019) Long-term mean, interannual and seasonal circulation in the Gulf of Finland – the wide salt wedge estuary or gulf type ROFI. J Mar Syst 195:1–19. https://doi.org/10.1016/j.jmarsys.2019.03.004
68. Lagemaa P (2012) Operational forecasting in Estonian marine waters. PhD thesis B128, TUT Press, 130 pp
69. Elken J, Zujev M, She J, Lagemaa P (2019) Reconstruction of large-scale sea surface temperature and salinity fields using sub-regional EOF patterns from models. Front Earth Sci 7:232. https://doi.org/10.3389/feart.2019.00232

70. Zujev M, Elken J, Lagemaa P (2021) Data assimilation of sea surface temperature and salinity using basin-scale reconstruction from empirical orthogonal functions: a feasibility study in the northeastern Baltic Sea. Ocean Sci 17(1):91–109. https://doi.org/10.5194/os-17-91-2021
71. Haapala J, Alenius P (1994) Temperature and salinity statistics for the northern Baltic Sea 1961–1990. Finnish Mar Res 262:51–121
72. Soomere T, Myrberg K, Leppäranta M, Nekrasov A (2008) The progress in knowledge of physical oceanography of the Gulf of Finland: a review for 1997–2007. Oceanologia 50 (3):287–362. https://www.iopan.pl/oceanologia/503myrbe.pdf
73. Liblik T, Lips U (2019) Stratification has strengthened in the Baltic Sea – an analysis of 35 years of observational data. Front Earth Sci 7:174. https://doi.org/10.3389/feart.2019.00174
74. Pitkänen H, Lehtoranta J, Peltonen H (2008) The Gulf of Finland. In: Schiewer U (ed) Ecology of Baltic coastal waters. Ecological studies (analysis and synthesis), vol 197. Springer, Berlin, pp 285–308. https://doi.org/10.1007/978-3-540-73524-3_13
75. Elken J, Raudsepp U, Lips U (2003) On the estuarine transport reversal in deep layers of the Gulf of Finland. J Sea Res 49:267–274. http://doi-org-443.webvpn.fjmu.edu.cn/10.1016/S1385-1101(03)00018-2
76. Meier HEM, Feistel R, Piechura J et al (2006) Ventilation of the Baltic Sea deep water: a brief review of present knowledge from observations and models. Oceanologia 48(S):133–164. https://www.iopan.pl/oceanologia/48Smeier.pdf
77. Westerlund A, Tuomi L, Alenius P, Miettunen E, Vankevich RE (2018) Attributing mean circulation patterns to physical phenomena in the Gulf of Finland. Oceanologia 60:16–31. https://doi.org/10.1016/j.oceano.2017.05.003
78. Andersen JG, Carstensen J, Conley DJ, Dromph K, Fleming-Lehtinen V, Gustafsson BG, Josefson AB, Norkko A, Villnäs A, Murray C (2017) Long-term temporal and spatial trends in eutrophication status of the Baltic Sea. Biol Rev 92:135–149. https://doi.org/10.1111/brv.12221
79. Pitkänen H, Lehtoranta J, Räike A (2001) Internal nutrient fluxes counteract decreases in external load: the case of the estuarial eastern Gulf of Finland, Baltic Sea. Ambio 30:195–201
80. Telesh IV, Golubkov SM, Alimov AF (2008) The Neva estuary ecosystem. In: Schiewer U (ed) Ecology of Baltic coastal waters. Ecological studies (analysis and synthesis), vol 197. Springer, Berlin, pp 259–284. https://doi.org/10.1007/978-3-540-73524-3_12
81. Ylöstalo P, Seppälä J, Kaitala S, Maunula P, Simis S (2016) Loadings of dissolved organic matter and nutrients from the Neva River into the Gulf of Finland – biogeochemical composition and spatial distribution within the salinity gradient. Mar Chem 186:58–71. https://doi.org/10.1016/j.marchem.2016.07.004
82. Pitkänen H, Tamminen T, Kangas P, Huttula T, Kivi K, Kuosa H, Sarkkula J, Eloheimo K, Kauppila P, Skakalsky B (1993) Late summer trophic conditions in the north-eastern Gulf of Finland and the river Neva estuary, Baltic Sea. Estuar Coast Shelf Sci 37(5):453–474. https://doi.org/10.1006/ecss.1993.1067
83. Fleming-Lehtinen V, Kaitala S, Haahti H, Maunula P (2008) Temporal and spatial variation of dissolved surface nutrients in the Gulf of Finland, The Northern Baltic Proper and the Bornholm Sea from 1951 to 2008. HELCOM Indicator Fact Sheets 2008. http://archive.iwlearn.net/helcom.fi/environment2/ifs/ifs2008/en_GB/DissolvedNutrients/index.html
84. Yurkovskis A (2004) Long-term land-based and internal forcing of the nutrient state of the Gulf of Riga (Baltic Sea). J Mar Syst 50(3–4):181–197. https://doi.org/10.1016/j.jmarsys.2004.01.004
85. Lainela S, Herkül K, Leito I, Jaanus A, Suursaar Ü (2020) Contemporary trends in hydrophysical and -chemical parameters in the NE Baltic Sea. Est J Earth Sci 69:91–108. https://doi.org/10.3176/earth.2020.06
86. Meier HEM (2005) Modeling the age of Baltic Seawater masses: Quantification and steady state sensitivity experiments. J Geophys Res 110:C02006. https://doi.org/10.1029/2004JC002607

87. Bērziņš V, Bethers U, Sennikovs J (1994) Gulf of Riga: bathymetric, hydrological and meteorological databases, and calculation of the water exchange. Proc Latv Acad Sci B7 (8):107–117
88. Raudsepp U (2001) Interannual and seasonal temperature and salinity variations in the Gulf of Riga and corresponding saline water inflow from the Baltic Proper. Nord Hydrol 32:135–160
89. Ojaveer E (ed) (1995) Ecosystem of the Gulf of Riga between 1920 and 1990. Estonian Academy Publishers, Tallinn, 277 pp
90. Yurkovskis A, Wulff F, Rahm L, Andruzaitis A, Rodriguez-Medina M (1993) A nutrient budget of the Gulf of Riga: Baltic Sea. Estuar Coast Shelf Sci 37:113–127. https://doi.org/10.1006/ecss.1993.1046
91. Mägi L, Lips U (1998) Budget components of water, salt, and nutrients in the Gulf of Riga in 1993–95. Proc Est Acad Sci Biol Ecol 47:14–31
92. Kotta J, Lauringson V, Martin G, Simm M, Kotta I, Herkül K, Ojaveer H (2008) Gulf of Riga and Pärnu Bay. In: Schiewer U (ed) Ecology of Baltic coastal waters. Ecological studies (analysis and synthesis), vol 197. Springer, Berlin, pp 217–243. https://doi.org/10.1007/978-3-540-73524-3_10
93. Suursaar Ü, Kullas T (2016) Water flows, nutrient exchange and excursions of fronts in the straits between the Baltic sub-basins. In: IEEE international geoscience and remote sensing symposium (IGARSS 2016), pp 4595–4598. https://doi.org/10.1109/IGARSS.2016.7730199
94. Aitsam AM (1981) Hydrodynamics as a limiting factor in the development of the Baltic Sea ecosystem. Elsevier Oceanogr Ser 32:165–189
95. Krauss W (1981) The erosion of a thermocline. J Phys Oceanogr 11(4):415–433. https://doi.org/10.1175/1520-0485(1981)011%3C0415:TEOAT%3E2.0.CO;2
96. Elken J, Pavelson J, Talpsepp L (1987) Dynamics and water mass distribution at the southern Gotland Basin (polygon studies). In: Proc 15th CBO, Copenhagen, pp 136–144
97. Elken J, Talpsepp L, Kõuts T, Pajuste M (1994) The role of mesoscale eddies and saline stratification in the generation of spring bloom heterogeneity in the southeastern Gotland Basin: an example from PEX'86. Coop Res Rep Int Counc Explor Sea 201:40–48. https://doi.org/10.17895/ices.pub.7981
98. Fennel K (2001) The generation of phytoplankton patchiness by mesoscale current patterns. Ocean Dyn 52(2):58–70. https://doi.org/10.1007/s10236-001-0007-y
99. Kahru M, Nõmmann S, Simm M, Vilbaste K (1986) Plankton distributions and processes in the Baltic boundary zones. In: Nihoul JCJ (ed) Marine interfaces ecohydrodynamics. Elsevier oceanography series, vol 42, pp 273–293. https://doi.org/10.1016/S0422-9894(08)71049-3
100. Fennel W, Seifert T, Kayser B (1991) Rossby radii and phase speeds in the Baltic Sea. Cont Shelf Res 11(1):23–36. https://doi.org/10.1016/0278-4343(91)90032-2
101. Moisander PH, Rantajärvi E, Huttunen M, Kononen K (1997) Phytoplankton community in relation to salinity fronts at the entrance to the Gulf of Finland, Baltic Sea. Ophelia 46 (3):187–203. https://doi.org/10.1080/00785326.1997.10432878
102. Demchenko N, Chubarenko I, Kaitala S (2011) The development of seasonal structural fronts in the Baltic Sea after winters of varying severity. Clim Res 48:73–84. https://doi.org/10.3354/cr01032
103. Cornillon P, Gilman C, Stramma L, Brown O, Evans R, Brown J (1987) Processing and analysis of large volumes of satellite-derived thermal infrared data. J Geophys Res 92:12993–13002. http://po.gso.uri.edu/poweb/PCC/pcornillon/1987-Cornillonb.pdf
104. Holyer RJ, Peckinpaugh SH (1989) Edge detection applied to satellite imagery of the oceans. IEEE Trans Geosci Remote Sens 21:46–56. https://doi.org/10.1109/36.20274
105. Simpson JJ (1990) On the accurate detection and enhancement of oceanic features observed in satellite data. Remote Sens Environ 33(1):17–33. https://doi.org/10.1016/0034-4257(90)90052-N
106. Belkin IM, O'Reilly JE (2009) An algorithm for oceanic front detection in chlorophyll and SST satellite imagery. J Mar Syst 78(3):319–326. https://doi.org/10.1016/j.jmarsys.2008.11.018

107. Uiboupin R, Laanemets J (2009) Upwelling characteristics derived from satellite sea surface temperature data in the Gulf of Finland, Baltic Sea. Boreal Environ Res 14(2):297–304. http://www.borenv.net/BER/archive/pdfs/ber14/ber14-297.pdf
108. Elken J (1994) Numerical study of fronts between the Baltic sub-basins. In: Proceedings of the 19th conference of the Baltic oceanographers, Sopot, Poland, vol 1, pp 438–446
109. Eilola K, Stigebrandt A (1998) Spreading of juvenile freshwater in the Baltic proper. J Geophys Res Oceans 103(C12):27795–27807. https://doi.org/10.1029/98JC02369
110. Hordoir R, Meier HM (2010) Freshwater fluxes in the Baltic Sea: a model study. J Geophys Res Oceans 115(C8). https://doi.org/10.1029/2009JC005604
111. Lehmann A, Hinrichsen HH (2000) On the wind driven and thermohaline circulation of the Baltic Sea. Phys Chem Earth B Hydrol Oceans Atmos 25(2):183–189. https://doi.org/10.1016/S1464-1909(99)00140-9
112. Meier HEM (2007) Modeling the pathways and ages of inflowing salt-and freshwater in the Baltic Sea. Estuar Coast Shelf Sci 74(4):610–627. https://doi.org/10.1016/j.ecss.2007.05.019
113. Elken J, Matthäus W (2008) Baltic Sea oceanography. In: The BACC Author Team (ed) Assessment of climate change for the Baltic Sea basin. Springer, Berlin, pp 379–386. https://link.springer.com/content/pdf/bbm%3A978-3-540-72786-6%2F1.pdf
114. Vankevich RE, Sofina EV, Eremina TE, Ryabchenko VA, Molchanov MS, Isaev AV (2016) Effects of lateral processes on the seasonal water stratification of the Gulf of Finland: 3-D NEMO-based model study. Ocean Sci 12(4):987–1001. https://doi.org/10.5194/os-12-987-2016
115. Zhurbas V, Väli G, Kuzmina N (2019) Rotation of floating particles in submesoscale cyclonic and anticyclonic eddies: a model study for the southeastern Baltic Sea. Ocean Sci 15 (6):1691–1705. https://doi.org/10.5194/os-15-1691-2019
116. Onken R, Baschek B, Angel-Benavides IM (2020) Very high-resolution modelling of submesoscale turbulent patterns and processes in the Baltic Sea. Ocean Sci 16(3):657–684. https://doi.org/10.5194/os-16-657-2020
117. Chrysagi E, Umlauf L, Holtermann P, Klingbeil K, Burchard H (2021) High-resolution simulations of submesoscale processes in the Baltic Sea: the role of storm events. J Geophys Res Oceans 126(3):e2020JC016411. https://doi.org/10.1029/2020JC016411
118. Andrejev O, Myrberg K, Lundberg PA (2004) Age and renewal time of water masses in a semi-enclosed basin - application to the Gulf of Finland. Tellus A 56(5):548–558. https://doi.org/10.1111/j.1600-0870.2004.00067.x
119. Elken J, Nõmm M, Lagemaa P (2011) Circulation patterns in the Gulf of Finland derived from the EOF analysis of model results. Boreal Env Res 16A:84–102
120. Soomere T, Delpeche N, Viikmäe B, Quak E, Meier HEM, Döös K (2011) Patterns of current-induced transport in the surface layer of the Gulf of Finland. Boreal Environ Res 16A:49–63. http://www.borenv.net/BER/archive/pdfs/ber16/ber16A-049.pdf
121. Kononen K, Nõmmann S (1992) Spatio-temporal dynamics of the cyanobacterial blooms in the Gulf of Finland, Baltic Sea. In: Carpenter EJ, Capone DG, Rueter JG (eds) Marine pelagic cyanobacteria: trichodesmium and other diazotrophs. NATO ASI series (C: mathematical and physical sciences), vol 362. Springer, Dordrecht, pp 95–113. https://doi.org/10.1007/978-94-015-7977-3_7
122. Pavelson J (2005) Mesoscale physical processes and the related impact on the summer nutrient fields and phytoplankton blooms in the western Gulf of Finland. PhD Thesis 39, TUT Press, 216 pp
123. Ryabchuk D, Zhamoida V, Orlova M, Sergeev A, Bublichenko J, Bublichenko A, Sukhacheva L (2017) Neva Bay: a technogenic lagoon of the eastern gulf of Finland (Baltic Sea). In: Kosyan R (ed) The diversity of Russian estuaries and lagoons exposed to human influence. Estuaries of the world. Springer, Cham, pp 191–221. https://doi.org/10.1007/978-3-319-43392-9_7
124. Kaukver K (ed) (2015) Estonian environmental monitoring 2013. Estonian Environmental Agency, Tallinn, 193 pp

125. Simis SGH, Ylöstalo P, Kallio KY, Spilling K, Kutser T (2017) Contrasting seasonality in optical-biogeochemical properties of the Baltic Sea. PLoS One 12(4):e0173357. https://doi.org/10.1371/journal.pone.0173357
126. Kratzer S, Moore G (2018) Inherent optical properties of the Baltic Sea in comparison to other seas and oceans. Remote Sens 10:418. https://doi.org/10.3390/rs10030418
127. Kyryliuk D, Kratzer S (2019) Summer distribution of total suspended matter across the Baltic Sea. Front Mar Sci 5:504. https://doi.org/10.3389/fmars.2018.00504
128. Lilover M-J, Lips U, Laanearu J, Liljebladh B (1998) Flow regime in the Irbe Strait. Aquat Sci 60(3):253–265. https://doi.org/10.1007/s000270050040
129. Lips U, Lilover MJ, Raudsepp U, Talpsepp L (1995) Water renewal processes and related hydrographic structures in the Gulf of Riga. EMI Rep Ser 1:1–34
130. Raudsepp U, Elken J (1999) Application of the Bryan-Cox-Type Ocean Model to reproduce synoptic and mesoscale variability of the Irbe Strait salinity front. Dtsch Hydrogr Z 51:477–488
131. Suursaar Ü, Kullas T, Otsmann M (2001) The influence of currents and waves on ecological conditions of the Väinameri. Proc Est Acad Sci Biol Ecol 50:231–247
132. Astok V, Otsmann M, Suursaar Ü (1999) Water exchange as the main physical process in semi-enclosed marine systems: the Gulf of Riga case. Hydrobiologia 393:11–18. https://link.springer.com/article/10.1023/A:1003517110726
133. Otsmann M, Suursaar Ü, Kullas T (2001) The oscillatory nature of the flows in the system of straits and small semi-enclosed basins of the Baltic Sea. Cont Shelf Res 21(15):1577–1603. https://doi.org/10.1016/S0278-4343(01)00002-4
134. Suursaar Ü, Kullas T, Aps R (2012) Currents and waves in the northern Gulf of Riga: measurement and long-term hindcast. Oceanologia 54(3):421–447. https://doi.org/10.5697/oc.54-3.421
135. Delpeche-Ellmann N, Soomere T, Kudryavtseva N (2018) The role of nearshore slope on cross-shore surface transport during a coastal upwelling event in Gulf of Finland, Baltic Sea. Estuar Coast Shelf Sci 209:123–135. https://doi.org/10.1016/j.ecss.2018.03.018
136. Haapala J (1994) Upwelling and its influence on nutrient concentration in the coastal area of the Hanko Peninsula, entrance of the Gulf of Finland. Estuar Coast Shelf Sci 38(5):507–521. https://doi.org/10.1006/ecss.1994.1035
137. Myrberg K, Andrejev O (2003) Main upwelling regions in the Baltic Sea – a statistical analysis based on three-dimentional modelling. Boreal Environ Res 8(2):97–112. http://www.borenv.net/BER/archive/pdfs/ber8/ber8-097.pdf
138. Vahtera E, Laanemets J, Pavelson J, Huttunen M, Kononen K (2005) Effect of upwelling on the pelagic environment and bloom-forming cyanobacteria in the western Gulf of Finland, Baltic Sea. J Mar Syst 58(1–2):67–82. https://doi.org/10.1016/j.jmarsys.2005.07.001
139. Suursaar Ü, Aps R (2007) Spatio-temporal variations in hydro-physical and -chemical parameters during a major upwelling event off the southern coast of the Gulf of Finland in summer 2006. Oceanologia 49(2):209–228. https://www.iopan.pl/oceanologia/492suurs.pdf
140. Lehmann A, Myrberg K, Höflich K (2012) A statistical approach to coastal upwelling in the Baltic Sea based on the analysis of satellite data for 1990–2009. Oceanologia 54:369–393. https://doi.org/10.5697/oc.54-3.369
141. Laanemets J, Zhurbas V, Elken J, Vahtera E (2009) Dependence of upwelling mediated nutrient transport on wind forcing, bottom topography and stratification in the Gulf of Finland: model experiments. Boreal Environ Res 14(1):213–225. http://www.borenv.net/BER/archive/pdfs/ber14/ber14-213.pdf
142. Talpsepp L (2008) On the influence of the sequence of coastal upwellings and downwellings on the surface water salinity in the Gulf of Finland. Est J Eng 14:29–41
143. Krezel A, Ostrowski M, Szymelfenig M (2005) Sea surface temperature distribution during upwelling along the Polish Baltic coast. Oceanologia 47(4):415–432. https://www.iopan.pl/oceanologia/474krez1.pdf

144. Suursaar Ü (2020) Combined impact of summer heat waves and coastal upwelling in the Baltic Sea. Oceanologia 62(4):511–524. https://doi.org/10.1016/j.oceano.2020.08.003
145. Lips I, Lips U, Liblik T (2009) Consequences of coastal upwelling events on physical and chemical patterns in the central Gulf of Finland (Baltic Sea). Cont Shelf Res 29:1836–1847. https://doi.org/10.1016/j.csr.2009.06.010
146. Suursaar Ü, Kullas T (2006) Influence of wind climate changes on the mean sea level and current regime in the coastal waters of west Estonia, Baltic Sea. Oceanologia 48(3):361–383. https://www.iopan.pl/oceanologia/483suurs.pdf
147. Meier HEM (2015) Projected change – marine physics. In: The BACC II Author Team (ed) Second assessment of climate change for the Baltic Sea basin. Springer, Cham, pp 243–252. https://doi.org/10.1007/978-3-319-16006-1_13
148. Kowalewska-Kalkowska H, Kowalewski M (2019) Combining satellite imagery and numerical modelling to study the occurrence of warm upwellings in the southern Baltic Sea in winter. Remote Sens 11:2982. https://doi.org/10.3390/rs11242982
149. Attila J, Koponen S, Kallio K, Lindfors A, Kaitala S, Ylöstalo P (2013) MERIS case II water processor comparison on coastal sites of the northern Baltic Sea. Remote Sens Environ 128:138–149. https://doi.org/10.1016/j.rse.2012.07.009
150. Siegel H, Gerth M, Mutzke A (1999) Dynamics of the Oder river plume in the southern Baltic Sea: satellite data and numerical modelling. Cont Shelf Res 19(9):1143–1159. https://doi.org/10.1016/S0278-4343(98)00114-9
151. Schernewski G, Neumann T, Podsetchine V, Siegel H (2001) Spatial impact of the Oder river plume on water quality along the south-western Baltic coast. Int J Hyg Environ Health 204 (2–3):143–155. https://doi.org/10.1078/1438-4639-00086
152. Soosaar E, Maljutenko I, Uiboupin R, Skudra M, Raudsepp U (2016) River bulge evolution and dynamics in a non-tidal sea–Daugava River plume in the Gulf of Riga, Baltic Sea. Ocean Sci 12(2):417–432. https://doi.org/10.5194/os-12-417-2016
153. Yankovsky AE, Chapman DC (1997) A simple theory for the fate of buoyant coastal discharges. J Phys Oceanogr 27(7):1386–1401. https://doi.org/10.1175/1520-0485(1997)027%3C1386:ASTFTF%3E2.0.CO;2
154. Avicola G, Huq P (2003) The characteristics of the recirculating bulge region in coastal buoyant outflows. J Mar Res 61(4):435–463. https://doi.org/10.1357/002224003322384889
155. Horner-Devine AR (2009) The bulge circulation in the Columbia River plume. Cont Shelf Res 29(1):234–251. https://doi.org/10.1016/j.csr.2007.12.012
156. Stokowski M, Winogradow A, Szymczycha B, Carstensen J, Kuliński K (2021) The CO2 system dynamics in the vicinity of the Vistula River mouth (the southern Baltic Sea): a baseline investigation. Estuar Coast Shelf Sci 258:107444. https://doi.org/10.1016/j.ecss.2021.107444
157. Salmela J, Kasvi E, Alho P (2020) River plume and sediment transport seasonality in a non-tidal semi-enclosed brackish water estuary of the Baltic Sea. Estuar Coast Shelf Sci 245:106986. https://doi.org/10.1016/j.ecss.2020.106986
158. Hofmeister R, Bolding K, Hetland RD, Schernewski G, Siegel H, Burchard H (2013) The dynamics of cooling water discharge in a shallow, non-tidal embayment. Cont Shelf Res 71:68–77. https://doi.org/10.1016/j.csr.2013.10.006
159. Lass HU, Mohrholz V, Seifert T (2005) On pathways and residence time of saltwater plumes in the Arkona Sea. J Geophys Res Oceans 110(C11). https://doi.org/10.1029/2004JC002848
160. Burchard H, Janssen F, Bolding K, Umlauf L, Rennau H (2009) Model simulations of dense bottom currents in the Western Baltic Sea. Cont Shelf Res 29:205–220. https://doi.org/10.1016/j.csr.2007.09.010
161. Gueguen C, Kowalczuk P (2012) Colored dissolved organic matter in frontal zones. In: Belkin IM (ed) Chemical oceanography of frontal zones, Hdb Env Chem. Springer, Berlin. https://doi.org/10.1007/698_2013_244
162. Attila J, Kauppila P, Kallio KY, Alasalmi H, Keto V, Bruun E, Koponen S (2018) Applicability of earth observation chlorophyll-a data in assessment of water status via MERIS — with

implications for the use of OLCI sensors. Remote Sens Environ 212:273–287. https://doi.org/10.1016/j.rse.2018.02.043
163. Kowalczuk P, Olszewski J, Darecki M, Kaczmarek S (2005) Empirical relationships between coloured dissolved organic matter (CDOM) absorption and apparent optical properties in Baltic Sea waters. Int J Remote Sens 26:345–370. https://doi.org/10.1080/01431160410001720270
164. Metsamaa L, Kutser T, Strömbeck N (2006) Recognising cyanobacterial blooms based on their optical signature: a modelling study. Boreal Environ Res 11(6):493–506. http://www.borenv.net/BER/archive/pdfs/ber11/ber11-493.pdf
165. Kutser T, Paavel B, Metsamaa L, Vahtmäe E (2009) Mapping coloured dissolved organic matter concentration in coastal waters. Int J Remote Sens 30:5843–5849. https://doi.org/10.1080/01431160902744837
166. Raag L, Sipelgas L, Uiboupin R (2014) Analysis of natural background and dredging-induced changes in TSM concentration from MERIS images near commercial harbours in the Estonian coastal sea. Int J Remote Sens 35:6764–6780. https://doi.org/10.1080/01431161.2014.963898
167. Woźniak SB, Darecki M, Sagan S (2019) Empirical formulas for estimating backscattering and absorption coefficients in complex waters from remote-sensing reflectance spectra and examples of their application. Sensors 19:4043. https://doi.org/10.3390/s19184043
168. Konik M, Kowalewski M, Bradtke K, Darecki M (2019) The operational method of filling information gaps in satellite imagery using numerical models. Int J Appl Earth Obs Geoinf 75:68–82. https://doi.org/10.1016/j.jag.2018.09.002
169. Cao Z, Ma R, Duan H, Pahlevan N, Melack J, Shen M, Xue K (2020) A machine learning approach to estimate chlorophyll-a from Landsat-8 measurements in inland lakes. Remote Sens Environ 248:111974. https://doi.org/10.1016/j.rse.2020.111974
170. Sipelgas L (2011) Monitoring the concentration, distribution, origin and size of suspended particles in the water column during harbour dredging in Pakri Bay, the Gulf of Finland. Boreal Environ Res 16(supp A):42–48. http://www.borenv.net/BER/archive/pdfs/ber16/ber16A-042.pdf
171. Suursaar Ü, Kutser T, Aps R, Kullas T, Vahtmäe E, Metsamaa L, Otsmann M (2009) Hydrodynamically induced or modified patterns derived from satellite images in the coastal waters of Estonia. J Coast Res Spec Issue 56:1602–1606. https://www.jstor.org/stable/pdf/25738060.pdf
172. Kutser T, Metsamaa L, Vahtmäe E, Aps R (2007) Operative monitoring of the extent of dredging plumes in coastal ecosystems using MODIS satellite imagery. J Coast Res Spec Issue 50:180–184
173. Sipelgas L, Uiboupin R, Arikas A, Siitam L (2018) Water quality near Estonian harbours in the Baltic Sea as observed from entire MERIS full resolution archive. Mar Pollut Bull 126:565–574. https://doi.org/10.1016/j.marpolbul.2017.09.058
174. Chelton DB, Schlax MG, Samelson RM (2011) Global observations of nonlinear mesoscale eddies. Prog Oceanogr 91(2):167–216. https://doi.org/10.1016/j.pocean.2011.01.002

The Kuroshio Nutrient Stream: Where Diapycnal Mixing Matters

Takeyoshi Nagai and Gloria Silvana Durán Gómez

Contents

Abstract While the Kuroshio and Gulf Stream carry a large amount of nutrients in the dark subsurface layers and supply them to the downstream regions, it remains unclear how the elevated nutrient concentrations along these nutrient streams are formed and/or maintained. This is particularly important as these positive nutrient anomalies on relatively less dense water along the nutrient streams can be injected

T. Nagai (✉) and G. S. Durán Gómez
Department of Ocean Sciences, Tokyo University of Marine Science and Technology, Tokyo, Japan
e-mail: takeyoshi@gmail.com

Igor M. Belkin (ed.), *Chemical Oceanography of Frontal Zones*,
Hdb Env Chem (2022) 116: 183–254, DOI 10.1007/698_2021_833,

Published online: 8 February 2022

more easily into the euphotic zone. The regions of anomalously high nitrate concentrations along the Kuroshio Extension coincide with one of the major net CO_2 sinks for the Earth's atmosphere. Although, in the earlier studies, the elevated nutrient concentrations on the density surfaces were attributed to diapycnal nutrient flux along the nutrient stream, recently it has been concluded that horizontal advection carries the preexisting anomaly from the upstream tropical regions, and that diapycnal mixing can be negligible in the Gulf Stream. Nonetheless, since the Kuroshio flows over more rough topography than the Gulf Stream, it remains elusive whether diapycnal mixing plays any role in the formation of a high nutrient concentration layer along the Kuroshio nutrient stream. In this chapter, by using a numerical model of the Kuroshio and direct microscale turbulence measurements, the importance of diapycnal mixing in the Kuroshio is examined. The time scale required for the diapycnal nitrate flux of 1 mmol m^{-2} day^{-1} to generate the observed positive anomaly of nitrate concentrations, which is estimated to be 26 Gmol over 5,000 km, is found to be as short as 50 days, assuming the nitrate injection area of 1,000 km^2. However, the question is whether there are such large-scale nutrient injection hot-spots, and where they may be formed. The tow-yo microstructure observations of the Kuroshio flowing over rough topography show that the large nitrate diffusive diapycnal flux of 1–10 mmol m^{-2} day^{-1} is frequently observed in the Tokara Strait, the Hyuganada Sea, and over the Izu Ridge. These observations suggest that the nitrate injection at 1 mmol m^{-2} day^{-1} over the 1,000 km^2 area is plausible if all the contributions in these mixing hot-spots along the Kuroshio are integrated. This time scale of 50 days is estimated to be comparable to the advection time scale over a distance of 5,000 km along the Kuroshio. Thus, unlike the Gulf Stream, the diapycnal turbulent diffusive flux can be important in the Kuroshio nutrient stream. From simulations, the eddy flux analyses on the density layer suggest that eddies can stir and dilute elevated nitrate concentrations, resulting in the decreasing trend of nitrate along the nutrient stream, but with increasing trends on both sides across the stream. Additionally, in the Kuroshio Extension region, rich high vertical wavenumber thermohaline interleaving structures are found to induce a large double-diffusive nitrate flux of 1–10 mmol m^{-2} day^{-1} in the density range of $\sigma_\theta = 26$–27 kgm^{-3}. Because these relatively dense layers in this region do not outcrop even during winter, whether this nitrate diffused upward by double-diffusion can be supplied to the euphotic zone depends on further stirring processes, such as the northward and upward eddy fluxes.

Keywords Baroclinic, Barotropic, Diapycnal mixing, Euphotic zone, Thermohaline

Key Point

Diapycnal mixing could have comparable importance to the advection for generation and maintenance of elevated nitrate concentration on the density surfaces along the Kuroshio.

1 Introduction

Fronts are ubiquitous features often formed at the boundaries of fluid properties in the ocean and the atmosphere. In the ocean, they typically separate water masses with different temperature, salinity, and therefore, density. When the front is formed with different seawater densities, it is no longer merely a boundary, it becomes a dynamic front [1, 2] through a thermal-wind relationship. The across frontal density gradient is accompanied by a jet parallel to it, which often results in growing meanders caused by baroclinic and barotropic instabilities. These instabilities lead to the generations of meso- and submesoscale eddies emanating from the meandering fronts which, at the same time, present their own fronts in their peripheries.

Fronts can strongly influence the distribution of marine organisms, from small phytoplankton to large migratory fish species. For instance, in situ high-resolution water samplings have shown that phytoplankton community structures are separated by mesoscale frontal features of the Kuroshio, with more abundant diatoms at the front, dinoflagellates near the surface of mesoscale upwelling regions, haptophytes in the subducting layers, and high abundance of picophytoplankton on the warm side of the Kuroshio Extension [3]. Moreover, Durán Gómez et al. [4] showed that high catch-per-unit-effort (CPUE) values of swordfish catches are found mostly in and near the warm-core eddies in the Kuroshio Extension region, by analyzing relatively long-term swordfish CPUE with an ocean reanalysis dataset. The high CPUE, associated with warm-core eddies, modulates interannually with the temporal variations of the Kuroshio Extension path and eddy activities.

The Kuroshio is a western boundary current originating from the North Equatorial Current (NEC), off the east coast of the Philippine Islands. Some of the Kuroshio water goes through the Luzon Strait and enters the South China Sea and moves north through the Taiwan Strait, but the majority of the water passes by the Luzon Strait and reaches the region east of Taiwan. Then, it flows through the Okinawa Trough, and finally enters the region off the south coast of Japan. The current separates from the Japanese coast near 35°N and flows eastward as the Kuroshio Extension. Previous studies have found that the Kuroshio transports a large amount of nutrients from the south, and partially from South China Sea [5, 6] and East China Sea (e.g., [7]) through regeneration processes, to the downstream regions [8–12], similar to the Gulf Stream [13, 14]. The subsurface nutrient conduits associated with these currents are, therefore, called nutrient streams. In addition, recent studies have found that the nutrient concentrations on the density layers are elevated in the subsurface of the Kuroshio [12], similar to the Gulf Stream [13–16]. However, the mechanisms to induce such an along isopycnal positive anomaly of nutrients in the Gulf Stream and the Kuroshio have been elusive. Since these anomalously high nutrient concentrations along the nutrient streams appear on the across-density surface, it has been plausibly explained by the diapycnal mixing processes caused by large vertical shear of the Gulf Stream [14, 17]. On the other hand, recent studies have concluded that as the positive nutrient anomaly is advected from the tropical regions, diapycnal mixing processes are not necessarily involved [15, 16, 18]. However, since in situ turbulence measurements with simultaneous samplings of nutrient concentrations along the strong western boundary currents are still limited, further observations are required to elucidate whether the diapycnal mixing matters.

These diapycnal mixing processes can be profoundly affected by a geostrophic current that is associated with a dynamic front. When the geostrophic currents encounter topographic features, strong mechanical turbulence can be generated directly. This turbulence can mix water masses and modify water properties and tracer distributions. Previous studies have pointed out that the lee wave generated by the geostrophic current over topography is one of the most important processes to mix water vertically in the abyssal ocean [19, 20]. Widespread large eddy diffusivity over rough topography, with a horizontal distance of 2,000–3,000 km, is found across the Antarctic Circumpolar Current, which could be attributed to the lee waves [21]. However, internal waves, including lee waves, can exchange energy with subinertial currents [22, 23]. Recent observations in the Drake Passage in the Southern Ocean showed much less depth-integrated energy dissipation rates than the vertical energy flux of the lee waves, suggesting that a large fraction of lee wave energy could be radiated away and reabsorbed into the mean flow [24]. On the other hand, several recent numerical studies have shown that the negative (positive) potential vorticity generation in the Northern (Southern) Hemisphere by subinertial currents on steep bottom slopes leads to centrifugal instability, followed by a forward energy cascade through three-dimensional turbulence [25–27]. Since the Kuroshio flows over many more topographic obstacles than the Gulf Stream, the contributions of the diapycnal mixing in the Kuroshio can be larger than that in the Gulf Stream.

In contrast to diapycnal mixing processes, along isopycnal stirring, caused by subinertial flows associated with meso- and submesoscale eddies, tends to dilute the positive anomaly of nutrient concentrations on the density surfaces. The eddy-induced dilution of the positive anomaly of nutrients occurs because both the eddy-induced advection, which tends to flatten the isopycnal, and the eddy-induced diffusion, which induces down-gradient tracer flux, reinforce and dilute away any tracer anomaly appearing at the front [28]. The pinched-off eddies that move to the dense side of the front undergo shoaling of the isopycnals, leading to upwelling; conversely, those emanated that move toward the less dense side of the front experience a deepening of isopycnals, resulting in subduction. Besides, the along isopycnal meso- and submesoscale stirring often generates interleaving structures of tracers including nutrients on the density surfaces, associated with an increase of their vertical gradient of concentrations. The nearly along isopycnal flow and shear associated with near-inertial internal waves have the same effect but with periodic shear flows; the tracer patches on the density surfaces are stretched and squeezed back and forth [29]. Nitrate concentrations measured across the Kuroshio Extension showed a nitrate interleaving structure extending southward over 30 km below the mixed layer on the less dense side of the front, which is attributed to the submesoscale stirring [3, 29].

These eddy-induced currents, associated with vertical shear, can prompt thermohaline filaments with no density signatures and are referred to as density-compensated filaments [30]. Because these filaments present large spatial variabilities in temperature and salinity, i.e., spiciness in their density range, the vertical stratification often satisfies double-diffusion favorable conditions [31]. Spiciness is defined as the orthogonal coordinate to the isopycnal in a temperature and salinity diagram that characterizes water masses [32]. Spiciness variations are more often expected after the Kuroshio flows away from the western boundary, i.e., the Kuroshio Extension region, where the warm salty Kuroshio water encounters the cold, fresh Oyashio-related waters in the same density layer. Therefore, the diapycnal fluxes caused by the double-diffusive processes may play a role in supplying nutrients in the Kuroshio Extension and in the Kuroshio-Oyashio Confluence regions.

As mentioned above, the physical mechanisms related to the generation, maintenance, and decay of the positive anomalies of nutrient concentrations along the nutrient streams are tightly connected to the frontal processes. One of the plausible generation mechanisms, diapycnal mixing along the Kuroshio, can be expected, for example, by the strong frontal jet of the Kuroshio flowing over topographic features. The advection of the preexisting nutrient positive anomaly on the density surfaces from the upstream to the downstream can also be caused by the frontal jet, while eddies and filaments emanating from the meandering front dilute the positive anomaly. In this chapter, the effects of diapycnal mixing processes on the generation of the positive nutrient anomaly on the density surfaces along the Kuroshio nutrient stream are quantified by using a series of direct turbulence and microstructure measurements with nutrient samplings. An eddy-resolving simulation is also conducted to understand lateral transports and the role of eddies generated from the Kuroshio front in distributing the nitrate.

2 Observations and Simulations

In this chapter, data obtained by the authors during many in situ observations are used. This section provides the information of these field surveys as well as numerical simulations conducted and analyzed in this study.

2.1 *In Situ Observations*

2.1.1 Observations in the Kuroshio Extension Using a Turbulence Profiler

In August 2008, in situ observations were carried out using the R.T.V. *Umitaka-Maru* (Tokyo University of Marine Science and Technology). During this experiment, 48 expendable bathythermographs (XBTs) were deployed every 3.7 km across the Kuroshio Extension front along 143°E from 37° to 35°N (vertical line marked

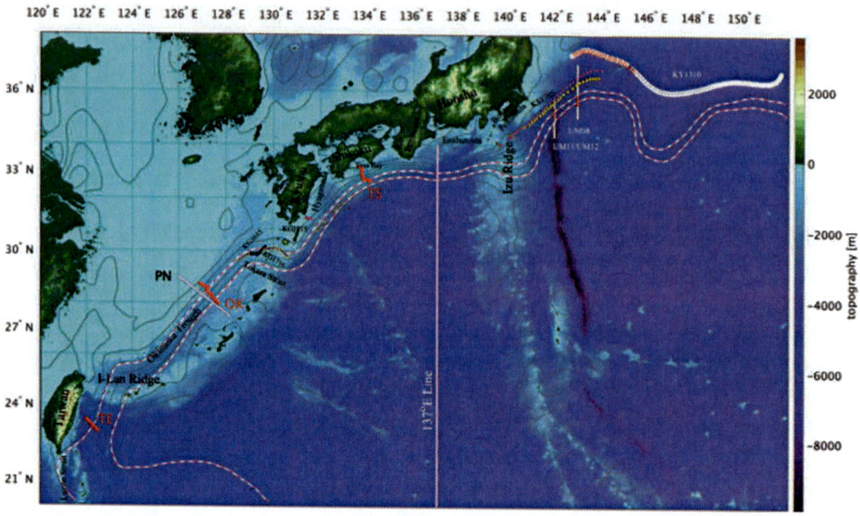

Fig. 1 Map of the Kuroshio region. Vertical lines in the Kuroshio Extension show the location of the microstructure surveys in August of 2008 (UM08), 2011 (UM11), and 2012 (UM12). Thick red lines indicate the observation lines (TE: East of Taiwan; OK: the Okinawa Trough; TS: Tosa Bay). The black contours show average sea surface height during the November 2015 cruise. The broken red-white lines show the approximate Kuroshio path during November 2015. White and red circles represent two autonomous profiling float trajectories during July 2013 cruise (KY1310). Magenta and yellow triangles also show two autonomous profiling float trajectories during June 2017 cruise (KS1705). Short black, blue, and magenta lines in the regions south of Kyushu show Tow-yo microstructure observation lines in the Tokara Strait during November 2016 (KG1615, black), November 2017 (KG1716, black), and in Hyuganada Sea during November 2018 (KG1815, magenta). Shading indicates the bathymetry

"UM08" in Fig. 1), while the shipboard acoustic Doppler current profiler (ADCP, 38.4 kHz Teledyne RD) measured current velocity every 16-m vertical bin. A microstructure profiler, Turbulence Ocean Microstructure Acquisition Profiler (TurboMAP, JFE-Advantech, [33]) was deployed at four stations across the main stream of the Kuroshio Extension (red line superimposed on white at UM08 in Fig. 1) to measure turbulent kinetic energy (TKE) dissipation rates (see [34]).

In August 2011 and 2012, similar in situ observations were conducted using the R.T.V. *Umitaka-Maru* with the shipboard ADCP, XBTs, and TurboMAP along 142°E (vertical line marked "UM11/UM12" in Fig. 1). For the August 2012 cruise, in addition to the XBTs profiling every 2.8 km, a tow-yo CTD (conductivity temperature depth), namely the Underway-CTD (UCTD, Teledyne Oceanscience) was deployed every 14.8 km to measure temperature and salinity down to 500 m depth.

2.1.2 Tow-Yo CTD Observations Across the Kuroshio in 2015

In November 2015, the UCTD was used to obtain temperature and salinity data across the Kuroshio front in three different vertical sections: the region off east of Taiwan, in the Okinawa Trough, and Tosa Bay (thick red lines in Fig. 1), using the R.V. *Hakuho-Maru* (Japan Agency for Marine-Earth Science and Technology, JAMSTEC). The UCTD was deployed down to 500 m roughly at 5 km resolution at all the observational lines (Figs. 2, 3, and 4). The shipboard ADCP (38.4 kHz Teledyne RD) measured the current velocity with 16-m bins.

2.1.3 Observations in the Izu Ridge Using Profiling Floats

In July 2013, the autonomous profiling floats were deployed in the Kuroshio Extension, using the R.V. *Kaiyo* (JAMSTEC, "KY1310" in Fig. 1). One of the

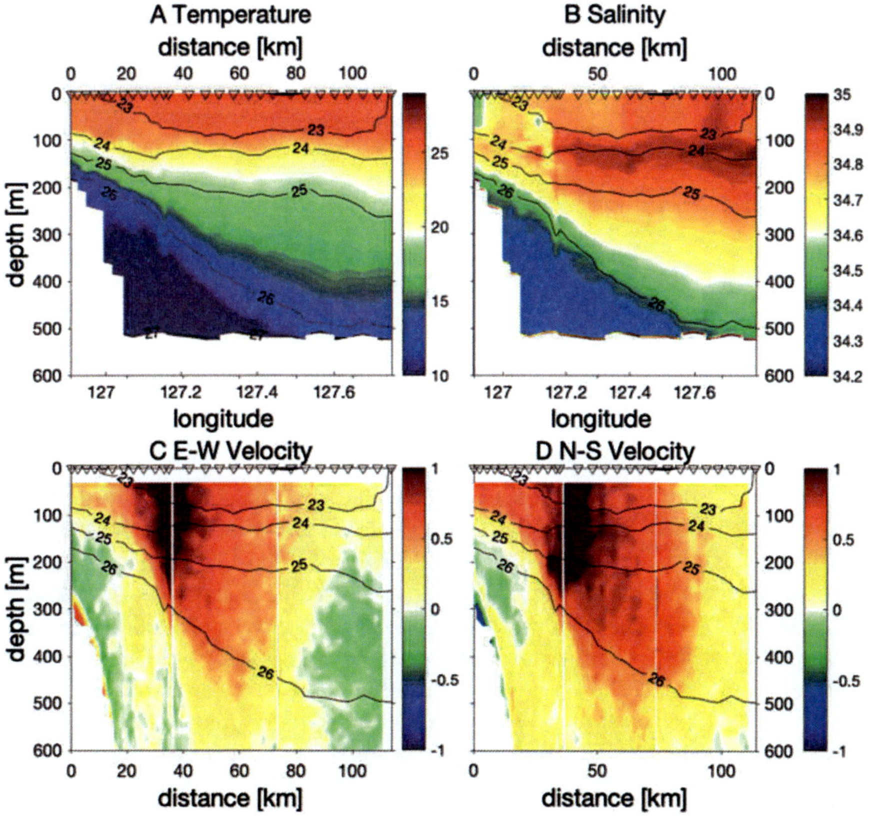

Fig. 2 Vertical sections for (**a**) temperature (°C), (**b**) salinity, (**c**) zonal, and (**d**) meridional velocity (ms^{-1}) along the OK line in the Okinawa Trough (see Fig. 1)

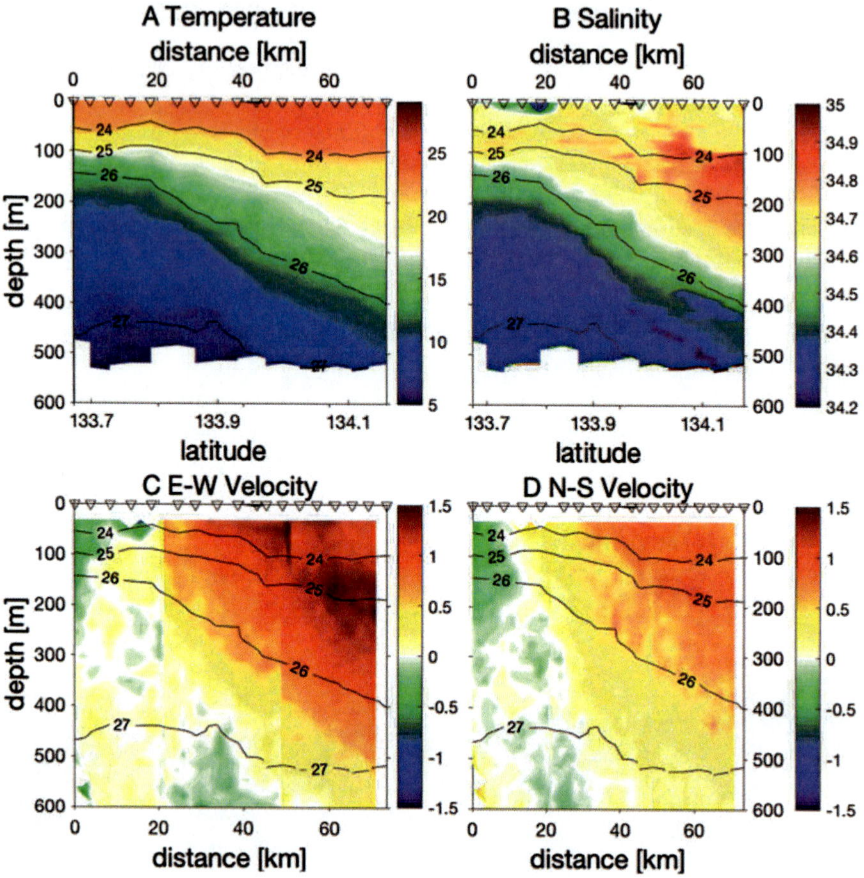

Fig. 3 Vertical sections for (**a**) temperature (°C), (**b**) salinity, (**c**) zonal and (**d**) meridional velocity (ms^{-1}) along the TS line off Tosa Bay (see Fig. 1)

floats, Navis-MR (SBE and Rockland Scientific), is equipped with turbulence sensors (MicroRider, Rockland Scientific) to profile microscale velocity shear and microscale temperature gradient to obtain TKE dissipation rates and microscale thermal variance dissipation rates. Other floats EM-APEX Floats (Teledyne Webb Research) carry a CTD sensor and electromagnetic sensors to vertically profile temperature, salinity, and absolute horizontal current velocity. One of these EM-APEX Floats and the Navis-MR float were deployed at the meander crest of the Kuroshio Extension around 142.5°E. After the float deployments, the UCTD was repeatedly dropped across the Kuroshio Extension front to obtain across frontal flow and density structures. The Navis-MR (red circles in Fig. 1) was advected by the Kuroshio Extension flow for 3 days before recovering at around 145.5°E. The EM-APEX float (white circles in Fig. 1) took almost the same trajectory as the Navis-MR and continued to travel along the Kuroshio Extension flow for 10 days.

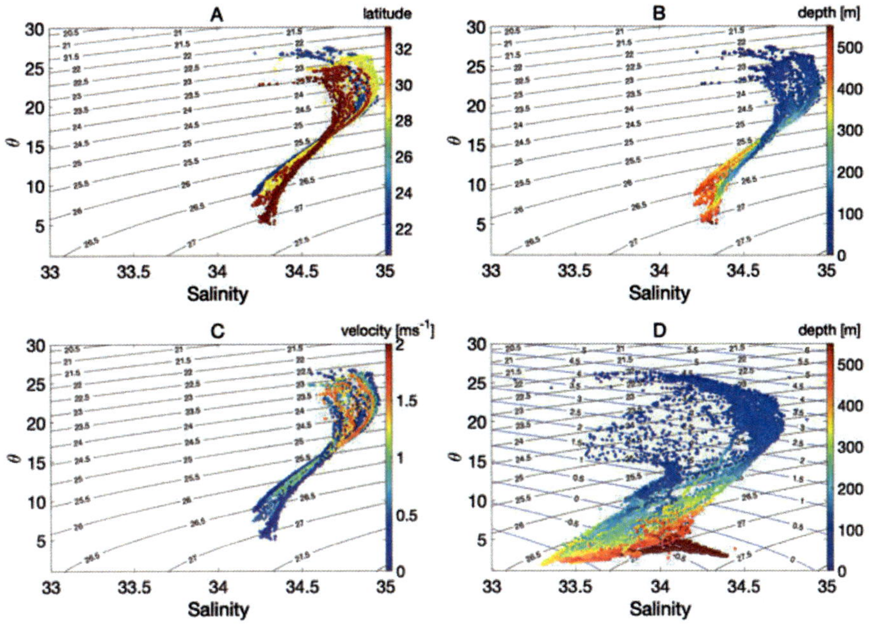

Fig. 4 Temperature-Salinity diagrams for across Kuroshio observations in November 2015 with color indicating (**a**) latitude, (**b**) depth (m), and (**c**) velocity magnitude (ms^{-1}). For comparison, a T-S diagram during the observations in July 2013 along the Kuroshio Extension axis ([31], Fig. 1) is shown in (**d**) with the color indicating depth (m). Longitude and latitude range for July 2013 data is 142–152°E and 36–38°N in the Kuroshio Extension region

While they were advected, the Navis-MR and the EM-APEX Float profiled the upper 500 m every 3 h and 2 h, respectively (see [35] for more details).

Another autonomous profiling float survey was carried out during June 2017 cruise, using the R.V. *Shinsei-Maru* (JAMSTEC, "KS1705" in Fig. 1). During the cruise, the Navis-MR float and a Navis-BGC float, which carried a CTD and chlorophyll-a, dissolved oxygen, backscatter and CDOM sensors, were deployed in the Kuroshio over the Izu Ridge (see [36] for more details). Both floats took very similar trajectories; however, the Navis-MR's trajectory shifted slightly more northward than that of the Navis-BGC. In addition to the float observations, 15 CTD casts were made near the Kuroshio, the Kuroshio Extension and at the edge of a cyclonic eddy generated at the Kuroshio Extension that simultaneously collected water samples at several depths to determine nutrient concentrations. Nutrient concentrations were measured using an AutoAnalyzer (QuAAtro 2-HR, BLTech) for relatively deeper layers (>100 m depth), and by a highly sensitive method [37] that can detect nanomolar nutrient concentrations for the upper 100 m depth. Nitrate and phosphate concentrations were bin-averaged as a function of water density to establish the empirical models of nitrate and phosphate concentrations (Fig. 2 of [36]). The empirical model for nitrate concentration was used to estimate its vertical gradient and vertical diffusive flux for the KS1705 cruise data and for similar profiling float observations during July 2013 (KY1310).

2.1.4 Tow-Yo Turbulence Observations in the Tokara Strait

To measure turbulence in the Kuroshio over the seamounts of the Tokara Strait south of Kyushu, two tow-yo microstructure surveys were conducted by the R.T.V. *Kagoshima-Maru* (Kagoshima University) using the Underway-VMP (see Sect. 2.2.1) in November 2016 (KG1615) and June 2017 (KG1716) (Fig. 1). One tow-yo survey was in the upstream side of the strait, while the other covered from the upstream to the downstream along the Kuroshio across the Tokara Strait (Fig. 1). During the November 2017 cruise, CTD casts were made at 12 stations along the Kuroshio across the Tokara Strait along with water samplings which were analyzed using an AutoAnalyzer (QuAAtro 2-HR, BLTech) to measure nitrate concentrations, and used to derive an empirical model of nitrate concentration as a function of water density. The empirical model of nitrate was then used to estimate the nitrate concentration and its vertical gradient to compute the vertical diffusive nitrate flux during the 2016 cruise.

2.1.5 Tow-Yo Turbulence Observations in the Hyuganada Sea

In November 2018, another tow-yo profiling survey ("KG1815" in Fig. 1) was conducted across the Kuroshio in Hyuganada Sea, southeast of Kyushu using the R.T.V. *Kagoshima-Maru*. Two tow-yo profilers, the Underway-VMP (see Sect. 2.3) and the Underway-RINKO, which uses a RINKO profiler (JFE Advantech) equipped with a CTD housing, an in situ nitrate sensor (SUNA, SBE), a dissolved oxygen sensor, and chlorophyll-a turbidity sensors, were deployed alternately to obtain turbulence and biogeochemical parameters quasi-simultaneously. During the cruise, 5 CTD casts were made near the tow-yo observation line. Water samples obtained at these stations were analyzed with an AutoAnalyzer to determine nitrate concentrations, which were then compared with the SUNA data. The calibrated SUNA data were then used to derive the empirical model of nitrate concentrations as a function of water density, which was used to estimate the nitrate concentration and its vertical gradient to obtain the vertical nitrate diffusive flux (see [12]).

2.2 *Instrumentation*

2.2.1 Tow-Yo Microstructure Profiler, Underway-VMP

The tow-yo turbulence and microstructure profiler, Underway-VMP (UVMP) used in this study consists of a vertical microstructure profiler, VMP-250 (Rockland Scientific International, Victoria, Canada) and a winch of the Underway-CTD (UCTD, Teledyne Oceanscience, USA). The UVMP carries two airfoil shear probes, two high-resolution thermistor sensors FP07, an accelerometer, and a CTD (JFE-Advantech). The shear probes and the high-resolution thermistor sensors measure microscale velocity shear and microscale temperature gradient, respectively, at 512 Hz. The accelerometer measures acceleration, which can be used to detect the instrument low frequency vibrations that could contaminate the shear signals. The sinking speed of the UVMP after the deployment is about 0.8 ms^{-1}, which slows down gradually to 0.3 ms^{-1} after deploying it for 7 min down to 300 m depth. The ship speed, which is relative to moving water while conducting the UVMP observations, is kept as slow as 1–2 ms^{-1}. With 7 min for deployment and 7 min for recovering, the lateral resolution of the turbulence data is 1–2 km.

2.2.2 Autonomous Microstructure Profiling Float, Navis-MR

An autonomous microstructure profiling float (Navis-MR) used in this study consists of a Navis Float (Sea-Bird Electronics, USA) which is equipped with a CTD sensor (SBE 41 N), a MicroRider (Rockland Scientific International, Victoria, Canada) and its external battery. The MicroRider also carries two airfoil shear probes, and two high-resolution thermistor sensors FP07 which measure the microscale velocity shear and microscale temperature gradient, respectively, at 512 Hz. The microstructure data are recorded internally. The microstructure sensors are mounted upward on the Navis Float so that the turbulence and microstructure data can be profiled while the float comes up to the surface. The rising up speed of the Navis-MR is about 0.2 ms^{-1}.

2.3 *P-N Line and 137°E Line Data*

In this study, hydrographic and nutrient concentration data measured in the East China Sea across the Kuroshio in the Okinawa Trough, known as the P-N Line [11] by the Japan Meteorological Agency (JMA), from 1997 to 2015 were used. The P-N Line data were obtained every season until 2014, and every two seasons year 2015 onwards. Similarly, JMA's hydrographic data and nutrient dataset along the 137°E line off Enshunada, known as the 137°E Line, from 1997 to 2016 were used. The 137°E Line data were taken mostly 3 to 4 times per year from 2000 to 2008, otherwise data were measured every two seasons. The data were optimally interpolated as a function of depth and distance from the Kuroshio axis with decorrelation scales of 100 km and 150 m in lateral and vertical directions, respectively. The Kuroshio axis is determined as the location of the maximum surface current, whenever the current data are available. When there are no surface current data, the Kuroshio axis is defined as the location of the maximum lateral density gradient.

2.4 *Numerical Simulation*

To reproduce the observed nitrate distributions in the North Pacific, and to investigate the role of eddies, a Regional Oceanic Modeling System (ROMS, [38]) coupled with a nitrogen-based, nitrate-ammonium-phytoplankton-zooplankton large and small detritus N_2PZD_2 model [29, 39] was used. The model was eddy resolving with lateral resolution of ~10 km in the Kuroshio region and eddy permitting for the other regions in the North Pacific. The number of the vertical levels was 64 with K-profile parameterization (KPP, [40]) for the sub-grid vertical mixing. The model was forced with monthly Comprehensive Ocean Atmosphere Data Set (COADS; [41]) climatology for the first 5 years. After these spin-up simulations, the model was forced with 6-hourly NCEP CFS Reanalysis from 1980 through 2015 [42]. The lateral boundary conditions were obtained from the World Ocean Atlas [43]. The simulation runs totalled 36 years; the output for the years 1985 to 2015 was used in this study.

3 Hydrographic and Nitrate Distributions along the Kuroshio

3.1 *Thermohaline Distributions*

As described above, the Kuroshio is a major ocean current associated with a dynamic front. To characterize the Kuroshio frontal structures, in situ data, which were obtained by UCTD (Teledyne Oceanscience) during a November 2015 cruise (thick red lines in Fig. 1), are used in this section. The warm salty current of the Kuroshio (depicted as a band bordered on either side by the red-white broken lines over the congested black contour lines of sea surface height in Fig. 1) originates from the North Equatorial Current (NEC) and starts its northward journey from south of the Luzon Strait. As the northward flowing Kuroshio is influenced by the westward propagating mesoscale eddies, the path and the strength of the Kuroshio can be profoundly altered [44], and the typical frontal structures can be unclear in some circumstances. This was the case during November 8–9, 2015 when the UCTD was deployed off Taiwan to the east. Mesoscale cyclonic circulation, caused by a cyclonic eddy, moved the Kuroshio path away from the Taiwan coast, which resulted in a flattening of isopycnals and no clear northward current (not shown). Chen et al. [45] showed that the separation of the flow at the ridge southeast of Taiwan provides multiple modes of the Kuroshio nutrient stream. This could affect the location of the Kuroshio axis off east of Taiwan as well.

In contrast to the region east of Taiwan, the Kuroshio in the Okinawa Trough flows along a relatively steady and straight path due to the characteristic half-pipe-like bottom topography in the trough, in which the Kuroshio flows parallel to the isobath in the northwestern shelf slope (Fig. 1). Vertical sections of temperature, salinity, and density show steep isothermal, isohaline, and isopycnal slopes across the front (Fig. 2). A strong northeastward current, with a speed exceeding 2 ms^{-1} in upper 200 m depth, is associated with the steep isopycnal tilt, while a moderately strong current, with speeds over 0.5 ms^{-1}, reaches deep layers down to 500 m depth (Fig. 2c–d).

After passing through the Okinawa Trough, the Kuroshio turns to flow through the Tokara Strait and reaches to the south of Honshu Island. The current of the Kuroshio typically becomes faster as it approaches closer to the coast near Tosa Bay due to the presence of a clockwise flowing Kuroshio recirculation gyre. The vertical sections across the Kuroshio south of Tosa Bay also show the tilted thermocline, halocline, and pycnocline shoaling toward the coast. At the northern edge of the recirculation gyre, the width of the Kuroshio becomes much wider (>50 km) compared to the ~50 km width in the Okinawa Trough, exceeding the length of the observation line off Tosa Bay (Fig. 3).

In the upstream Kuroshio regions east of Taiwan, the Okinawa Trough, and south of Tosa Bay, the temperature shows a monotonic decrease with depth, and salinity presents a subsurface maximum. Because of these thermohaline distributions, all the profiles of temperature and salinity exhibit inverted S-shapes in the temperature-salinity (T-S) diagrams, reflecting the subsurface salinity maximum and minimum, although most of our UCTD profiles did not reach the depths at the salinity minimum (Fig. 4).

Properties of water denser than $\sigma_\theta = 25.5$ are similar across the front, and among the three transect surveys compared to that for lighter water. However, water denser than $\sigma_\theta = 25.5$ shows higher temperature and are less dense in the region east of Taiwan, which then becomes denser and colder with higher latitude. Also, as the Kuroshio flows to higher latitudes, the denser water appears in shallower depths (Fig. 4b). For example, density of $\sigma_\theta \sim 26.5$ can be observed at 200 m depth south of Tosa Bay, at latitude 33°N, while density of $\sigma_\theta \sim 25.5$ appears at the same depth east of Taiwan, at latitude 23°N (Fig. 4a, b). This isopycnal shoaling, toward the downstream of the Kuroshio, may support the induction process similar to the Gulf Stream which is the dominant supplier of subsurface nutrients to the upper mixed layer [46].

For lighter water ($\sigma_\theta < 25.5$) in the Kuroshio during the November 2015 cruise, the T-S profiles show wider salinity ranges on the same density surfaces, than that for dense water, especially for the observation lines that went across the front successfully. The strong current (>1 m s^{-1}) of the Kuroshio carries mostly the water with different salinity of the lighter water ($\sigma_\theta < 26$, Fig. 4c).

The various salinity values along the isopycnal suggest that across the front there is certain spiciness anomaly. The spiciness variations along the same density surface are dramatically increased when the Kuroshio reaches the Kuroshio Extension region (Fig. 4d). The T-S profile drawn for a single vertical CTD profile exhibits zigzag structures in the T-S space, reflecting the influence probably from along isopycnal stirring. The bin-averaged salinity as a function of σ_θ and the distance from the Kuroshio axis, which is defined as the position of the maximum mean velocity at 150 m depth, shows salinity variations along the same density surface (Fig. 5).

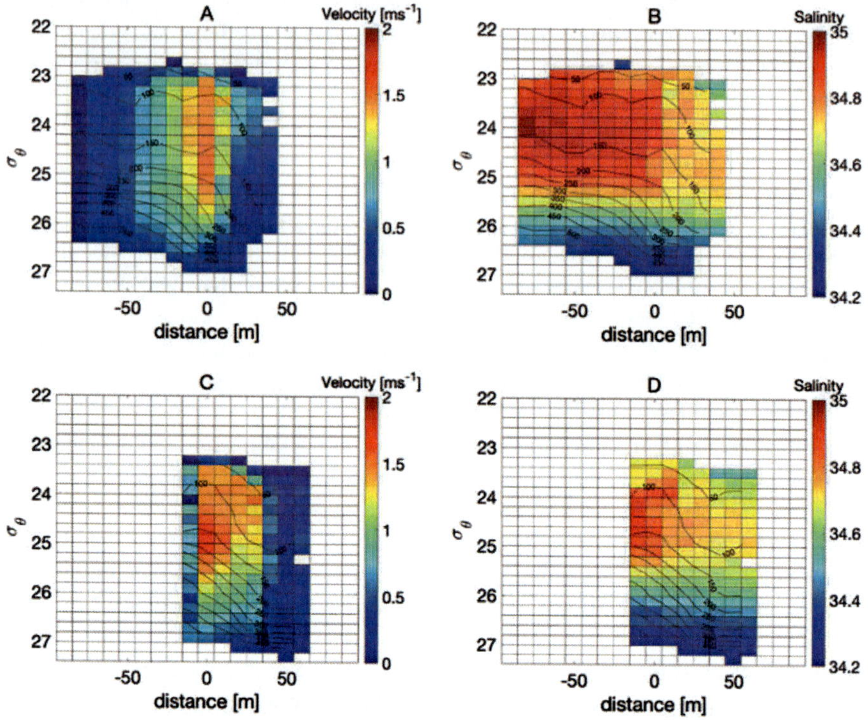

Fig. 5 Bin-averaged velocity magnitude (ms^{-1}) for (**a**) OK Line and (**c**) TS Line, and that for salinity for (**b**) OK Line and (**d**) TS Line

Higher salinity is observed on the offshore side of the front (negative distance) for both regions, the Okinawa Trough (Fig. 5b), and the off Tosa Bay (Fig. 5d). It should be noted that salinity shows a convex upward structure across density surfaces with low salinity values, in the range of $\sigma_\theta = 24.5$–25.75 and distance of 0–50 km (Fig. 5b, d). Moreover, the low salinity values appearing across density surfaces are found in the light water ($\sigma_\theta < 24$) for both the regions. The latter low salinity water in the light water ($\sigma_\theta < 24$) can be attributed to the freshwater flux at the surface, which is transported to the subsurface by diapycnal mixing processes driven by surface forcing, such as wind and cooling. However, the cause of the former dense water freshening is unclear. This could be just along isopycnal advection of low salinity water from the Kuroshio's upstream, and/or caused by the diapycnal mixing near the front, that diffuses up fresher intermediate water near the salinity minima. Because this dense low salinity anomaly is located on the onshore side, at a few tens of kilometers from the Kuroshio axis, the advection speeds that transport this anomaly are 0.1–1 ms^{-1}, not at the peak current of the Kuroshio (Fig. 5a–d).

3.2 *Nitrate Distributions*

Although the amount of available light for photosynthesis depends mostly on the depth and latitude without being influenced directly by the flow, nutrients, vital elements for phytoplankton growth are advected mainly along density surfaces by subinertial flows. These subinertial along isopycnal flows tend to homogenize any tracer anomalies such as the nutrient concentrations and potential vorticity within the density layers [47], leading to higher nutrient concentrations in denser layers in which bacteria remineralize sinking particulate organic matter. Isolines of nutrient concentrations in the ocean interior with weak diapycnal fluxes tend to be parallel to isopycnals. Isopycnals that shoal from the subtropical ocean toward subpolar regions, therefore, result in higher values of the upper 100-m average nitrate concentrations in subpolar regions (Fig. 6a). On the other hand, nitrate distributions on the density layer have less spatial gradients compared to those averaged over the range of 0–100 m depth, caused by the along isopycnal homogenization (Fig. 6b). Yet, the nitrate concentrations sliced at $\sigma_\theta = 25$ kgm^{-3} show distinct higher values along the western boundary where the Kuroshio flows (Figs. 1 and 6b). This positive anomaly of nitrate concentration along the Kuroshio subsurface has been recently reported in a number of studies [12, 49, 50]. The climatological mixed layer depth (MLD; GLORYS12V1, [51]) and the σ_θ at the mixed layer base during winter suggest that the nitrate positive concentration anomaly appearing on $\sigma_\theta = 25$ kgm^{-3} cannot be entrained into the surface mixed layer until it reaches the Kuroshio Extension (Fig. 7a, b). This also rules out the possibility of the surface mixing processes within the mixed layer to be the generation mechanism of this positive nitrate anomaly. In the following sections, the mechanisms to form the elevated subsurface nitrate concentrations on the density surfaces along the Kuroshio are addressed using a series of direct turbulence and nitrate measurements.

Similar to salinity, nitrate concentrations measured along the P-N Line and 137°E Line are bin-averaged as a function of σ_θ and the distance from the Kuroshio axis; the latter is defined as the position of the maximum surface current or lateral density gradient when the surface current data are not available. Results show higher nitrate concentrations near the Kuroshio front, on the shore side (−20 to +40 km) along the

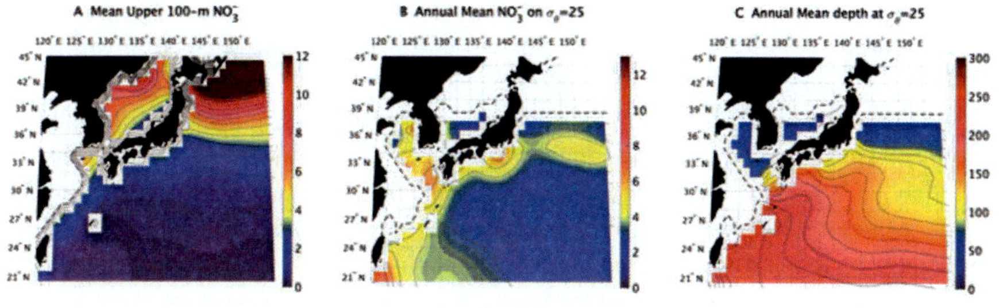

Fig. 6 Climatological nitrate concentration (μM) averaged within (**a**) upper 100 m depth and sliced at (**b**) $\sigma_\theta = 25$ kgm^{-3}. The depth at which $\sigma_\theta = 25$ kgm^{-3} is shown in (**c**). Data from [48]

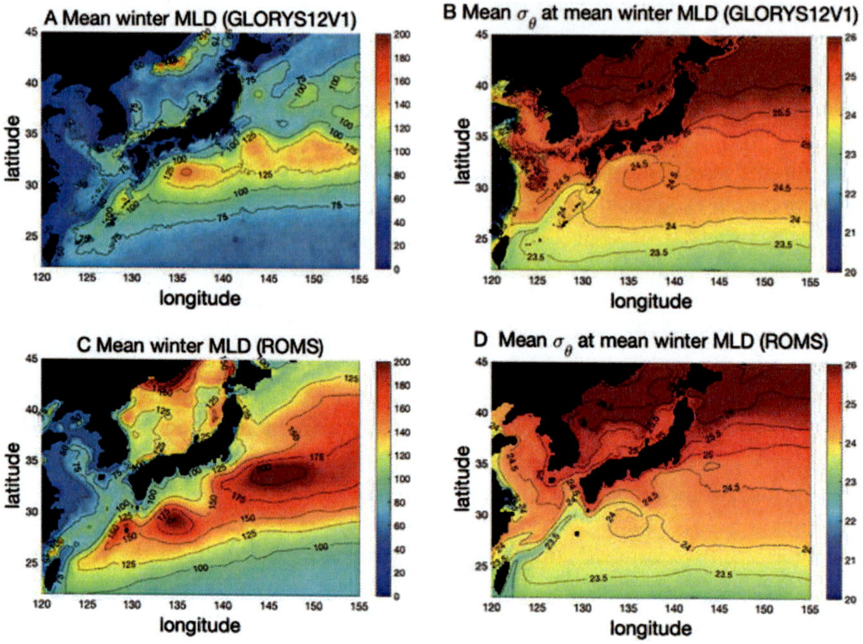

Fig. 7 Climatological mixed layer depth during winter (Dec.–Feb.) shown for (**a**) Mercator global ocean reanalysis (GLORYS12V1) with (**b**) σ_θ at the wintertime mixed layer depth (MLD). Mean wintertime (Dec.–Feb.) MLD for 31-year ROMS simulation and σ_θ are shown in (**c**) and (**d**), respectively

P-N Line in the density range σ_θ = 25.5–24.25 kgm^{-3} (Fig. 8c). The nitrate concentrations extracted along two density layers at σ_θ = 25 and 24.6 kgm^{-3} clearly show peaks at the distance of 20 km with the values of 8.5 and 5 μM, respectively (Fig. 8a). The elevated nitrate concentrations are found mostly between 100 and 250 m depth. However, on the shore side, high concentrations ~5 μM are also found in the density range σ_θ = 25.25–24.75 kgm^{-3}, which corresponds with the depth range as shallow as ~75–100 m. The mean MLD in the P-N Line during winter suggests that the higher nitrate concentration found near the Kuroshio axis in the density range σ_θ = 25.5–24.25 kgm^{-3} is not entrained into the surface mixed layer (Fig. 8c).

Similar to the P-N Line data, the nitrate concentrations along the 137°E Line show elevated values on the density surfaces in the slightly denser layers σ_θ = 25.5–24.5 kgm^{-3} (Fig. 8b, d). It should be noted that the depths of the density surfaces are shallower in the 137°E Line compared to that along the P-N Line, especially in the less dense water on the shore side (0–80 km). This reflects the shoaling of the isopycnals toward the downstream along the Kuroshio. More importantly, the high nitrate concentrations found on the density layers σ_θ = 25.25–24.75-kgm^{-3}, on the shore side of the P-N Line (on the right in Fig. 8c), decrease in the downstream along the 137°E Line (on the right in Fig. 8d).

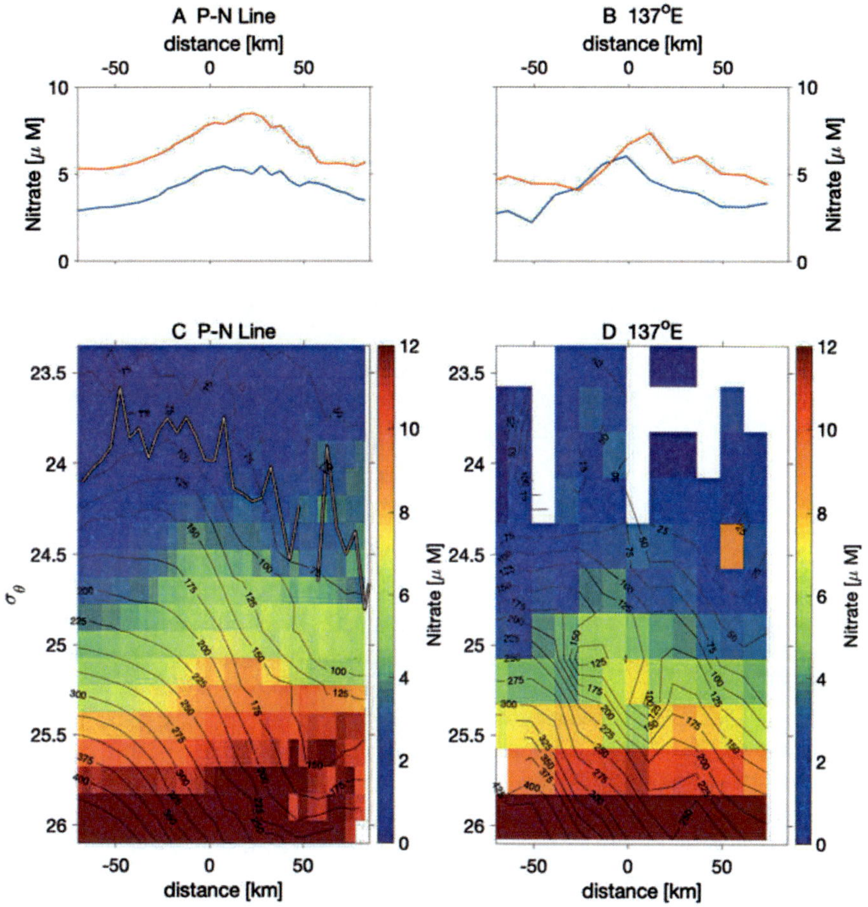

Fig. 8 Nitrate concentration (μM) are bin-averaged as a function of σ_θ and the distance from the Kuroshio axis for (**c**) P-N Line in the Okinawa Trough and (**d**) 137°E Line off Enshunada. Black contours in (**c**) and (**d**) are average depths. Nitrate concentrations extracted along two density layers at $\sigma_\theta = 25$ and $24.6\ \mathrm{kgm^{-3}}$ and $\sigma_\theta = 25$ and $24.75\ \mathrm{kgm^{-3}}$ are shown in (**a**) and (**b**) for the P-N Line and the 137°E Line, with blue and red lines, respectively. The thick white line in (**c**) is mean σ_θ at the mixed layer depth (MLD) from December to February for the P-N Line. Density at the mixed layer depth in the 137°E Line is not shown due to small number of available data

The less nitrate concentrations in the downstream on the shore side compared to the upstream P-N Line revealed by the long-term mean field indicate that a certain fraction of the nitrate, probably transported by the Kuroshio, is supplied to the phytoplankton in the regions south of Honshu, before reaching the Kuroshio Extension region. Even with this decreased nitrate, elevated concentrations remain near the Kuroshio axis. This is consistent with the mean plane view of the nitrate on the density surface, which shows that streaks of elevated values along the Kuroshio persist even after it reaches the Kuroshio Extension (Fig. 6b). The amount of vertically integrated nitrate lost between these two observation lines on the shore side (10–100 km), between 50 and 150 m depth, is 40 mmol m^{-2}. As well, the time scale that the nitrate would require to travel the distance of roughly 1,000 km between these two lines is 30–100 days [12], yielding a vertical nitrate flux of 0.4–1.3 mmol m^{-2} day^{-1}. This flux value is consistent with the corresponding net primary production, estimated by using satellite chlorophyll and sea surface temperature data on the shore side of the Kuroshio over the continental shelf (e.g., [52]).

In addition to the nitrate lost to the sunlit layers on the shoaled shore side layers, the elevated nitrate on the density surfaces can be stirred and diluted along the isopycnal direction caused by eddies. The depth range of these elevated concentrations is 100–250 m depth, which can be largely affected by meso- and submesoscale eddy stirring. Despite this expected lateral stirring, the elevated nitrate concentrations seen along the P-N Line persist in the downstream region along the 137°E Line, suggesting that some other processes preserve the higher concentrations of nitrate.

4 Nitrate Transport by the Kuroshio Nutrient Stream

Despite its oligotrophic upper layers, previous studies have found that the Kuroshio carries a large amount of nutrients in the dark subsurface layer [8–11]. Guo et al. [11] estimated the mean nitrate transport across the P-N Line as 170.8 k $mols^{-1}$ by integrating nitrate lateral advection flux with the geostrophic flow referenced at 700 m depth. The velocity magnitude as a function of σ_θ and the distance from the Kuroshio axis along the OK Line shows that its maximum values are located at the Kuroshio axis (Fig. 5a), while the elevated nitrate concentrations on the density surfaces near the Kuroshio along the P-N Line are misaligned slightly on the onshore side with respect to the core of the nutrient stream (Fig. 8a, c), suggesting that the streaks of high nitrate concentrations on the density surfaces are not entirely advected by the fastest velocity at the Kuroshio axis.

Because the Oyashio, rather than the Kuroshio, has been long recognized as the nutrient-rich cold southward current that can transport nutrients to the Kuroshio-Oyashio Interfrontal Zone, these nutrient transports caused by the oligotrophic Kuroshio are somewhat counterintuitive. Komatsu and Hiroe [49] investigated the nitrate budget in the Kuroshio Extension and in the Kuroshio-Oyashio Interfrontal Zone and reported that along isopycnal transport of nitrate leads to convergence (increasing trend) in these regions, with the dominant contribution coming from the Kuroshio rather than from the Oyashio in the density layers lighter than 26.5 kgm^{-3}. The convergent nitrate advective flux per unit area within these density layers is estimated as +4.4 mmol m^{-2} day^{-1}, with additional nitrification flux of +1.8 mmol m^{-2} day^{-1} which can be balanced by nitrate consumption by phytoplankton in the same regions.

The core of the Kuroshio nutrient stream is in the density range of $\sigma_\theta = 25.5$–26.5-kgm^{-3} [11], which lies in the above-mentioned density layers lighter than 26.5 kgm^{-3}. Part of this density layers outcrops in the Kuroshio Extension and in this Kuroshio-Oyashio Interfrontal Zone during the winter season (Fig. 7b, d), implying that without strong subsurface diapycnal mixing and the upwelling caused by flow-topography interactions, most of these nutrients are merely advected away by the along isopycnal current to the downstream regions without being supplied to phytoplankton in the regions south of Honshu, while sub- and mesoscale eddies stir and dilute the subsurface nutrients along isopycnal direction. In the following sub-sections, nitrate transport is further investigated using the numerical simulations.

4.1 Mean Along Isopycnal Transport

The 31-year average nitrate transport computed for the ROMS simulation within the density range $\sigma_\theta = 25\text{–}26.5$ kgm^{-3}, which is at the core of the nutrient stream, shows that the model reproduces the Kuroshio nutrient stream with an integrated transport magnitude of 150–400 kmols^{-1}, increasing from 125 to 135°E (Fig. 9b). This transport is consistent with the previous estimations of 100–1,000 kmols^{-1} by Guo et al. [10, 11]. Previous studies have shown that there are higher values of nitrate transport in the downstream region, south of Honshu, that reach ~1,000 k mols^{-1}. In the area off Tosa Bay, Guo et al. [10] reported that the merging of the Kuroshio nutrient stream with the nutrients circulating clockwise, around the recirculation gyre, is responsible for this increase. The width and total transport of the large values of the model nitrate transport keep increasing from the upstream off the east coast of Taiwan to the south of Kii Peninsula. The expansion in stream width is found after the merger with other nutrient streams, such as the Ryukyu Current, and the recirculation gyre south of Tosa Bay (Fig. 9b). The divergence of the transport shows large magnitude mostly near the model Kuroshio nutrient stream, with positive and negative values associated with increase and decrease of nitrate concentration within these density layers (not shown). Positive values (increasing trend) appear in the region southeast of Kyushu (133°E, 32°N) and south of Enshunada (140°E, 34°N). The positive divergence suggests that the nitrate of the Kuroshio nutrient stream is supplied to these coastal regions, which is consistent with the recent view that the Kuroshio brings a large amount of nutrients to the southern coast of Japan [12].

For the North Atlantic Gulf nutrient stream, it has been reported that nutrients, which arrive at its downstream subpolar regions, are supplied to the mixed layer through induction processes [46]. Thirty-one-year model output from an eddy-resolving ROMS simulation, coupled with a N_2PZD_2 ecosystem model, shows that the horizontal component of induction is one of the dominant nitrate supply mechanisms for the Kuroshio-Oyashio Interfrontal Zone, with a flux of >5 mmol m^{-2} day^{-1}, strongly influenced by mesoscale structures in the mixed layer depth ([53], Fig. 10b). Nevertheless, the large lateral induction flux is limited to the region relatively close to the east coast of Honshu Island, 140–147°E, which is much narrower compared to that estimated for the Gulf Stream [46] and is consistent with the results of Qiu and Huang [54]. With vertical advective (induction) and

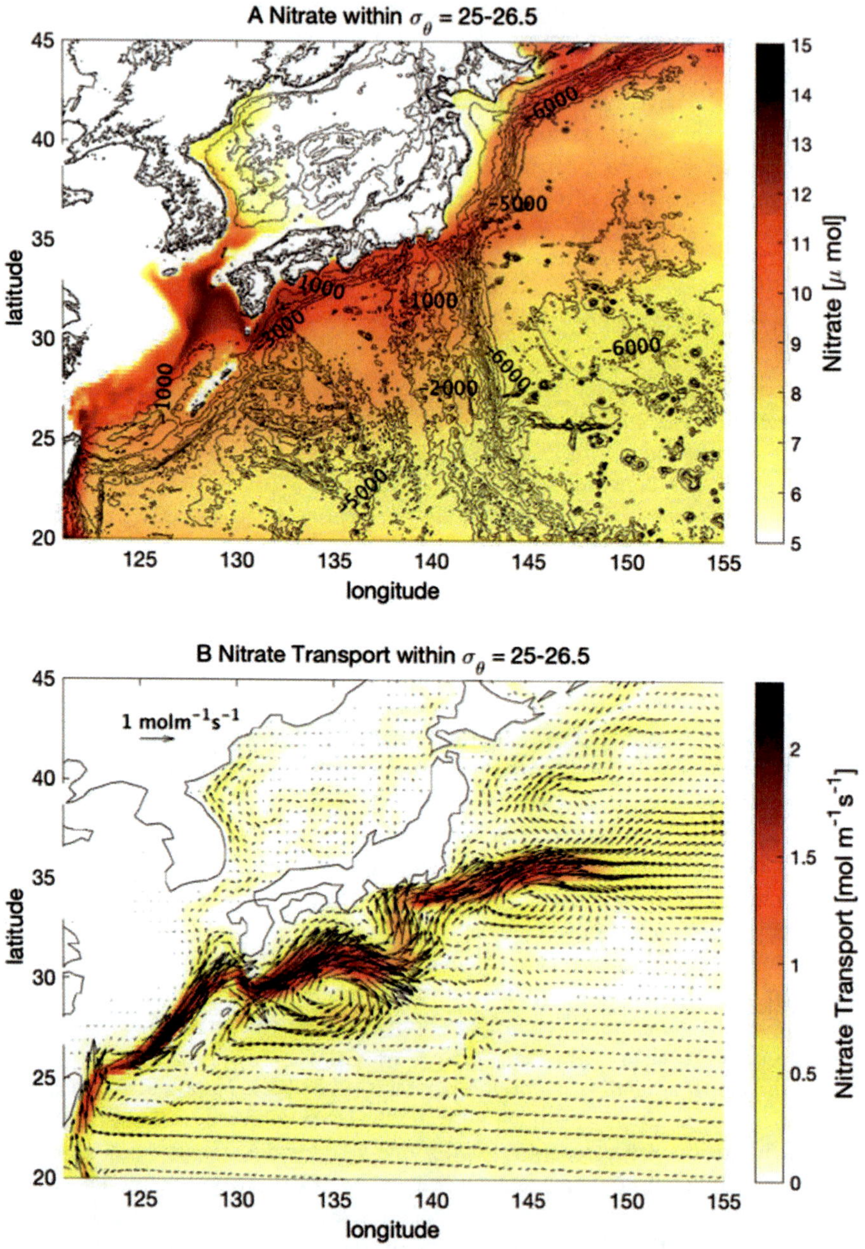

Fig. 9 The 31-year average (**a**) nitrate concentration (μM), (**b**) nitrate flux (mol m^{-1} s^{-1}) integrated over the density range σ_θ = 25–26.5 kgm^{-3} from ROMS. The color shading in (**a**) shows concentrations and that in (**b**) represents flux magnitudes. Black contours in (**a**) denote topography

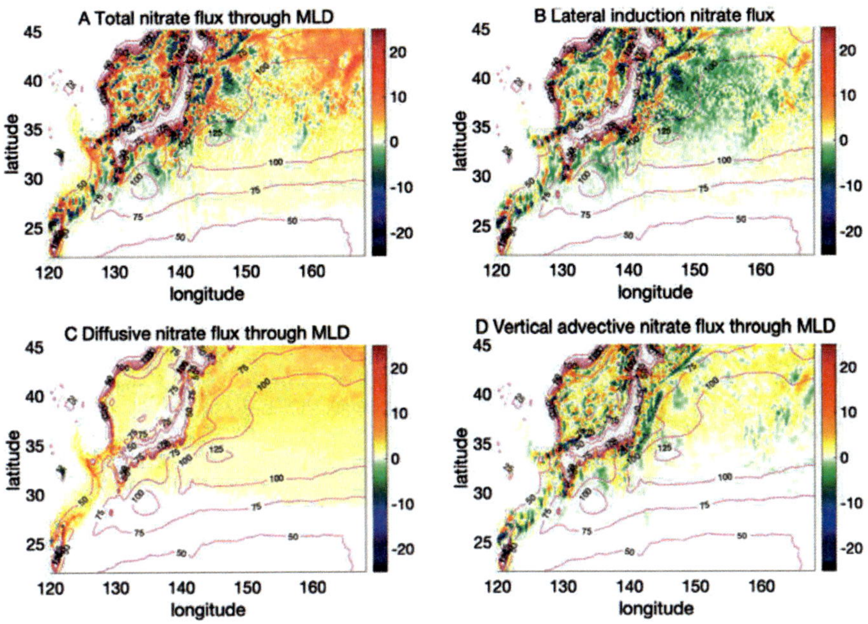

Fig. 10 The 31-year mean nitrate flux through mixed layer depth (MLD) with positive and negative values indicating in and out flux to the mixed layer. (**a**) Total nitrate flux at the MLD (m mol m^{-2} day^{-1}), which consists of (**b**) lateral induction flux, (**c**) vertical diffusive flux, and (**d**) vertical advective (induction) flux. Dashed contours represent mean sea surface height every 0.25 m, and magenta contours indicate mean MLD (m)

diffusive fluxes through the MLD, the transported nitrate by the Kuroshio nutrient stream is supplied to the Kuroshio Extension and the Kuroshio-Oyashio mixed water region at 1–10 mmol m^{-2} day^{-1} (Fig. 10a). The magnitude of the simulated nitrate flux, through the mixed layer in the Kuroshio Extension and in the Kuroshio-Oyashio mixed water region, is consistent with the convergent nitrate flux (increasing trend) estimated by Komatsu and Hiroe [49].

The plane view of the 31-year average nitrate concentration within the density range $\sigma_\theta = 25$–26.5 kgm^{-3} shows elevated values along the western boundary where the Kuroshio flows (Fig. 9a). Interestingly, relatively larger values, >12 μM, are found in several locations where the Kuroshio encounters shallow topography such as to the east of Taiwan, the northern Okinawa Trough, and the Tokara Strait.

4.2 *Eddy-Induced Along Isopycnal Nutrient Flux*

Ocean mesoscale eddies have been thought to increase the nutrient upward flux caused by the uplifted pycnocline at the center of cyclonic eddies, which often coincides with the large nutrient vertical gradient, i.e., nutricline. However, because these eddies are often nonlinear and can hold water inside traveling a long distance, whether the uplift of the pycnocline can contribute directly to the net nutrient upward flux is unclear. Also, because these eddies induce anomalous thickness flux at the front, which tends to flatten the front when they are generated from a meandering free jet, the uplift of the pycnocline does not necessarily mean upwelling. Thus, the influence of eddies needs to be investigated by integrating the effects of the eddy-induced flux.

Using the Reynolds decomposition, velocity **u** and tracer N can be separated into their mean and deviation parts, i.e.,

$$u = \overline{u} + u' \tag{1}$$

$$N = \overline{N} + N', \tag{2}$$

where $\overline{(\ )}$ represents mean and $(\ )'$ deviation from the mean. Note that the mean of these terms satisfies the following:

$$\overline{\overline{u}} = \overline{u}$$

$$\overline{\overline{N}} = \overline{N}.$$

$$\overline{u'} = 0$$

$$\overline{N'} = 0, \tag{3}$$

Substituting these expressions into the tracer conservation Eq. (4),

$$\frac{\partial N}{\partial t} + u \cdot \nabla N = D\nabla^2 N, \tag{4}$$

where ∇ is the derivative operator, and D is the molecular diffusivity for N as a constant. Here, we consider that all the advective tracer fluxes are resolved by the second term on the left-hand side, and that diffusion, the first term on the right-hand side, is driven entirely by molecular diffusion. Thus the mean tracer equation can be written as

$$\frac{\partial \overline{N}}{\partial t} + \overline{u} \cdot \nabla \overline{N} + \frac{\partial \overline{u'_j N'}}{\partial x_j} = D\nabla^2 \overline{N}, \tag{5}$$

where the third term on the left-hand side written in the tensor form with $j = 1, 2, 3$, which newly appears in the mean tracer Eq. (5) compared to the instantaneous tracer Eq. (4). This term is called Reynolds flux divergence term, which represents the effects of fluctuations. This fluctuating effect can include mesoscale eddies and filaments by choosing a sufficiently long averaging period, months or years, or large spatial scales, to obtain the mean variables in the Reynolds decomposition.

This term can also represent microscale turbulence induced fluxes with shorter averaging periods and/or distances.

Using eddy flux analyses for the results from eddy-resolving numerical simulations, Gruber et al. [55] showed counterintuitively that the mesoscale eddies off the California coast increase the subduction of upwelled nitrate and induce biological production drawdown. The effects of eddy-induced tracer flux depend on how the tracers are distributed with respect to the front. Supposing there is a zonal front when the tracer has higher concentrations on the northern side where the isopycnal shoals northward, the eddy-induced flux makes the tracers subduct to deeper layers. On the contrary, when the higher concentrations of tracers locate on the southern side of the front, the eddy-induced flux causes them to upwell [28]. The former corresponds to the coastal upwelling regions with higher nutrient concentrations on the dense side of the front [35, 55].

In this study, eddy flux analysis is conducted for the ROMS simulation outputs for the density range $\sigma_\theta = 25$–$26.5\ \mathrm{kgm}^{-3}$, where the core of the Kuroshio nutrient stream lies. The mean tracer equation on a homogeneous density layer can be written as

$$\frac{\partial \overline{N}}{\partial t} + \overline{\boldsymbol{u}}_{\boldsymbol{h}} \cdot \nabla_h \overline{N} + \overline{\boldsymbol{u}}_{\boldsymbol{h}}^{*} \cdot \nabla_h \overline{N} + \frac{1}{\overline{h}} \nabla_h \cdot \overline{(h\boldsymbol{u}_h)'N'} = D\nabla^2 \overline{N}, \tag{6}$$

where $\boldsymbol{u}_h$ is the along isopycnal velocity vector, ∇_h is the along isopycnal derivative operator, h is the thickness of the density layer with the mean $\overline{h}$, the fluctuation h' and $(h\boldsymbol{u}_h)'$ are the fluctuating components of the thickness flux, and $\overline{u}_h^* = \overline{h'\boldsymbol{u}_h'}/\overline{h}$ is the bolus velocity [56]. In this study, the isopycnal velocity and the derivative operator are replaced with the ones along horizontal directions for simplicity. It is also assumed that isopycnals are flat and the vertical flow is negligible.

The mean nitrate flux $\overline{u}_h\overline{N}$ shows very similar patterns to the 31-year average nitrate transport (Figs. 10b and 11a), suggesting that mean transport dominates the total nitrate transport by the Kuroshio nutrient stream. The bolus flux of nitrate $\overline{u}_h^*\overline{N}$ exhibits northeastward flux in nearshore regions south of Honshu Island, 132–140°E (Fig. 11b), as well as along the east coast of the Tohoku and Hokkaido regions, 35–45°N. On the other hand, the eddy flux $\overline{(hu_h)'N'}/\overline{h}$ is directed southwestward, as opposed to the mean nitrate flux in the region south of Honshu, 132–140°E (Fig. 11c). The southwestward eddy nitrate flux also appears along the east coast of the Tohoku and Hokkaido regions, 35–45°N, which is directed opposite to the bolus flux. Note that in Fig. 11 the color scales have different ranges for mean, bolus, and eddy flux, with the largest magnitude found for the mean flux.

However, the divergence of these fluxes shows similar magnitudes (Fig. 12). The bolus flux divergence shows minimum contributions to the local nitrate variations (not shown). The mean flux divergence mostly shows negative values (decreasing trend) along the Kuroshio nutrient stream, except in the region south of Kii Peninsula, 32°N, 137°E, where the mean model Kuroshio presents a meandering path, and in the region east of Boso Peninsula, 35°N, 141°E (Fig. 12c). While this decreasing

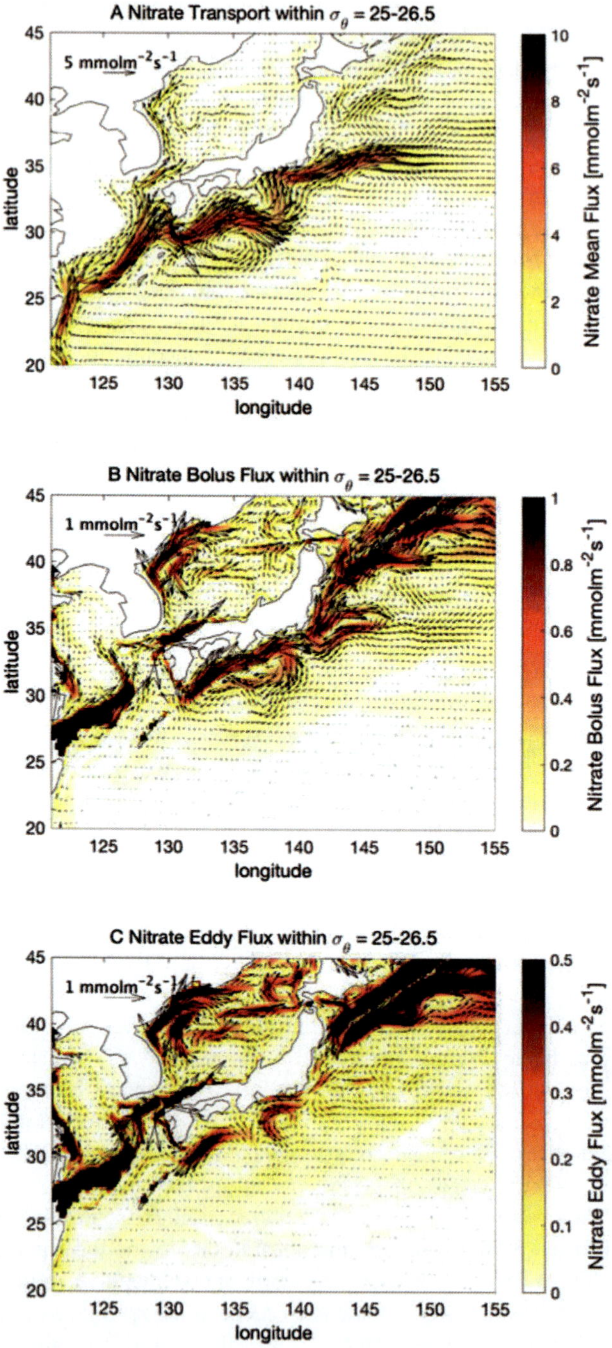

Fig. 11 The 31-year average (**a**) nitrate mean flux (m mol m^{-2} s^{-1}), (**b**) bolus flux and (**c**) eddy flux within the density range $\sigma_\theta = 25$–26.5 kgm^{-3}. Color shading shows flux magnitudes

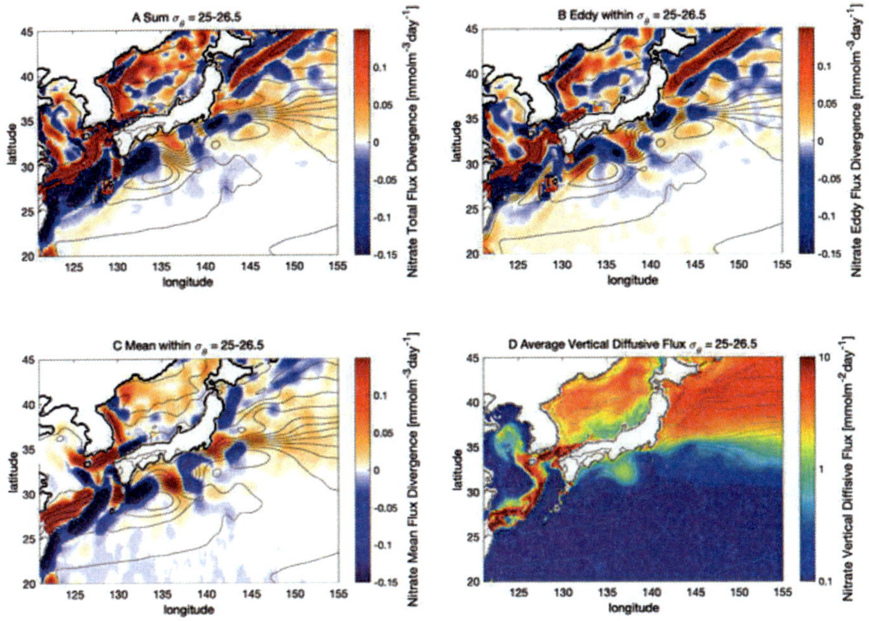

Fig. 12 The 31-year average nitrate flux divergence within density range $\sigma_\theta = 25\text{–}26.5\ \text{kgm}^{-3}$. Color shading shows (**a**) total nitrate flux divergence (m mol $\text{m}^{-3}\ \text{day}^{-1}$), (**b**) eddy flux divergence, and (**c**) mean flux divergence within the density range $\sigma_\theta = 25\text{–}26.5\ \text{kgm}^{-3}$. The divergence is moving averaged over 100 km distance. The vertically averaged 31-year mean vertical diffusive flux within the density range $\sigma_\theta = 25\text{–}26.5\ \text{kgm}^{-3}$ is also shown in (**d**) with color shading in log scale. Gray contours represent model mean sea surface height (SSH, m). Congested SSH contours indicate the Kuroshio

trend along the nutrient stream may imply the removal of nitrate from the stream by physical and biological processes, it could also be by the increased transport toward the downstream, caused by the merger between nutrient streams, as mentioned earlier. Furthermore, there are slightly positive values of mean flux divergence in the regions between the Kuroshio and the south coast of Japan, and in the Kuroshio Extension region. The positive values in these regions suggest a mean flux and/or local vertical injection supply of nitrate to the coastal seas and to the downstream. Remineralization could have also increased the nitrate locally.

On the other hand, eddy flux divergence shows negative values near the Kuroshio nutrient stream, but on the slightly dense side of the front (Fig. 12b). A sign of eddy flux divergence indicates a mirror image of the mean flux divergence in the region from Kii Peninsula to the east of Boso Peninsula. Positive eddy flux divergence south of Enshunada, 137–138°E, where the mean flux divergence shows a blob of negative values, suggests that the eddy flux provides nitrate in this region. Except for the opposing trends of the eddy to the mean flux divergence, the eddy-induced flux seems to remove nitrate from the Kuroshio, suggesting that eddy stirring dilutes the elevated nitrate concentrations along the nutrient stream. The positive eddy flux divergence next to the negative values along the Kuroshio supports this interpretation that the eddies inject nitrate along isopycnal direction to the onshore and offshore sides of the nutrient stream. In the Kuroshio Extension, positive divergence values found on the northern and southern sides of the negative values are also consistent with this eddy dilution process. A band of large positive eddy flux divergences extends southwestward along the east coast of Hokkaido and Tohoku (Fig. 12b). As the positive divergence appears in the regions where the Oyashio extends southward, this is probably caused by mesoscale Oyashio water intrusion in the denser layer. Although the bolus flux divergence is found to have the smallest contributions, it shows a large increasing trend in East China Sea (not shown), which is also in the mean and eddy flux divergence, suggesting that the Kuroshio nutrient stream is also an important nutrient supplier for the East China Sea.

5 Diapycnal Nutrient Flux

While along isopycnal flow and eddy stirring can advect and mix the nutrients, resulting in their uniform distribution on the density surfaces [47], the diapycnal fluxes caused by microscale vertical mixing processes such as turbulence and double-diffusion generate along isopycnal anomalies of tracer concentrations on the density surface, as opposed to the lateral stirring. As shown in the earlier sections, the long-term mean nitrate concentrations, as a function of σ_θ and the distance from the Kuroshio axis, in the Okinawa Trough, in the upstream Kuroshio region (P-N Line), and in the Kuroshio south of Honshu (137°E Line), are slightly higher on the shore side of the Kuroshio front. Although these elevated nitrate concentrations on the density surface of the nutrient stream have been already reported in the Gulf Stream [13], the same structures have recently been discovered for the Kuroshio [49, 52]. The mechanisms to enhance the nutrient concentrations along the nutrient streams have been under debate. While earlier studies pointed out the importance of diapycnal mixing [13, 14, 17], recent studies consider that the diapycnal flux is negligible to form the elevated concentrations, compared to the advection of the preexisting nutrient anomalies in the tropical ocean by the Gulf Stream [15, 16]. The comparison of transport-weighted nutrient concentrations in light water between the upstream and the downstream consistently shows no increase, suggesting that there is no large diapycnal flux along the Gulf Stream [18]. In contrast, the importance of diapycnal mixing in the Kuroshio nutrient stream has been unclear due to the lack of sufficient direct turbulence and nutrient measurements. Thus, the present study attempts to fill this gap for the Kuroshio nutrient stream.

In the numerical model of the present study, the 31-year mean diffusive flux averaged over the density layers, $\sigma_\theta = 25\text{–}26.5\ \text{kgm}^{-3}$, shows 1–10 mmol m^{-2} day^{-1} along the Kuroshio flowing through Okinawa Trough and the Tokara Strait (Fig. 12d). Importantly, these density layers do not outcrop even in winter in both the numerical model and in situ observations (Fig. 7b, d), indicating that enhanced model vertical diffusive nitrate flux is not driven by surface boundary layer mixing but by mixing near bottom. This result implies the importance of topographically induced turbulence for the nutrient supply in the upstream Kuroshio. However, because this is parameterized diffusive flux using the KPP scheme [40] in a rather coarse resolution (~10 km) numerical model, it is not clear whether the enhanced diffusive nutrient flux can be found in in situ observations.

5.1 Challenge to Measure Diapycnal Nutrient Flux

Diapycnal or vertical mixing occurs at the smallest microscale, ranging from millimeters to a few meters in the ocean. The corresponding time scales are also short. Using the upper bounds of spatial and temporal scales for the Reynolds decomposition, the flow and tracer variations can be separated into the non-turbulent mean and the fluctuating turbulent components.

Assuming no lateral gradients in the mean tracer concentration $\overline{N}$ and in the mean velocity $\overline{u}$, Eq. (5) for the mean tracer concentration $\overline{N}$ for a one-dimensional water column can be written as

$$\frac{\partial \overline{N}}{\partial t} + \overline{w}\frac{\partial \overline{N}}{\partial z} + \frac{\partial \overline{w'N'}}{\partial z} = D\frac{\partial^2 \overline{N}}{\partial z^2}. \tag{7}$$

Equation (7) is same as (5), except that the spatial gradients remain only along the vertical direction z. The third term on the left-hand side is the divergence of vertical Reynolds flux $\overline{w'N'}$. With the analogy to the molecular diffusion, this flux is modeled similarly as

$$\overline{w'N'} = -K_v\frac{\partial \overline{N}}{\partial z}, \tag{8}$$

where K_v is the vertical eddy diffusivity, representing any vertical mixing processes, such as turbulence and double-diffusive convection. Note that the negative sign on the right-hand side is because flux is down-gradient, i.e., directed from high to low concentration of $\overline{N}$. Because the isopycnals lie flat, even in the steep frontal region, fluctuating vertical velocity w' and the associated Reynolds flux $\overline{w'N'}$ can be considered as in diapycnal direction. When the fluctuating vertical flow is caused by turbulent velocity, the resulting Reynolds flux represents turbulent diapycnal flux of the tracer, which we aim to measure. However, as can be seen on the left-hand side of (8), to obtain the turbulent tracer flux, fluctuating velocity and tracer concentration must be measured simultaneously. Although it is possible to obtain only the former, the measurements of the fluctuating tracer filed, e.g., nitrate concentration, is still not feasible due to the slow sensor response speed. Thus, instead of directly measuring the Reynolds flux on the left-hand side of (8), obtaining the eddy diffusivity K_v with the mean vertical gradient of tracer concentration on the right-hand side has been a challenge.

5.1.1 Turbulent Kinetic Energy Dissipation Rate and Eddy Diffusivity

The successful measurements of microscale temperature gradients or turbulent shear in the ocean allowed us to quantify the turbulent kinetic energy (TKE) dissipation rate. To measure the TKE dissipation rate, free fall or free rising instruments equipped with airfoil shear probes and high-resolution thermistors have been most frequently utilized in the ocean. Assuming isotropic turbulence, the TKE dissipation rate ε can be determined by quantifying the turbulent shear variance $\overline{(du/dz)^2}$, i.e.,

$$\epsilon = \frac{15}{4}\nu\left[\overline{\left(\frac{\partial u}{\partial z}\right)^2} + \overline{\left(\frac{\partial v}{\partial z}\right)^2}\right], \tag{9}$$

where u and v are the turbulent velocities, normal to each other and to the profiling direction, and ν is the kinematic viscosity. The quantification of the turbulent shear variance can be achieved by integrating the measured turbulent shear spectra, i.e.,

$$\overline{\left(\frac{\partial u}{\partial z}\right)^2} = \int_{k_1}^{k_2} \phi_{u_z} dk, \tag{10}$$

where k is the vertical wavenumber, ϕ_{u_z} is the shear spectrum for u_z, measured by one of the two shear probes mounted on the profiler; and k_1 and k_2 are the lower and the upper bounds of the wavenumber to integrate the shear spectrum. The wavenumber range for the integration is adjusted to avoid integrating the signals from mechanical noises and low frequency vibrations of the instrument. The Nasmyth spectrum is integrated for the outside of this integration range instead [57].

Using the measured TKE dissipation rate ε, the eddy diffusivity K_ρ can be estimated, assuming steady state in the TKE equation. Neglecting all the lateral gradients in the mean variables and advection of TKE, the TKE equation can be simplified as

$$\overline{u'w'}\frac{\partial \overline{u}}{\partial z} = -\nu\overline{\frac{\partial u'_i}{\partial x_j}\frac{\partial u'_j}{\partial x_i}} - \frac{g\overline{w'\rho'}}{\overline{\rho}}, \tag{11}$$

where the first term on the right-hand side is in the tensor form $i = 1, 2, 3, j = 1, 2, 3$, and ρ is density with its mean $\overline{\rho}$ and fluctuation ρ'. The term $P = \overline{u'w'}\partial\overline{u}/\partial z$ on the left-hand side is the shear production of TKE, and the first term $\epsilon = \overline{\nu(u'_i/\partial x_j)(u'_j/\partial x_i)}$ and the second term $B = \frac{g\overline{w'\rho'}}{\overline{\rho}}$ on the right-hand side are TKE dissipation rate and buoyancy destruction, respectively. With these expressions, the equation can be simply written as

$$P = -\epsilon - B. \tag{12}$$

Since Reynolds flux of density $\overline{w'\rho'}$ can be modeled using the eddy diffusivity K_ρ similar to (8),

$$\overline{w'\rho'} = -K_\rho \frac{\partial \overline{\rho}}{\partial z}. \tag{13}$$

Then K_ρ can be written as

$$K_\rho = \frac{g\overline{w'\rho'}}{\overline{\rho}N^2} = -\frac{B}{N^2}, \tag{14}$$

where $N^2 = -g\partial\overline{\rho}/\partial z/\overline{\rho}$ is the buoyancy frequency square. The flux Richardson number R_f (also known as the mixing efficiency) is the ratio of the buoyancy production (– buoyancy destruction), $-B$ to the shear production, P, defined as

$$R_f = \frac{-B}{P} = -\frac{K_\rho N^2}{P}. \tag{15}$$

Dividing both sides of (12) by P results in

$$1 = -\frac{\epsilon R_f}{K_\rho N^2} + R_f, \tag{16}$$

which immediately leads to the following equation [58],

$$K_\rho = \left(\frac{R_f}{1 - R_f}\right)\frac{\epsilon}{N^2}. \tag{17}$$

By defining the mixing efficiency factor, $\gamma = R_f/(1 - R_f)$, the eddy diffusivity is finally written as

$$K_\rho = \gamma\frac{\epsilon}{N^2}, \tag{18}$$

From the laboratory experiment results, the maximum value of γ is estimated to be 0.2.

Once turbulence occurs, the large eddies break into smaller ones until the inertia force balances with the viscous force, stirring the heat, momentum, and tracers almost equally. Therefore, the turbulent eddy diffusivity K_ρ can be used to quantify the diffusive flux of other tracers, such as salinity and nutrients. While these turbulent eddies stir the tracers, tracer streaks are mingled and thinned, which facilitates the molecular diffusion with large spatial gradients of tracer concentrations. Thus, the vertical diffusive flux of nutrients can be quantified by obtaining the vertical eddy diffusivity and the vertical gradient of mean nutrient concentration. This method has been used since the 1980s after the development of microstructure profilers that measure TKE dissipation rates in the ocean [59], and are now widely employed in many observational studies (e.g., [52, 60–63]).

5.1.2 Microscale Thermal Dissipation Rate and Effective Diffusivity for Heat

Another turbulent quantity that can be measured by vertical or horizontal profilers is the microscale thermal variance dissipation rate, χ. Similar to the TKE Eq. (11), neglecting the lateral gradients of the mean field and transport terms, the microscale temperature variance equation can be written as

$$\frac{\partial \overline{T'^2}}{\partial t} + 2\overline{T'w'}\frac{\partial \overline{T}}{\partial z} = -2k_t\left[\overline{\left(\frac{\partial T'}{\partial x}\right)^2} + \overline{\left(\frac{\partial T'}{\partial y}\right)^2} + \overline{\left(\frac{\partial T'}{\partial z}\right)^2}\right], \tag{19}$$

where T' is the fluctuating temperature and k_t is the molecular diffusivity for heat. The first term is the local time derivative, the second term represents the variance production term, and the right-hand side term is the thermal variance dissipation rate, χ.

Assuming isotropy, the thermal variance dissipation rate, χ can be simplified as

$$\chi = 2k_t\left[\overline{\left(\frac{\partial T'}{\partial x}\right)^2} + \overline{\left(\frac{\partial T'}{\partial y}\right)^2} + \overline{\left(\frac{\partial T'}{\partial z}\right)^2}\right] = 6k_t\overline{\left(\frac{\partial T'}{\partial z}\right)^2}, \tag{20}$$

which allows us to obtain χ by measuring only one component of the microscale temperature gradient, which is typically the vertical one. The thermal variance can be obtained by integrating the temperature gradient spectrum

$$\overline{\left(\frac{\partial T'}{\partial z}\right)^2} = \int_{k_1}^{k_2} \phi_{T'_z} dk, \tag{21}$$

where $\phi_{T'_z}$ is the temperature gradient spectrum. Thus, the high-resolution temperature measurements, which allow us to resolve the microscale temperature gradient, can be used to obtain χ. Similar to the TKE equation, assuming the steady state, the thermal variance Eq. (19) can be simplified further as

$$2\overline{T'w'}\frac{\partial \overline{T}}{\partial z} = -\chi. \tag{22}$$

In Eq. (22), the vertical Reynolds flux of heat, $\overline{T'w'}$ can also be written with the eddy diffusivity for heat, K_θ, and mean temperature gradient, i.e.,

$$\overline{T'w'} = -K_\theta \frac{\partial \overline{T}}{\partial z}. \tag{23}$$

Substituting Eq. (23) into Eq. (22), the eddy diffusivity for heat can be obtained with measured microscale thermal variance dissipation rate [64] with O(100 m) mean background temperature gradient as,

$$K_\theta = \frac{\chi}{2\left(\frac{\partial \overline{T}}{\partial z}\right)^2}. \tag{24}$$

As mentioned earlier, when mechanical turbulence is the dominant mixing agent, eddy diffusivities for density K_ρ and heat K_θ should be equal. In this case, the mixing efficiency factor or the dissipation ratio [65] can be written using (18) and (24) as

$$\gamma = \frac{\chi N^2}{2\epsilon\left(\frac{\partial \overline{T}}{\partial z}\right)^2}, \tag{25}$$

which has been examined by measuring ε, χ and O(100 m) mean buoyancy frequency and temperature vertical gradient simultaneously to quantify γ from in situ microstructure observations.

When microscale mixing is dominated by double-diffusion, γ (in this case, dissipation ratio) becomes as large as 1 on average for salt fingers [65], and 4 for diffusive convection [31], because $K_\rho \neq K_\theta$ in double-diffusive layers.

6 Mixing near the Kuroshio

6.1 Turbulence Induced by the Kuroshio Flowing Over Topographic Features

The Kuroshio is a unique western boundary current because along its path it encounters many topographic obstacles, such as the I-Lan Ridge east of Taiwan, the Okinawa Trough in the East China Sea, the Tokara Strait south of Kyushu, and the Izu Ridge south of Tokyo (Fig. 1). Therefore, strong turbulence in these regions is inevitable in the subsurface layers along the Kuroshio, which can mix the subsurface nutrients into the sunlit surface layers. Previous studies have shown that when the current flows over the topographic features, lee wave generation is one of the mechanisms responsible for such strong mixing near the bottom [19]. Nikurashin and Ferrari [20] showed that the energy input from the general circulation to lee waves is estimated to be 0.2 TW. However, it remains unclear how much energy transferred to the lee waves can ultimately dissipate to heat [24]. As internal waves can exchange their energy with the mean flow, a large fraction of the radiated lee waves from the generation sites could be reabsorbed into the mean flow [31]. On the other hand, recent high-resolution numerical studies have suggested that turning the sign of potential vorticity, caused by the subinertial flows at steep bottom slopes, can trigger inertial-symmetric instability [25–27]. Ertel's potential vorticity, PV,

$$PV = -\frac{1}{\rho}\left[\left(\frac{\partial w}{\partial y} - \frac{\partial v}{\partial z}\right)\frac{\partial \rho}{\partial x} + \left(\frac{\partial u}{\partial z} - \frac{\partial w}{\partial x}\right)\frac{\partial \rho}{\partial y} + \left(f + \frac{\partial v}{\partial x} - \frac{\partial u}{\partial y}\right)\frac{\partial \rho}{\partial z}\right], \quad (26)$$

where f is the Coriolis parameter, is conserved following the fluid, $\frac{DPV}{Dt} = 0$ when there is no friction. However, at the surface and the bottom boundaries, the frictional force is inevitable, which can alter the PV. Two extreme cases that can lead to negative PV in the Northern Hemisphere are when the vertical component of the relative vorticity becomes smaller than $-f$, i.e., negative absolute vorticity, and when vertical density inversion occurs. The flow in the former is unstable for inertial instability, while the latter is known to induce gravitational instability. But before reaching gravitational instability, when the isopycnal tilt of the front becomes steeper than the isoline of absolute momentum, $v + fx$ for a zonal front (here x is the zonal direction), PV turns its sign and the flow becomes unstable for symmetric instability [66], leading to secondary Kelvin-Helmholtz instability and turbulence. Previous theoretical and observational studies showed that the downfront wind (wind that blows in the same direction to the frontal jet) can induce PV flux, which can change the sign of the PV [66–68]. When the PV changes its signs from that of the ambient water near the surface, fluctuation grows due to symmetric instability, producing turbulent motions along the isopycnals to reset the PV to zero. On the other hand, when the current flows over a steep bottom slope, the enhancement of the vertical shear of the current occurs because of the weakened current by bottom friction, which in turn produces lateral shear. When this lateral shear changes the relative vorticity sufficiently, the PV with different sign is generated, resulting in strong turbulence by inertial and symmetric instability [25–27]. Assuming a straight front where all the along frontal gradients in (26) have disappeared, neglecting small vertical velocity terms, and assuming the thermal-wind balance, $-g\partial\rho/\partial x/\rho_o = f\partial v/\partial z$, the two-dimensional potential vorticity PV_{2D} multiplied with f can be written as

$$fPV_{2D} = \frac{1}{g}\left[f\left(f + \frac{\partial v}{\partial x}\right)N^2 - f^2\left(\frac{\partial v}{\partial z}\right)^2\right], \quad (27)$$

where x in (27) is the cross-frontal direction and v is the along frontal velocity. When the wind blows in the same direction to the frontal jet, the wind stress at the surface, greater than geostrophic shear stress, can produce ageostrophic shear to enhance the preexisting shear [69]. This increased shear contributes to the reduction of the fPV_{2D} through the second term. Also, if the downfront wind blows on the anticyclonic side of the front, where $f\partial v/\partial z$ is negative, it becomes more effective to turn the sign of fPV_{2D} to negative. Similarly, when the current flows over a steep bottom slope, increased shear can reduce the fPV_{2D}, and if this enhanced vertical shear on the sloping bottom can produce anticyclonic vorticity, both terms in (27) decrease, leading to the negative fPV_{2D} generation. The Tokara Strait is one of such regions where a strong current, the southeastward flowing Kuroshio, encounters many seamounts. The Kuroshio can produce the negative fPV_{2D} on the northeastern slopes of the seamounts that could induce inertial-symmetric instability followed by strong turbulence.

Even with the positive fPV_{2D} , lowered fPV_{2D} close to zero, induced by the frictional force at the boundaries, can draw down the minimum frequency of the internal waves, ω_{min}. The minimum internal wave frequency for a two-dimensional front is exactly equal to $\sqrt{gfPV_{2D}/N^2}$, i.e.,

$$\omega_{\text{min}} = \sqrt{f\left(f + \frac{\partial v}{\partial x}\right) - f^2 \frac{\left(\frac{\partial v}{\partial z}\right)^2}{N^2}} \tag{28}$$

$$\omega_{\text{min}} = f\sqrt{R_o - \frac{1}{R_i}}, \tag{29}$$

where $R_o = (f + \partial v/\partial x)/f$ is the Rossby number and $R_i = N^2/(\partial v/\partial z)^2$ is the Richardson number [70]. When fPV_{2D} is negative, then $gfPV_{2D}/N^2$ is also negative with $N^2 > 0$, which results in a complex ω_{min}, suggesting that the internal waves propagating in the x-z plane are not allowed with the negative PV but allowed in the positive low fPV_{2D} water. Previous studies showed that the regions of lowered minimum internal wave frequency can trap near-inertial internal waves [70–72]. Thus, in the flow over the seamounts, regions of the lowered minimum internal wave frequency associated with the low fPV_{2D} can be formed, which may enhance the vertical mixing caused by the trapped near-inertial waves. Therefore, the currents over the topographic features not only can induce lee waves but also can turn the signs of fPV_{2D} over the steep bottom slopes, accompanied by strong turbulence. Furthermore, regions of low fPV_{2D} in the stratified layers are associated with the lowered minimum internal wave frequency, which can trap near-inertial internal waves, leading to a secondary turbulence. If this is the case, the flow over steep slopes can provide very efficient mixing hot-spots.

Despite the expected mixing hot-spots, the microstructure surveys in these regions, where the Kuroshio flows over the topographic features, are rather limited. The first such microstructure observations, made near one of the islands in Izu Ridge by Hasegawa et al. [73], showed that the Kuroshio induced intense TKE dissipation rates of O(10^{-4} W kg^{-1}), which seemed to increase nitrate concentrations near the surface. Recent observational study by Tanaka et al. [63] found that strong turbulence and a large nitrate diffusive flux occur in the Kuroshio near a seamount in the Izu Ridge, by using quasi-simultaneous profiling of turbulence and nitrate concentrations. However, as the Kuroshio changes its path frequently near the Izu Ridge, how frequently such intense turbulence is generated by the Kuroshio still remains unclear. Also, as these previous observations were limited to the regions in close proximity to the topographic features, it is not clear how far the strong turbulence continues toward the downstream. On the other hand, the Kuroshio flowing through the Tokara Strait always encounters many seamounts and islands. Recent microstructure observations in the strait have found very strong turbulence and high vertical wavenumber shear, which is most likely caused by near-inertial internal waves [74, 75]. However, how much nutrients can be supplied to the euphotic layers has still been elusive, as the number of simultaneous observations of nutrient concentration and turbulence is still very limited [62]. In the following subsections, the nutrient diffusive flux is quantified in the Kuroshio flowing in these regions, the Tokara Strait and the Izu Ridge, using a tow-yo microstructure profiler and an autonomous microstructure float.

6.1.1 Tow-Yo Microstructure Surveys in the Tokara Strait

In November 2016, the tow-yo microstructure observations were conducted using the R.T.V *Kagoshima-Maru*. The Underway-VMP was deployed in the Kuroshio along Leg A and B in the Tokara Strait on November 14, and 17–18, respectively (Fig. 13, Legs A and B overlap the observation line above the seamounts). The tidal phase during the tow-yo surveys was in the spring tide in Leg A and at the middle of the transition from spring to neap tide for Leg B. Despite the differences in the regions and in the tidal phase, the ADCP vertical shear exhibited banded structures roughly along the isopycnal, with a small vertical wavelength of ~100 m especially above the rough topography (black shading in Fig. 14b, c, e, f). The angle of the shear banding was found to be closer to the near-inertial ray-paths rather than that for M_2 semidiurnal tidal frequency, suggesting that the banded shear is caused by near-inertial internal waves. The rotary spectral analysis showed that these shear bands rotate both clockwise and anticlockwise directions with depth [74]. This result indicates that these shear layers are caused by upward and downward propagating near-inertial internal waves. The minimum frequency of the internal waves, ω_{min} (28), was found to be lower in the region with the large amplitude banded shear [74], suggesting trapped near-inertial internal waves.

The microstructure observations at a high lateral resolution of 1–2 km reveal that strong turbulent layers also formed banded structures associated with bands of high vertical wavenumber shear (Fig. 15b, c). The large TKE dissipation rates in the banded layers show values of O(10^{-7} W kg^{-1}), sometimes reaching O(10^{-6} W kg^{-1}). The enhanced dissipation rates associated with the banded shear, which is probably caused by the trapped near-inertial internal waves, suggest that the

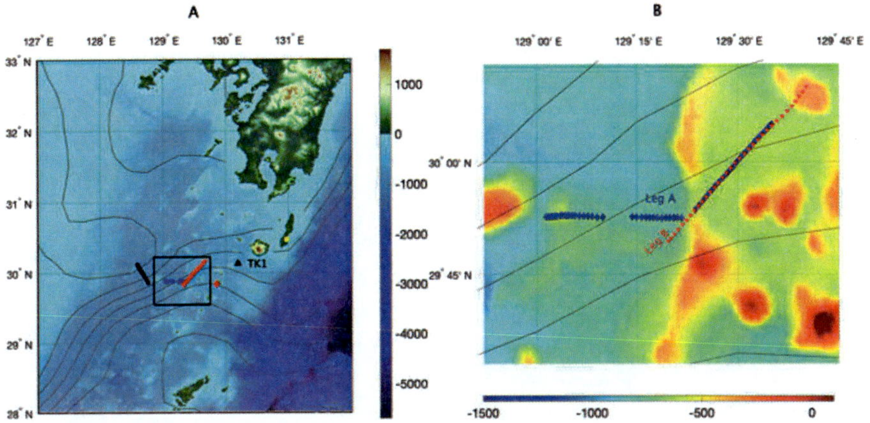

Fig. 13 Observation site in the Tokara Strait near the Tokara Island chain off Kyushu Island. (**a**) Underway-VMP (UVMP) observation lines are in blue for Leg A and red for Leg B. CTD stations are denoted by solid white circles. Black contours are AVISO sea surface height (m) averaged from November 14–19, 2016. (**b**) Enlarged map of the UVMP observation site. UVMP in Leg A is denoted by blue diamonds and Leg B by red triangles. Color shadings indicate depth (m). From Nagai et al. [74]

efficient route that leads to primary and secondary turbulence with low fPV_{2D} water, as mentioned earlier, could be the case in the upstream Kuroshio region.

Using the nitrate concentration data from the bottle samples taken in the Tokara Strait during the November 2017 cruise, the empirical model of nitrate concentration as a function of the water density is first derived (Fig. 15a). This empirical model is then used to estimate the nitrate concentrations and their vertical gradients for the same vertical sections measured by the Underway-VMP. The eddy diffusivity is computed by (18) with measured TKE dissipation rates and buoyancy frequency, and the nitrate flux is estimated as the right-hand side of (8). The estimated nitrate vertical diffusive flux shows O(1 mmol m^{-2} day^{-1}) in the banded shearing and turbulent layers (Fig. 15d, e). These large fluxes are found in the pycnocline at $\sigma_\theta = 25$–26 kg m^{-3}, where elevated nitrate concentrations are on the density layers near the Kuroshio axis (Figs. 6 and 8). The lateral scale of the large nitrate flux is about several tens of kilometers. Considering that the observations in November 2016 were only limited on the upstream side of the Tokara Strait, the lateral scale of the enhanced turbulence and diffusive nitrate flux could be much larger. In fact, observations in later years, with wider spatial coverage along the Kuroshio from the upstream to the downstream side across the Tokara Strait, reveal that strong turbulent layers persist over a lateral scale of O(100 km) (Fig. 16, [76]). Since the Kuroshio flowing through the Tokara Strait is steadily passing over several seamounts, this large-scale strong turbulence with wide spatial scales suggests that the diffusive flux in the Kuroshio over the topography plays an important role in forming and/or maintaining enhanced nitrate concentrations in the Kuroshio nutrient stream.

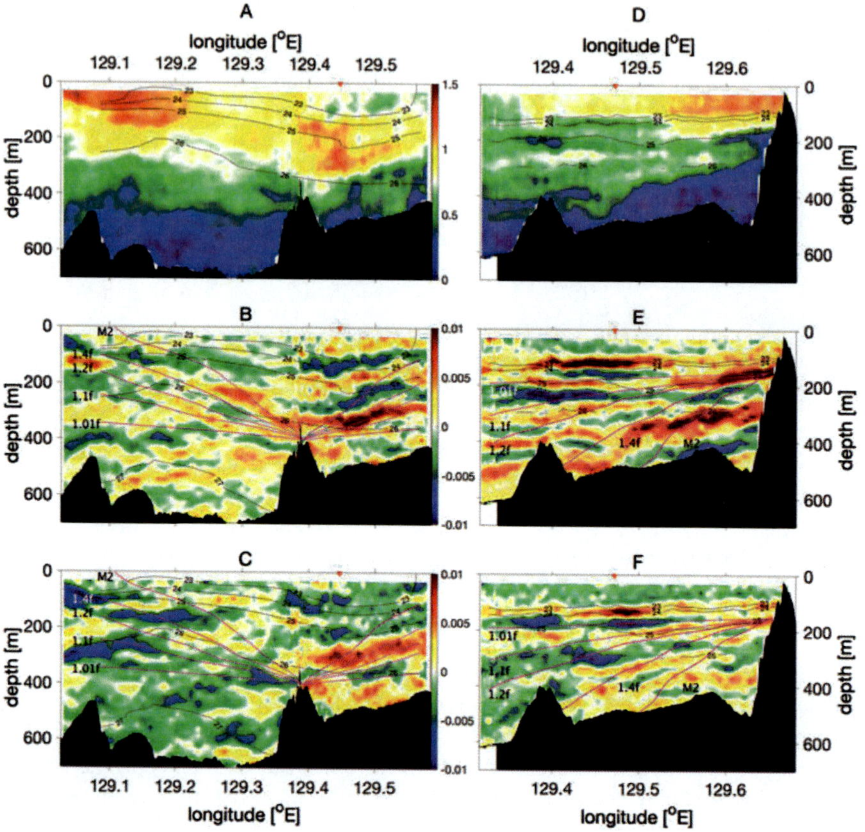

Fig. 14 Vertical sections of the current velocity in the Tokara Strait. (**a–c**) are for Leg A (blue in Fig. 13b) and (**d–f**) are for Leg B (red in Fig. 13b). Horizontal absolute current velocity (m s^{-1}) is shown in (**a**) and (**d**). Back-rotated shear [s^{-1}] is shown for zonal shear $u_z(t_0)$ (**b** and **e**), and for meridional shears $v_z(t_0)$ (**c**) and (**f**) (see Nagai et al. [74]). Black contours are σ_θ (kgm^{-3}). Magenta curves in (**b–f**) are internal-wave ray paths at frequencies of 1.01 *f*, 1.1 *f*, 1.2 *f*, 1.4 *f*, and M_2 tidal frequency, where *f* is the Coriolis frequency, as indicated in the panels. From Nagai et al. [74]

6.1.2 Tow-Yo Microstructure Surveys in the Hyuganada Sea

After the Kuroshio leaves the Tokara Strait, it changes the current direction from southeast to northeast due to the abrupt deepening of the bottom topography in the region south of Kyushu. The northeastward flowing Kuroshio sometimes approaches the eastern coast of Kyushu and follows along the southern coast of Shikoku Island. The tow-yo microstructure surveys and the water samplings for the nutrient concentrations were conducted across the Kuroshio in the Hyuganada Sea, southeast of Kyushu, in November 2018 (Fig. 17). The figure shows the observational line oriented in the oblique direction with respect to the Kuroshio flow. The 200-m average current shows a large across frontal gradient of the Kuroshio flow

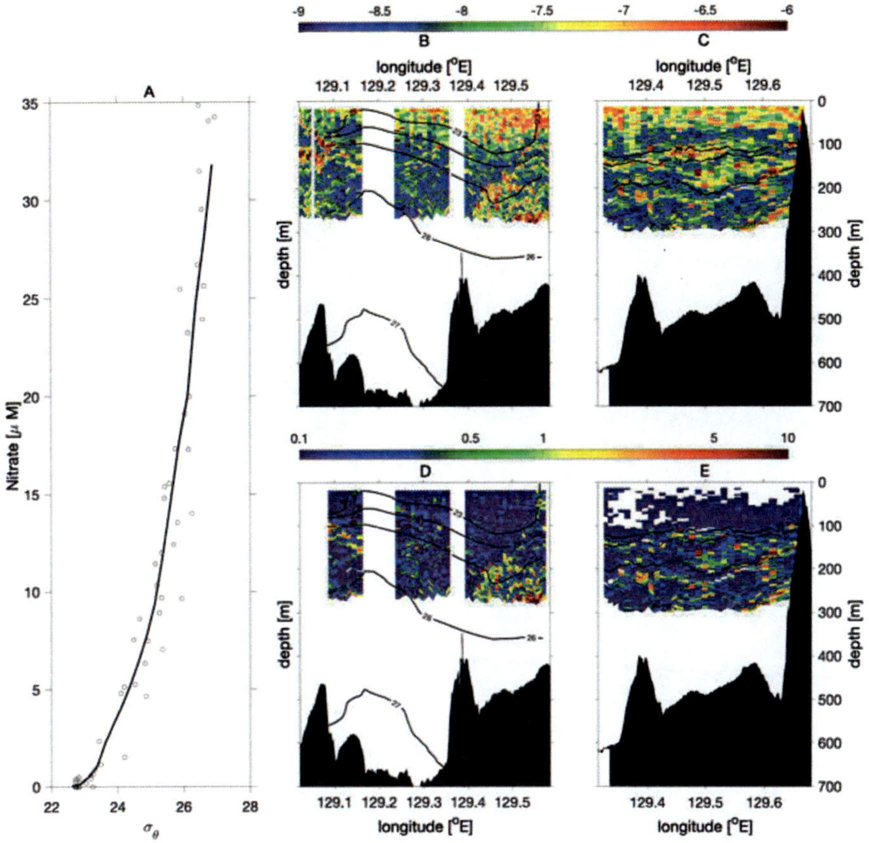

Fig. 15 Turbulent kinetic energy dissipation rates and associated diffusive nitrate flux in the Tokara Strait during the November 2016 cruise. (**a**) Empirical model of nitrate as a function of potential density based on water samples taken in the strait during the November 2017 cruise. (**b, c**) Turbulent kinetic energy (TKE) dissipation rates $\log_{10}$ ε (W kg^{-1}) and (**d, e**) nitrate vertical diffusive flux (m mol $m^{-2}day^{-1}$). Black contours are σ_θ at every 1 kgm^{-3}

speed, resulting in a large anticyclonic vorticity, approximately $-2f$ (Fig. 17c). The northern tip of the observation line seems to be influenced by the south-westernmost tip of a warm water filament (Fig. 17a, c).

The measured ADCP current velocity decomposed into the parallel (Fig. 18a) and normal (Fig. 18b) flows, with respect to the observation line, show banded velocity for the parallel flow, while the strong Kuroshio has normal flow. The vertical shear, computed for both the velocity components, shows banded structures, with a small vertical wavelength ~100 m, roughly align with the isopycnal, similar to those observed in the Tokara Strait (Fig. 18e, f). The ratio of anticlockwise rotating shear variance to that of clockwise rotation with depth shows that both the upward and the downward propagations are also equally found in this region (Fig. 18c).

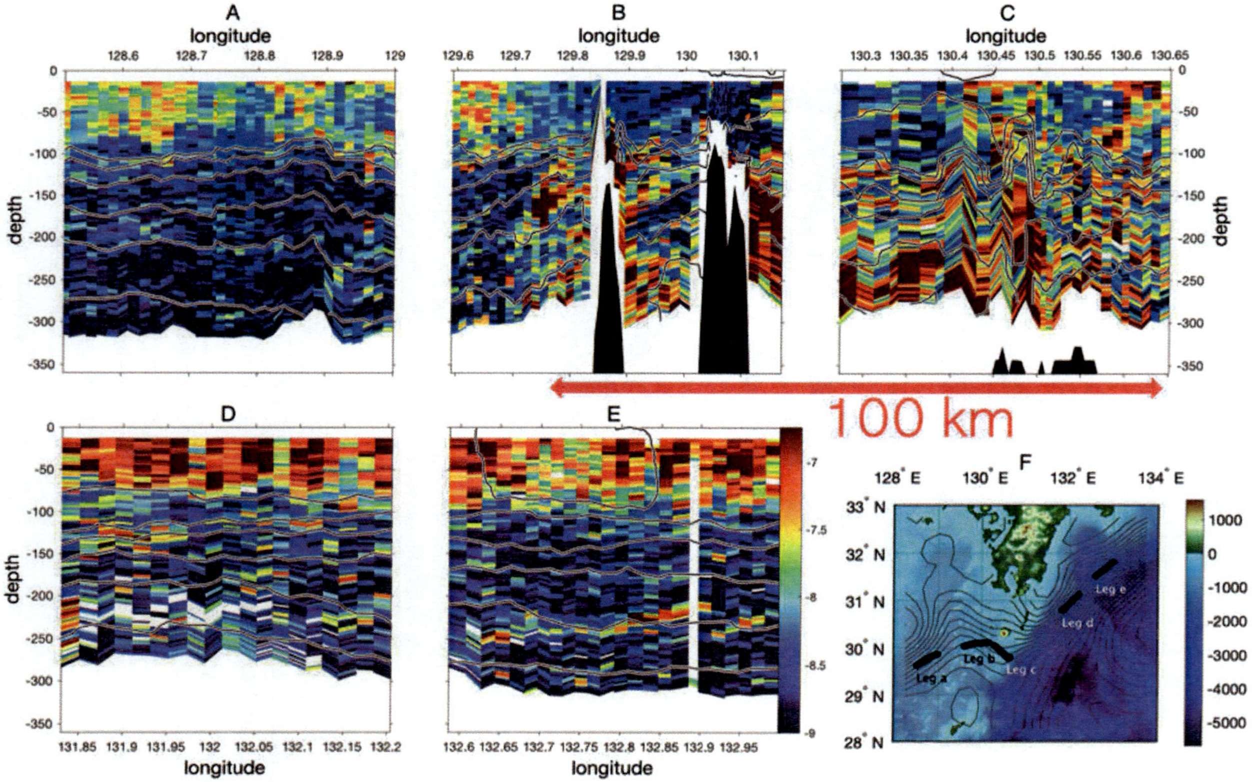

Fig. 16 Turbulent kinetic energy (TKE) dissipation rates ε ($\log_{10}$ W kg^{-1}) in the Tokara Strait during the November 2017 cruise for **(a)** Leg a, **(b)** Leg b, **(c)** Leg c, **(d)** Leg d, and **(e)** Leg e. Location of each leg is shown in **(f)**. Contours in **(a–e)** represent isopycnals. Modified from Nagai et al. [76]

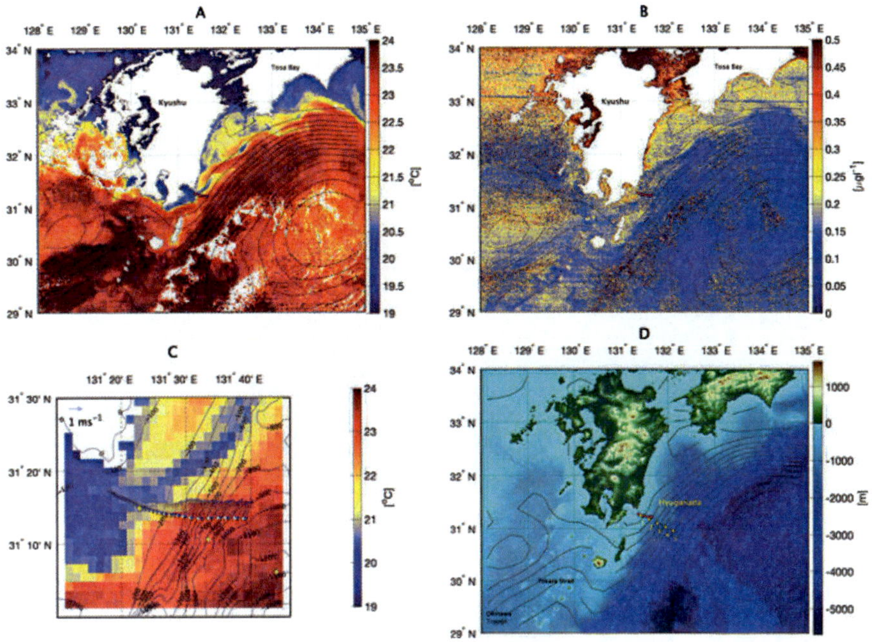

Fig. 17 (**a**) Daily Sea surface temperature (SST, °C) and (**b**) daily surface chlorophyll-a concentration (μgL^{-1}) taken on November 25, 2018, in the Hyuganada Sea, measured by Himawari-8 (Japan Meteorological Agency). Data obtained from P-Tree System. (**c**) Expanded SST off Hyuganada Sea with stations for (blue) Underway-VMP, (white) Underway-RINKO, and (yellow) CTD. Blue vectors indicate current velocity. Topography (m) with (yellow) CTD stations is shown in (**d**). Black contours in (**a, b**) and (**d**) show average sea surface height (m) during the observations. Black contours in (**c**) show topography (m). From Nagai et al. [52]

The internal-wave ray path computed, assuming quiescent condition with a near-inertial frequency $1.1f$ (solid yellow line in Fig. 18e, f), shows a similar angle from the horizon to that of the observed vertical shear away from the frontal structure, but the angle is much smaller compared to the shear angle near the front. Contrarily, the ray path with an M_2 tidal frequency shows a much closer angle to the observed one at the front. Because the observed density and flow clearly show the frontal structures of the Kuroshio, the influence of the front should be considered in the dispersion relation for the ray tracing. The ray paths, considering the effect of the front with a near-inertial frequency, $1.1f$, show a much steeper angle near the front, and become flatter away from the front, similar to the ADCP shear angle distributions across the section. It should be noted that these ray paths bounced back and forth in the regions with a large shear amplitude of the parallel flow component (Fig. 18e). When these ray paths are reflected, as the group velocity approaches zero, the speed of the propagation becomes very slow. Interestingly, the convex downward profiles of the computed minimum frequency of the internal waves, ω_{min}, and those for the

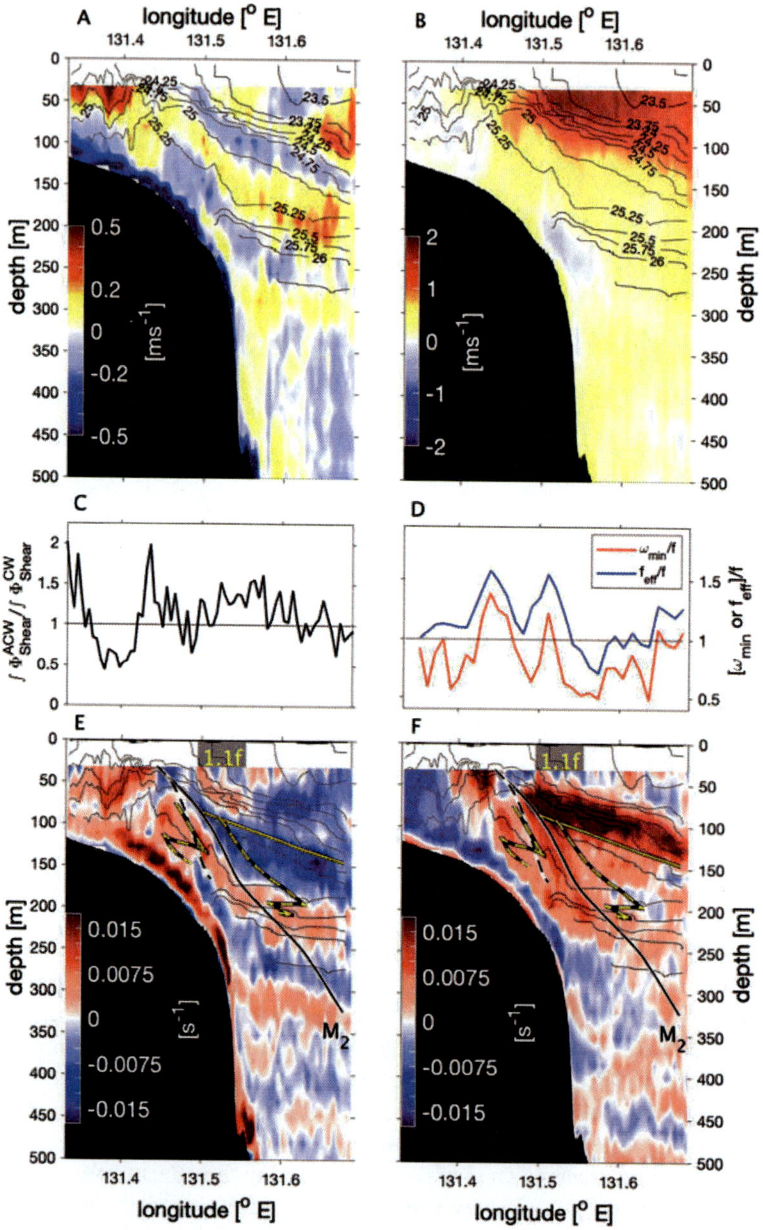

Fig. 18 ADCP velocity for the along (**a**) and across (**b**) component to the observation line, and back-rotated zonal (**e**) and meridional (**f**) shear is shown with black contours for density. (**e, f**), internal-wave ray-paths are shown for (solid) quiescent condition and for (broken) frontal case with (black-yellow) near-inertial frequency 1.1f, and (white-black) M_2 tidal frequency. The ratio of shear variance for anticlockwise rotating component with depth to that for the clockwise rotating component is shown as a black line in (**c**). (**d**) the effective Coriolis parameter f_{eff}/f and minimum internal-wave frequency ω_{min}/f (28) are shown as blue and red lines, respectively. The location of the transect observation is shown in Fig. 17d. From Nagai et al. [52]

barotropic front, i.e., without vertical shear, $f_{eff} = \sqrt{f\left(f + \frac{\partial v}{\partial x}\right)}$, are seen where the reflections of the near-inertial ray paths occur (Fig. 18d), consistent with the near-inertial wave trapping.

The measured TKE dissipation rates across the Kuroshio in the Hyuganada Sea show large values of >O(10^{-6} W kg^{-1}) over 5–10 km, near the Kuroshio axis at 50–150 m depth. Also, large dissipation rates of >O(10^{-7} W kg^{-1}) are found over 30–40 km across the Kuroshio [52]. These results suggest that the trapped near-inertial waves are causing the observed intense turbulent mixing in the Kuroshio.

Nitrate concentrations are estimated using an empirical model as a function of seawater density, based on the water samples and the in situ nitrate sensor deployed near the observation line during the same 2018 cruise (Fig. 19a). The estimated nitrate concentration shows climbing tongue-like nitrate-rich water above the continental slope (Fig. 19c). Using these estimated nitrate distributions and the eddy diffusivities based on the measured TKE dissipation rates (18), the vertical diffusive flux of nitrate is estimated. The estimated diffusive flux shows widespread large values from the surface down to 225 m depth centered at the Kuroshio axis (Fig. 19d). The average diffusive flux in the upper 100 m is 1–6 mmol m^{-2} day^{-1} over 20 km across the Kuroshio (Fig. 19b).

Further analyses by Durán Gómez and Nagai [77] suggest that lowered minimum internal wave frequency is caused by the generation of negative and/or low PV in the region slightly upstream of the observation line, where the Kuroshio flows over a small bump like a topographic obstacle. Also, their results show that these primary and secondary mixings caused by negative and low PV water extend further downstream over 50–100 km along the Kuroshio, accompanied by large diffusive nitrate fluxes of O(1 mmol m^{-2} day^{-1}) along the same distance.

6.1.3 Profiling Float Surveys in the Kuroshio Over the Izu Ridge During June 2017

The Kuroshio south of Honshu is known to take a large meandering path non-periodically, and has been doing so since September 2017. Also, it frequently undergoes transient non-large meanders. When the Kuroshio takes a non-large meandering path, chances of the Kuroshio flowing over the shallow ridges, sills, and seamounts in the Izu Ridge increase (Fig. 1). However, even with the large meandering path, it tends to flow along the western edge of the ridge. Small fluctuations in the path, therefore, may be able to provide an opportunity for the flows run over shallow topographic features.

In June 2017, the autonomous microstructure profiling float observations were conducted along the Kuroshio flowing near the Izu Ridge [36]. Although it was just a few months before the occurrence of the Kuroshio large meander, the Kuroshio had already started to take a very similar path to the large meander near the Izu Ridge during the observation period (Fig. 20). The microstructure float (hereinafter Navis-MR) and a biogeochemical float (hereinafter Navis-BGC) were deployed in the

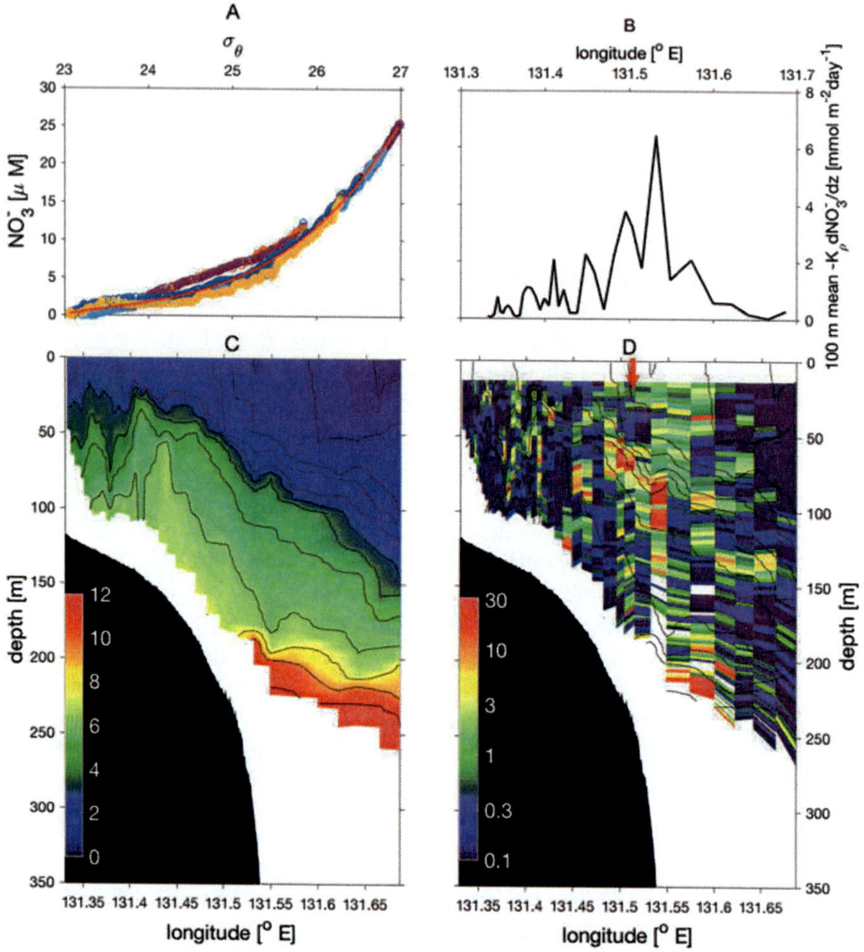

Fig. 19 (**a**) Nitrate concentrations measured at 4 CTD stations as a function of σ_θ, and the third order polynomial fitting function as a red curve. The vertical section of (**c**) is estimated nitrate concentration (mmol N m^{-3}), based on the UVMP measured σ_θ, and vertical section of (**d**) is the estimated vertical diffusive nitrate flux (m mol m^{-2}day^{-1}). Black contours in (**c**) and (**d**) represent isopycnals. In (**b**) the diffusive flux within the upper 100 m in (**d**) is averaged as a function of latitude. Location of the transect observation is shown in Fig. 17d. Red arrow in (**d**) shows the Kuroshio axis. Modified from Nagai et al. [52]

Kuroshio above the ridge. These floats profiled down to only 200 m for a few days due to improper buoyancy settings, but after the first few profiles they are adjusted to dive down to 450 m. Both floats were rapidly advected northeastward by the Kuroshio, profiling the water columns every 3–4 h, and then were recovered at the northernmost portion of the first meander crest of the Kuroshio Extension front.

The microstructure data obtained by the Navis-MR show that the TKE dissipation rates were larger, with values O(10^{-7} W kg^{-1}) over 20–40 km away near the Izu

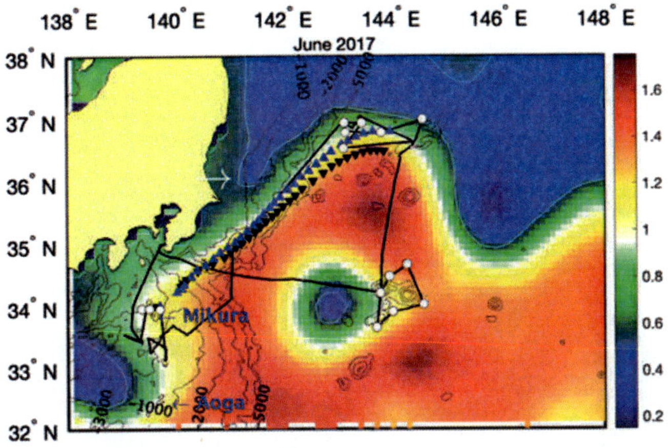

Fig. 20 Map of the surveys for the June 2017 cruise. Color shading shows average satellite sea surface height (m) during June 24–29, 2017. Black line indicates the ship track, blue and black triangles are Navis-MR microstructure float and Navis-BGC float profiling positions, respectively. Thin black contours show the topography of the Izu Ridge. Solid white circles are the CTD-rosette samplings for nutrients. From Nagai et al. [36]

Ridge [36]. The corresponding eddy diffusivities for density (18) and heat (24) show elevated values of 10^{-3}–10^{-2} m^2s^{-1} in the same region (Fig. 21). Similar to the analyses above, nitrate concentrations were estimated with an empirical model as a function of seawater density (see Fig. 2 of [36]), using the bottle sampled data. The computed nitrate diffusive fluxes, estimated with the nitrate concentrations (Fig. 22a) and the obtained diffusivities, show that the nitrate flux is enhanced as $>O(1\ \mathrm{mmol\ m^{-2}\ day^{-1}})$ above the Izu Ridge. The enhanced nitrate flux of $>O(1\ \mathrm{mmol\ m^{-2}\ day^{-1}})$, averaged between 50 and 125 m depth, extends over 20–40 km from the southwesternmost part of the observations (Fig. 23), and that of $O(0.1\ \mathrm{mmol\ m^{-2}\ day^{-1}})$ extends further downstream up to 100 km away. The similar orders of magnitudes found for vertical nitrate diffusive flux, using eddy diffusivity for density and for heat, suggest that the dominant mechanism driving the diapycnal mixing is mechanical turbulence.

The data obtained by the Navis-BGC float show a subsurface increase in chlorophyll-a concentrations from 35.5°N, about 200 km away from the shallowest part of the Izu Ridge, toward the downstream (see Fig. 4 of [36]). With the Kuroshio speed of 1–2 ms^{-1}, the time required to travel over the 200 km distance is about 1–2 days, which is similar to the phytoplankton doubling time. Thus, the enhanced nutrient supply, caused by the Kuroshio flowing over the Izu Ridge, is a plausible source of the increased subsurface chlorophyll-a concentrations.

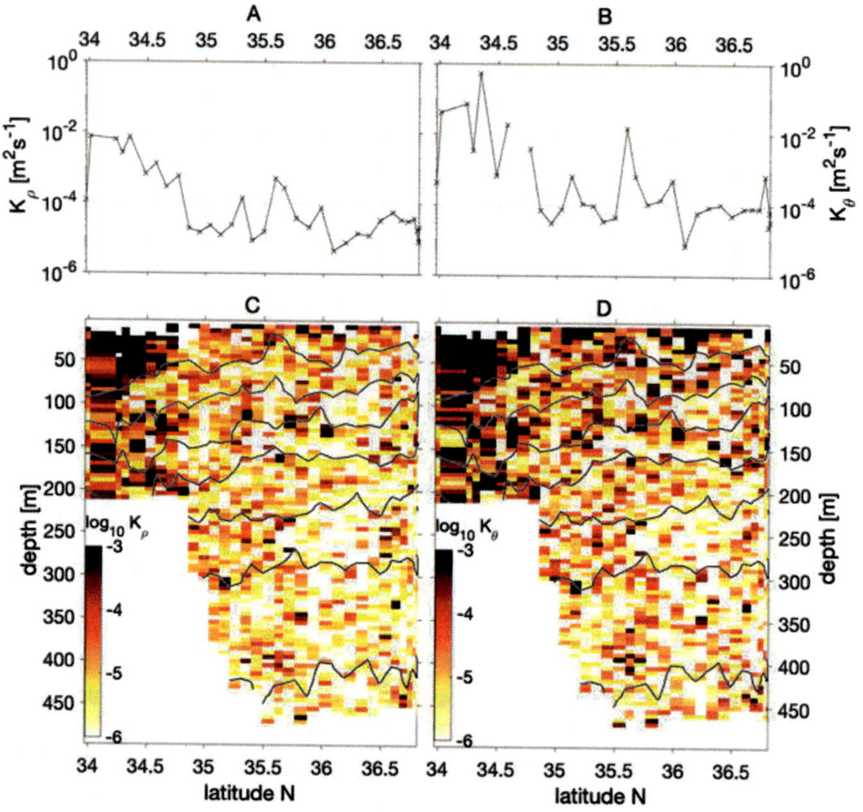

Fig. 21 Vertical eddy diffusivity, K_ρ (m^2s^{-1}) and effective thermal diffusivity K_θ (m^2s^{-1}). Mean (**a**) K_ρ and (**b**) K_θ averaged between 40 and 200 m depth as a function of latitude. Vertical sections for (**c**) K_ρ and (**d**) K_θ. Values are shown in log scale. Black contours show isopycnals. From Nagai et al. [36]

6.2 *Turbulence Near the Kuroshio Away from the Topographic Features*

After the Kuroshio passes by the Izu Ridge and the Boso Peninsula, it turns direction eastward and continues as the Kuroshio Extension. Because there is no shallow topography along the Kuroshio Extension, intense turbulence is not expected as in the upstream regions and over the Izu Ridge. Nevertheless, several studies have shown an enhanced nitrate diffusive flux of O(0.1 mmol m^{-2} day^{-1}) in the Kuroshio Extension thermocline, which has been one of the possible explanations for the missing nutrients source to sustain biological production in the nutrient-depleted subtropical gyre [60, 61]. However, the number of microstructure profiles used in these observational studies was very limited and they were made with rather coarse lateral resolutions to support this claim. Here, additional microstructure data

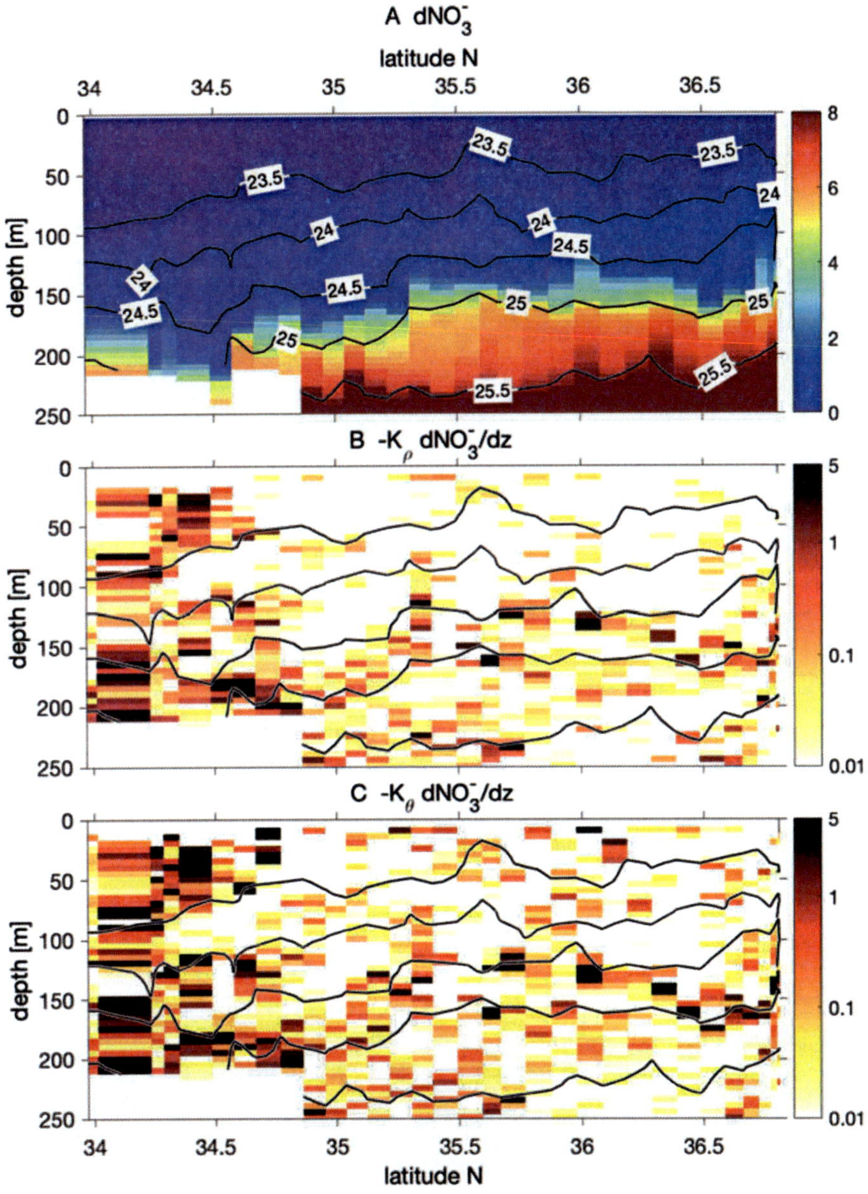

Fig. 22 Vertical sections of (**a**) nitrate concentration (μM), diffusive flux with (**b**) eddy diffusivity K_ρ and with (**c**) effective thermal diffusivity K_θ, $\log_{10}$(mmol N m^{-2}day^{-1}) over the Izu Ridge

obtained in the Kuroshio Extension in different years are summarized to back this conclusion.

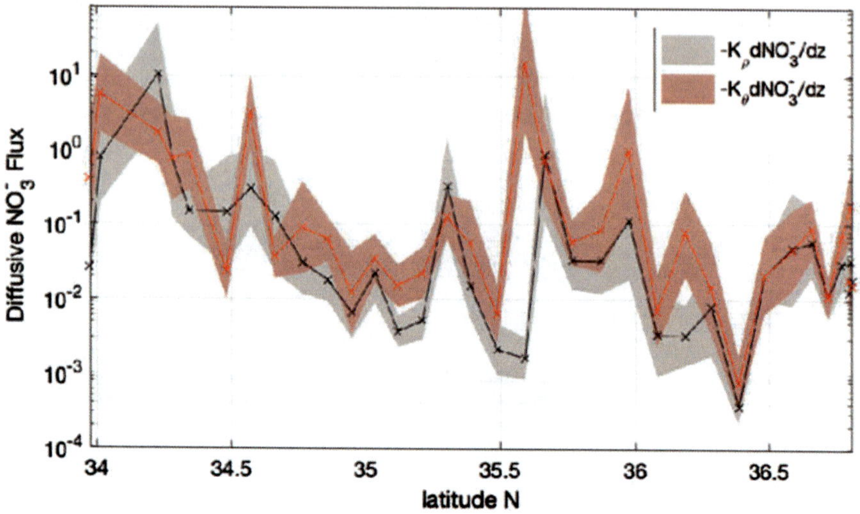

Fig. 23 Nutrient diffusive flux (mmol N m^{-2}day^{-1}) averaged at 50–125 m depth as a function of latitude for nitrate. Black line indicates diffusive flux computed with the eddy diffusivity K_ρ for density and red line represents that with the effective thermal diffusivity K_θ

6.2.1 Microstructure Observations in the Kuroshio Extension

The microstructure observations in the Kuroshio Extension were conducted along 143°E during August 2008 in the first meander of the Kuroshio Extension front [34]. The zonal thermal-wind shear is subtracted from that measured by the shipboard ADCP to obtain non-geostrophic zonal shear, i.e.,

$$\frac{\partial u_n}{\partial z} = \frac{\partial u_A}{\partial z} + \frac{g}{\rho_o}\frac{\partial \rho}{\partial y}, \tag{30}$$

where the subscript n stands for non-geostrophic, and A stands for ADCP. The obtained non-geostrophic shear shows banded structure near the region with the sharpest lateral temperature gradient formed at the Kuroshio Extension thermocline. The banded structures appear nearly along isothermal (Fig. 24a). Because the stratification is formed dominantly by temperature, these banded shear structures also align along the isopycnals, similar to those observed in the Tokara Strait and the Hyuganada Sea. The ray paths computed with several frequencies from near-inertial frequencies (1.01f, 1.1f, 1.2f and 1.5f) show angles that agree with the observed angle of the shear, suggesting they are caused by near-inertial internal waves (Fig. 24a). A microstructure profile in the Kuroshio Extension axis shows large TKE dissipation rates of O(10^{-7} W kg^{-1}) right at these shear banding layers [34]. Although Nagai et al. [34] considered frontogenesis to be the cause of the strong turbulence, these results suggest that it is likely caused also by the trapped

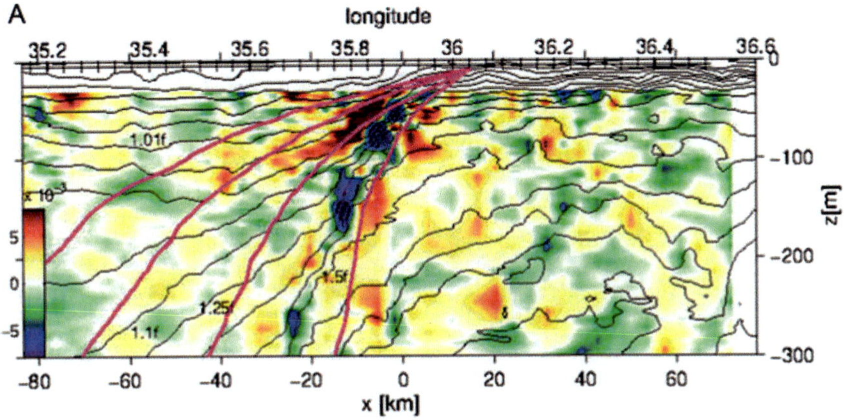

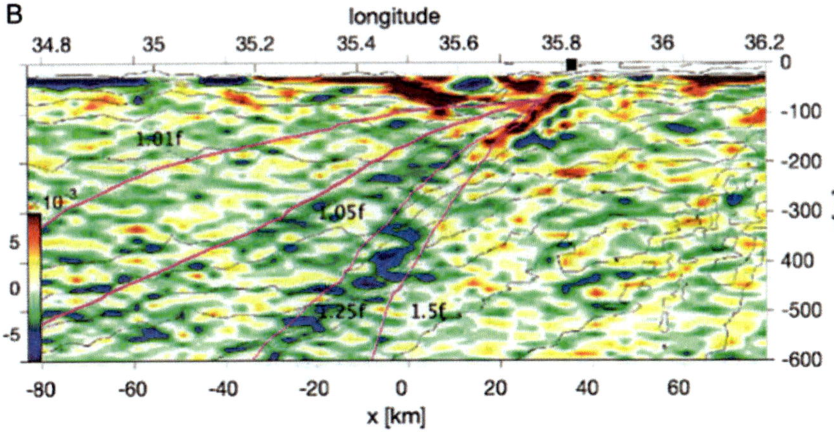

Fig. 24 Vertical sections of zonal ageostrophic shear (s^{-1}) for observations during (**a**) August 2008 (see Fig. 1 "UM08") and (**b**) August 2011 (see Fig. 1 "UM11") in the Kuroshio Extension

near-inertial internal waves. These banded non-geostrophic shear layers are frequently observed in the Kuroshio thermocline. During another cruise in August 2011, the banded non-geostrophic shear structures were also observed roughly along tilted isopycnals of the Kuroshio Extension (Fig. 24b). The angles of the same near-inertial ray paths are in good agreement with those for the observed banded shear. Several microstructure profiles at these banded shear layers also show large TKE dissipation rates of O(10^{-7} W kg^{-1}) (Fig. 25a).

Other microstructure observations in August 2012 also show large TKE dissipation rates, with the same order of magnitude O(10^{-7} W kg^{-1}) at the tilted sharp Kuroshio thermocline (Fig. 25b), suggesting that the banded shear structures accompanied by O(10^{-7} W kg^{-1}) TKE dissipation rates are commonly observable features in the first meander of the Kuroshio Extension. The mean TKE dissipation rates below the mixed layer for these observation data during 2008, 2011, 2012, as well as

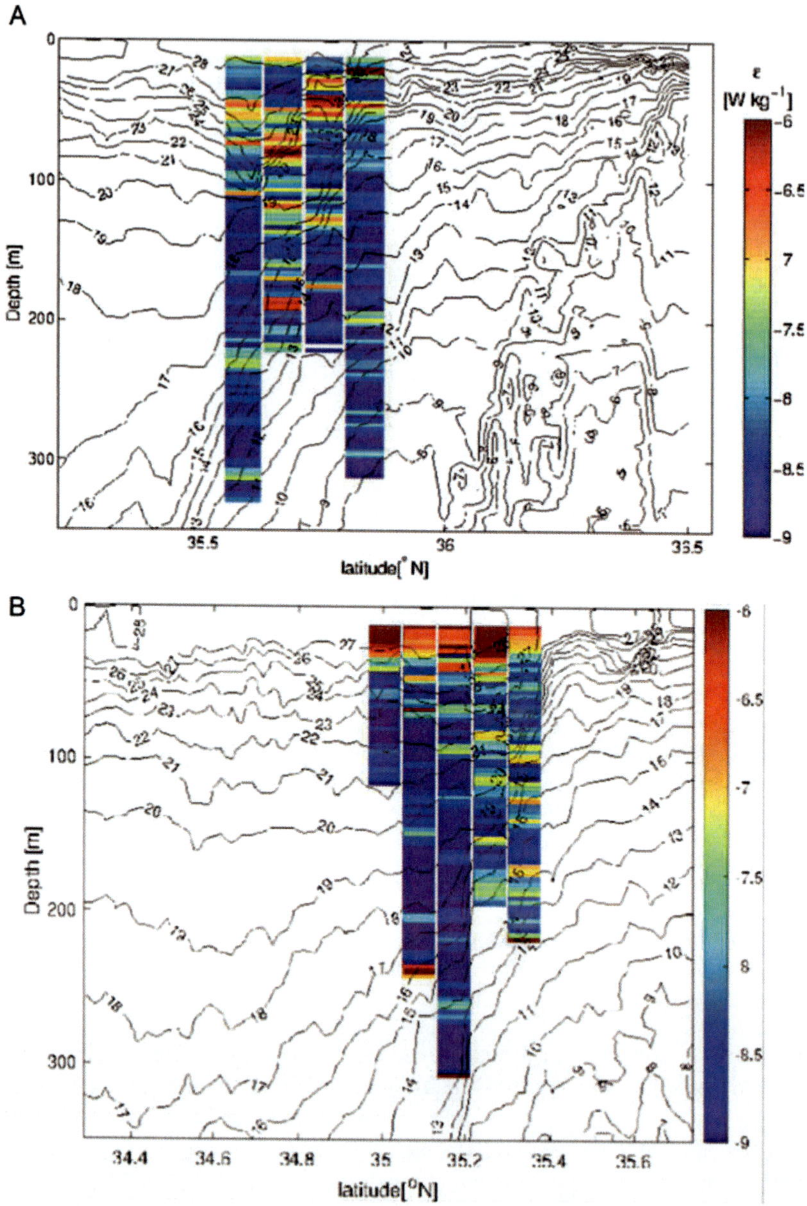

Fig. 25 TKE dissipation rates $\log_{10}\varepsilon$ (W kg^{-1}) during (**a**) the August 2011 cruise (see Fig. 1 "UM11") and (**b**) the August 2012 cruise (see Fig. 1 "UM12") in the Kuroshio Extension, along 142° and 143°E, respectively

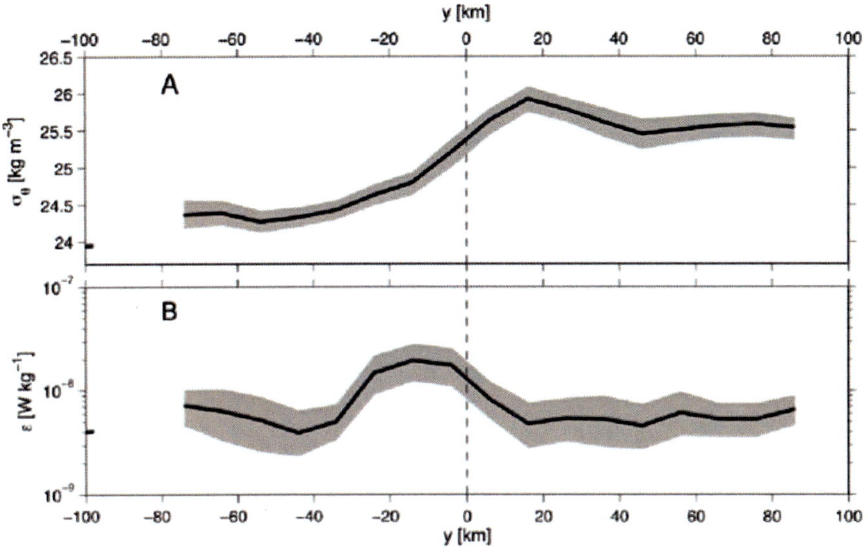

Fig. 26 Mean (**a**) σ_θ (kg m^{-3}) and (**b**) TKE dissipation rates ε (W kg^{-1}) averaged below the mixed layer for August 2008, 2011, 2012, and October 2009 [78] cruises in the Kuroshio Extension as a function of distance from the Kuroshio Extension axis. Shaded area is 95% confidence interval

those taken during October 2009 [78], as a function of the distance from the Kuroshio Extension axis, show that they are elevated as O(10^{-8} W kg^{-1}) from 0 to 30 km on the north side of the Kuroshio Extension axis (Fig. 26). These elevated dissipation rates in the Kuroshio Extension are, however, one order smaller than those obtained in the upstream Kuroshio regions, such as the Tokara Strait, the Hyuganada Sea, and the Izu Ridge, where average dissipation rates are O (10^{-7} W kg^{-1}). Since the large amplitude high vertical wavenumber banded shear structures are frequently seen in many different regions along the entire Kuroshio path near its current axis, such as the I-Lan Ridge (not shown), the Okinawa Trough (not shown), the Tokara Strait (Fig. 14), the Hyuganada Sea (Fig. 18), and the Kuroshio Extension (Fig. 24), it is suggested that the trapping of the near-inertial internal waves in the Kuroshio by geostrophic lateral and vertical shear plays a very important role to increase the base level of the TKE dissipation in the Kuroshio.

6.3 *Double-Diffusive Convection in the Kuroshio Extension*

Even without strong mechanical turbulence, the ocean has a functionality to mix tracers vertically through double-diffusive processes, i.e., salt fingering and diffusive convection. These are microscale vertical mixing processes which can occur in fluids whose buoyancy is affected by these two constituents that diffuse at different rates [79], exchanging heat, salt, and tracers and even forming new water masses in the case of ocean [80].

Salt fingering (SF) occurs when the stable temperature stratification overcompensates for the unstable salinity stratification. It can be triggered by small vertical displacements of a fluid parcel. Since the heat diffuses 100 times faster than salt, the parcel brought downward becomes heavier and therefore, keeps moving downward, and vice versa for the parcel displaced upward (e.g., [81]). Diffusive convection (DC) occurs when a cold-fresh layer is introduced on top of a warmer saltier layer; the fluid parcel undergoes growing oscillatory motions because of the differing diffusivities [82].

In the following subsections, the influence and the temporal modulations of double-diffusive convection on the diapycnal nutrient flux in the Kuroshio Extension and in the Kuroshio-Oyashio Confluence region are analyzed.

6.3.1 Profiling Float Surveys in the Kuroshio Extension During July 2013

In the Kuroshio flowing over the Izu Ridge, the diffusive nutrient fluxes have mostly similar orders of magnitudes with eddy diffusivities for both density and heat. This is a clear indication that the diffusive fluxes are dominated by mechanical turbulence, i.e., $K_\theta \sim K_\rho$. However, after the Kuroshio flows away from the Boso Peninsula (Fig. 1), it encounters the cold-fresh Oyashio-related waters transported from the Kuroshio–Oyashio Confluence region, along the Kuroshio Extension front (Fig. 4d). In July 2013, Nagai et al. [31] deployed a microstructure float (Navis-MR) and an E-M (Electromagnetic) APEX float in the Kuroshio Extension axis to measure turbulence, and lateral velocity with temperature and salinity ([31], Fig. 27). The

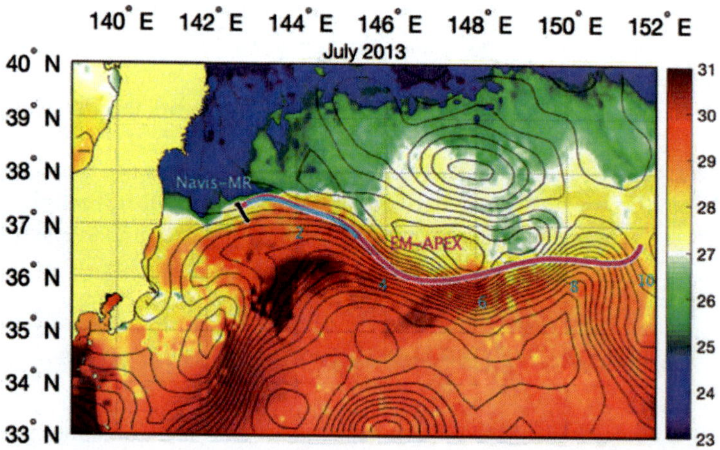

Fig. 27 Map of the July 2013 survey. Color shading shows satellite sea surface temperature (°C) on July 18, 2013. Magenta and cyan curves are trajectories of the EM-APEX and Navis-MR floats, respectively. Numbers along the EM-APEX trajectory indicate days after the deployment. The thick black line represents the across-front Underway-CTD transect shown in Fig. 28a. Contours are sea surface height

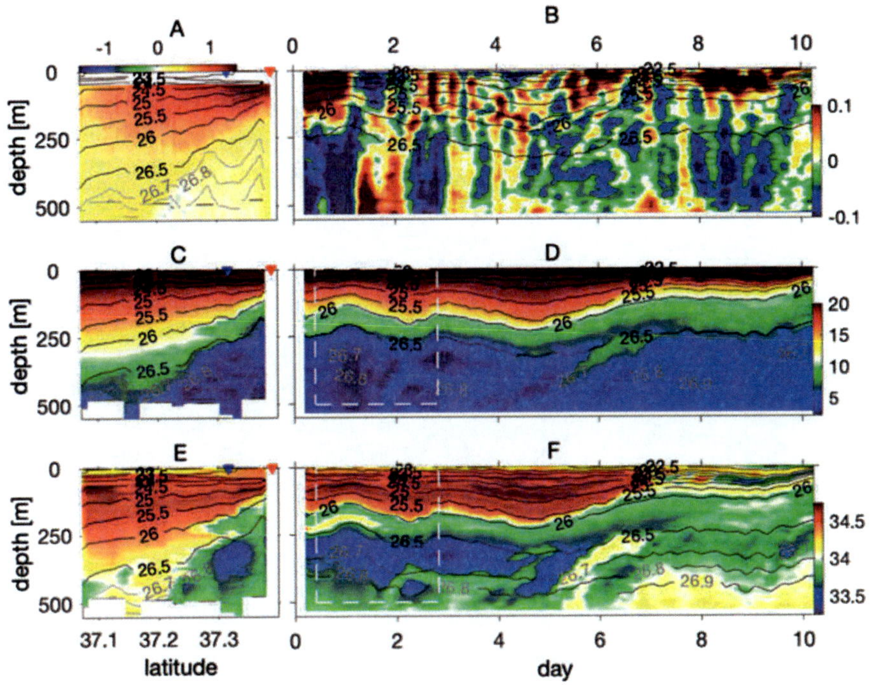

Fig. 28 Vertical sections of (**a**) along Kuroshio Front flow, (**b**) across-front flow, (**c**, **d**) temperature (°C), and (**e**, **f**) salinity. (**a**, **c**, **e**) show across-frontal sections obtained by the Underway-CTD observations. (**b**, **d**, **f**) are along frontal sections measured by the EM-APEX float. Black contours are isopycnals. White dashed rectangle in (**d**) and (**f**) is the region measured by the Navis-MR microstructure float during the July 2013 survey. Blue and red triangles in (**a**, **c**, **e**) indicate the deployment positions for the Navis-MR and the Navis-BGC floats, respectively

results showed vertical high wavenumber thermohaline interleaving structures in and below the thermocline of the Kuroshio Extension ([31]; Fig. 28c–f). Further analyses suggested that along isopycnal stirring and shearing caused by the meso- and submesoscale eddies and propagating near-inertial internal waves (Fig. 28b) are responsible for the generation of the interleaving with small vertical wavelengths [31]. To measure the susceptibility to double-diffusive processes, the Turner angle Tu was computed [83], identifying the values of $45° < Tu < 90°$ for SF and $-90° < Tu < -45°$ for DC. Upper and lower part of the interleaving layers were found to be in SF and DC favorable stratification, respectively, being stable for the Kelvin-Helmholtz instability with gradient Richardson number $\mathrm{Ri} > 0.25$ (Fig. 29c, d).

To compare the diffusivities between the empirical models for the double-diffusion and those measured by the microstructure float (Navis-MR), a salt-finger induced thermal diffusivity, K_θ^{SL} parameterized by Radko and Smith [84], and that for diffusive convection, K_θ^{DC} by Fedorov [1], are employed as follows:

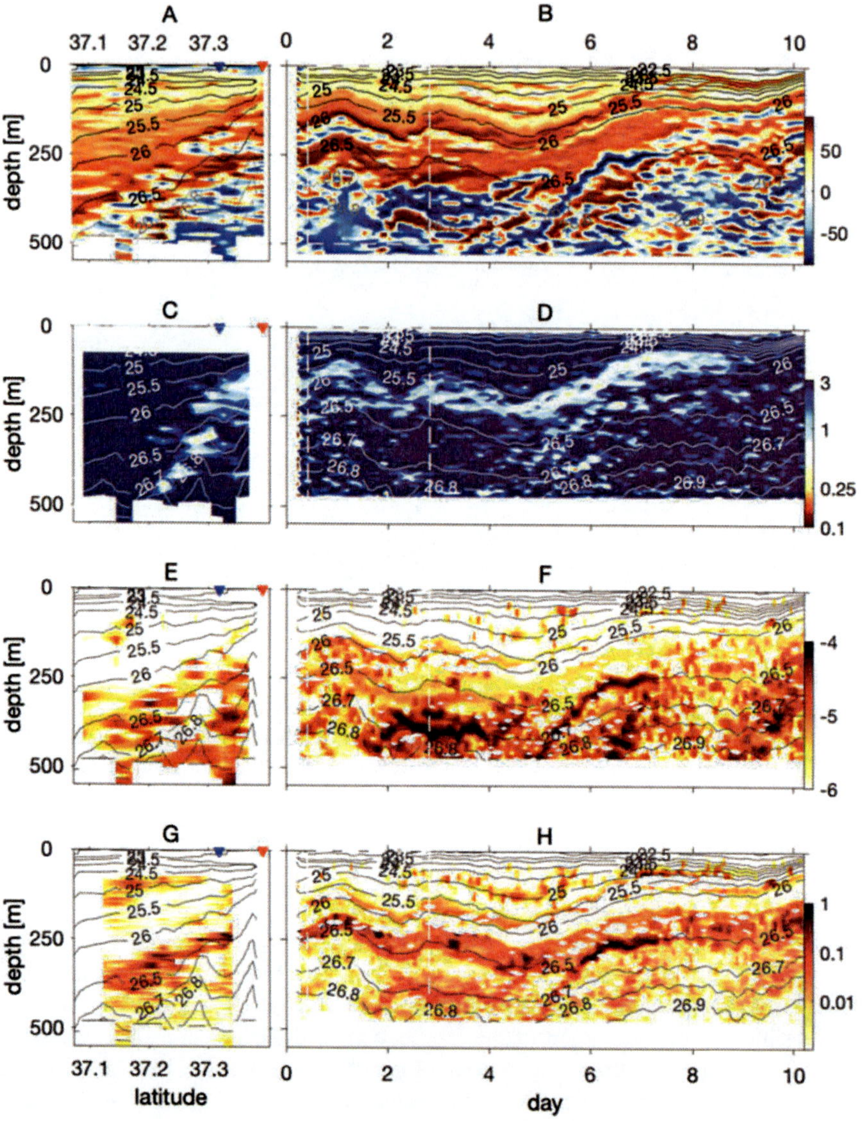

Fig. 29 Vertical sections of (**a**, **b**) Turner angle, *Tu* (°), (**c**, **d**) gradient Richardson number, (**e**, **f**) parameterized thermal diffusivity for double-diffusion using (31–32) (m^2s^{-1}), and (**g**, **h**) parameterized double-diffusive nitrate flux [mmol N $m^{-2}day^{-1}$]. Blue and red triangles in (**a**, **c**, **e**, **g**) indicate the deployment positions for the Navis-MR and the Navis-BGC floats, respectively

$$K_\theta^{\mathrm{SL}} = \left(\frac{a_s}{\sqrt{R_\rho - 1}} + b_s\right) k_t a_g \left[\exp\left(-b_g R_\rho\right) + c_g\right], \tag{31}$$

where $a_s = 135.7, b_s = -62.75, a_g = 2.709, b_g = 2.513, c_g = 0.5128$ and $R_\rho > 1$ for salt finger, and

$$K_\theta^{\mathrm{DC}} = 0.909\nu \exp\left\{4.6 \exp\left[-0.54\left(\frac{1}{R_\rho} - 1\right)\right]\right\}, \tag{32}$$

where ν is the kinematic viscosity ($\mathrm{m^2s^{-1}}$) and $0 < R_\rho < 1$ for diffusive convection. R_ρ is the density ratio defined as $\alpha\Theta_z/\beta S_z$, where α and β are the thermal and salt expansion coefficients, respectively, Θ_z and S_z are background vertical gradients of temperature and salinity, respectively.

Using these parameterizations for measured temperature and salinity, enhanced diffusivity of $O(10^{-5}\ \mathrm{m^2s^{-1}})$ for heat is found below the layer denser than $\sigma_\theta = 26\ \mathrm{kgm^{-3}}$, where temperature and salinity inversions are frequently seen (Fig. 29e, f). Some distinct layers of large thermal diffusivity, $O(10^{-4}\ \mathrm{m^2s^{-1}})$, are associated with vertical high wavenumber thermohaline interleaving. Additionally, Nagai et al. [36] estimated nitrate concentrations using an empirical model of nitrate as a function of seawater density (see Fig. 2 of [36]), derived from water samples in the Kuroshio Extension front. The nitrate vertical diffusive flux was computed with the estimated nitrate concentration and parameterized double-diffusion induced thermal diffusivity. Here, the thermal diffusivity was used for comparing the parameterized fluxes with those based on the measured thermal diffusivity using temperature microstructure data shown below. Although the molecular diffusivity for nitrate is closer to that of salinity, $O(10^{-9}\ \mathrm{m^2s^{-1}})$, than to that of heat, $O(10^{-7}\ \mathrm{m^2s^{-1}})$, parameterizations of double-diffusion (e.g., [85]) provide very similar effective diffusivities for heat and salt as a function of density ratio. Thus, employing the effective thermal diffusivity to infer nitrate flux can still be useful. The parameterized nitrate diffusive flux was mostly ~$O(0.1\ \mathrm{mmol\ m^{-2}\ day^{-1}})$, with $O(1\ \mathrm{mmol\ m^{-2}\ day^{-1}})$ values at the density layer along $\sigma_\theta = 26.5\ \mathrm{kgm^{-3}}$ around 6–8 days after the EM-APEX float deployment (Fig. 29h).

Because one (Navis-MR) of the two deployed autonomous floats carried microstructure sensors, it was able to measure the eddy diffusivity for density and effective thermal diffusivity directly without any parameterization, but with some assumptions described in the previous section. The measured eddy diffusivity, K_ρ was found to be moderately large in the layer denser than $\sigma_\theta = 26.5\ \mathrm{kgm^{-3}}$, with values of $O(10^{-5}\text{–}10^{-4}\ \mathrm{m^2s^{-1}})$ (Fig. 30a). On the other hand, effective thermal diffusivity K_θ was one order larger, $O(10^{-4}\text{–}10^{-3}\ \mathrm{m^2s^{-1}})$, in the layer denser than $\sigma_\theta = 26\ \mathrm{kgm^{-3}}$ (Fig. 30b). The effective thermal diffusivity larger than the eddy diffusivity for density is an indicator of active double-diffusion. The same empirical nitrate model (Fig. 2 of [36]), utilized to estimate nitrate concentration (Fig. 30c), is used with the measured diffusivities, K_ρ and K_θ, to compute diffusive nitrate fluxes. The obtained fluxes were also one order magnitude larger, 1–10 $\mathrm{mmol\ m^{-2}\ day^{-1}}$, with

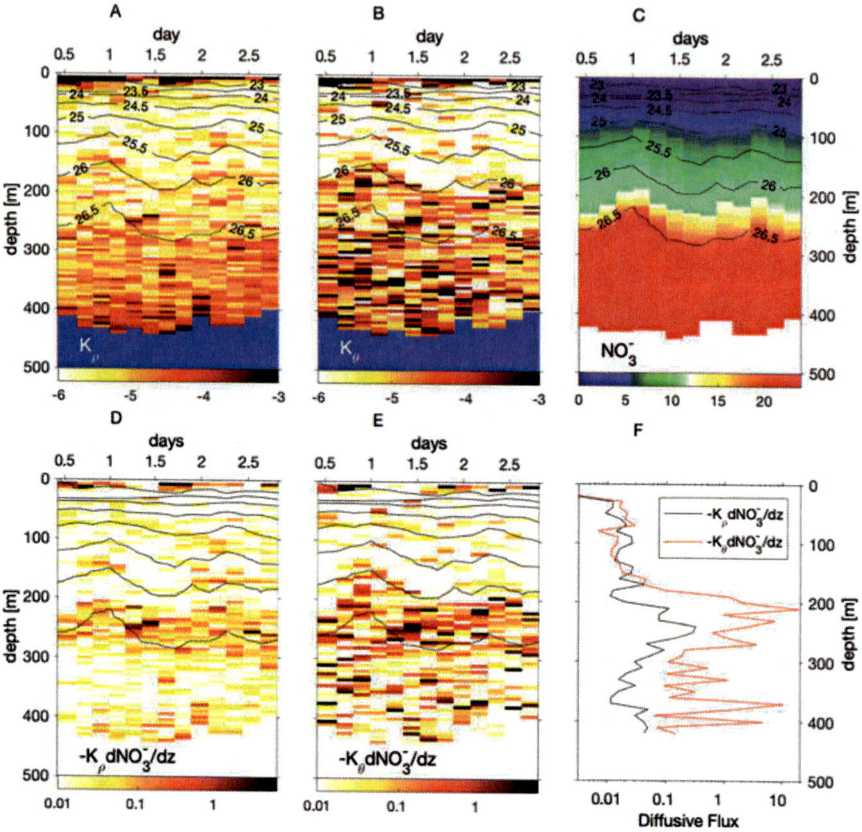

Fig. 30 Microstructure and turbulence data obtained by the Navis-MR float deployed along the Kuroshio during the July 2013 cruise. The trajectory and the measured depth range of the float is shown in Fig. 27. (**a**) Eddy diffusivity for density K_ρ $\log_{10}$ (m^2s^{-1}) using TKE dissipation rates ε (W kg^{-1}), (**b**) effective thermal diffusivity K_θ $\log_{10}$ (m^2 s^{-1}), (**c**) estimated nitrate concentration (μM), (**d**) nitrate diffusive flux by turbulence and (**e**) double-diffusive nitrate flux (mmol N $m^{-2}day^{-1}$), (**f**) average diffusive nitrate flux with the eddy diffusivity for density K_ρ (in black) and that with effective thermal diffusivity K_θ (in red) (mmol N $m^{-2}day^{-1}$)

K_θ compared with K_ρ (Fig. 30d–f). These results suggest that the double-diffusion is the dominant agent to diffuse nutrients upward along the Kuroshio Extension in the layer denser than $\sigma_\theta = 26$ kgm^{-3} [31]. It should be noted that direct estimation of the nitrate diffusive flux using high-resolution thermistor data, 1–10 mmol m^{-2} day^{-1}, is one order larger than that parameterized for the double-diffusion, 0.1–-1 mmol m^{-2} day^{-1}. A recent glider hydrographic survey by Jan et al. [86] has also found salinity interleaving layers in the same density range, $\sigma_\theta > 26.5$ kgm^{-3} below the upstream Kuroshio core east of Taiwan, which are found to be driven primarily by double-diffusive processes, implying that the effects of double-diffusion on the nutrient flux are more widespread in the subsurface layers of the Kuroshio.

6.3.2 Interannual Variabilities of Double-Diffusive Favorable Stratification Revealed by Argo Float Data

The double-diffusive processes can be present ubiquitously in ocean fronts where the along isopycnal variations of spiciness are large, such as in the Kuroshio Extension (KE) and the Kuroshio-Oyashio Confluence regions. It has been shown that double diffusion, dominated by SF, has significant influence in setting the water mass properties within the Kuroshio-Oyashio Mixed Water region [87, 88].

Since double-diffusion processes can also play a role in supplying nutrients to the upper layers [89], the temperature and salinity profiled vertically by Argo floats are analyzed in this section. The analyses are carried out for the Kuroshio Extension and the Kuroshio-Oyashio Confluence region, located between 33°N – 39°N and 140°E – 155°E. For the latitude range, the Kuroshio Extension axis is known to be between 33°N – 36°N, while the Kuroshio-Oyashio Confluence region locates between 36°N – 39°N [90]. As for the longitude range, since two quasi-steady meander crests are located at 144°E and 150°E and their presence is related to the Kuroshio Extension state [4, 91, 92], the zonal range is determined to cover these crests. The study period 2006–2014 is divided into two (unstable (2006–2009) and stable (2010–2014)) Kuroshio Extension times according to the Kuroshio Extension state described and analyzed in previous studies [92, 93].

First, the Turner angle, *Tu* [83] was computed for the Argo data to investigate the occurrences of the SF and DC. Furthermore, the study region was diced into 9 regions (3 by 3 in longitude and latitude), and an empirical regression model between nitrate and density was determined for each region, using nitrate, temperature and salinity data obtained from World Ocean Atlas 2018 ([94], WOA18 1°× 1° grid, https://www.nodc.noaa.gov/OC5/woa18/woa18data.html). Using this empirical model and the parameterizations for the double-diffusion induced diffusivities for SF (31) and DC (32), the nitrate flux (mmol m^{-2} day^{-1}) from Argo float data was estimated.

In the Kuroshio Extension region, salinity exhibits a subsurface maximum since past evaporation exceeds the precipitation; this condition provides a basic stratification above the subducted low saline North Pacific Intermediate Water (NPIW), more favorable for the SF process. The SF favorable stratification seems to be persistent throughout the unstable period, in the density range $\sigma_\theta = 26$–26.5 kgm^{-3} (Fig. 31a), where there is a decrease in its occurrence from 2010, followed by the increase in DC favorable conditions, which coincides with the beginning of the Kuroshio Extension stable period. The increase in DC favorable conditions is accompanied by a relatively large nitrate flux, ~0.5 mmol m^{-2} day^{-1}, especially from 2012 (Fig. 31b). These increases coincide with the large amplitude negative values of the vertical gradients of spiciness τ_z [32], approximately −0.03 m^{-1} at low *Tu* (Fig. 31c), suggesting that the increased thermohaline inversion results in a large nitrate flux during the stable Kuroshio Extension period. It should be noted that the number of the Argo data obtained for the years, 2013–2014 accounts for 62% of the total number of data, which would produce a bias in the results. However, the increase in

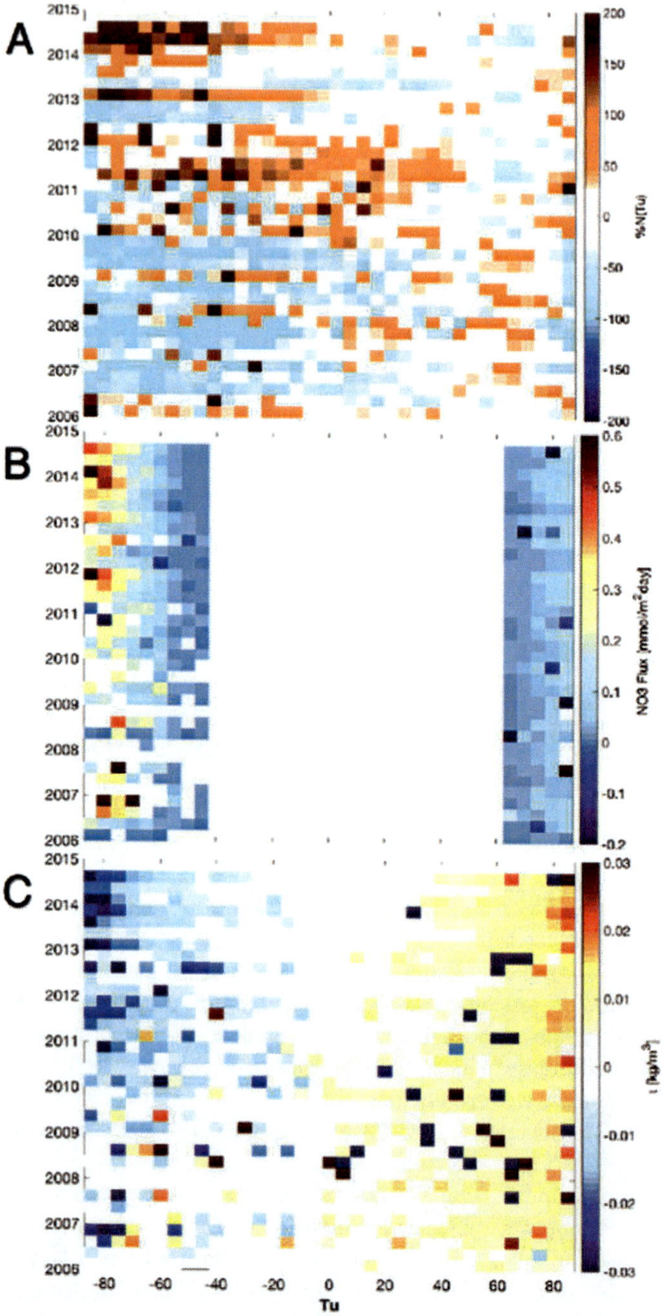

Fig. 31 Hovmöller diagram of anomaly for (**a**) the normalized frequency of Turner angle N(*Tu*)%, (**b**) average nitrate diffusive fluxes (mmol $m^{-2}day^{-1}$), (**c**) spiciness vertical gradient τ_z (m^{-1}) in the density range $\sigma_\theta = 26$–26.5 kgm^{-3}. For (**a**), the normalized histogram of *Tu* is computed for every

the DC favorable conditions, the nitrate flux, and the amplitude of negative vertical spiciness gradient started from 2010, suggesting that these are not caused by the difference in the amount of data.

To evaluate the possible influence of the Oyashio intrusion on the analyzed results, the fraction of the Oyashio water is quantified by extracting the water with a temperature range 0–7°C, salinity range of 33–33.7, and a density criterion, $\sigma_\theta < 26.7$ kgm^{-3} [95]. The calculated fraction of the Oyashio water is found to be only 4% and 2% of the total Argo floats data for unstable and stable period, respectively, indicating that the results of the analyses are not largely affected by the contribution from nutrient-rich Oyashio water.

The more favorable conditions for DC during the Kuroshio Extension stable period are also seen in the temperature-salinity diagram, especially in the water denser than $\sigma_\theta = 25.5$ kgm^{-3}, with lower values of *Tu* (Fig. 32b). Similarly, a larger magnitude of vertical gradient of spiciness was also observed in the same dense waters (Fig. 32d); suggesting that high vertical wavenumber thermohaline interleaving is more frequently formed in the layers $\sigma_\theta > 25.5$ kgm^{-3} during the stable period. Note that the lower *Tu* and larger spiciness gradient during the stable period are found with slightly lighter water than the Oyashio water range mentioned above. To identify where the DC favorable stratification during the stable period is formed in the density range $\sigma_\theta = 26$–26.5 kgm^{-3}, the *Tu* for only the DC favorable range and the corresponding depth are bin-averaged onto the grid with 0.25° resolution (Fig. 33). The results show that the DC favorable stratification is more frequently found during the stable period compared to the unstable period, especially at the first meander trough of the Kuroshio Extension (Fig. 33b). Furthermore, the depths of these DC favorable stratifications during the stable Kuroshio Extension period are relatively shallow, 100–200 m, in the Kuroshio-Oyashio Confluence region (Fig. 33b, d), suggesting that the DC may be able to supply nutrients to an even shallower depth which can reach to the euphotic zone by additional mixing during wintertime. On the other hand, the average depths of density layers $\sigma_\theta = 26$–26.5-kgm^{-3} in the unstable period are much deeper than those in stable period due to the northward spread of thick warm salty Kuroshio Extension water, implying that DC-induced nitrate flux is hard to reach the euphotic zone.

Although the interannual variations in double-diffusive favorable stratification are observed in the Argo float data, it is not clear what the driving mechanisms are. Warm streamers from the Kuroshio Extension are more abundant during the unstable period, fueling warm salty water to warm-core eddies [96] which move generally northward [91] to the confluence region, covering the cold-fresh Oyashio-related water, and possibly providing stratification more favorable for the SF process. In contrast, during the stable period, the Kuroshio Extension tends to present a straight path [93], making the waters from the Oyashio Current to be located right next to the

Fig. 31 (continued) 3 months, and the time mean normalized histogram of *Tu* is subtracted to obtain the anomaly of the normalized frequency and is represented as the percentage increase with respect to the time mean

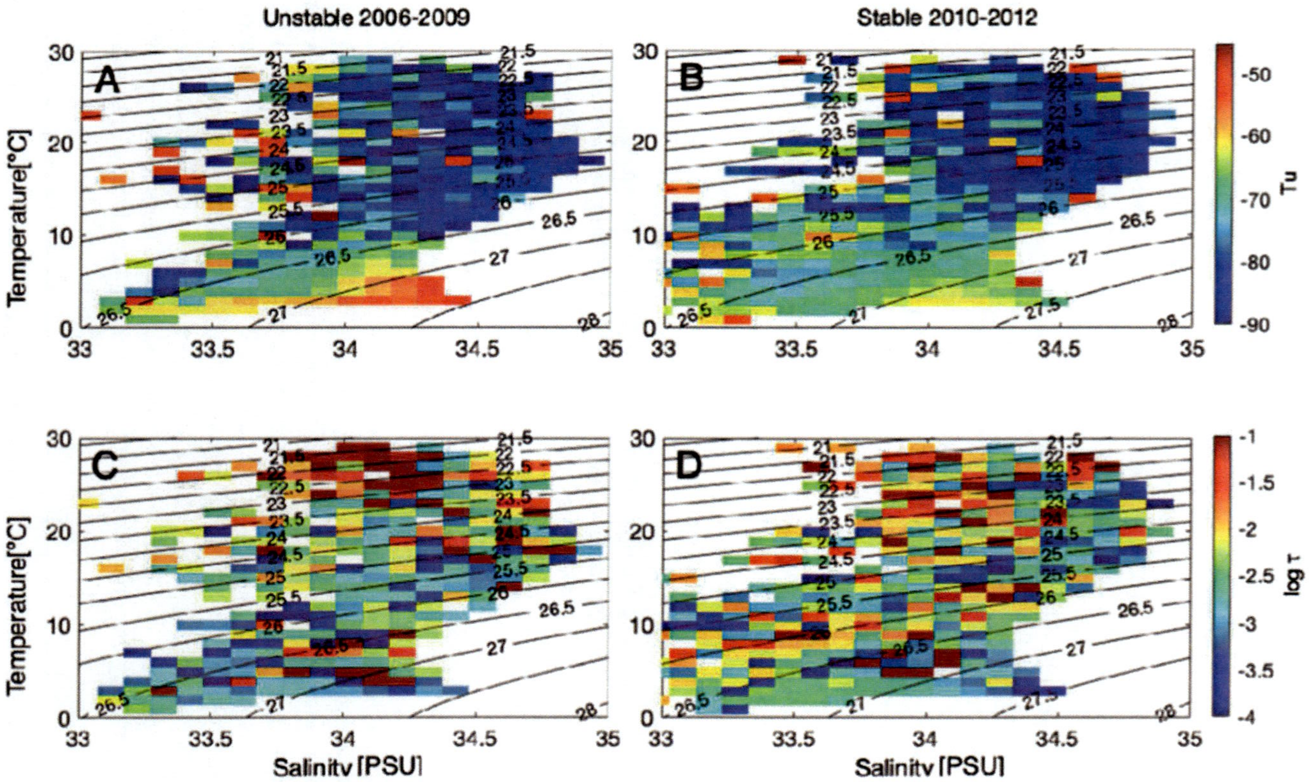

Fig. 32 T-S diagrams with the color indicating bin-averaged (**a**, **b**) Turner angle (°) only for diffusive convection (DC) range and (**c**, **d**) corresponding $\log_{10}$ spiciness vertical gradient τ_z (m^{-1}) for the Kuroshio Extension unstable period, 2006–2009 (on the left, **a** and **c**), and stable period, 2010–2012 (on the right, **b** and **d**). Note that year 2013–2014 is excluded to avoid bias from the large difference in the number of available profiles obtained from Argo float data

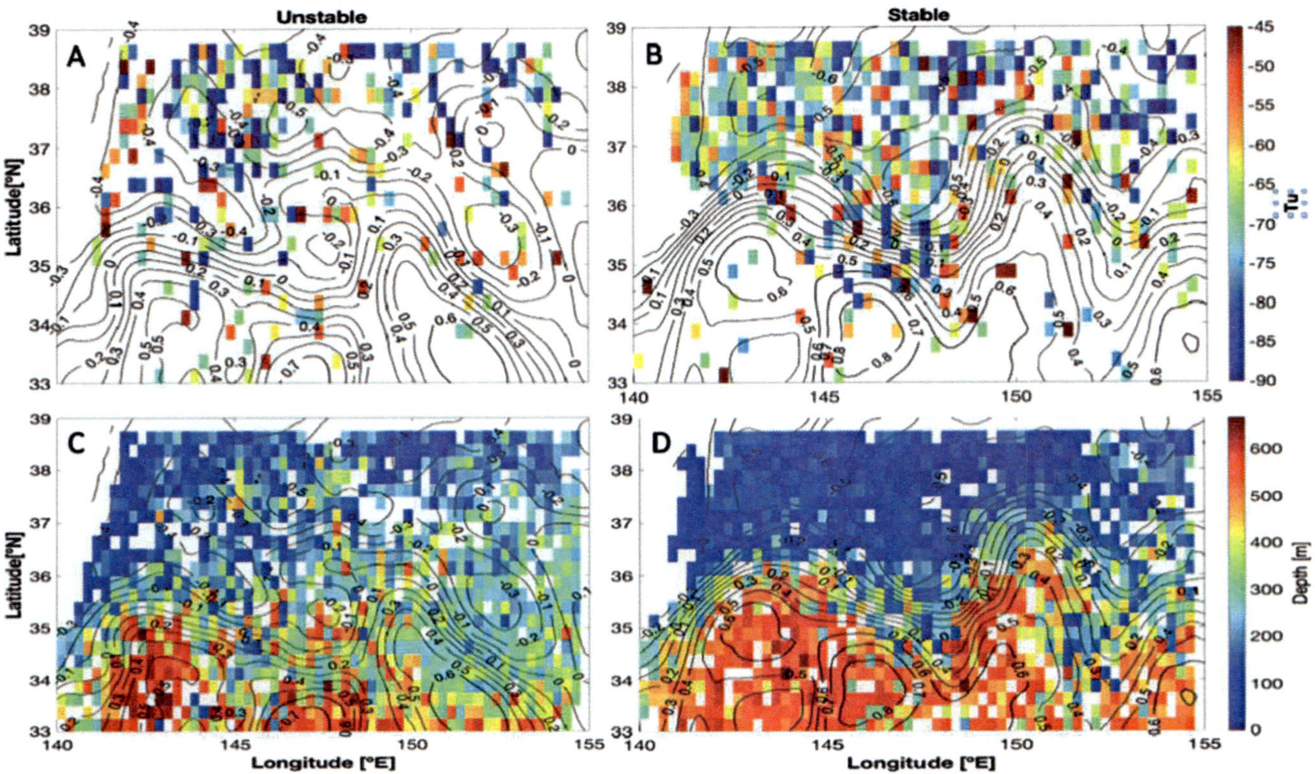

Fig. 33 Location of bin-averaged (**a**, **b**) $Tu < -45°$ (for diffusive convection process only) and (**c**, **d**) depth (m) for (on the left, **a** and **c**) the Kuroshio Extension unstable period, 2006–2009, and (on the right, **b** and **d**) stable period, 2010–2012, within the density range $\sigma_\theta = 26$–$26.5\ kgm^{-3}$. Contours represent annual averaged sea surface height (SSH) values (m). Note that average depths are computed for both diffusive convection and salt-finger cases

strong Kuroshio Extension jet, with warmer salty waters along 36–37°N, and increasing the chance to form favorable conditions for the DC process. Further analyses are still needed to understand the mechanisms in detail, considering, for example, external forcing at the surface, as previous studies reported that the wind power input to the near-inertial energy also modulates interannually [97], and that near-inertial shear can be effective to form thermohaline interleaving layers [31].

7 Lateral Advection Versus Diapycnal Nutrient Flux

Since Pelegri and Csanady [13] discovered elevated nutrient concentrations along the Gulf Stream on density surfaces, the mechanisms to form higher nutrient concentrations have been under debate. Pelegri and Csanady [17] and Pelegri et al. [14] proposed that the vertical shear of the Gulf Stream could induce strong diapycnal mixing that could generate the characteristic nutrient distributions. On the other hand, Palter and Lozier [15] analyzed the source of the Gulf Stream nutrients, using the historical hydrographic data, and concluded that the majority of the nutrients is advected from the tropical regions, and that diapycnal mixing is not important. Similarly, Whitt [16] deduced the time scale of advection, diapycnal mixing and isopycnal eddy stirring to make a rough estimation of their relative importance, assuming that the time scale for diffusion can be scaled by $\Delta D^2/k$, where D and k are the spatial scale and the diffusivity, respectively. With a thickness of the nutrient elevation ΔH ~100 m and a diapycnal eddy diffusivity K_ρ ~ 10^{-5}–10^{-4} m^2s^{-1}, $\Delta H^2/K_\rho$ approximately 1,000 to 10,000 days are required for diapycnal diffusion to generate elevated nutrient concentrations. Contrarily, the effects of along isopycnal eddy stirring can be represented with the effective horizontal eddy diffusivity of K_h ~ 10^2–10^4 m^2s^{-1}, which yields a time scale of 10 to 1,000 days to diffuse the elevated concentrations in a 100 km width. With the advection time scale of a few weeks that the Gulf Stream travels between the Strait of Florida and the Grand Banks (~3,000 km), Whitt [16] concluded that the diapycnal diffusion is too slow compared to the advection. However, the question remains on whether the diapycnal mixing has only negligible effects to form and/or maintain the higher concentrations of nutrients on the density surface. Since the nutrients in the Gulf Stream and in the Kuroshio lie in the subsurface layer, the time scale of the advection could be longer than these previous estimates. If this is the case, the time scale of the along isopycnal eddy stirring, 10 to 100 days with K_h ~ 10^3–10^4 m^2s^{-1}, is similar or shorter to the advection time scale of, for example, 100 days. Nevertheless, the high concentration of nutrients seems to persist over the long distance, without being completely diluted by the eddy stirring. Nagai et al. [12] conducted passive tracer release experiments with the same initial distribution as model nitrate in the western North Pacific, and showed that the released tracer in the western tropical region takes O(100 days) to travel from east of Taiwan to the south of Japan. This is consistent with the results shown in the previous sections, where the elevated nitrate concentrations are slightly on the onshore side of the Kuroshio, with slower advection

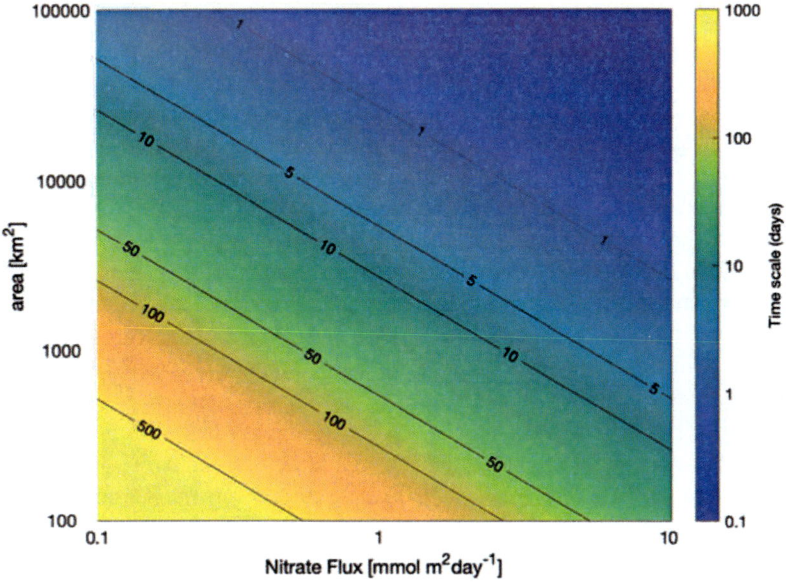

Fig. 34 Time scale required to generate nitrate anomaly of 26 Gmol as a function of nitrate flux (mmol m^{-2}day^{-1}) along the Kuroshio and the area over which the nitrate diffusive flux occurs

speeds. Furthermore, as mentioned in the previous sections, the Kuroshio flows over rough topography more frequently compared to the Gulf Stream. Thus, in this section, the effects of diapycnal mixing to form and/or maintain the elevated nitrate concentrations in the Kuroshio nutrient stream are examined.

To this end, the time scale to form the amount of positive nitrate anomaly found in the subsurface layers of the Kuroshio is estimated. First, positive anomalies of nitrate concentration at the P-N Line and 137°E Line are computed by subtracting the mean nitrate concentration averaged along the cross-frontal direction as a function of only density, from that as a function of both density and distance from the Kuroshio axis. Then, only positive anomalies are integrated along the cross-frontal direction and the vertical direction over a limited distance range of ±30 km from the Kuroshio axis, and over a density range of $\sigma_\theta = 24.25–25.5$ kgm^{-3}. The resultant nitrate amounts per unit length along the Kuroshio are 4.8 kmol m^{-1} and 5.2 kmol m^{-1} for the P-N Line and 137°E Line, respectively. Assuming an along frontal scale of 5,000 km of the meandering Kuroshio from off east coast of Taiwan through the Kuroshio Extension up to 160°E, the amount of the positive anomaly is estimated to be 5.2 kmol m^{-1} × 5,000 km = 26 Gmol. The vertical diffusive flux of nitrate estimated in the previous section shows values in the range of 0.1 to 10 mmol m^{-2} day^{-1}. Assuming a large nitrate flux of 1 mmol m^{-2} day^{-1}, similar to those found in the Tokara Strait on average [62] and an area of 1,000 km^2 where this large flux keeps injecting nitrate through the density layer $\sigma_\theta = 25.5$ kgm^3, the time scale to generate the observed nitrate anomaly is estimated to be <50 days, which is comparable to or less than the nitrate advection time scale of O(100 days) (Fig. 34).

The question is, however, whether there is a diapycnal nitrate diffusive flux of 1 mmol m^{-2} day^{-1} over an area of 1,000 km^2. According to our direct turbulence measurements and the estimations of the nitrate diffusive flux, there are most likely such large-scale nitrate injections. In the Kuroshio flowing through the Tokara Strait, even in the upstream side of the strait the nitrate flux of O(1 mmol m^{-2} day^{-1}) is found near the density layer $\sigma_\theta \sim 25$ kgm^3 over several tens of kilometers (Fig. 15d, e). A more extensive tow-yo microstructure survey during November 2017 from the upstream to the downstream across the Tokara Strait by Nagai et al. [76] showed that a 100- to 1,000-fold enhancement of turbulent dissipation rates spreads over 100 km to the downstream. In 2020 and 2021, the authors repeated similar microstructure observations in the Tokara Strait over 100 km distance along the Kuroshio. In these surveys, a similar enhancement of turbulent dissipation rate was found over 100 km, suggesting the plausibility of a large nitrate flux of O(1 mmol m^{-2} day^{-1}) over a 100 km scale in the Tokara Strait.

Also, in the Hyuganada Sea Nagai et al. [52] reported an intense nitrate diffusive flux of 1–10 mmol m^{-2} day^{-1} over 20–30 km distance across the Kuroshio (Fig. 19). Durán Gómez and Nagai [77], using a numerical simulation, showed that this elevated nitrate flux of O(1 mmol m^{-2} day^{-1}) extends further, over 50–100 km downstream. Furthermore, Nagai et al. [36], using an autonomous microstructure float, showed that a large nitrate flux of 1–10 mmol m^{-2} day^{-1} occurs in the Kuroshio within 20–40 km from the Izu Ridge (Fig. 22). These results suggest that a nitrate flux of O(1 mmol m^{-2} day^{-1}) over an area of 1,000 km^2 is plausible in the Kuroshio. These observations, showing enhanced turbulence along the Kuroshio, are consistent with the estimated global map of turbulent kinetic energy dissipation rates and vertical eddy diffusivity by Whalen et al. [98], using Argo float data in which large dissipation rates of O(10^{-8} W kg^{-1}) and high vertical eddy diffusivities of $>10^{-5}$ m^2s^{-1} are found at 250–500 m depth along the western boundary of the North Pacific where the Kuroshio flows.

Accordingly, the time scale of the diapycnal diffusion can be as short as ~50 days. Because this is comparable to that of the advective time scale of ~100 days, and because the time scale of along isopycnal stirring is ~100 days with 10^3 m^2s^{-1} of lateral diffusivity which acts to dilute the nitrate positive anomaly at the same time, it is unlikely that diapycnal diffusion has only negligible effects on the formation and/or maintenance of the elevated nitrate concentrations along the Kuroshio.

Uchiyama et al. [99], using submesoscale eddy-permitting simulations of the Kuroshio, showed that the eddy kinetic energy along the Kuroshio is continuously high from south of Kyushu to the Kuroshio Extension, suggesting that the level of eddy lateral dilution effects is similar along the Kuroshio and the Kuroshio Extension. Nevertheless, the elevated nitrate concentrations along the Kuroshio are not diluted away until they reach ~1,000 km away from the east coast of Japan in the Kuroshio Extension (Figs. 6b and 10a, [12]). This indicates that nitrate is supplied to form its positive anomaly along the Kuroshio when it flows along the western boundary, but this does not happen in the Kuroshio Extension. Because the Kuroshio encounters topographic features frequently in the upstream and regions south of

Shikoku Island and Honshu Island, the Kuroshio–topography interaction is a plausible candidate as the formation mechanism of the nitrate positive anomalies.

8 Conclusions and Open Questions

In this study, using a numerical model of the Kuroshio nutrient stream and direct microscale turbulence measurements, the long-lasting question on how the layer of elevated nutrient concentrations is formed and/or maintained has been addressed. For this purpose, three important processes are investigated, such as the nutrient advection toward the downstream, the effects of eddy stirring, and the diapycnal mixing processes. This chapter also introduced some of the required tools or theories that were necessary for the investigations conducted in this study.

First, the hydrographic data shows that the salinity distribution, as a function of density and distance from the Kuroshio axis, exhibits subsurface freshening on the slightly shore side of the Kuroshio, implying a diapycnal freshwater flux from the salinity minimum layer (Fig. 5b, d). Similar along isopycnal variations were found in the nitrate concentrations using the hydrographic data by JMA and World Ocean Atlas 2018 data. When these variations on the density surface are visualized in a plane view, elevated nitrate concentrations on the density surfaces along the Kuroshio are observed (Fig. 6b). These enhanced nutrient concentrations along the Kuroshio are very similar to those along the Gulf Stream [14].

Second, to investigate the lateral advection of the nitrate and the role of eddy stirring, an eddy-resolving numerical simulation was conducted. The results obtained from the model showed realistic nitrate distributions with elevated concentration along the Kuroshio on the density surfaces, which is consistent with the observations. The transport of model nitrate and its divergence on the density layers $\sigma_\theta = 25$–26.5 near the core of the Kuroshio nutrient stream suggest that the Kuroshio carries a large amount of nitrate and brings a large fraction to the subsurface layers of the southern coast of Japan (Fig. 12). Further investigations, using an eddy flux analysis on the density layer, showed that the mean flux acts to transport nitrate of the nutrient stream to the Kuroshio Extension regions, where the induction and vertical diffusion supply nitrate to the mixed layer (Fig. 9a). The eddy flux also acts to remove the nitrate from the nutrient stream, while supplying it to the onshore and offshore side of the Kuroshio, consistent with the eddy isopycnal stirring (Fig. 12b). However, as these results are obtained with the assumptions of a homogeneous density layer and flat isopycnals, more careful analysis is needed for accurate quantification of the eddy effects (e.g., [100]).

Third, the in situ turbulence data obtained with a tow-yo microstructure profiler are utilized to estimate the nitrate diffusive flux in several different locations along the Kuroshio. The largest nitrate injection site for the Kuroshio is found in the Tokara Strait, where the Kuroshio flows steadily over many seamounts. Even with slight temporal shifts of its path near the strait, it always flows over some of the seamounts. The measured TKE dissipation rates show large values of O

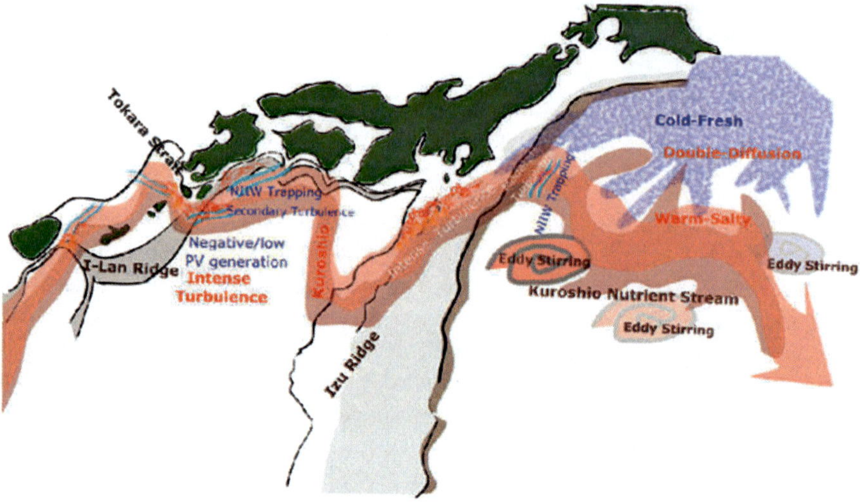

Fig. 35 Summary of this study. The Kuroshio flows over many topographic features in the upstream regions and in the Izu Ridge, where the nitrate diffusive flux is driven mostly by mechanical turbulence. Integrating these mixing hot-spots, the diffusive flux of 1 mmol $m^{-2}day^{-1}$ over 1000 km^2, which leads to a time scale of 50 days to generate the elevated nutrient anomaly of 26 Gmol over 5000 km along the stream, is plausible. The advection needs O(100 days) to transport nutrients over the same distance, suggesting that the diapycnal diffusive flux is important in the Kuroshio nutrient stream. In the dense water $\sigma_\theta > 26$ kgm^{-3} of the Kuroshio Extension and the Kuroshio-Oyashio Confluence region, the dominant agent for the diffusive nitrate flux is the double-diffusion process. PV denotes potential vorticity, NIIW is near-inertial internal wave

(10^{-7} W kg^{-1}) in the banded shear layers, suggesting that the near-inertial internal waves induce the strong turbulent banded layers. The estimated nitrate diffusive flux in these banded layers showed large values of O(1 mmol m^{-2} day^{-1}) over several tens of kilometers. Recent observations by the authors confirm that the elevated turbulent dissipation of O(10^{-7} W kg^{-1}) spreads further downstream over 100 km [76], suggesting that the associated large nitrate flux could extend further downstream along the Kuroshio (Fig. 35). In the Hyuganada Sea southeast of Kyushu, slightly downstream with respect to the Tokara Strait, the same observations also show a large nitrate diffusive flux of 1–10 mmol m^{-2} day^{-1} over 20–30 km across the Kuroshio (Fig. 19, [52]). The further analyses by Durán Gómez and Nagai [77], using a numerical model, suggest that this elevated nitrate flux continues 50–100 km further downstream. Moreover, in the Izu Ridge, the autonomous microstructure profiler data showed a large nitrate diffusive flux of 1–10 mmol m^{-2} day^{-1} over 20–40 km away near the Izu Ridge in the Kuroshio.

By integrating these large diapycnal nitrate fluxes of O(1 mmol m^{-2} day^{-1}) in different locations along the Kuroshio, it is estimated that only about 50 days are required to generate the observed positive anomaly of the nitrate concentrations of 26 Gmol on the density surfaces along the Kuroshio nutrient stream. This time scale for the diapycnal diffusion is comparable or less than that required for the nitrate advection over a distance of 5,000 km, O(100 days). Furthermore, with the effective horizontal diffusivity of O(10^3 m^2s^{-1}), the time required for erasing the nitrate positive anomaly on the density surface with a 100 km width is also 100 days. These results suggest that diapycnal mixing of nutrients in the Kuroshio plays an important role in forming and/or maintaining the elevated nitrate concentrations along the Kuroshio (Figs. 34 and 35).

However, obtaining observations of the diapycnal diffusive flux of nutrients is still insufficient. The nitrate vertical diffusive fluxes in the in situ observations in this study are estimated using turbulence data and the empirical model of nitrate concentration. The latter is as a function of water density derived from the water samples within the same regions where the turbulence data are collected. Although the derived empirical relations between nitrate concentrations and water density for each region are relatively tight (Figs. 15a and 19a; Fig. 2 of [36]), the estimation of the diffusive nitrate flux is done somewhat indirectly. The accurate quantification of the scale over which large diffusive fluxes occur is required to elucidate whether the diffusive flux of nutrients is indeed important.

The autonomous microstructure profiling float data showed that the tracers in the thermocline of the Kuroshio Extension are mixed dominantly by the double-diffusive processes. However, these large double-diffusive nitrate fluxes are found in relatively deep layers of σ_θ = 26–27 kgm^{-3}, which are not outcropped even during winter in this region. Thus, whether the double-diffusion induced upward nitrate flux in the Kuroshio Extension is ultimately supplied to the euphotic zone depends on additional mixing or stirring processes, such as eddy flux toward the shallower northern directions. Further investigations are needed to clarify the importance of the double-diffusive nutrient fluxes.

Also, the analyses of the Argo float data from 2006 through 2014 showed interannual modulations of double-diffusive nitrate flux in the density range σ_θ = 26–26.5 kgm^{-3}, with more diffusive convection flux in the stable period of the Kuroshio Extension in the Kuroshio-Oyashio Confluence region. In this region, the depths of σ_θ = 26–26.5 kgm^{-3} layers during the stable period are 100–200 m, suggesting that the nitrate injected to the shallower depths from these layers by double-diffusion can be entrained into the surface mixed layer during winter. However, it remains unclear what the driving mechanism is of this modulations. Further analyses are required, including the contributions of decadal variability in the wind energy input to the near-inertial energy [97], as the previous study suggested that near-inertial internal wave shear is partly responsible for generating double-diffusion favorable thermohaline interleaving structures [29, 31].

In addition, the diapycnal fluxes of nitrate on the shore side of the front in the regions south of Japan, such as Hyuganada Sea at relatively shallow depths close to the euphotic zone (Figs. 17, 18, and 19), indicate that some fractions of nutrients of the Kuroshio nutrient stream are directly supplied to the primary producers in the coastal regions (c.f. Sect. 7, Fig. 7). However, the measurements of the diffusive nitrate flux are still very scarce in these regions. Therefore, the direct nutrient supplies from the Kuroshio nutrient stream to the south coast of Japan should be quantified, using simultaneous observations of turbulence and nutrient concentrations.

Lastly, it should be noted that the climate models (e.g., [101]) predict a slower Gulf Stream and a faster Kuroshio several decades from now caused by global warming. At the same time, stratification in the upper layers of the Kuroshio and the Kuroshio Extension is expected to be enhanced, preventing subsurface nutrients along the Kuroshio from being supplied through induction and diffusion, with shallower mixed layer depths. Importantly, the Kuroshio Extension and the Gulf Stream Extension are two major net CO_2 sinks for the Earth's atmosphere [102]. Thus, changes caused by global warming in physical and biogeochemical processes leading to the nutrient supply in the Kuroshio and the Gulf Stream regions, and their impacts on the biological pump and how the ocean ecosystem responds, need to be monitored with continuous interdisciplinary observations and accurately predicted to mitigate the impacts on the society.

Acknowledgments Nagai thanks SKED (JPMXD0511102330), OMIX (KAKENHI16H01590 and 18H04914), and KAKENHI (19H01965) and Capt. Ukekura, Capt. Inoue, Capt. Noda, Capt. Uchiyama, Capt. Ryono, and Capt. Tsugio Nagai. Durán Gómez thanks JASSO and MEXT. Nagai and Durán Gómez thank Sandy and Ponta, respectively, and the Kuroshio and Cerdo Florentino.

References

1. Fedorov KN (1988) Layer tickles and effective diffusivities in diffusive thermohaline convection in the ocean. In: Nihoul J, Jamart B (eds) Small-scale turbulence and mixing in the ocean. Elsevier
2. Uda M (1936) Locality of fishing Centre and shoals of "Katsuwo", Euthynnus Vagans (lesson), correlated with the contact zone of cold and warm currents. Bull Jap Soc Sci Fish 4:385–390
3. Clayton S, Nagai T, Follows MJ (2014) Fine scale phytoplankton community structure across the Kuroshio Extension Front. J Plankton Res 36:1017–1030. https://doi.org/10.1093/plankt/fbu020
4. Durán Gómez GS, Nagai T, Yokawa K (2020) Mesoscale warm-core eddies drive interannual modulations of swordfish catch in the Kuroshio Extension System. Front Mar Sci 7. https://doi.org/10.3389/fmars.2020.00680
5. Chen C-TA, Huang M-H (1996) A mid-depth front separating the South China Sea water and the Philippine Sea water. J Oceanogr 52:17–25
6. Cheng Z (1995) Fluxes of seawater and some dissolved chemicals in the Bashi Strait. Chin J Oceanol Limnol 13:240–246. https://doi.org/10.1007/BF02850525
7. Chen C-TA, Wang S-L (1999) Carbon, alkalinity and nutrient budgets on the East China Sea continental shelf. J Geophys Res 104:20675–20686. https://doi.org/10.1029/1999JC900055
8. Chen CTA (1996) The Kuroshio intermediate water is the major source of the nutrients on the East China Sea continental shelf. Oceanol Acta 19:523–527
9. Chen CTA, Liu CT, Pai SC (1995) Variations in oxygen, nutrient and carbonate fluxes of the Kuroshio Current. La Mer 33:161–175
10. Guo X, Zhu XH, Long Y, Huang D (2013) Spatial variations in the Kuroshio nutrient transport from the East China Sea to south of Japan. Biogeosciences 10:6403–6417. https://doi.org/10.5194/bg-10-6403-2013

11. Guo X, Zhu XH, Wu QS, Huang D (2012) The Kuroshio nutrient stream and its temporal variation in the East China Sea. Global Biogeochem Cycles 117:C01026. https://doi.org/10.1029/2011JC007292
12. Nagai T, Clayton S, Uchiyama Y (2019b) Multiscale routes to supply nutrients through the Kuroshio nutrient stream. In: Nagai T, Saito H, Suzuki K, Takahashi M (eds) Kuroshio Current: physical, biogeochemical and ecosystem dynamics. AGU geophysical monograph series. AGU-Wiley. https://doi.org/10.1002/9781119428428.ch6
13. Pelegri JL, Csanady GT (1991) Nutrient transport and mixing in the Gulf Stream. J Geophys Res 96:2577–2583. https://doi.org/10.1029/90JC02535
14. Pelegri JL, Csanady GT, Martins A (1996) The North Atlantic nutrient stream. J Oceanogr 52: 275–299. https://doi.org/10.1007/BF02235924
15. Palter JB, Lozier MS (2008) On the source of Gulf Stream nutrients. J Geophys Res Oceans 113(C6). https://doi.org/10.1029/2007JC004611
16. Whitt DB (2019) On the role of the Gulf Stream in the changing Atlantic nutrient circulation during the 21st century. In: Kuroshio Current. American Geophysical Union (AGU), pp 51–82. https://doi.org/10.1002/9781119428428.ch4
17. Pelegri JL, Csanady GT (1994) Diapycnal mixing in western boundary currents Gulf Stream. J Geophys Res 99:18275–18304
18. Williams RG, McDonagh E, Roussenov VM, Torres-Valdes S, King B, Sanders R, Hansell DA (2011) Nutrient streams in the North Atlantic: advective pathways of inorganic and organic nutrients. Glob Biogeochem Cycles 25:GB4008. https://doi.org/10.1029/2010GB003853
19. Nikurashin M, Ferrari R (2010) Radiation and dissipation of internal waves generated by geostrophic motions impinging on small-scale topography: theory. J Phys Oceanogr 40:1055–1074. https://doi.org/10.1175/2010JPO4315.1
20. Nikurashin M, Ferrari R (2011) Global energy conversion rate from geostrophic flows into internal lee waves in the deep ocean. Geophys Res Lett 38:L08610. https://doi.org/10.1029/2011GL046576
21. Naveira Garabato AC, Polzin KL, King BA, Heywood KJ, Visbeck M (2004) Widespread intense turbulent mixing in the Southern Ocean. Science 303:210–213. https://doi.org/10.1126/science.1090929
22. Nagai T, Tandon A, Kunze E, Mahadevan A (2015c) Spontaneous generation of near-inertial waves by the Kuroshio front. J Phys Oceanogr 45:2381–2406. https://doi.org/10.1175/JPO-D-14-0086.1
23. Polzin KL (2010) Mesoscale eddy-internal wave coupling. Part II: energetics and results from PolyMode. J Phys Oceanogr 40:789–801. https://doi.org/10.1175/2009JPO4039.1
24. Cusack JM, Naveira Garabato AC (2017) Observation of a large lee wave in the Drake Passage. J Phys Oceanogr 47:793–810. https://doi.org/10.1175/JPO-D-16-0153.1
25. Dewar W, Molemake MJ, McWilliams JC (2015) Centrifugal instability and mixing in the California Undercurrent. J Phys Oceanogr 45:1224–1241. https://doi.org/10.1175/JPO-D-13-0269.1
26. Gula J, Molemaker MJ, McWilliams JC (2016) Topographic generation of submesoscale centrifugal instability and energy dissipation. Nat Commun. https://doi.org/10.1038/ncomms12811
27. Molemaker MJ, McWilliams JC, Dewar WK (2015) Submesoscale instability and generation of mesoscale anticyclones near a separation of the California Undercurrent. J Phys Oceanogr 45:613–629. https://doi.org/10.1175/JPO-D-13-0225.1
28. Lee M-M, Williams RG (2000) The role of eddies in the isopycnic transfer of nutrients and their impact on biological production. J Mar Res 58:895–917. https://doi.org/10.1357/002224000763485746
29. Nagai T, Clayton S (2017) Nutrient interleaving below the mixed layer of the Kuroshio Extension Front. Ocean Dyn 67:1027–1046. https://doi.org/10.1007/s10236-017-1070-3

30. Smith KS, Ferrari R (2009) The production and dissipation of compensated thermohaline variance by mesoscale stirring. J Phys Oceanogr 39(10):2477–2501. https://doi.org/10.1175/2009JPO4103.1
31. Nagai T, Inoue R, Tandon A, Yamazaki H (2015b) Evidence of enhanced double-diffusive convection below the main stream of the Kuroshio Extension. J Geophys Res 120:8402–8421. https://doi.org/10.1002/2015JC011288
32. Flament P (2002) A state variable for characterizing water masses and their diffusive stability: spiciness. Progr Oceanogr 54(1):493–501. https://doi.org/10.1016/S0079-6611(02)00065-4
33. Wolk F, Yamazaki H, Seuront L, Lueck RG (2002) A new free-fall profiler for measuring biophysical microstructure. J Atmos Ocean Technol 19:780–793. https://doi.org/10.1175/1520-0426(2002)019<0780:ANFFPF>2.0.CO;2
34. Nagai T, Tandon A, Yamazaki H, Doubell MJ (2009) Evidence of enhanced turbulent dissipation in the frontogenetic Kuroshio front thermocline. Geophys Res Lett 36:L12609. https://doi.org/10.1029/2009GL038832
35. Nagai T, Gruber N, Frenzel H, Lachkar Z, McWilliams JC, Plattner G-K (2015) Dominant role of eddies and filaments in the offshore transport of carbon and nutrients in the California Current System. J Geophys Res. https://doi.org/10.1002/2015JC010889
36. Nagai T, Rosales Quintana GM, Durán Gómez GS, Hashihama F, Komatsu K (2021a) Elevated turbulent and double-diffusive nutrient flux in the Kuroshio over the Izu Ridge and in the Kuroshio Extension. J Oceanogr 77:55–74. https://doi.org/10.1007/s10872-020-00582-2
37. Hashihama F (2013) A biogeochemical study in surface waters of the subtropical ocean using a highly-sensitive method for measuring nanomolar nutrients. Oceanogr Jpn 22:169–185. https://doi.org/10.5928/kaiyou.22.5_169
38. Shchepetkin AF, McWilliams JC (2005) The regional oceanic modeling system (ROMS): a split-explicit, free-surface, topography-following-coordinate ocean model. Ocean Model 9: 347–404. https://doi.org/10.1016/j.ocemod.2004.08.002
39. Gruber N, Frenzel H, Doney SC, Marchesiello P, McWilliams JC, Moisan JR, Oram JJ, Plattner G-K, Stolzenbach KD (2006) Eddy-resolving simulation of plankton ecosystem dynamics in the California Current System. Deep Sea Res I 53:1483–1516. https://doi.org/10.1016/j.dsr.2006.06.005
40. Large WG, McWilliams JC, Doney SC (1994) Oceanic vertical mixing: a review and a model with a nonlocal boundary layer parameterization. Rev Geophys 32(4):363–403. https://doi.org/10.1029/94RG01872
41. da Silva A, Young AC, Levitus S (1994) Atlas of surface marine data 1994 (Algorithms and procedures no. tech. rep. 6). Department of Commerce, NOAA, NESDIS
42. NCEP (National Centers for Environmental Prediction) (2015)
43. Garcia HE, Locarnini RA, Boyer TP, Antonov JI, Baranova O, Zweng M, Reagan JR, Johnson D (2014) World ocean atlas 2013, volume 4: dissolved inorganic nutrients (phosphate, nitrate, silicate). In: Levitus S (ed), Mishonov A (Technical ed) NOAA atlas NESDIS, 76, 25 pp
44. Qiu B (2019) Kuroshio and Oyashio Currents. In: Cochran JK, Bokuniewicz H, Yager P (eds) Encyclopedia of ocean sciences, 3rd edn. Academic Press, pp 384–394
45. Chen C-TA, Huang T-H, Wu C-H, Yang H, Guo X (2021) Variability of the nutrient stream near Kuroshio's origin. Sci Rep 11:5080. https://doi.org/10.1038/s41598-021-84420-5
46. Williams RG, Roussenov V, Follows M (2006) Induction of nutrients into the mixed layer and maintenance of high latitude productivity. Global Biogeochem Cycles 20:GB1016. https://doi.org/10.1029/2005GB002586
47. Rhines PB, Young W (1982) Homogenization of potential vorticity in planetary gyre. J Fluid Mech 122:347–368. https://doi.org/10.1017/S0022112082002250
48. Garcia HE, Boyer TP, Baranova OK, Locarnini RA, Mishonov AV, Grodsky A, Paver CR, Weathers KW, Smolyar IV, Reagan JR, Seidov D, Zweng MM (2019) World ocean atlas 2018. In: Mishonov A (Technical ed) Product documentation

49. Komatsu K, Hiroe Y (2019) Structure and impact of the Kuroshio nutrient stream. In: Nagai T, Saito H, Suzuki K, Takahashi M (eds) Kuroshio Current: physical, biogeochemical and ecosystem dynamics. AGU geophysical monograph series. AGU-Wiley. https://doi.org/10.1002/9781119428428.ch5
50. Komatsu K, Yasuda I, Itoh S, Ikeya T, Hidaka K, Yagi M et al (Dec 2010) Nutrient stream in the Kuroshio region. AGU fall meeting abstracts, OS51A-1273
51. Lellouche J-M, Bourdalle-Badie R, Greiner E, Garric G, Melet A, Bricaud C, Legalloudec O, Hamon M, Candela T, Regnier C, Drevillon M (2021) The Copernicus global 1/12° oceanic and sea ice reanalysis, EGU general assembly 2021, online, 19–30 Apr 2021, EGU21-14961. https://doi.org/10.5194/egusphere-egu21-14961
52. Nagai T, Durán Gómez GS, Otero DA, Mori Y, Yoshie N, Ohgi K, Hasegawa D, Nishina A, Kobari T (2019a) How the Kuroshio Current delivers nutrients to sunlit layers on the continental shelves with aid of near-inertial waves and turbulence. Geophys Res Lett 46(12):6726–6735. https://doi.org/10.1029/2019GL082680
53. Nagai T (2019) Diapycnal mixing processes and their impacts along the Kuroshio from its upstream to the Kuroshio Extension. Bull Coast Oceanogr 57:43–58 (in Japanese)
54. Qiu B, Huang RX (1995) Ventilation of the North Atlantic and North Pacific: subduction versus obduction. J Phys Oceanogr 25:2374–2390. https://doi.org/10.1175/1520-0485(1995)0252.0.CO;2
55. Gruber N, Lachkar Z, Frenzel H, Marchesiello P, Münnich M, McWilliams JC, Nagai T, Plattner G-K (2011) Eddy-induced reduction of biological production in Eastern Boundary Upwelling Systems. Nat Geosci 4(11):787–792. https://doi.org/10.1038/ngeo1273
56. Williams RG, Follows M (2011) Ocean dynamics and the carbon cycle. Cambridge University Press, p 404. https://doi.org/10.1017/CBO9780511977817
57. Nasmyth PW (1970) Oceanic turbulence. Unpublished doctoral dissertation, University of British Columbia
58. Osborn T (1980) Estimates of the local rate of vertical diffusion from dissipation measurements. J Phys Oceanogr 10:83–89. https://doi.org/10.1175/1520-0485(1980)010<0083:EOTLRO>2.0.CO;2
59. Lewis MR, Harrison WG, Oakey NS, Hebert D, Platt T (1986) Vertical nitrate fluxes in the oligotrophic ocean. Science 234:870–873. https://doi.org/10.1126/science.234.4778.870
60. Kaneko H, Yasuda I, Komatsu K, Itoh S (2012) Observations of the structure of turbulent mixing across the Kuroshio. Geophys Res Lett 39. https://doi.org/10.1029/2012GL052419
61. Kaneko H, Yasuda I, Komatsu K, Itoh S (2013) Observations of vertical turbulent nitrate flux across the Kuroshio. Geophys Res Lett 40:3123–3127. https://doi.org/10.1002/grl.50613
62. Kobari T, Honma T, Hasegawa D, Yoshie N, Tsutsumi E, Matsuno T, Nagai T, Kanayama T, Karu F, Suzuki K, Tanaka T, Guo X, Kume G, Nishina A, Nakamura H (2020) Phytoplankton growth and consumption by microzooplankton stimulated by turbulent nitrate flux suggest rapid trophic transfer in the oligotrophic Kuroshio. Biogeosciences 17(9):2441–2452. https://doi.org/10.5194/bg-17-2441-2020
63. Tanaka T, Hasegawa D, Yasuda I, Tsuji H, Fujio S, Nishioka J (2016) Enhanced vertical turbulent nitrate flux in the Kuroshio across the Izu Ridge. J Oceanogr 75:195–203. https://doi.org/10.1007/s10872-018-0500-2
64. Osborn TR, Cox CS (1972) Oceanic fine structure. Geophys Fluid Dyn 3:321–345
65. St. Laurent LC, Schmitt RW (1999) The contribution of salt fingers to vertical mixing in the North Atlantic tracer release experiment. J Phys Oceanogr 29:1404–1424. https://doi.org/10.1175/1520-0485(1999)029<1404:TCOSFT>2.0.CO;2
66. Thomas LN, Lee CM (2005) Intensification of ocean fronts by down-front winds. J Phys Oceanogr 35:1086–1102. https://doi.org/10.1175/JPO2737.1
67. D'Asaro E, Lee C, Rainville L, Harcourt L, Thomas L (2011) Enhanced turbulence and energy dissipation at ocean fronts. Science 332:318–322. https://doi.org/10.1126/science.1201515
68. Thomas L, Ferrari R (2008) Frontogenesis, and the stratification of the surface mixed layer. J Phys Oceanogr 38:2501–2518. https://doi.org/10.1175/2008JPO3797.1

69. Nagai T, Tandon A, Rudnick DL (2006) Two dimensional ageostrophic secondary circulation at ocean fronts due to vertical mixing and large-scale deformation. J Geophys Res 111. https://doi.org/10.1029/2005JC002964
70. Whitt D, Thomas L (2013) Near-inertial waves in strongly baroclinic currents. J Phys Oceanogr 43:706–725. https://doi.org/10.1175/JPO-D-12-0132.1
71. Kunze E (1985) Near-inertial wave propagation in geostrophic shear. J Phys Oceanogr 15: 544–565. https://doi.org/10.1175/1520-0485(1985)015<0544:NIWPIG>2.0.CO;2
72. Mooers CN (1975) Several effects of baroclinic current on the cross-stream propagation of inertial-internal waves. Geophys Fluid Dyn 6:245–275. https://doi.org/10.1080/03091927509365797
73. Hasegawa D, Yamazaki H, Lueck RG, Seuront L (2004) How islands stir and fertilize the upper ocean. Geophys Res Lett 31(16). https://doi.org/10.1029/2004GL020143
74. Nagai T, Hasegawa D, Tanaka T, Nakamura H, Tsutsumi E, Inoue R, Yamashiro T (2017) First evidence of coherent bands of strong turbulent layers associated with high-wavenumber internal-wave shear in the upstream Kuroshio. Sci Rep. https://doi.org/10.1038/s41598-017-15167-1
75. Tsutsumi E, Matsuno T, Lien RC, Nakamura H, Senjyu T, Guo X (2017) Turbulent mixing within the Kuroshio in the Tokara Strait. J Geophys Res Oceans 122(9):7082–7094. https://doi.org/10.1002/2017JC013049
76. Nagai T, Hasegawa D, Tsutsumi E, Nakamura H, Nishina A, Senjyu T, Endoh T, Matsuno T, Inoue R, Tandon A (2021b) The Kuroshio flowing over seamounts and associated submesoscale flows drive 100-km-wide 100-1000-fold enhancement of turbulence. Comm Earth Environ 2:170. https://doi.org/10.1038/s43247-021-00230-7
77. Durán Gómez GS, Nagai T (2020) Elevated nutrient supply caused by the Kuroshio approaching to the continental shelf associated with submesoscale flows and near-inertial waves. Presented at 2020 Fall meeting, AGU, San Francisco, CA, 1–17 Dec. Abstract, OS011-0009
78. Nagai T, Tandon A, Yamazaki H, Doubell MJ, Gallager S (2012) Direct observations of microscale turbulence and thermohaline structure in the Kuroshio front. J Geophys Res 117: C08013. https://doi.org/10.1029/2011JC00722
79. Radko T (2013) Double-diffusive convection. Cambridge University Press. https://doi.org/10.1017/CBO9781139034173
80. Schmitt RW, Ledwell J, Montgomery E, Polzin K, Toole J (2005) Enhanced diapycnal mixing by salt fingers in the thermocline of the tropical Atlantic. Science 308(5722):685–688. https://doi.org/10.1126/science.1108678
81. Kundu PK, Cohen IM (2001) Fluid mechanics. Academic Press, p 730
82. Kantha LH, Clayson CA (2000) Small scale processes in geophysical fluid flows. Academic Press
83. Ruddick B (1983) A practical indicator of the stability of the water column to double-diffusive activity. Deep Sea Res Part A Oceanogr Res Pap 30(10):1105–1107. https://doi.org/10.1016/0198-0149(83)90063-8
84. Radko T, Smith DP (2012) Equilibrium transport in double-diffusive convection. J Fluid Mech 692:5–27. https://doi.org/10.1017/jfm.2011.343
85. Zhang JR, Schmitt W, Huang RX (1998) Sensitivity of the GFDL Modular Ocean model to parameterization of double-diffusive processes. J Phys Oceanogr 28:589–605. https://doi.org/10.1175/1520-0485(1998)028<0589:SOTGMO>2.0.CO;2
86. Jan S, Wang SH, Yang KC, Yang YJ, Chang MH (2019) Glider observations of interleaving layers beneath the Kuroshio primary velocity core east of Taiwan and analyses of underlying dynamics. Sci Rep 9(1):1–11. https://doi.org/10.1038/s41598-019-47912-z
87. Inoue R, Yamazaki H, Wolk F, Kono T, Yoshida J (2007) An estimation of buoyancy flux for a mixture of turbulence and double diffusion. J Phys Oceanogr 37(3):611–624. https://doi.org/10.1175/JPO2996.1

88. Talley LD, Yun J-Y (2001) The role of cabbeling and double diffusion in setting the density of the North Pacific Intermediate Water salinity minimum. J Phys Oceanogr 31:1538–1549. https://doi.org/10.1175/1520-0485(2001)031<1538:TROCAD>2.0.CO;2
89. Oschlies A, Dietze H, Kahler P (2003) Salt-finger driven enhancement of upper ocean nutrient supply. Geophys Res Lett 30(23). https://doi.org/10.1029/2003GL018552
90. Seo Y, Sugimoto S, Hanawa K (2014) Long-term variations of the Kuroshio Extension path in winter: meridional movement and path state change. J Clim 27(15):5929–5940. https://doi.org/10.1175/JCLI-D-13-00641.1
91. Mizuno K, White WB (1983) Annual and interannual variability in the Kuroshio Current system. J Phys Oceanogr 13(10):1847–1867. https://doi.org/10.1175/1520-0485(1983)013<1847:AAIVIT>2.0.CO;2
92. Qiu B, Chen S (2005) Variability of the Kuroshio Extension jet, recirculation gyre, and mesoscale eddies on decadal time scales. J Phys Oceanogr 35(11):2090–2103. https://doi.org/10.1175/JPO2807.1
93. Qiu B, Chen S, Schneider N, Taguchi B (2014) A coupled decadal prediction of the dynamic state of the Kuroshio Extension System. J Clim 27(4):1751–1764. https://doi.org/10.1175/JCLI-D-13-00318.1
94. Boyer TP, García HE, Locarnini RA, Zweng, Melissa M, Mishonov AV, Reagan JR, Weathers KA, Baranova OK, Paver CR, Seidov D, Smolyar IV (2018) World ocean database 2018. In: Mishonov AV (Technical ed) NOAA atlas NESDIS 87, 207 p
95. Hanawa K, Mitsudera H (1986) Variation of water system distribution in the Sanriku coastal area. J Oceanogr 42(6):435–446. https://doi.org/10.1007/BF02110194
96. Sugimoto T, Kawasaki Y, Li J (1992) A description of the time-dependent hydrographic structure of the warm streamer around the Kuroshio warm-core ring 86b. Deep Sea Res Part A Oceanogr Res Pap 39:S77–S96. https://doi.org/10.1016/S0198-0149(11)80006-3
97. Inoue R, Watanabe M, Osafune S (2017) Wind-induced mixing in the North Pacific. J Phys Oceanogr 47:1587–1603. https://doi.org/10.1175/JPO-D-16-0218.1
98. Whalen CB, Talley LD, MacKinnon JA (2012) Spatial and temporal variability of global ocean mixing inferred from Argo profiles. Geophys Res Lett 39:L18612. https://doi.org/10.1029/2012GL053196
99. Uchiyama Y, Suzue Y, Yamazaki H (2017) Eddy-driven nutrient transport and associated upper-ocean primary production along the Kuroshio. J Geophys Res Oceans 122:5046–5062. https://doi.org/10.1002/2017JC012847
100. Wolfe CL, Cessi P (2010) What sets the strength of the middepth stratification and overturning circulation in eddying ocean models? J Phys Oceanogr 40(7):1520–1538. https://doi.org/10.1175/2010jpo4393.1
101. Chen C, Wang G, Xie S, Liu W (2019) Why does global warming weaken the Gulf Stream but intensify the Kuroshio? J Clim 32(21):7437–7451. https://doi.org/10.1175/JCLI-D-18-0895.1
102. Takahashi T, Sutherland SC, Wanninkhof R, Sweeney C, Feely RA, Chipman DW, Hales B, Friederich G, Chavez F, Watson A, Bakker DCE, Schuster U, Metzl N, Yoshikawa-Inoue H, Ishii M, Midorikawa T, Nojiri Y, Sabine C, Olafsson J, Arnarson TS, Tilbrook B, Johannessen T, Olsen A, Bellerby R, Körtzinger A, Steinhoff T, Hoppema M, de Baar HJW, Wong CS, Delille B, Bates NR (2009) Climatological mean and decadal changes in surface ocean pCO_2, and net sea-air CO_2 flux over the global oceans. Deep-Sea Res II 56:554–577. https://doi.org/10.1016/j.dsr2.2008.12.009

Front-Driven Physical–Biogeochemical–Ecological Interactions in the Yellow Sea Large Marine Ecosystem

Qin-Sheng Wei, Ming-Zhu Fu, Xian-Sen Li, Jun-Chuan Sun, Bao-Dong Wang, and Zhi-Gang Yu

Contents

Q.-S. Wei (✉), M.-Z. Fu, and B.-D. Wang
Key Laboratory of Marine Eco-Environmental Science and Technology, First Institute of Oceanography, Ministry of Natural Resources, Qingdao, China

Laboratory for Marine Ecology and Environmental Science, Qingdao National Laboratory for Marine Science and Technology, Qingdao, China
e-mail: weiqinsheng@fio.org.cn

X.-S. Li
Key Laboratory for Sustainable Utilization of Marine Fisheries Resource, Ministry of Agriculture, Yellow Sea Fisheries Research Institute, Chinese Academy of Fishery Sciences, Qingdao, China

J.-C. Sun
Key Laboratory of Marine Science and Numerical Modeling, First Institute of Oceanography, Ministry of Natural Resources, Qingdao, China

Z.-G. Yu
Laboratory for Marine Ecology and Environmental Science, Qingdao National Laboratory for Marine Science and Technology, Qingdao, China

Frontiers Science Center for Deep Ocean Multispheres and Earth System, and Key Laboratory of Marine Chemistry Theory and Technology, Ministry of Education, Ocean University of China, Qingdao, China

Igor M. Belkin (ed.), *Chemical Oceanography of Frontal Zones*,
Hdb Env Chem (2022) 116: 255–282, DOI 10.1007/698_2021_832,

Published online: 21 December 2021

Abstract The Yellow Sea (YS) is among the most important large marine ecosystems (LMEs). By synthesizing a variety of physicochemical and biological data in the western SYS (Southern Yellow Sea), we provide an overview of front-driven physical–biogeochemical–ecological interactions in the Yellow Sea Large Marine Ecosystem (YSLME), mainly addressing (1) thermohaline, density, turbidity, and nutrient fronts in relation to the water-mass structure; (2) nutrient transport and light conditions associated with the fronts; (3) front-driven primary production regime; and (4) anchovy distribution and other ecological processes in the frontal zone. Generally, forced by complicated hydrodynamics under coastal water/currents, estuarine runoff, the Yellow Sea Warm Current (YSWC), and the Yellow Sea Cold Water Mass (YSCWM), robust thermohaline and density fronts are formed. Moreover, a prominent surface turbidity front forms year-round near the 30 m isobath. Three types of nutrient fronts, i.e., estuarine, coastal, and offshore, exist and change seasonally in this region; their locations generally represent the range of substance transport along with different water masses. The frontal system largely regulates the primary production regime in the SYS. Specifically, the front between the southward cold-water belt and eastern YSWC roughly shapes the western boundary of the central phytoplankton-blooming area in spring. In summer, upwelling is observed near the bottom YSCWM boundary, leading to surface cold patches, nutricline shoaling, and upward transport of nutrients. This upwelling system can extract nutrients from the YSCWM to the euphotic layer, resulting in a spatial shift in phytoplankton blooms from the central SYS in spring to the YSCWM frontal region in summer. Phytoplankton biomass also tends to peak near the autumn YSCWM front and winter YSWC front. Furthermore, anchovy distribution is closely related to the frontal system. In winter, the front adjacent to the YSWC core area at the YS entrance acts as an overwintering ground for anchovy; in spring, anchovy migrates northward for feeding, and a high biomass forms in the YSWC-affected bloom area; during summer and autumn, the YSCWM frontal region is an important spawning and nursery ground for anchovy due to food availability and suitable temperature. Overall, the frontal system may play a substantial role in shaping the YSLME via close physical–biogeochemical–ecological interactions.

Keywords Anchovy, Front, Nutrient, Physical–biogeochemical–ecological interactions, Primary production, Upwelling, Yellow Sea Large Marine Ecosystem

1 Introduction

Oceanic fronts are physical boundaries between adjacent water masses that have different properties. These fronts manifest as sharp gradients in environmental variables in horizontal or vertical directions [1]. In particular, alongshore winds [2], tidal mixing [3], river plumes [4, 5], eddies [6], and upwelling [7] can contribute to the formation of fronts. Frontogenesis is fundamentally important to ocean dynamic processes and marine ecosystems, separating water masses that have distinct characteristics and forming a barrier/pathway that influences the exchange/transport of energies and materials, such as heat, solutes, and organisms. As highly active interfaces and unique venues, fronts also generally involve complicated physical–biogeochemical–ecological relationships [8–10].

Marginal seas are among the most important regions that feature multiple-scale physical and biogeochemical processes [11, 12]. As a result, these seas are hot spots for the occurrence of fronts; therefore, investigating front-driven physical–biogeochemical–ecological interactions in these seas has been a primary focus in marine sciences [13]. The Yellow Sea (YS), which is bounded by mainland China and the Korean Peninsula, is a typical semi-enclosed marginal sea of the northwest Pacific Ocean. The YS is one of the important large marine ecosystems (LMEs) [14] and is formally known as the Yellow Sea Large Marine Ecosystem (YSLME) [15, 16]. This marginal sea is characterized by complicated hydrodynamics [17, 18] and biogeochemical-ecological processes [19–22]. In winter, the general circulation over this basin-scale shelf is primarily controlled by the northward Yellow Sea Warm Current (YSWC) [23–25] in the central area and southward coastal current on both sides of the inshore area. In summer, the Yellow Sea Cold Water Mass (YSCWM) [26, 27], which is entrenched at the bottom in the central YS, dominates the current system. Moreover, the Lubei Coastal Current (LCC) and Subei Coastal Water (SCW) exhibit seasonal variations in the western YS [22, 28–32]. The Changjiang Diluted Water (CDW) [33–35] can influence the southwestern YS in summer due to its northeastward expansion under the southerly monsoon; the northward Taiwan Warm Current (TWC) [36, 37] may also reach the southeastern area of the YS. Notably, tidal mixing is intense in the YS and can control the YSCWM boundary during the stratified warm season [38]. Forced by these physical dynamics, robust fronts [22, 39–47] form in this LME.

In the past few decades, thermal fronts in the YS have been examined using multiple tools, including satellite remote sensing, in situ observations, and numerical models. As revealed by previous studies [39, 40, 42, 48–57], fronts are primarily observed off the China and Korea coasts. Significant fronts generally form around the boundaries of the YSWC and YSCWM, leading to sharp temperature gradients [21, 58–60]. The frontal region with a water depth of 20–40 m constitutes a transitional zone between coastal and offshore waters [61], and it may act as a link between the stratified central YS and the well-mixed nearshore area in summer [62]. Pronounced physical fronts also exist around the YSWC core area at the YS entrance during winter [23, 43, 45], accompanying the temperature inversion

[63, 64]. Notably, during summer, cold patches often occur on the surface in the region corresponding to the YSCWM boundary under tidal mixing, especially off the Subei Shoal, the eastern tip of the Shandong Peninsula, and the southwest coast of Korea, resulting in strong thermal fronts [59, 65–71]. Most of these fronts exhibit remarkable features in terms of their spatial scale and temporal variability in the YS [13, 39, 40]. The front system off the Subei Shoal (i.e., the Subei Shoal front), which is attributed to the shear between the cold China Coastal Current and the central YSWC during winter [39, 42, 43] and tidal mixing around the YSCWM in summer [59], expands southeastward in winter and northeastward in summer, showing an arc and "S" shape off the shoal, respectively [71]. The winter thermal front off Shidao, which forms due to convergence of the YSWC and LCC, exhibits an ear shape [56]. Based on multiscale ultrahigh-resolution (~1 km) SST (sea-surface temperature) analysis data [72], a summertime double front pattern in the YS was suggested by Lin et al. [73]. Moreover, salinity fronts (especially in the subsurface) and nutrient fronts were identified in the YS using in situ data [53]. In addition, influenced by the CDW plume, an estuarine front also exists in the southwestern YS [22]. Considering the significant roles of fronts in hydrographic, chemical, and biological interactions [8, 10, 74], some studies have focused on front-driven physical and biogeochemical processes in the YS. Specifically, tidal-induced upwelling occurs near the YSCWM frontal region during summer [21, 59, 65, 66, 75], leading to vertical transport of nutrients and phytoplankton patches [22, 76]. The tidal mixing front contributes to the distribution and productivity of phytoplankton in the YS [21, 77–80], and it may act as an oasis [71]. A correlation also exists between quasi-stationary mesoscale fronts and the spatial distribution of microfossils (diatoms, dinoflagellates, and silicoflagellates) in the YSLME [81]. The YS fronts may greatly influence the transport of suspended particulate matter [71, 82–85] and play a role in mud deposition [86, 87]. Moreover, due to moderate temperatures and high phytoplankton biomass (rich food), the summer YSCWM front and its offshore adjacent area may provide a habitat for *C. sinicus* [88–90], where it can avoid high temperature and actively feed to meet its metabolic needs. The YS fronts also influence the distribution of important fishes. Anchovy is one of the most critical fish species in the YSLME [91, 92], playing a key "trophic linkage role" in the ecosystem structure, as it is both a plankton feeder and the primary food source for various predatory fishes [93]. The migration and distribution of anchovies are closely related to the YSWC and YSCWM fronts [21, 42, 94–98]. Consequently, oceanic fronts have been recognized as important environmental factors in the YS, resulting in highly complex physical and biogeochemical-ecological dynamics in this LME.

In this chapter, mainly based on our in situ observations and previous studies, we attempt to assemble an overview of front-driven physical–biogeochemical–ecological interactions in the YSLME, providing a comprehensive and systematic framework. Detailed information on the sampling and associated analyses can be found in the cited papers [21, 22, 99–104]. By synthesizing a variety of physicochemical and biological data in this context, we mainly address (1) thermohaline, density, turbidity, and nutrient fronts in relation to the water-mass structure; (2) nutrient transport and light conditions associated with fronts; (3) front-driven primary production regime; and (4) anchovy distribution and other ecological processes in the frontal zone. This review may improve our insights into the synergies among physical, chemical, and biological variables near frontal regions in marginal sea ecosystems from a multidisciplinary perspective.

2 Study Area and Data

The YS is surrounded by the Jiangsu Province and the Shandong, Liaodong, and Korean Peninsulas (Fig. 1), with an area of ~381,000 km^2 and a mean depth of ~44 m. The YS is a crucial region for the marine-based economic development of China and Korea, being under the influence of multiple stressors (e.g., human activities and climate change) [105, 106]. Generally, this semi-enclosed marginal sea has curved shorelines and complex topography, with a deep trough (>70 m) generally extending northward in the central area. This sea is connected to the Bohai Sea (BS) through a narrow strait, and it is more open to the East China Sea (ECS) (Fig. 1). The line connecting Chengshantou on the Shandong Peninsula of China and Chang San-got on the Korean Peninsula divides the YS into two parts, the Northern Yellow Sea (NYS) and the Southern Yellow Sea (SYS). There are several special subregions in the SYS, including the Subei Shoal and Haizhou Bay. The Subei Shoal is linked to the Changjiang Bank and forms a large, tongue-shaped shoal extending to the southeast (Fig. 1b). The YSWC (strongest in winter) [23, 107], YSCWM (strongest in summer) [26], and East Asian monsoons are the main components affecting the current system in the YS. The YSWC generally enters the YS from southwestern Cheju Island (Fig. 1a). Moreover, the TWC, which mainly originates from the Kuroshio Branch Current [108], and the CDW [33–35] affect the

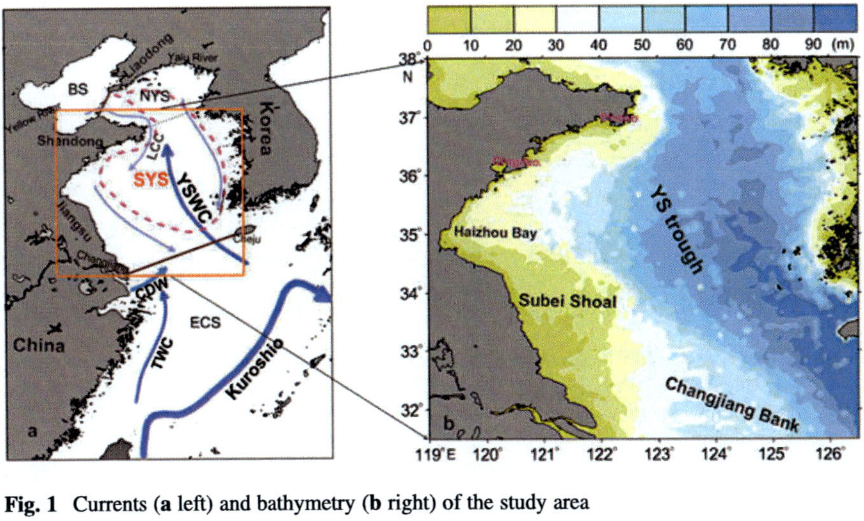

Fig. 1 Currents (a left) and bathymetry (b right) of the study area

southwestern region of the SYS. In winter, when strong northerly winds prevail, the water column is generally homogeneous in the YS, while in summer, the water column is strongly stratified because of intense solar radiation [17]. Due to relatively intense land–ocean interactions, large amounts of nutrients are transported into the YS directly or indirectly via rivers, such as the Changjiang, Yellow and Yalu Rivers [109], and atmospheric deposition [110, 111]. As a result, the YS is known as one of the most productive LMEs in the world, with high biological diversity and a complicated food web structure [15]. Front-related processes play an important role in regulating the physical and biogeochemical regimes in the YS and are highly ecologically significant. Here, we take the western SYS as an example to integrate the physical–biogeochemical–ecological interactions in the frontal region of the YSLME, providing a comprehensive and systematic framework for marginal sea ecosystems.

The in situ data were obtained by four seasonal surveys in the western SYS during 2006–2007 in the region of 32°20′–35°30′N and west of 124°00′E. The observed physicochemical and biological variables include temperature, salinity, density, turbidity, dissolved inorganic nitrogen (DIN), phosphate (PO_4-P), silicate (SiO_3-Si), chlorophyll a (Chl-a), primary productivity (PP), and anchovy biomass. Details of the cruises, sampling, analyses, and data processing can be found in numerous papers [21, 22, 62, 71, 99–104], in which the data were analyzed from different perspectives. Moreover, based on the climatological monthly mean SST from MODIS (Moderate Resolution Imaging Spectroradiometer)/Aqua for the 2003–2018 period, the winter thermal fronts in the SYS are detected by gradient analysis [112]. The satellite data were downloaded from the U.S. National Aeronautics and Space Administration (NASA) website (https://oceancolor.gsfc.nasa.gov), covering an area between 30.5 and 37.5°N with a spatial resolution of 4 km × 4 km. The SST gradient was calculated using the oceanic front detection algorithm developed by Belkin and O'Reilly [112], and the patterns of thermal fronts were produced according to the gradient magnitude. In this study, an SST image from February was chosen to represent the observations in winter. A gradient value of 0.02°C km^{-1} was selected for the occurrence of the SST front, as adopted in previous studies [71, 73]. The names of thermal fronts generally follow the frontal nomenclature by Hickox et al. [39].

3 Physicochemical Regimes and Frontogenesis in the Southwestern YSLME

3.1 Thermohaline and Density Fronts in Relation to the Water-Mass Structure

Based on the spatiotemporal distributions of temperature, salinity, and potential density in the western SYS (Fig. 2), the current regime and water-mass structure

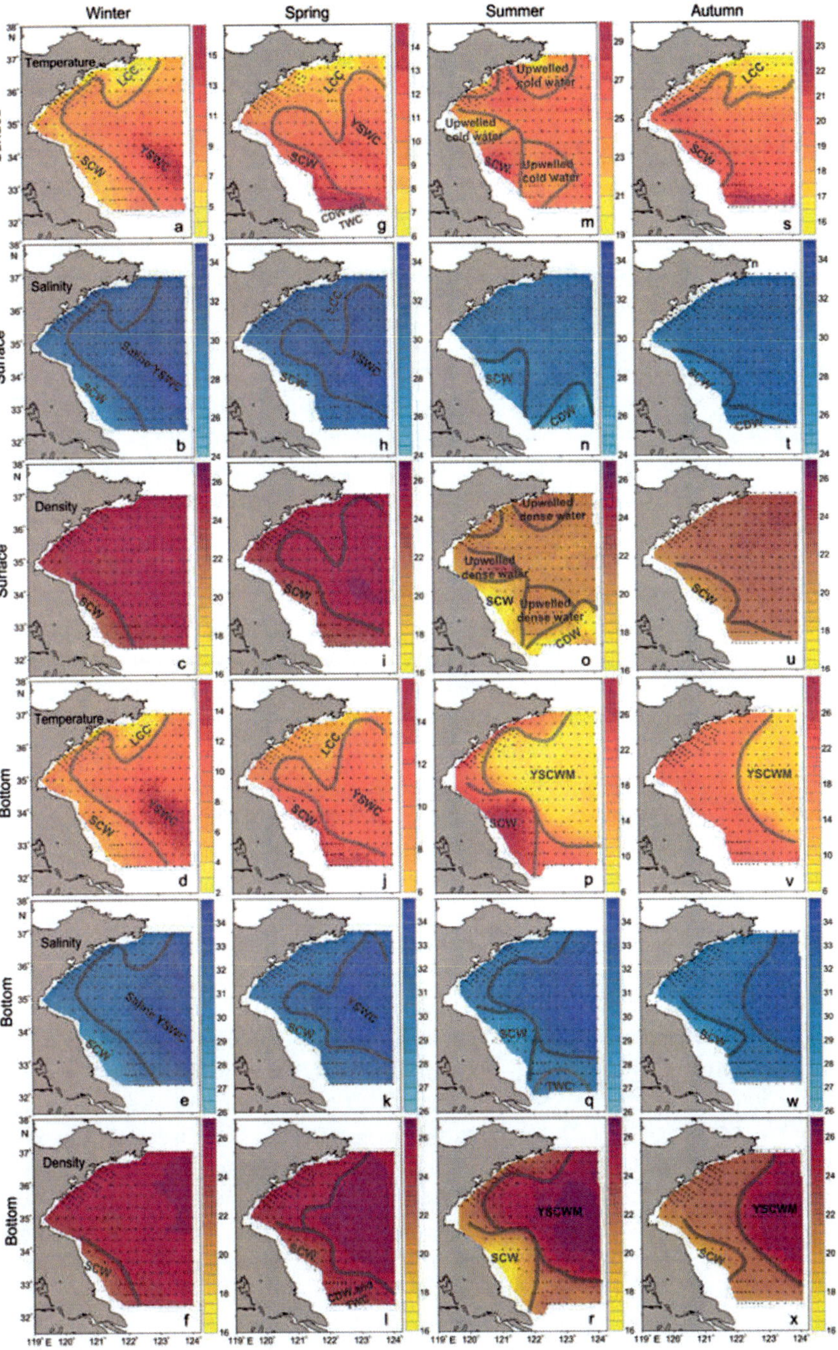

Fig. 2 (a, g, m, s, d, j, p, v) Temperature (°C), (c, i, o, u, f, l, r, x) salinity, and (b, h, n, t, e, k, q, w) potential density (kg m^{-3}) in the western SYS. Black dots are stations. Gray bands are fronts. The temperature and salinity maps are modified after [22]

are distinguished. In winter and spring, the saline YSWC, which intrudes from the southeast to northwest, dominates the central SYS; the SCW, which is characterized by low salinity and low density, expands toward the southeastern area along the Subei coast; and the cold LCC is observed off the eastern tip of the Shandong Peninsula (i.e., Shidao) and constitutes the southward-moving Yellow Sea Coastal Current (YSCC). In summer, several surface cold patches with low temperature and high density exist off the Shandong Peninsula tip (off Shidao), Qingdao coast, Haizhou Bay, and Subei Shoal, potentially indicating the presence of upwelling. The extension direction of the SCW in summer is different from that in winter and spring, being generally northeastward. The tongue-shaped CDW with low salinity and low density emerges in the northeastern region off the Changjiang Estuary (CE). The bottom YSCWM with low temperature (<10°C) (Fig. 2k), high salinity (Fig. 2k), and high density (Fig. 2k) occupies most of the central SYS area. The bottom saline TWC is observed off the CE. In autumn, the bottom YSCWM scope is much smaller than in summer, and the CDW in the southwestern SYS becomes weaker; the SCW and LCC seem to expand southeastward again. The presence of these water masses is confirmed by temperature-salinity plots [22].

According to the current regime and water-mass structure in the western SYS, different thermohaline and density fronts are identified. Their locations are determined by sharp gradients in temperature, salinity, and density (Fig. 2). A dominant "S"-shaped front appears around the YSWC boundary in winter and spring due to the interaction of central warm water and coastal cold water. In summer, remarkable fronts emerge around the surface cold patches (Fig. 2m). A salinity-density front is induced by the northeastward CDW plume (Fig. 2n, o). An offshore bottom front is observed around the YSCWM boundary in summer and autumn (Fig. 2p–r, v–x), and this YSCWM front (Fig. 2p) generally corresponds to the surface cold patches (Fig. 2m) in summer. Tidal mixing coupled with sloping topography plays an important role in establishing summer fronts at a critical bottom depth, as described by Simpson and Hunter [3] in the Irish Sea. Such a front has been verified around the YSCWM [57, 59], and its location is mainly controlled by tidal mixing [38]. The tide-induced process can initiate homogenization of the water column inshore of some critical depth, creating a bottom thermal front [57]. Thus, rapid spatial changes in bottom water temperature can be observed near the YSCWM boundary (Fig. 2p), and temperature oscillations may occur in this area due to tidal currents variability [113]. Surface cold patches are closely related to tidal mixing near the bottom YSCWM boundary [59]. In addition, an SCW-related coastal front off the Subei Shoal is present year-round, and it is roughly consistent with the outer boundary of the YSWC in winter and spring. A coastal front off Shidao, which is associated with the southward LCC, is distinct in winter, spring, and autumn. A bottom TWC-induced salinity front is observed off the CE in summer (Fig. 2q). These results based on intensive sampling data sets revealed a detailed pattern of surface and bottom physical fronts in the western SYS [22].

The abovementioned surface thermal fronts stand out in satellite SST imagery in winter, when five major fronts with high gradients (>0.02°C km^{-1}) can be clearly observed (Fig. 3), including the Shandong Peninsula Front (front 1), Subei (Jiangsu)

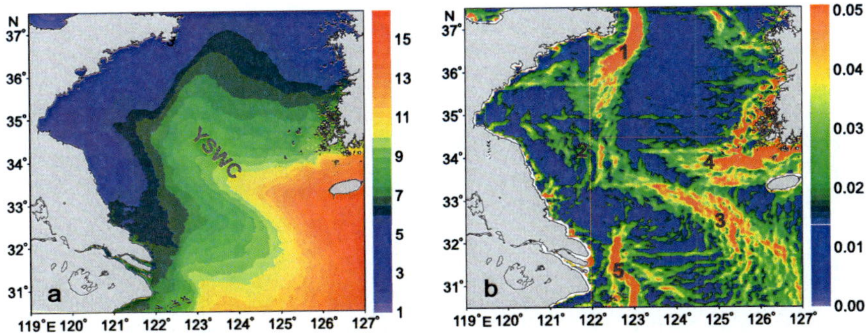

Fig. 3 Satellite SST (°C) (**a** left) and SST gradient (°C km^{-1}) (**b** right) in the SYS and northern ECS in winter (February). Major fronts are (1) Shandong Peninsula Front, (2) Subei (Jiangsu) Shoal Front, (3) Changjiang Bank Ring Front, (4) Western Cheju Front, and (5) Changjiang Estuary Front

Shoal Front (front 2), Changjiang Bank Ring Front (front 3), Western Cheju Front (front 4), and Changjiang Estuary Front (front 5). Fronts (1), (2), and (3) are closely related to the YSWC expansion, and linkages among these fronts constitute the "S"-shaped frontal structure near the YSWC western boundary (Fig. 3b). Fronts (3) and (4) are sometimes connected at their western tips to form a unique tongue-like front around the YSWC core area at the YS entrance [23]. Front (5) is the northern part of the Zhejiang–Fujian Front [39, 114]; thus, it is beyond the scope of the present chapter. Surface cold patches and associated thermal fronts in the YS in summer stand out in satellite SST [59]. The surface thermal front in summer generally corresponds to the strong subsurface front near the YSCWM boundary, which is controlled by tidal mixing [59, 115]. Overall, the thermal fronts and their structures in the YSLME are relatively spatially stable during particular seasons, as revealed by previous studies [39–41, 43, 59].

3.2 *Turbidity Front*

Figure 4 shows seasonal distributions of surface turbidity in the western SYS. Generally, the turbidity in the nearshore region is higher than offshore; three turbid areas are located off the Shandong Peninsula tip (off Shidao), Qingdao coast, and Subei Shoal. As a result, a prominent turbidity front forms between the nearshore turbid water and offshore clear water, as indicated by the 2 NTU isoline (Fig. 4). Moreover, turbidity over the Subei Shoal is the highest, and the expansion range of turbid water is the largest year-round, exhibiting the strongest turbidity front; in winter, autumn, and spring, the turbid waters (>2 NTU) in this region mainly expand southeastward; in summer, the turbidity is greatly reduced, and the turbid waters are mostly confined within the shoal [71]. In contrast, the scope of the turbid zone off the Shidao and Qingdao coasts is much smaller, and its intensity is weaker. In fact,

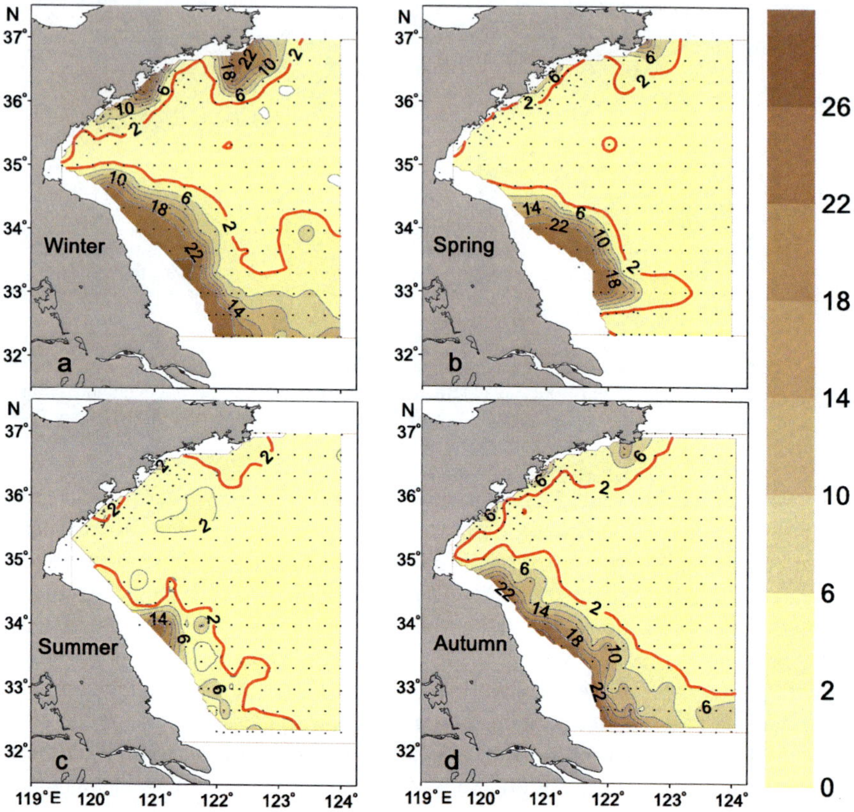

Fig. 4 Seasonal distributions of surface turbidity (NTU) during (a) winter, (b) spring, (c) summer and (d) autumn in the western SYS. The red isoline of 2 NTU generally indicates the prominent turbidity front between the turbid coastal water and offshore clear water

influenced by the "S"-shaped thermohaline front around the YSWC boundary in winter (Fig. 2a, b) and the YSCWM-related front in summer (Fig. 2m, p), the suspended sediments in the SYS can be confined to the nearshore region; therefore, the turbid water plume becomes trapped along the coast [71]. This process may be responsible for the formation of the turbidity front. In fact, the results by Zhong et al. [85] confirmed that the thermal front of the YSCWM can block the offshore transport of coastal sediments, particularly in summer.

3.3 Nutrient Fronts

Figure 5 shows the nutrient distributions in the western SYS, including DIN, PO_4-P, and SiO_3-Si. In the surface layer, nutrient concentrations are high in the

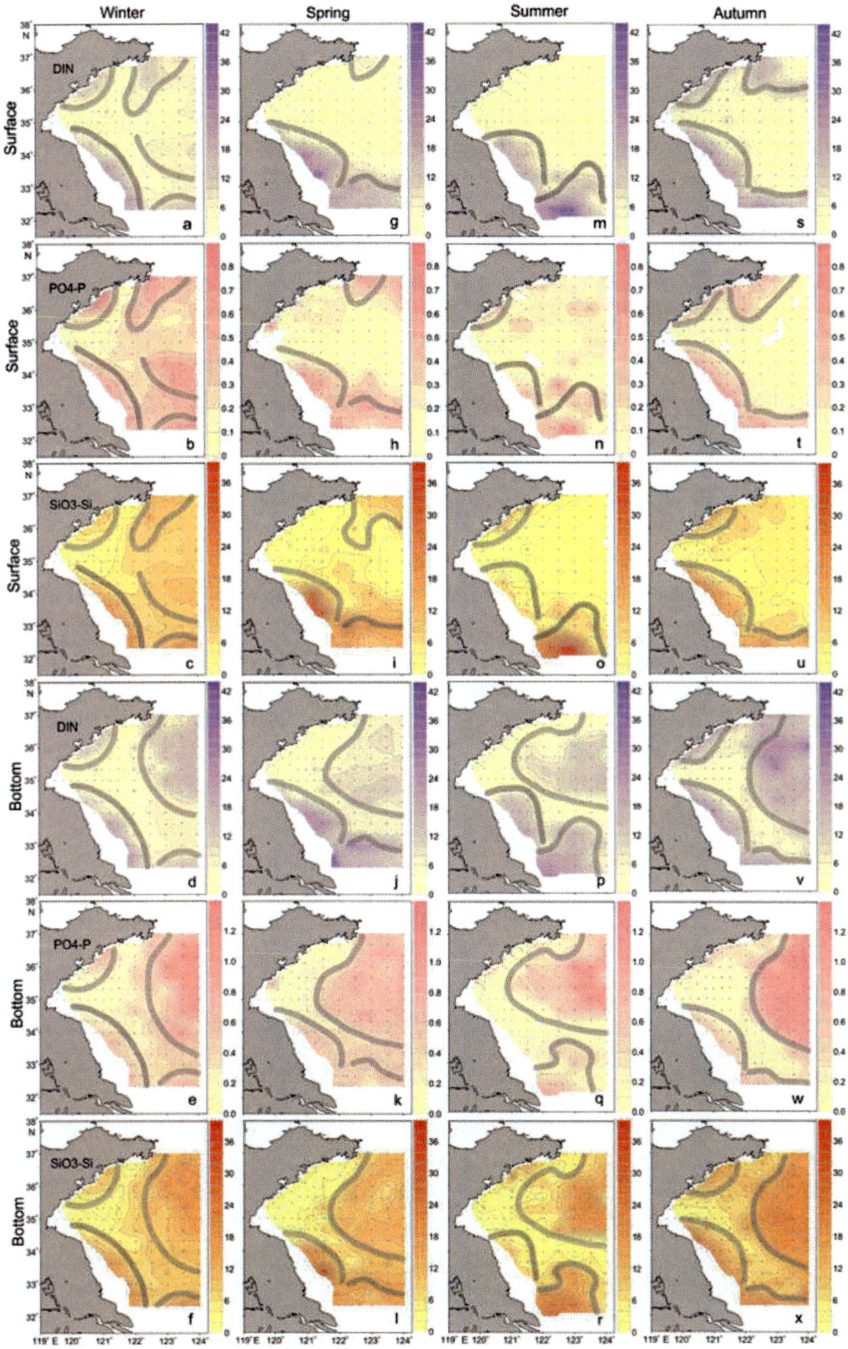

Fig. 5 Seasonal distributions of nutrients (μmol dm^{-3}) during (a–f) winter, (g–l) spring, (m–r) summer and (s–x) autumn in the western SYS. Gray bands are nutrient fronts. The DIN maps are modified after [22]

southwestern and northern areas and low in the central SYS year-round. The highest nutrient concentrations are observed from the northeastern area off the CE to the Subei Shoal, decreasing seaward and showing significant gradients. Another high-nutrient region off Shidao generally expands southward by passing the Shandong Peninsula tip, especially in winter, spring, and autumn. At the bottom, high-nutrient concentrations persist in the southeastern SYS, especially over the Subei Shoal. A high-nutrient region exists in the central SYS, and its scope is in accordance with the bottom YSCWM in summer and autumn.

Based on the annual cycle of nutrient distributions, nutrient fronts and their locations in the western SYS were identified (Fig. 5). Coastal nutrient fronts exist off the Subei Shoal and Shidao, and they are closely associated with SCW and LCC expansion (Fig. 2). Another coastal nutrient front is formed off the Qingdao coast. The bottom nutrient front in the central SYS follows the YSCWM boundary in summer and moves eastward in autumn, while it was consistent with the outer edge of the YSWC in winter and spring, suggesting a role played by the YSCWM/YSWC boundary in the nutrient front formation. A surface nutrient front, which is related to the CDW plume, exhibits a tongue-shaped pattern in summer and retreats southward in autumn. A bottom nutrient front, which is induced by the northward TWC expansion, is present near 123°E in the southern area of the SYS, especially in summer (Fig. 5p–r). A circular nutrient front is observed in the southeastern SYS in winter (Fig. 5a–f), implying a cross-shelf transport resulting from western YS coastal water under the northwesterly monsoon [22, 116]. Accordingly, three types of nutrient fronts were determined in the western SYS: estuarine (due to the CDW plume), offshore (mainly formed around the boundaries of the YSWC, YSCWM, and TWC), and coastal (induced by coastal water extension) [22].

4 Front-Driven Physical–Biogeochemical–Ecological Interactions in the Southwestern YSLME

4.1 Nutrient Transport and Light Conditions Associated with the Fronts

The distribution and transport of nutrients in the western SYS are closely related to the current regime and water-mass structure. The horizontal distributions of nutrients in the Subei Shoal and adjacent area are mainly controlled by the SCW year-round; the nutrients off Shidao are primarily delivered by the southward LCC, especially in winter and spring; from spring to autumn (especially in summer), the surface high-nutrient region in the northeastern CE is due to the transport associated with the CDW expansion. Statistical analyses further show that nutrient concentration decreases as salinity increases in the SCW-dominated area (characterized by low salinity), and a similar relationship exists between nutrient concentration and temperature in the LCC-dominated area (characterized by low temperature); surface nutrient concentration is also negatively correlated with salinity in the CDW-affected area, especially in summer [22]. These results suggest the two-end mixing of high-nutrient coastal water and offshore seawater with low nutrients. A positive relationship is found between nutrient concentration and salinity at the bottom off the CE in summer, indicating that the saline TWC may transport nutrients northward to the southern YS [22]. The locations of nutrient fronts generally represent the range of material transport along with different water masses, and they roughly agree with the positions of related thermohaline and density fronts. During the stratified warm season, nutrients can accumulate within the YSCWM due to decomposition of organic matter under the pycnocline, resulting in a nutrient pool [62]. Therefore, a bottom nutrient front is shaped by the YSCWM boundary in the central SYS during summer and autumn.

Turbidity levels, which are related to the amount of suspended sediments, can partially determine the light conditions in marine environments. Previous studies have shown that the suspended sediment amount in the nearshore area of the YS is high year-round [117, 118]. Undoubtedly, the coastal thermohaline and density fronts may influence the offshore transport of suspended sediments and thereby restrict the turbid coastal water plume within a certain area in the western SYS (Fig. 4). Because three main turbid areas exist year-round off the Subei Shoal, Qingdao coast, and Shidao, turbidity frontal patterns are most conspicuous in these regions (Fig. 4). The turbidity front is collocated with the 30 m isobath year-round, indicating a region with improved light conditions. The vertically averaged irradiance, which can be regarded as a light index denoting the average irradiance, is much higher near the frontal region than in nearshore and offshore areas in the SYS [119].

Vertical transport of nutrients exists in the frontal region of the YSCWM in the stratified summer. As shown by the nutrient distributions along typical sections across the YSCWM-dominated area (Fig. 6a–c), a nutrient deficit prevails in the upper mixed layer in the central SYS, and a notable nutricline occurs at the 30 m depth. A nutrient pool forms within the YSCWM, and nutrient contours tend to be significantly uplifted along the western sloping area of the SYS, which leads to nutricline shoaling and vertical transport of nutrients near the YSCWM frontal region. This process is driven by upwelling, which mainly results from tidal mixing around the YSCWM front [59]. As a result, cold bottom water near the YSCWM boundary can be pumped upward, and cold patches form at the surface with distinct thermal fronts (i.e., upwelling fronts) (Fig. 2m). By multiplying the nutrient concentration with upward velocity in the upwelling area, the estimated upwelled nutrient fluxes in summer were $1.4 \pm 0.9 \times 10^3$, $0.1 \pm 0.1 \times 10^3$, and $2.0 \pm 1.3 \times 10^3$ μmol m^{-2} day^{-1} for DIN, PO_4-P, and SiO_3-Si, respectively [22]. Consequently, this upwelling system plays a crucial role in extracting nutrients from the YSCWM to the euphotic layer. During the eastward movement of the YSCWM boundary from summer to autumn, nutrients that accumulate inside the cold-water mass can also be released to the nearshore side of the front [21]. The front may prevent the offshore transport of suspended sediments; as a result, the turbid water becomes coastally trapped (Fig. 4). The sharp reduction in turbidity in the

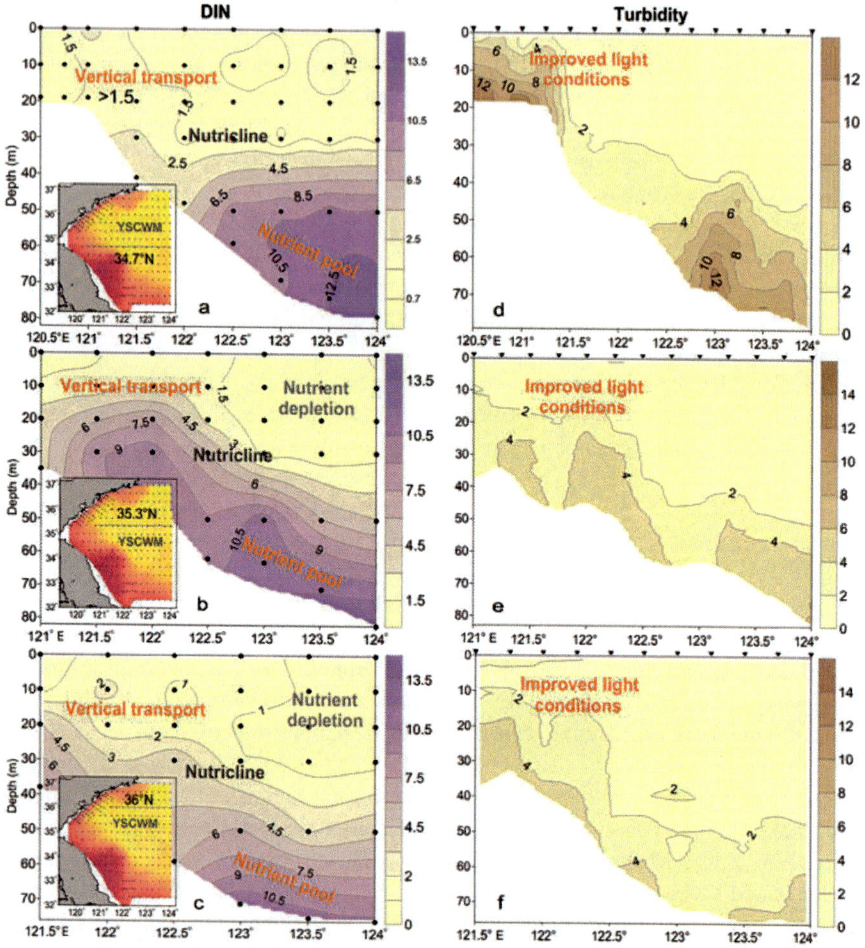

Fig. 6 Vertical distributions of (a–c) DIN (μmol dm^{-3}) and (d–f) turbidity (NTU) along typical sections crossing the YSCWM-dominated area in summer. Insets in panels **a**, **b**, and **c** show the section locations and the summer bottom temperature (°C) distribution in the SYS; panels **a** and **d** are modified after [21]

frontal region of the YSCWM provides a rapid improvement in light conditions (Fig. 6d–f).

4.2 Front-Driven Primary Production Regime

In spring, patch-like phytoplankton blooms with high Chl-a content can be observed in the central SYS (Fig. 7b–d). Their western boundaries are approximately

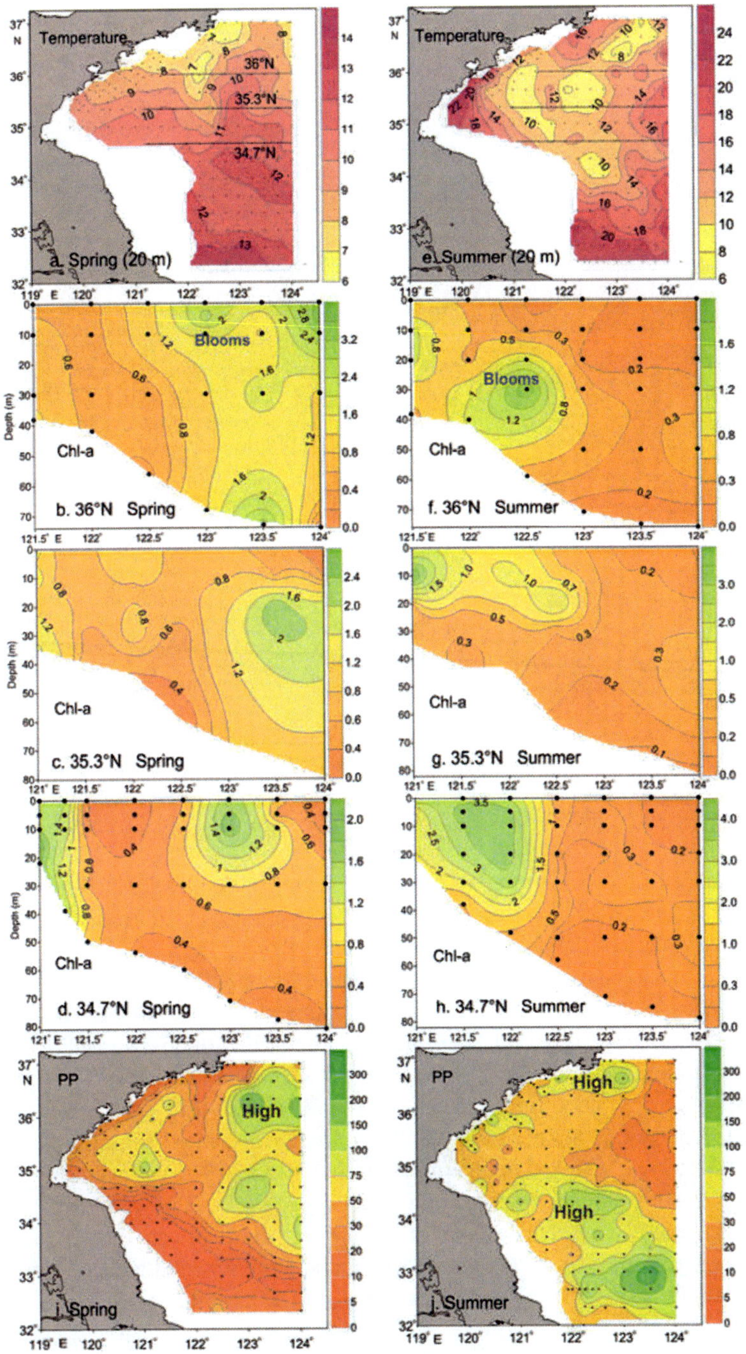

Fig. 7 (**a**, **e**) Temperature (°C); (**b–d** and **f–h**) Chl-a (μg m^{-3}) along typical sections; (**i**, **j**) PP (mg C m^{-2} day^{-1}) in spring and summer. The left-/right-hand side of each comparative image is summer/winter. The section locations are shown in panels **a** and **e**; panels **d** and **h** are modified after [21, 76]

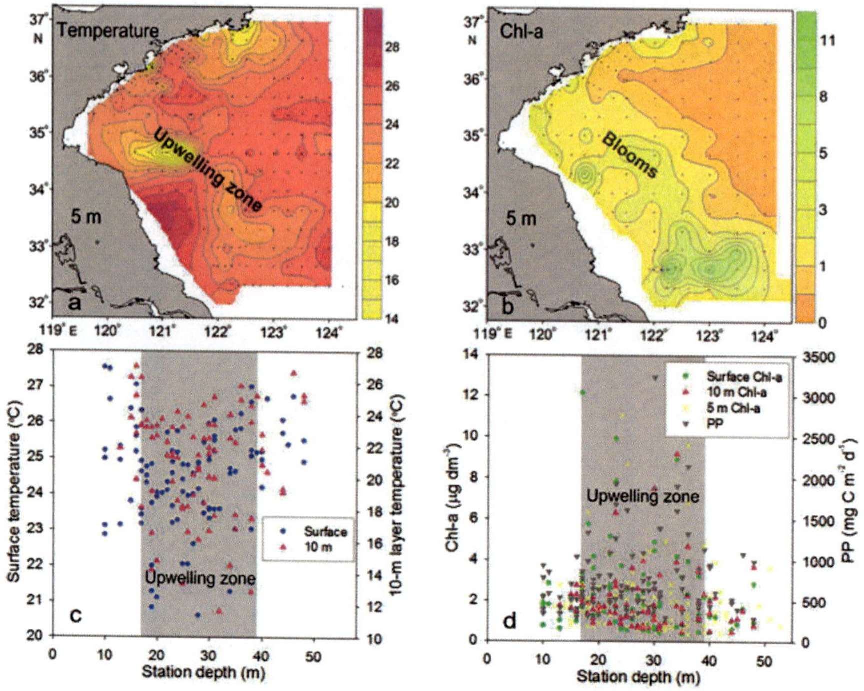

Fig. 8 (**a**, **b**) Temperature (°C) and Chl-a (μg m^{-3}) distributions in the surface 0–5-m layer and (**c**, **d**) statistical correlations between temperature, Chl-a, PP, and water depth in the western area (station depth < 50 m) of the SYS in summer. Panel **a** is modified after [21]; panels c and d are modified after [62] by adding 5-m and 10-m data

collocated with the cold-water belt (i.e., the result of southward-expanding LCC) (Fig. 7a). Therefore, the presence of this cold belt and its eastern front can shape the western boundary of the phytoplankton-blooming area in the central SYS [21]. Phytoplankton patchiness mainly occurs in the western sloping region of the SYS in summer (Fig. 7f–h), and its location roughly corresponds to the western boundary of the low-temperature area (Fig. 7e), which is associated with tidal upwelling near the YSCWM boundary [59]. This result is related to the vertical transport of nutrients and improved light conditions around the YSCWM front (Fig. 6). Therefore, Chl-a enhancement can occur near the YSCWM frontal region in summer. Horizontal distributions of Chl-a and temperature in the surface 0–5-m layer (Fig. 8a, b) illustrate this phenomenon (i.e., the high-Chl-a area is roughly consistent with the upwelling zone). Statistical analyses also show that the high values of Chl-a and PP in the western SYS correspond to the low-temperature upwelling zone, both occurring in the water depth range of approximately 15–35 m (Fig. 8c, d). In surface cold patches caused by upwelling, Chl-a and nutrients are positively correlated [22]. Thus, the Chl-a content and PP in the frontal region in summer are generally higher than those in the offshore region and coastal area (Fig. 7f–h, j). These results confirm the important role of the upwelling system in supplying nutrients and thus maintaining primary production in the western SYS in summer. Consequently, there is a spatial shift for phytoplankton blooms from the central SYS in spring to the YSCWM frontal region in summer [22], and the seasonal variations in the high-PP region also exhibit similar patterns (Fig. 7i, j).

In autumn, forced by enhanced wind stress and vertical mixing, the bottom YSCWM and its front retreat eastward to the deep area (Fig. 2v–x) when compared with that in summer (Fig. 2p–r). This process may cause a nutrient exchange between the inside and outside areas of the YSCWM, leading to a release of nutrients accumulated inside the YSCWM to the exterior. Thus, a continuous supply of nutrients to the euphotic layer can be achieved around the YSCWM front from summer to autumn, which may play an essential role in maintaining the relatively high Chl-a near the frontal region [21]. Due to the upward transport of nutrients induced by deepening stratification in the YSCWM-dominated area, the Chl-a content in the central SYS in autumn is slightly higher than that in summer, as shown by Wei et al. [21].

In winter, the YSCWM front disappears, and a thermohaline front occurs around the YSWC boundary (Fig. 2a, b). The reverse flows on both sides of this front may be favorable for the convergence of plankton near the frontal zone [71]. The sharp reduction in turbidity (compared with the nearshore) (Fig. 4a) also provides improved light conditions. Consequently, the frontal region features high Chl-a and PP (Fig. 9a, b). In contrast, Chl-a and PP are relatively low on both sides of the frontal zone. This phenomenon can be further illustrated by vertical distributions of Chl-a along the cross-front sections off the Subei Shoal [71]. As shown by the relationships between Chl-a/PP and temperature/salinity in the western SYS (Fig. 9c, d), high Chl-a/PP values generally occur at stations characterized by medium temperatures and salinities, forming a Chl-a/PP peak near the region with temperatures of ~5–9°C and salinities of ~31–32; this region corresponds to the frontal zone formed by the offshore YSWC and coastal water (Fig. 2a, b). These patterns are consistent with previous in situ observations [120] and satellite-derived results [119].

4.3 *Anchovy Distribution and Other Ecological Processes in Relation to the Fronts*

Anchovies are critically important fishery species; therefore, exploring their spawning, feeding, migration, and overwintering is highly important for understanding the ecological dynamics in the YSLME. As revealed by previous studies, anchovies are sensitive to seawater temperature (they prefer a relatively narrow range of 11–13°C), and their distribution is closely related to physical conditions [95, 121–123]. In winter, high anchovy biomass mainly exists near the front adjacent to the YSWC core area (Fig. 10a, b) at the YS entrance. The appropriate temperature in this region (Fig. 2a, d) provides a favorable overwintering environment for

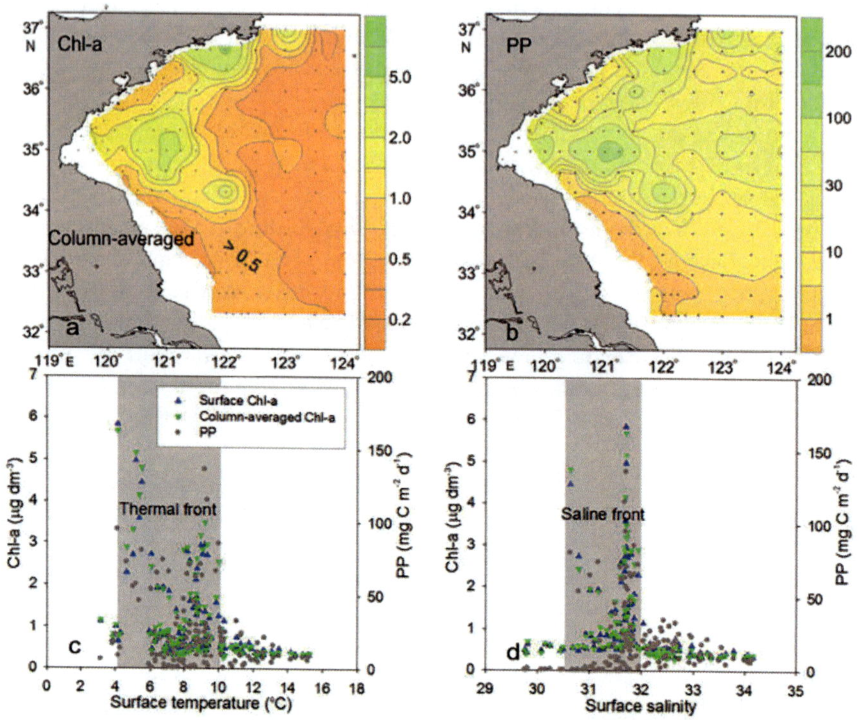

Fig. 9 (**a**, **b**) Horizontal distributions of column-averaged Chl-a and PP; (**c**, **d**) relationships between Chl-a, PP, surface temperature, and surface salinity in the SYS in winter. Panels **c** and **d** are modified after [21] by adding PP data

anchovies. In spring, the feeding demands of anchovies significantly increase [95]. The abundant plankton in the YSWC-affected blooming area [98], whose boundary is shaped by the southward LCC front (Fig. 7a–d), may provide sufficient food for anchovies. The anchovy biomass increases with column-averaged Chl-a content in spring (Fig. 10c). The seawater temperature (10–12°C) in this region (Fig. 2g, j) is also suitable. Therefore, anchovies leave the southern overwintering ground and migrate northward to feed [95]. As a result, high concentrations of anchovies are formed in the YSWC-affected bloom area in spring [21].

During the stratified summer, due to a nutrient deficit in the upper mixed layer, primary production can be limited in the central SYS, while the vertical transport of nutrients by upwelling near the YSCWM (Figs. 6a–c, 8a) triggers Chl-a increase in the frontal region (Figs. 7f–h, 8b). Therefore, the organisms that anchovies feed upon may shift from the central SYS in spring to the YSCWM frontal region in summer. An appropriate temperature (~10°C) for the anchovy is also achieved near this thermohaline front (Fig. 2p, q). Statistical analyses show that high anchovy biomass is found in a region with a temperature of ~10°C and salinity of ~33 (Fig. 10d, e), which is approximately collocated with the YSCWM boundary. The

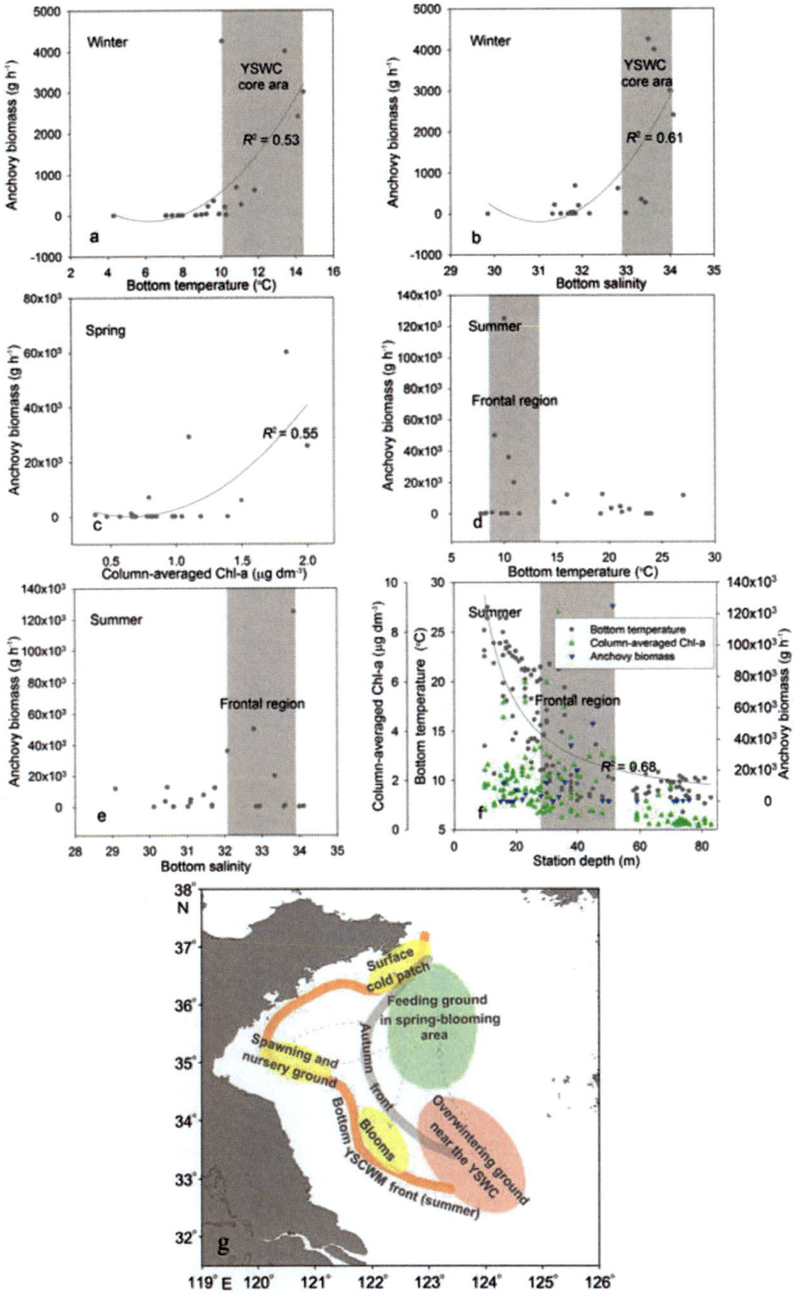

Fig. 10 **(a–f)** Relationships between anchovy biomass and environmental variables; **(g)** schematic diagram illustrating seasonal migration routes of anchovies in the western SYS. Panels **(a–f)** are modified after [21]. Arrows in panel **(g)** indicate possible migration paths of anchovy

area with high column-averaged Chl-a and high anchovy biomass roughly corresponds to a high bottom temperature gradient zone (i.e., YSCWM front) (Fig. 10f). Correspondingly, anchovies migrate from the central SYS in spring to the YSCWM frontal region in summer [21], which can lead to high concentrations of anchovy eggs and larvae near the front [90, 124]. In autumn, the YSCWM shrinks (Fig. 2v) and thus the distribution area of anchovies around the YSCWM front, where available food and suitable temperature are found, shifts seaward. Thus, the frontal region around the YSCWM may generally act as an important spawning, feeding, and nursery ground for anchovies in summer and autumn [21]. A reasonable schematic diagram generally illustrating the seasonal migration routes of anchovies in the western SYS is given in Fig. 10g.

The western SYS has been well known for its extraordinary green tides in recent years, which are caused by massive blooms of *Ulva prolifera* [125, 126]. Other ecological disasters, such as jellyfish blooms [127–129] and Sargassum [130], have also been observed. Due to convergences associated with horizontal thermohaline and density fronts [131], frontogenesis in the western SYS is favorable for accumulation of these organisms. The tidally induced upwelling around the YSCWM front in summer can provide nutrients to support algae growth [71]. Thus, the convergence and propagation of *Ulva prolifera* blooms near the upwelling front may have a synergistic effect over the Subei Shoal, leading to the large-scale development of green tides, as indicated by Wei et al. [132]. Consequently, the frontal system in the western SYS may potentially influence and shape the distribution of algal blooms and other ecological disasters.

5 Concluding Remarks

Based on in situ observations and previous studies, this chapter provides a comprehensive and systematic overview of front-driven physical–biogeochemical–ecological interactions in the YSLME. The overview reveals that the fronts provide unique scenarios for physical-biogeochemical processes and interactions in the western SYS, laying a foundation for front-related studies in marginal sea ecosystems. The front-related physical-chemical processes and associated ecological effects determine the complexity of ecosystem dynamics in the YS. In future studies, special efforts should be dedicated to integrating high-resolution simulations and observations to better understand and quantitatively evaluate the diverse functions of fronts in shaping YSLME.

Acknowledgments We thank the book editor, Igor M. Belkin, for the invitation to prepare this chapter on physical–biogeochemical–ecological interactions in frontal zones. Support from the National Natural Science Foundation of China under Grant Nos. U1906210 and 41876085; the Basic Scientific Fund of the National Public Research Institutes of China under Grant No. 2020S03; and the National Project of Comprehensive Investigation and Research of Coastal Seas in China under Grant No. 908-01-ST03 are gratefully acknowledged.

References

1. Belkin IM (2002) Front. In: Nybakken JW, Broenkow WW, Vallier TL (eds) Interdisciplinary encyclopedia of marine sciences. Grolier, Danbury, pp 433–436
2. Wang Y, Castelao RM, Yuan Y (2015) Seasonal variability of alongshore winds and sea surface temperature fronts in eastern boundary current. J Geophys Res 120:2385–2400
3. Simpson JH, Hunter JR (1974) Fronts in the Irish Sea. Nature 250:404–406. https://doi.org/10.1038/250404a0
4. Geyer WR, Ralston DK (2015) Estuarine frontogenesis. J Phys Oceanogr 45:546–561. https://doi.org/10.1175/JPO-D-14-0082.1
5. Cao L, Tang R, Huang W, Wang Y (2021) Seasonal variability and dynamics of coastal sea surface temperature fronts in the East China Sea. Ocean Dyn 71:237–249. https://doi.org/10.1007/s10236-020-01427-8
6. Jing Z, Wang S, Wu L, Chang P, Zhang Q, Sun B, Ma X, Qiu B, Small J, Jin FF, Chen Z, Gan B, Yang Y, Yang H, Wan X (2020) Maintenance of mid-latitude oceanic fronts by mesoscale eddies. Sci Adv 6:eaba7880
7. Zeng X, Belkin IM, Peng S, Li Y (2014) East Hainan upwelling fronts detected by remote sensing and modelled in summer. Int J Remote Sens 35(11–12):4441–4451
8. Woodson CB, Litvin SY (2015) Ocean fronts drive marine fishery production and biogeochemical cycling. PNAS 112(6):1710–1715
9. Stukel MR, Aluwihare LI, Barbeau KA, Chekalyuk AM, Goericke R, Miller AJ, Ohman MD, Ruacho A, Song H, Stephens BM, Landry MR (2017) Mesoscale ocean fronts enhance carbon export due to gravitational sinking and subduction. PNAS 114(6):1252–1257
10. Belkin IM (2021) Remote sensing of ocean fronts in marine ecology and fisheries. Remote Sens 13:883. https://doi.org/10.3390/rs13050883
11. Gattuso JP, Frankignoulle M, Wollast R (1998) Carbon and carbonate metabolism in coastal aquatic ecosystems. Annu Rev Ecol Syst 29:405–434
12. Simpson JH, Sharples J (2012) Introduction to the physical and biological oceanography of shelf seas. Cambridge University Press, Cambridge
13. Belkin IM, Cornillon PC, Sherman K (2009) Fronts in large marine ecosystems. Prog Oceanogr 81(1–4):223–236
14. Sherman K (1990) Productivity, perturbations, and options for biomass yields in large marine ecosystems. In: Sherman K, Alexander LM, Gold BD (eds) Large marine ecosystems: patterns, processes and yields. AAAS Publications, Washington, pp 206–219
15. Tang QS, Ying YP, Wu Q (2016) The biomass yields and management challenges for the Yellow Sea Large Marine Ecosystem. Environ Dev 17:175–181
16. Zhang ZH, Qu FY, Wang SQ (2019) Sustainable development of the Yellow Sea Large Marine Ecosystem. Deep-Sea Res II 163:102–107
17. Lie HJ, Cho CH (2016) Seasonal circulation patterns of the Yellow and East China Seas derived from satellite-tracked drifter trajectories and hydrographic observations. Prog Oceanogr 146:121–141. https://doi.org/10.1016/j.pocean.2016.06.004
18. Liu ZQ, Gan JP, Hu JY, Wu H, Cai ZY, Deng YF (2021) Progress on circulation dynamics in the East China Sea and southern Yellow Sea: origination, pathways, and destinations of shelf current. Prog Oceanogr 193:102553. https://doi.org/10.1016/j.pocean.2021.102553
19. Liu X, Chiang KP, Liu SM, Wei H, Zhao Y, Huang BQ (2015) Influence of the Yellow Sea warm current on phytoplankton community in the Central Yellow Sea. Deep-Sea Res I 106:17–29
20. Liu X, Huang BQ, Huang Q, Wang L, Ni XB, Tang QS, Sun S, Wei H, Liu SM, Li CL, Sun J (2015) Seasonal phytoplankton response to physical processes in the southern Yellow Sea. J Sea Res 95:45–55
21. Wei QS, Li XS, Wang BD, Fu MZ, Ge RF, Yu ZG (2016) Seasonally chemical hydrology and ecological responses in frontal zone of the central southern Yellow Sea. J Sea Res 112:1–12

22. Wei QS, Yu ZG, Wang BD, Fu MZ, Xia CS, Liu L, Ge RF, Wang HW, Zhan R (2016) Coupling of the spatial-temporal distributions of nutrients and physical conditions in the southern Yellow Sea. J Mar Syst 156:30–45
23. Lie HJ, Cho HC, Lee S (2009) Tongue-shaped frontal structure and warm water intrusion in the southern Yellow Sea in winter. J Geophys Res 114:C01003. https://doi.org/10.1029/2007JC004683
24. Lie HJ, Oh KH, Cho CH, Moon JH (2019) Wintertime large temperature inversions in the Yellow Sea associated with the Cheju and Yellow Sea warm currents. J Geophys Res Oceans 124:4856–4874. https://doi.org/10.1029/2019JC015180
25. Lin XP, Yang JY, Guo JS, Zhang ZX, Yin YQ, Song XZ, Zhang XH (2011) An asymmetric upwind flow, Yellow Sea warm current: 1. New observations in the western Yellow Sea. J Geophys Res 116:C04026. https://doi.org/10.1029/2010JC006513
26. Ho CB, Wang YX, Lei ZY, Xu S (1959) A preliminary study of the formation of Yellow Sea cold water mass and its properties. Oceanol Limnol Sin 2(1):11–15 (in Chinese with English Abstract)
27. Yu F, Zhang ZX, Diao XY, Guo JS, Tang YX (2006) Analysis of evolution of the Huanghai Sea cold water mass and its relationship with adjacent water masses. Acta Oceanol Sin 28:26–34 (in Chinese with English Abstract)
28. Guo BH (1993) Major features of the physical oceanography in the Yellow Sea. J Oceanogr Huanghai Bohai Seas 11(3):7–18 (in Chinese)
29. Zhang QL, Weng XC, Yang YL (1996) Analysis of water masses in the South Yellow Sea in spring. Oceanol Limnol Sin 27(4):421–428 (in Chinese with English abstract)
30. Zhang ZX, Guo JS, Qiao FL, Liu YY, Guo BH (2016) Whereabouts and freshwater origination of the Subei coastal water. Oceanol Limnol Sin 47(3):527–532 (in Chinese with English abstract)
31. Zhu P, Wu H (2018) Origins and transports of the low-salinity coastal water in the southwestern Yellow Sea. Acta Oceanol Sin 37(4):1–11
32. Wu H, Gu JH, Zhu P (2018) Winter counter-wind transport in the inner southwestern Yellow Sea. J Geophys Res Oceans 123:411–436. https://doi.org/10.1002/2017JC013403
33. Mao HL, Kan TC, Lan SF (1963) A preliminary study of the Yangtze diluted water and its mixing processes. Oceanol Limnol Sin 5(3):183–206 (in Chinese with English abstract)
34. Beardsley RC, Limeburner R, Yu H, Cannon GA (1985) Discharge of the Changjiang (Yangtze River) into the East China Sea. Cont Shelf Res 4:57–76
35. Lie HJ, Cho CH, Lee JH, Lee S (2003) Structure and eastward extension of the Changjiang River plume in the East China Sea. J Geophys Res 108(C3):3077. https://doi.org/10.1029/2001JC001194
36. Guan BX (1994) Patterns and structures of the currents in Bohai, Huanghai and East China seas. In: Zhou D, Liang YB, Zeng CK (eds) Oceanology of China seas. Kluwer Academic Publishers, pp 17–26. https://doi.org/10.1007/978-94-011-0862-1_3
37. Qi JF, Yin BS, Zhang QL, Yang DZ, Xu ZH (2017) Seasonal variation of the Taiwan warm current water and its underlying mechanism. Chin J Oceanol Limnol 35(5):1045–1060
38. Zhao BR (1985) The fronts of the Huanghai Sea cold water mass induced by tidal mixing. Oceanol Limnol Sin 16:451–459 (in Chinese with English abstract)
39. Hickox R, Belkin IM, Cornillon P, Shan Z (2000) Climatology and seasonal variability of ocean fronts in the East China, Yellow and Bohai Seas from satellite SST data. Geophys Res Lett 27(18):2945–2948
40. Belkin IM, Cornillon PC (2003) SST fronts of the Pacific coastal and marginal seas. Pacific Oceanogr 1(2):90–113
41. Park S, Chu PC (2006) Thermal and haline fronts in the Yellow/East China Seas: surface and subsurface seasonality comparison. J Oceanogr 62(5):617–638
42. Huang DJ, Zhang T, Zhou F (2010) Sea-surface temperature fronts in the Yellow and East China seas from TRMM microwave imager data. Deep-Sea Res II 57(11):1017–1024

43. Wang F, Liu CY, Meng QJ (2012) Effect of the Yellow Sea warm current fronts on the westward shift of the Yellow Sea warm tongue in winter. Cont Shelf Res 45:98–107
44. Lie HJ, Cho CH, Lee S (2013) Frontal circulation and westward transversal current at the Yellow Sea entrance in winter. J Geophys Res Oceans 118(8):3851–3870. https://doi.org/10.1002/jgrc.20280
45. Lie HJ, Cho CH, Jung KT (2015) Occurrence of large temperature inversion in the thermohaline frontal zone at the Yellow Sea entrance in winter and its relation to advection. J Geophys Res Oceans 120(1):417–435. https://doi.org/10.1002/2014JC010653
46. Oh KH, Lee S, Lie HJ, Jung KT, Qiao F (2015) Study on the current structure of the thermohaline front in the southeastern entrance of the Yellow Sea during winter. Acta Oceanol Sin 34(12):29–36
47. Sun JC, Yang DZ, Yin BS, Chen HY, Feng XR (2018) Onshore warm tongue and offshore cold tongue in the western Yellow Sea in winter: the evidence. Chin J Oceanol Limnol 36: 1475–1483. https://doi.org/10.1007/s00343-018-7021-0
48. Huh OK (1976) Detection of oceanic thermal fronts off Korea with the defense meteorological satellite. Remote Sens Environ 5:191–213
49. Zheng QA, Klemas V (1982) Determination of winter temperature patterns, fronts, and surface currents in the Yellow Sea and East China Sea from satellite imagery. Remote Sens Environ 12(3):201–218
50. Lie HJ (1989) Tidal fronts in the southeastern Hwanghae (Yellow Sea). Cont Shelf Res 9(6): 527–546
51. Ning X, Liu Z, Cai Y, Fang M, Chai F (1998) Physicobiological oceanographic remote sensing of the East China Sea: satellite and in situ observations. J Geophys Res 103(C10): 21623–21635
52. Xie SP, Hafner J, Tanimoto Y, Liu WT, Tokinaga H, Xu HM (2002) Bathymetric effect on the winter sea surface temperature and climate of the Yellow and East China seas. Geophys Res Lett 29(24):2228. https://doi.org/10.1029/2002GL015884
53. Chen CTA (2009) Chemical and physical fronts in the Bohai, Yellow and East China seas. J Mar Syst 78:394–410
54. Wang F, Liu CY (2009) An N-shape thermal front in the western South Yellow Sea in winter. Chin J Oceanol Limnol 27(4):898–906
55. Huang RH, Wang QM, Zhao BR (1992) Satellite observations of continental shelf front in northern China Sea. Adv Space Res 13(5):79–83
56. Tana FY, Liu BC, Sun SW, Wang HW (2017) Dramatic weakening of the ear-shaped thermal front in the Yellow Sea during 1950s–1990s. Acta Oceanol Sin 36(5):51–56
57. Lin F, Asplin L, Budgell WP, Wei H, Fang J (2019) Currents on the northern shelf of the Yellow Sea. Reg Stud Mar Sci 32:100821. https://doi.org/10.1016/j.rsma.2019.100821
58. Ma J, Qiao FL, Xia CS, Yang YZ (2004) Tidal effects on temperature front in the Yellow Sea. Chin J Oceanol Limnol 22(3):314–321
59. Lü XG, Qiao FL, Xia CS, Wang GS, Yuan YL (2010) Upwelling and surface cold patches in the Yellow Sea in summer: effects of tidal mixing on the vertical circulation. Cont Shelf Res 30:620–632
60. Xia Y, Bao X, Song D, Ding Y, Wan K, Yan Y (2019) Tidal effects on the bottom thermal front of north Yellow Sea cold water mass near Zhangzi Island in summer 2009. J Ocean Univ China 18(4):751–760
61. Liu DY, Wang YQ (2013) Trends of satellite derived chlorophyll-a (1997-2011) in the Bohai and yellow seas, China: effects of bathymetry on seasonal and inter-annual patterns. Prog Oceanogr 116:154–166
62. Wei QS, Wang BD, Fu MZ, Sun JC, Yao QZ, Xin M, Yu ZG (2020) Spatiotemporal variability of physical-biogeochemical processes and intrinsic correlations in the semi-enclosed South Yellow Sea. Acta Oceanol Sin 39(10):11–26. https://doi.org/10.1007/s13131-020-1656-3

63. Hao J, Chen Y, Wang F (2010) Temperature inversion in China seas. J Geophys Res 115: C12025. https://doi.org/10.1029/2010JC006297
64. Pang IC, Moon JH, Lee JH, Hong JS, Pang SJ (2017) Modeling temperature inversion in southeastern Yellow Sea during winter 2016. J Geophys Res Oceans 122:3848–3860. https://doi.org/10.1002/2017JC012718
65. Xia ZW, Guo BH (1983) The cold water and upwelling in the tip areas of Shandong peninsula and Liaodong peninsula. J Oceanogr Huanghai Bohai Seas 1(1):13–19 (in Chinese)
66. Lie HJ (1986) Summertime hydrographic features in the southeastern Huanghai. Prog Oceanogr 17:229–242
67. Seung YH, Chung JH, Park YC (1990) Oceanographic studies related to the tidal front in the mid-Yellow Sea off Korea: physical aspect. J Soc Oceanogr Korea 25:84–95
68. Zou EM, Guo BH, Tang YX, Lee JH, Lie HJ (2001) An analysis of summer hydrographic features and circulation in the southern Yellow Sea and the northern East China Sea. Oceanol Limnol Sin 32:340–348 (in Chinese with English Abstract)
69. Ren SH, Xie JP, Zhu J (2014) The roles of different mechanisms related to the tide-induced fronts in the Yellow Sea in summer. Adv Atmos Sci 31:1079–1089
70. Liang XS, Huang M, Wu H, Wang Y (2018) The Yellow Sea surface cold patches in warm seasons. In: Liang XS, Zhang Y (eds) Coastal environment, disaster, and infrastructure: a case study of China's coastline. IntechOpen, pp 27–45. https://doi.org/10.5772/intechopen.80732
71. Wei QS, Fu MZ, Sun JC, Yao QZ, Wang BD, Liu XH, Yu ZG (2020) Seasonal physical fronts and associated biogeochemical-ecological effects off the Jiangsu Shoal in the Western Yellow Sea, China. J Geophys Res Oceans 125:e2020JC016304. https://doi.org/10.1029/2020JC016304
72. Chin TM, Vazquez-Cuervo J, Armstrong EM (2017) A multi-scale high-resolution analysis of global sea surface temperature. Remote Sens Environ 200:154–169
73. Lin L, Liu DY, Luo CX, Xie L (2019) Double fronts in the Yellow Sea in summertime identified using sea surface temperature data of multi-scale ultra-high resolution analysis. Cont Shelf Res 175:76–86
74. Sun YJ, Cho YK (2010) Tidal front and its relation to the biological process in coastal water. Ocean Sci J 45(4):243–251
75. Zhao BR (1987) A preliminary study of continental shelf fronts in the western part of southern Huanghai Sea and circulation structure in the front region of the Huanghai Cold Water Mass (HCWM). Oceanol Limnol Sin 18(3):217–226 (in Chinese with English Abstract)
76. Fu MZ, Wang ZL, Sun P, Li Y, Li RX (2009) Spatial distribution characteristics and the environmental regulation mechanisms of phytoplankton chlorophyll a in the southern Yellow Sea during summer 2006. Acta Ecol Sin 29(10):5366–5375 (in Chinese with English abstract)
77. Choi JK (1991) The influence of the tidal mixing front on primary productivity and distribution of phytoplankton in the mid-eastern coast of Yellow Sea. J Soc Oceanogr Korea 26:223–241
78. Li HB, Xiao T, Lv RH, Ding T, Lin YA (2007) Impact of tidal front on the distribution of bacterioplankton in the southern Yellow Sea, China. J Mar Syst 67:263–271
79. Lee Y, Choi JK, Shon JK (2012) Phytoplankton distribution in the eastern part of the Yellow Sea by the formation of tidal mixing front and upwelling during summer. Ocean Polar Res 34: 111–123
80. Choi BJ, Lee JA, Choi JS, Park JG, Lee SH, Yih W (2017) Influence of the tidal front on the three-dimensional distribution of spring phytoplankton community in the eastern Yellow Sea. Chemosphere 173:299–306
81. Liu DY, Wang YN, Wang YQ, Keesing JK (2018) Ocean fronts construct spatial zonation in microfossil assemblages. Glob Ecol Biogeogr 27:1225–1237. https://doi.org/10.1111/geb.12779
82. Gao F, Qiao LL, Li GX (2016) Winter meso-scale shear front in the Yellow Sea and its sedimentary effects. J Ocean Univ China 15(1):50–56. https://doi.org/10.1007/s11802-016-2668-7

83. Li WJ, Wang ZY, Huang HJ (2019) Relationship between the southern Yellow Sea cold water mass and the distribution and composition of suspended particulate matter in summer and autumn seasons. J Sea Res 154:101812. https://doi.org/10.1016/j.seares.2019.101812
84. Shi Y, Gao JH, Sheng H, Du J, Jia JJ, Wang YP, Li J, Bai FL, Chen YN (2019) Cross-front sediment transport induced by quick oscillation of the Yellow Sea warm current: evidence from the sedimentary record. Geophys Res Lett 46:226–234. https://doi.org/10.1029/2018GL080751
85. Zhong Y, Qiao LL, Song DH, Ding Y, Xu JS, Xue WJ, Xue C (2020) Impact of cold water mass on suspended sediment transport in the South Yellow Sea. Mar Geol 428:106244. https://doi.org/10.1016/j.margeo.2020.106244
86. Wang A, Ralston DK, Bi N, Cheng Z, Wu X, Wang H (2019) Seasonal variation in sediment transport and deposition on a muddy clinoform in the Yellow Sea. Cont Shelf Res 179:37–51
87. Zhong F, Xiang R, Zhang L, Yang Y, Zhao M, Zhou L (2021) A synthesized study of the spatiotemporal evolution of central Yellow Sea mud depositional processes during the holocene. Front Earth Sci 9:687344. https://doi.org/10.3389/feart.2021.687344
88. Liu GM, Sun S, Wang H, Zhang Y, Yang B, Ji P (2003) Abundance of *Calanus sinicus* across the tidal front in the Yellow Sea, China. Fish Oceanogr 12:291–298
89. Wang R, Zuo T, Wang K (2003) The Yellow Sea cold bottom water—an oversummering site for *Calanus sinicus* (Copepoda, crustacea). J Plankton Res 25:169–183
90. Zhou F, Huang DJ, Wan RJ, Su JL, Zhang T, Ding T, Zhou BF (2008) Observation and analysis of tidal fronts in the northwestern Huanghai Sea. Acta Oceanol Sin 30(3):9–15 (in Chinese with English abstract)
91. Tang QS, Su JL, Sun S, Zhang J, Huang DJ, Jin XS, Tong L (2005) A study of marine ecosystem dynamics in the coastal ocean of China. Adv Earth Sci 20(12):1288–1299 (in Chinese with English abstract)
92. Zhao X (2006) Population dynamic characteristics and sustainable utilization of the anchovy stock in the Yellow Sea. Doctor dissertation, Ocean University of China, Qingdao (in Chinese with English abstract)
93. Yu HQ, Yu HM, Ito SI, Tian YJ, Wang H, Liu Y, Xing QW, Bakun A, Kelly RM (2020) Potential environmental drivers of Japanese anchovy (Engraulis japonicus) recruitment in the Yellow Sea. J Mar Syst 212:103431
94. Ma SS (1989) Relationship between distribution and hydrological conditions of the wintering anchovy in the Yellow Sea and East China Sea. J Fish China 13(3):201–206 (in Chinese with English abstract)
95. Zhu DS, Iversen SA (1990) Anchovy and other fish resources in the Yellow Sea and East China Sea November 1984–April 1988: fisheries investigations by R/V "Beidou". Mar Fish Res 11:1–143 (in Chinese with English abstract)
96. Chen JF, Ma SS, Wang WL (1997) Variation characteristics of three large water systems influencing the behavior and distribution of anchovy Engraulis japonicus in the Yellow Sea and East China Sea in autumn and winter. Mar Fish Res 18(1):54–59 (in Chinese with English abstract)
97. Wei H, Su J, Wan RJ, Wang L, Lin YA (2003) Tidal front and the convergence of anchovy (*Engraulis japonicus*) eggs in the Yellow Sea. Fish Oceanogr 12(4/5):434–442
98. Li Y, Zhao XY, Zhang T, Li XS, Wei H (2007) Wintering migration and distribution of anchovy in the Yellow Sea and its relation to physical environment. Mar Fish Res 28:104–112 (in Chinese with English abstract)
99. Fu MZ, Wang ZL, Li Y, Li RX, Sun P, Wei XH, Lin XZ, Guo JS (2009) Phytoplankton biomass size structure and its regulation in the southern Yellow Sea (China): seasonal variability. Cont Shelf Res 29:2178–2194
100. Fu MZ, Wang ZL, Sun P, Li Y, Li RX (2009) Size structure and potential export of phytoplankton primary production in the southern Huanghai (Yellow) Sea. Acta Oceanol Sin 31(6):100–109 (in Chinese with English abstract)

101. Wei QS, Liu L, Zhan R, Wei XH, Zang JY (2010) Distribution features of the chemical parameters in the southern Yellow Sea in summer. Period Ocean Univ China 40(1):82–88 (in Chinese with English abstract)
102. Wei QS, Zhan R, Zang JY, Li RX (2010) Distributions and influence factors of the chemical parameters in the southern Yellow Sea in spring. Mar Sci 34(3):52–60 (in Chinese with English abstract)
103. Wei QS, Zhou M, Wei XH, Zang JY, Zhan R (2010) Distribution features and variation tendency of chemical elements in the southern Yellow Sea in winter. Adv Mar Sci 28(3): 353–363 (in Chinese with English Abstract)
104. Wei QS, Zang JY, Wei XH, Liu L (2011) The distribution of nutrients and the relationship of them with the circulation condition in the western southern Huanghai Sea in autumn. Acta Oceanol Sin 33(1):74–82 (in Chinese with English abstract)
105. Kong CE, Yoo S, Jang CJ (2019) East China Sea ecosystem under multiple stressors: heterogeneous responses in the sea surface chlorophyll-a. Deep-Sea Res I 151:103078. https://doi.org/10.1016/j.dsr.2019.103078
106. Yoo S, Kong CE, Son YB, Ishizaka J (2019) A critical re-assessment of the primary productivity of the Yellow Sea, East China Sea and sea of Japan/East Sea large marine ecosystems. Deep-Sea Res II 163:6–15
107. Teague WJ, Jacobs GA (2000) Current observations on the development of the Yellow Sea warm current. J Geophys Res 105(C2):3401–3411
108. Yang D, Yin B, Chai F, Feng X, Xue H, Gao G, Yu F (2018) The onshore intrusion of Kuroshio subsurface water from February to July and a mechanism for the intrusion variation. Prog Oceanogr 167:97–115
109. Liu SM, Hong GH, Zhang J, Ye XW, Jiang XL (2009) Nutrient budgets for large Chinese estuaries. Biogeosciences 6:2245–2263
110. Zhang J, Liu MG (1994) Observations on nutrient elements and sulphate in atmospheric wet depositions over the Northwest Pacific coastal oceans–Yellow Sea. Mar Chem 47(2):173–189
111. Shi JH, Gao HW, Zhang J, Tan SC, Ren JL, Liu CG, Liu Y, Yao XH (2012) Examination of causative link between a spring bloom and dry/wet deposition of Asian dust in the Yellow Sea, China. J Geophys Res 117:D17304. https://doi.org/10.1029/2012JD017983
112. Belkin IM, O'Reilly JE (2009) An algorithm for oceanic front detection in chlorophyll and SST satellite imagery. J Mar Syst 78(3):319–326
113. Asplin L, Lin F, Budgell WP, Strand Ø (2021) Rapid water temperature variations at the northern shelf of the Yellow Sea. Aquacult Env Interac 13:111–119. https://doi.org/10.3354/aei00394
114. He S, Huang D, Zeng D (2016) Double SST fronts observed from MODIS data in the East China Sea off the Zhejiang-Fujian coast, China. J Mar Syst 154:93–102
115. Liu G, Wang H, Sun S, Han B (2003) Numerical study on the velocity structure around tidal fronts in the Yellow Sea. Adv Atmos Sci 20(3):453–460
116. Yuan DL, Zhu JR, Li CY, Hu DX (2008) Cross-shelf circulation in the Yellow and East China seas indicated by MODIS satellite observations. J Mar Syst 70:134–149
117. Bian CW, Jiang WS, Greatbatch RJ (2013) An exploratory model study of sediment transport sources and deposits in the Bohai Sea, Yellow Sea, and East China Sea. J Geophys Res Oceans 118:5908–5923
118. Wei QS, Liu L, Zang JY, Ran XB (2012) The distribution and transport of suspended matter in the southern Huanghai Sea. Acta Oceanol Sin 34(2):73–84 (in Chinese with English abstract)
119. Lin L, Wang YQ, Liu DY (2019) Vertical average irradiance shapes the spatial pattern of winter chlorophyll-a in the Yellow Sea. Estuar Coast Shelf Sci 224:11–19
120. Zhu MY, Mao XH, Lü RH, Sun MH (1993) Chlorophyll a and primary productivity in the Yellow Sea. J Oceanogr Huanghai Bohai 11(3):38–51 (in Chinese with English abstract)
121. Wan RJ, Wei H, Sun S, Zhao XY (2008) Spawning ecology of the anchovy *Engraulis japonicus* in the spawning ground of the southern Shandong Peninsula I. abundance and

distribution characters of anchovy eggs and larvae. Acta Zool Sin 4(5):785–797 (in Chinese with English abstract)

122. Li XS, Ma J, Zhao XY, Sun S, Huang DJ (2013) Spawning aggregation characteristics and batch fecundity of anchovy in the Yellow Sea. Curr Dev Oceanogr 6(1):15–39
123. Niu MX, Jin XS, Li XS, Wang J (2014) Effects of spatio-temporal and environmental factors on distribution and abundance of wintering anchovy Engraulis japonicus in central and southern Yellow Sea. Chin J Oceanol Limnol 32(3):565–575
124. Wan RJ, Huang DJ, Zhang J (2002) Abundance and distribution of eggs and larvae of *Engraulis japonicus* in the northern part of East China Sea and the southern part of Yellow Sea and its relationship with environmental conditions. J Fish China 26(4):321–330 (in Chinese with English abstract)
125. Liu DY, Keesing JK, Xing QG, Shi P (2009) World's largest macroalgal bloom caused by expansion of seaweed aquaculture in China. Mar Pollut Bull 58:888–895
126. Zhou MJ, Liu DY, Anderson DM, Valiela I (2015) Introduction to the special issue on green tides in the Yellow Sea. Estuar Coast Shelf Sci 163:3–8
127. Dong Z, Liu D, Keesing JK (2010) Jellyfish blooms in China: dominant species, causes and consequences. Mar Pollut Bull 60:954–963
128. Sun S (2012) New perception of jellyfish bloom in the East China Sea and Yellow Sea. Oceanol Limnol Sin 43(3):406–410 (in Chinese with English abstract)
129. Zhang F, Sun S, Jin XS, Li CL (2012) Associations of large jellyfish distributions with temperature and salinity in the Yellow Sea and East China Sea. Hydrobiologia 690:81–96. https://doi.org/10.1007/s10750-012-1057-5
130. Xiao J, Wang ZL, Song HJ, Fan SL, Yuan C, Fu MZ, Miao XX, Zhang XL, Su RG, Hu CM (2020) An anomalous bi-macroalgal bloom caused by *Ulva* and sargassum seaweeds during spring to summer of 2017 in the western Yellow Sea, China. Harmful Algae 93:101760. https://doi.org/10.1016/j.hal.2020.101760
131. D'Asaro EA, Shcherbina AY, Klymak JM, Molemaker J, Novelli G, Guigand CM, Haza AC, Haus BK, Ryan EH, Jacobs GA, Huntley HS, Laxague NJM, Chen S, Judt F, McWilliams JC, Barkan R, Kirwan Jr AD, Poje AC, Özgökmen TM (2018) Ocean convergence and the dispersion of flotsam. PNAS 115(16):1162–1167. https://doi.org/10.1073/pnas.1718453115
132. Wei QS, Wang BD, Yao QZ, Fu MZ, Sun JC, Xu BC, Yu ZG (2018) Hydro-biogeochemical processes and their implications for *Ulva prolifera* blooms and expansion in the world's largest green tide occurrence region (Yellow Sea, China). Sci Total Environ 645:257–266

Colored Dissolved Organic Matter in Frontal Zones

Céline Guéguen and Piotr Kowalczuk

Abstract Dissolved organic matter (DOM) includes a broad range of organic molecules of various sizes and composition that are released by all living and dead plants and animals. Measuring the fraction of DOM that absorbs light (colored or chromophoric DOM; CDOM) and fluoresces (referred to as CDOM fluorescence or FDOM) at specific wavelengths is diagnostic of DOM source and amount. The composition and dynamics of CDOM and FDOM across estuarine and coastal mixing zones, eddies, upwelling, and nepheloid layers are discussed in relation to the anomalies in physical (e.g., salinity and temperature), chemical (e.g., nutrients, $\delta^{18}O$, dissolved oxygen), and biological properties (e.g., chlorophyll-a, primary production) reported in the frontal zone. In situ observations using profiling sensors and gliders, and remote sensing across coastal and oceanic fronts are described.

Keywords CDOM, Dissolved organic matter, FDOM, Mixing process, Water circulation

Contents

C. Guéguen (✉)
Department of Chemistry, Trent University, 1600 West Bank Drive, Peterborough, ON, Canada K9J 7B8
e-mail: celinegueguen@trentu.ca

P. Kowalczuk
Institute of Oceanology Polish Academy of Sciences, ul, Powstańców Warszawy 55, PL-81-712 Sopot, Poland
e-mail: piotr@iopan.gda.pl

Igor M. Belkin (ed.), *Chemical Oceanography of Frontal Zones*,
Hdb Env Chem (2022) 116: 283–318, DOI 10.1007/698_2013_244,

Published online: 15 April 2013

1 Introduction

Physical oceanic boundaries are formed by horizontal and vertical density gradients, at the interface of water masses with different temperatures and salinities. Typical examples of these frontal zones are buoyant riverine plumes where the lenses of fresh, and thus lighter, water float over more saline waters. The riverine plume and estuarine mixing can form physical boundaries hindering surface mixing and/or vertical mixing in coastal waters [1, 2]. The extent and vertical thickness of such frontal zones depend mainly on the fresh water discharge, wind direction and speed, and tidal mixing. The presence of eddies, upwelling, gyres, and bottom nepheloid layers can also affect the ocean circulation features, resulting in significant biological enhancements [3, 4]. In regions not influenced by riverine plumes, the vertical boundaries are usually associated with the radiative heat transfer and development and disappearance of the seasonal thermocline. Frontal zones separate water masses with distinct physical and chemical properties and form a barrier that limits transfer of heat, solutes, and plankton. For example, the advection of iron- and nutrient-enriched water masses has been documented to significantly enhance primary production at the polar front in the Southern Ocean (e.g., [5, 6]).

Changes in the physical properties of seawater in frontal zones due to biological activity can be easily detected by remote sensing at visible and infrared wavelengths. The change of inherent and apparent optical properties (IOP and AOP, respectively) caused by phytoplankton often abundant in frontal zones can be detected by satellite using optical ocean color scanners such as SeaWiFS, MODIS, and MERIS. Especially important are the light-absorbing properties of chromophoric (or colored) dissolved organic matter (CDOM) [7, 8], which absorbs light in the UV–visible range (Fig. 1a) and represents one fraction of the DOM pool. CDOM is one of the primary absorbers of sunlight that determines the optical properties of natural waters and directly affects both the availability and the spectral quality of light. Because it absorbs UV light, CDOM can strongly limit the penetration of damaging UVA/UVB radiation and thus acts to protect aquatic organisms including phytoplankton [10]. CDOM is thought to be important for most photochemically mediated processes in surface waters as they can stimulate and/or hinder biological activity [11].

A variable fraction of the absorbing chromophores that make up CDOM are also fluorescent and often named fluorescent DOM (or FDOM; Fig. 1b).

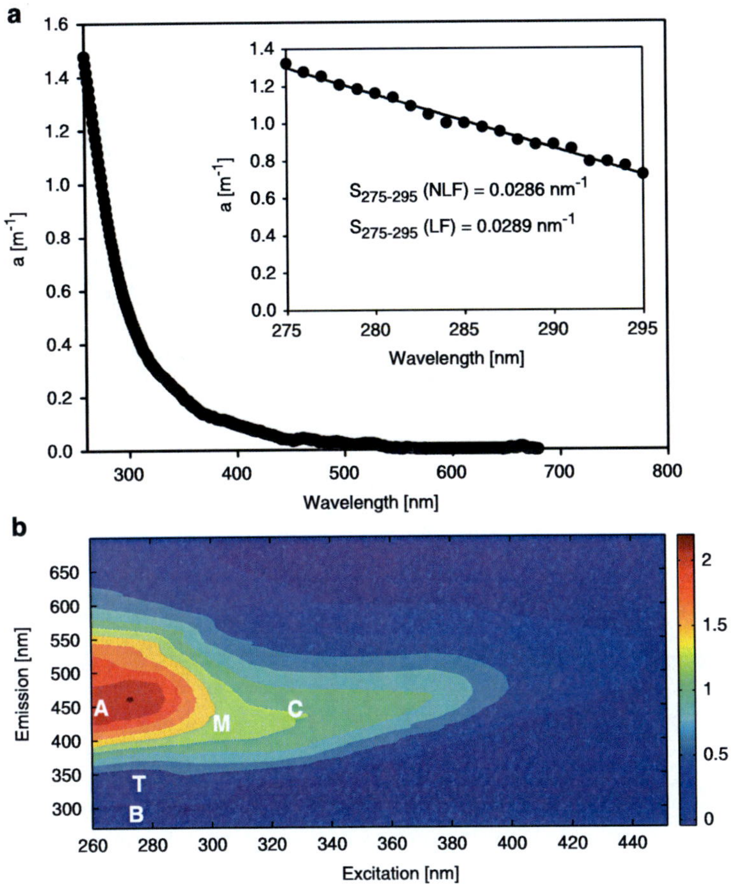

Fig. 1 Example of (**a**) CDOM absorption spectrum obtained in the Canadian Arctic Archipelago in July 2008 and (**b**) EEM spectrum and contour plot of FDOM in Beaufort Sea (Adapted from [102]). The major peak regions for FDOM are labeled according to previous convention (Table 1; [9])

Empirical relationships between FDOM and CDOM have been developed (Fig. 2), but they seem to be system-specific. This decoupling is mainly apparent in coastal and estuary regions where different water mass sources having distinct CDOM composition and optical properties (e.g., terrestrial, wetland, and wastewater inputs; [12, 13]) are mixed. This decoupling limits the application of FDOM for the assessment of CDOM distribution and dynamics. Patterns of FDOM distribution can be elucidated using excitation–emission matrix (EEM; [9]) spectroscopy. EEM datasets are extensive in nature, each typically containing >3,000 fluorescence intensities, which may be sufficiently informative for the differentiation of source-specific FDOM [9]. The overwhelming nature of the task of developing a statistically relevant deconstruction of fluorescence data contained in EEMs has been circumvented

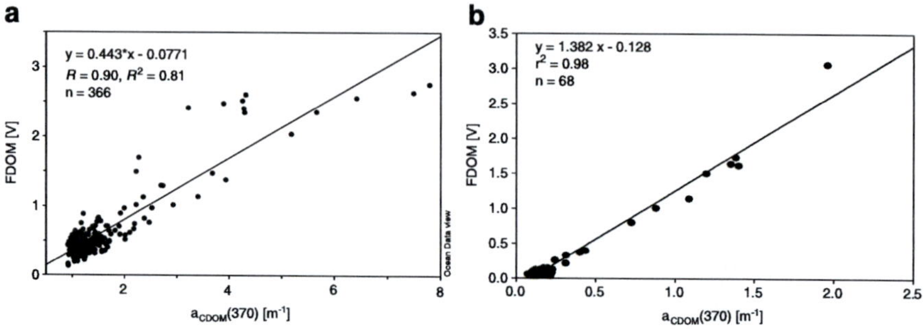

Fig. 2 Relationships between DOM fluorescence intensity, FDOM (expressed as fluorometer voltage output), and CDOM absorption coefficient $a_{CDOM}(370)$ in (**a**) Baltic Sea based on data set collected between 2008 and 2012 and (**b**) Beaufort Sea (JOIS2012 cruise; Dainard and Guéguen, unpublished)

Table 1 Spectral characteristics of the main FDOM components identified in marine waters

Component	Ex/Em	[9]	[19]	Description and probable source
Tyrosine-like	270–280/300–310	B	γ	Protein-like, autochthonous
Tryptophan-like	270–280/340–360	T	δ	Protein-like, autochthonous
UVC humic-like	230–260/380–460	A	α'	Terrestrial humic-like, autochthonous
UVA marine/microbial humic-like	290–310/370–420	M	β	Humic-like, marine, biological, and/or microbial origin
UVA humic-like	320–360/420–480	C	α	Terrestrial humic-like

through the application of parallel factor analysis (PARAFAC) [14–17]. PARAFAC enables the identification of statistically independent components that are the product of significant covariance amongst the respective fluorescence data [18]. Fluorescent components have been characterized based on positions and magnitudes of peaks corresponding to excitation–emission fluorescence maxima derived for each CDOM component (Table 1; [9, 19]). PARAFAC is now ubiquitous in investigations where the identity and source of chemically distinct CDOM components are of interest (e.g., [14, 17, 20, 21]). Changes in PARAFAC modeled FDOM components reflect the effects of physical and chemical processes that occur in the water column as well as variations in CDOM composition from different sources (e.g., [22, 23]).

This chapter gives an overview of CDOM and FDOM using field- and satellite-based observations of coastal and offshore frontal regions where physical, chemical, and biological properties of water change dramatically over very short distances. The change in CDOM and FDOM composition and dynamics will be related to the anomalies in physical (e.g., salinity and temperature), chemical (e.g., nutrients, $\delta^{18}O$, dissolved oxygen), and biological properties (e.g., chlorophyll-a, primary production) reported in the frontal zone.

2 Characterization of CDOM and FDOM

CDOM is defined operationally as the optically active fraction of dissolved material isolated by filtration. Typical filter types and sizes include glass fiber filter (GF/F 0.7 μm) and polycarbonate, mixed ester polysulfone filter (0.2 μm). The first step, filtration through a precombusted (500°C for 4 h) GF/F filter, is needed to remove most of the suspended solids and plankton cells. However, the GF/F filters do not separate bacteria from CDOM, which may cause biodegradation, if CDOM samples are stored for an extended period of time prior to analysis. Filtration through 0.2-μm membrane filters minimizes the presence of most bacteria and fine-sized particles in filtrates. The recommended 0.2-μm filter type includes polysulfone, polycarbonate, and nylon. These latter filters have to be solvent-rinsed thoroughly with methanol and Milli-Q water and then rinsed with the sample water. Samples processed through the two-step filtration should be stored in sterilized amber glass sample vials or bottles in the dark at 4°C. This method of storage has little to no effect on CDOM's absorption spectrum [24].

2.1 CDOM Absorption Measurements

The absorption of light in water together with scattering contributes to attenuation and penetration of the solar radiation through the water column. In all natural waters the spectral absorption coefficient of water, $a_{tot}(\lambda)$ is defined as a sum absorption coefficient of the pure water, $a_w(\lambda)$, and all optically significant water constituents. The absorption of pure water [25] is almost constant in natural waters and may be omitted in further analysis because it does not contribute to the variability of the total absorption coefficient of seawater. For practical reasons, the four most significant water constituents that contribute the most to the variability of $a_{tot}(\lambda)$ were defined: phytoplankton pigments contained in algae cells, non-algal particulate material, CDOM, and inorganic salts dissolved in seawaters. The total absorption is a sum of those components:

$$a_{tot}(\lambda) = a_w(\lambda) + a_{ph}(\lambda) + a_{NAP}(\lambda) + a_{CDOM}(\lambda) + a_s(\lambda) \quad (1)$$

where $a_w(\lambda)$ is the absorption coefficient of pure water, $a_{ph}(\lambda)$ is the absorption coefficient of phytoplankton pigments, $a_{NAP}(\lambda)$ is the absorption coefficient of non-algal particles, $a_{CDOM}(\lambda)$ is the absorption coefficient of colored or chromophoric dissolved organic matter, and $a_s(\lambda)$ is absorption of inorganic sea salts. The inorganic salts that contribute the most to sea salts absorption are nitrates and bromides [26], the ions of which absorb light at wavelengths shorter than 240 nm. When dissolved in marine waters, transition metals, such as manganese and iron, may create inorganic and organic complexes that may influence CDOM absorption measurements. The absorption of manganese and iron inorganic and organic complexes may be significant in UVA and visible spectral domains.

The effect of the metal–organic complexes on CDOM absorption may be significant; Xiao et al. [27] have estimated that iron ions associated with humic substances contribute 0.6% to the $a_{CDOM}(410)$ in Mississippi River waters and 56.4% in the Finish humic lake Löytynlähde.

The CDOM absorption curve can be parameterized using the following exponential function [28, 29]:

$$a_{CDOM}(\lambda) = a_{CDOM}(\lambda_{ref}) \times e^{-S(\lambda-\lambda_{ref})} + K \tag{2}$$

where $a_{CDOM}(\lambda)$ is the absorption coefficient at wavelength λ (m^{-1}), λ_{ref} is a reference wavelength, and S is the CDOM absorption spectrum slope coefficient (nm^{-1}). K is a background constant that accounts for attenuation caused by residual scattering by fine size particles, micro-air bubbles, or colloidal material present in the sample, as well as refractive index differences between the sample and the reference, or other attenuation not due to organic matter.

A comprehensive protocol for measurements of the CDOM absorption coefficient has been recommended by Mitchell et al. [30] for the field validation of satellite ocean color sensors. The CDOM absorption measurements are usually conducted with the use of research-grade dual beam spectrophotometers in the spectral range 240–700 nm. Purified water (Milli-Q or double distilled water) is used as the reference. Measured absorbance is converted to the Napierian absorption coefficient (m^{-1}) using the equation

$$a_{CDOM}(\lambda) = 2.303 \times \frac{A(\lambda)}{L} \tag{3}$$

where $A(\lambda)$ is the absorbance and L the pathlength of the optical cell in meters. The accuracy of the measurements is determined as the standard deviation of multiple scans of a blank sample (i.e., Milli-Q water). The blank spectra shall be spectrally flat over the whole spectral range, with no obvious spectral features below 0.001 (dimensionless). This absorbance value corresponds to a detection level for the absorption coefficient. The detection level is a function of pathlength and is 0.046 and 0.023 m^{-1} for a 5- and 10-cm cell, respectively. The application of longer pathlengths decreases the detection limit. For example, the detection limit of a spectrophotometer equipped with a 1-m capillary wave guide is 0.0023 m^{-1}, a 10-fold improvement relative to a 10-cm cell. In estuaries, marginal seas, and coastal ocean regions that are under influence of riverine outflow (i.e., CDOM-rich waters), the use of a cuvette with a 1- or 5-cm pathlength is sufficient to measure the CDOM absorption with the accuracy required by the measurement protocol for field calibration and validation of satellite ocean color sensors (at least 5% or better for given spectral values; [30]). However, the CDOM absorption can be very low in surface oligotrophic waters; therefore, the use of cuvettes with longer pathlengths is necessary. Most commercially available research-grade spectrophotometers do not offer either measurement cuvettes longer than 10 cm or measurement chambers that can fit longer cuvettes, and therefore an alternative approach for CDOM measurements is required. This involves using a liquid capillary wave guide cell system (LWCC) with the variety of optical pathlengths (from 0.25 up to 3 m). The capillary waveguide has the liquid forming the optical core contained in rigid quartz capillary tubing that is coated by an amorphous polymer optical cladding with a refractive index less than that of an aqueous solution [31]. Source light that is axially introduced into the waveguide, via an optical fiber, is transmitted and constrained within the capillary cell by total internal reflection because of the higher refractive index of the seawater in relation to the cell wall. At the opposite end of the waveguide, a detection fiber conducts the light that is not absorbed by the aqueous medium to a fiber-optics-based spectrometer that uses a diffraction grating to disperse the transmitted light into a CCD detector array. There is an inlet or outlet connection at each end of the waveguide for injecting filtered seawater samples or any other aqueous solution. The injected volume of sample needed for measurement is usually less than 4–5 ml. After injection of the sample, the capillary waveguide cells are flushed with a small aliquot of the Milli-Q water. The instrument readings require corrections for differences between the refraction index of pure and seawater and instrument drift [32]. This measurement methodology offers the highest precision and accuracy, as well as the full spectral resolution, and has been successfully applied in studies on CDOM properties and distribution in the open ocean [32–34].

A promising technique to measure CDOM absorption of filtered water samples is the integrating sphere. The instrument called PSICAM (point-source integrating-cavity absorption meter) enables measurements of the true absorption coefficient not disturbed by the presence of small particles, colloids, and viruses or small bacteria that can pass through the 0.22-μm filter. After initial testing [25, 35], Röttgers and Doerffer [36] have applied this technique in the field and proved that measurements of CDOM absorption in the integrated cavity guarantee a better accuracy when compared with the commercially available research-grade bench spectrophotometers. The CDOM absorption coefficients derived from PSICAM readings are lower than those obtained from spectrophotometric scans in the spectral range 350–442 nm. In the spectral range between 442 and 500 nm, results between the two techniques were comparable, and at longer wavelengths, the PSICAM enabled CDOM absorption measurements below detection limits of the spectrophotometer. Röttgers and Doerffer [36] have developed instrument-specific correction factors designed to account for subtle changes in the absorption of seawater caused by variation in temperature and salinity [37].

Another possible solution is to use the absorption and beam attenuation meter: the ac-9 in situ spectral and attenuation meter developed by WET Labs Inc., USA (presently part of the SeaBird Scientific group), fitted with 25-cm pathlength flow tubes [38, 39]. This instrument is capable of measuring CDOM absorption in situ when a 0.2-μm flow-through filter is attached to the water intake. Nine wavelengths – 412, 440, 488, 532, 560, 620, 650, 676, and 715 nm – can be measured with the accuracy of $\pm$ 0.005 m^{-1} [40]. The ac-9 field spectrophotometer is capable of online recording of multispectral CDOM absorption coefficients with very high spatial and temporal resolution in the vertical and horizontal directions.

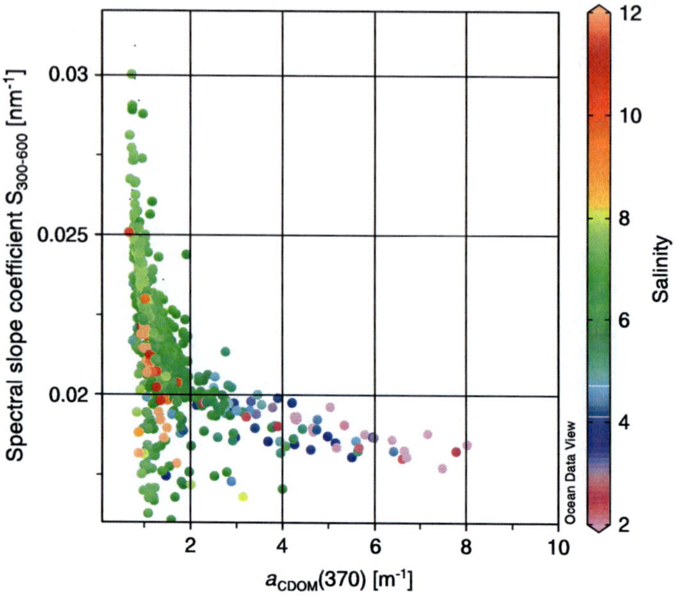

Fig. 3 Distribution of CDOM absorption spectrum slope $S_{300-600}$ as a function of CDOM absorption coefficient, $a_{CDOM}(370)$, and salinity in the Baltic Sea in 2002–2012

Furthermore, deployment of the ac-9 instrument on buoys and moorings allows acquisition of high-resolution time series data that enables observation of the temporal evolution of spectral CDOM absorption coefficients. The only disadvantage of the ac-9 in situ spectrophotometer is its coarse spectral resolution (i.e., 10 nm) which makes it difficult to calculate accurately the CDOM absorption slope coefficient, *S* [41]. This limitation has been overcome with the introduction of the hyperspectral version of the marine spectrophotometer, the ac-s instrument (4-nm resolution; [42]). These in situ instruments have been used by many researchers in numerous oceanographic cruises around the world in various marine regions producing extremely valuable data sets (e.g., [43–49]). For example, Nencioli et al. [47] showed the senescent and healthy diatom layers shared the same optical properties below an eddy-induced bloom using the ac-s in situ spectrophotometer. Berthon and Zibordi [49] showed that the general correlation between absorption by CDOM and non-algal particles was not observed in the southern Bothnian Bay due to significant input of CDOM-rich and suspended particle poor waters.

The spectral slope *S* (Figs. 1 and 3) can be calculated through a linear fitting (LF) of the log-linearized absorption spectra [50, 51] or through a nonlinear fitting (NLF; [50, 52, 53]). The NLF method of estimation of the spectral slope tends to produce consistently higher values of *S* relative to LF (~0.002 nm^{-1} higher with NLF; [23]), which is consistent with previous work [50]. This effect could be explained by the fact that log transformation puts more weight on samples with low absorption values at longer wavelengths which are less precisely determined [52]. However, spectral slope values can be highly variable depending on the wavelength range over which the values are estimated. Twardowski et al. [41] estimated that almost 75% variability in CDOM absorption spectrum spectral slope coefficient values reported in literature were due to different spectral ranges used for calculation of S as well as the differences in methodology. Stedmon et al. [52] have tested the differences in the spectral slope estimation calculated over different spectral ranges. They found that uncertainty of the spectral slope estimation increases with increasing wavelengths and the $S_{300–400}$ value was 40% higher than the $S_{400–500}$ value. Kitidis et al. [54] estimated that $S_{290–350}$ differed from $S_{250–650}$ by as much as 58%, highlighting the importance of the wavelength range.

Overall, in the transition from terrestrial to marine and oceanic environments during rapid mixing, the spectral slope values increase with decreasing absorption coefficient and increasing salinity (Fig. 2) [22, 50, 55–57]. The absorption spectrum spectral slope coefficient is often regarded as a proxy for the CDOM composition; typically a steeper S indicates low molecular weight material or lower aromaticity and a shallower S indicates DOM with a higher aromatic content and higher molecular weight [50, 58, 59]. In the Gulf of Mexico, Carder et al. [60] showed that different classes of organic compounds that compose the CDOM have distinctly different absorption spectral slope values. For example, the range of spectral slopes of fulvic and humic acids were 0.016–0.022 and 0.011–0.012 nm^{-1}, respectively. The relative contribution of those two major classes of organic compounds in the mixture of soluble organic compounds composing CDOM affects the spectral slope values (S). The significant changes in S values can also result from photodegradation of CDOM. Numerous irradiation experiments have shown a reduction in $a_{CDOM}(\lambda)$ associated with an increase in S upon solar exposure [61–63]. The change in S values has also been successfully modeled using a simple mixing model based on two end-members in a laboratory experiment [64, 65], suggesting that physical mixing is an important factor. Therefore, the distribution of the spectral slope coefficient in natural aquatic environments is the combined effect of the source of the precursory organic material that has formed the CDOM, mixing of water bodies with distinctly different CDOM optical properties (two or multiple end-members [66], parallel processes of autochthonous production [67], and decomposition of CDOM [45, 68]. Kowalczuk et al. [57] have proposed a simple method to discriminate the conservative and non-conservative processes that influence the spectral properties of CDOM absorption. The theoretical mixing line derived from the Stedmon and Markager's mixing model [64] has been superimposed over the empirical distribution of the spectral slope values in the function of the $a_{CDOM}(375)$. A large number of the $S_{300–650}$ values fell within brackets of the mixing model confidence interval, and therefore, those points were not considered to deviate significantly from the modeled conservative mixing of the two CDOM end-members. Values outside these confidence limits were indicative of the presence of additional sources or sinks for CDOM, e.g., the autochthonous production of CDOM or degradation of terrestrial CDOM. The majority of points that deviated from the mixing model were found at the high-salinity end of the mixing line. Field studies that present the distribution of the S in the function of $a_{CDOM}(\lambda)$ and salinity suggest that mixing is the dominant process that controls the distribution of the S values in the salinity gradient [22, 53, 56, 69]. The salinity at which a sudden increase in the spectral slope values is observed is called the inflection point [22]. Blough et al. [53] and Del Castillo et al. [56] have reported similar behavior in the Orinoco River. The position of the inflection point on the salinity scale depends on the initial concentration of CDOM in the freshwater end-member and the salinity gradient over which the mixing occurs. The influence of non-conservative processes, such as production and decomposition of CDOM, on its optical properties could become apparent only if the freshwater is sufficiently diluted [70, 71].

Recently, Helms et al. [58] showed that the greatest variations in S values in contrasted CDOM samples (river, estuary, coastal, and open ocean) were found over two spectral ranges (275–295 nm and 350–400 nm). The increase in MW shifts the absorption spectrum towards longer wavelengths [72] causing steeper $S_{350\text{–}400}$ and decreased spectral slope ratio S_R ($S_R = S_{275\text{–}295}/S_{350\text{–}400}$). The spectral slope values were highly correlated to MW, as measured using size exclusion chromatography [58] and flow field flow fractionation [59]. The S_R parameter was linearly correlated with MW ranging from ~800 [59] to 3,000 Da [58]. The loss in absorption upon CDOM photodegradation causes greater absorption at lower wavelengths which would induce steeper $S_{275\text{–}295}$, shallower $S_{350\text{–}400}$, and greater S_R.

2.2 FDOM Fluorescence Measurement

DOM excited with ultraviolet light fluoresces at wavelengths between 300 and 600 nm. Typically, the fluorescence intensity is Raman calibrated (Raman unit or r.u. [73, 74]; and sometimes standardized to quinine sulfate equivalents (QSE in ppb). The fluorescent properties of DOM (FDOM) have been known for a long time [75, 76] and the fluorescence signal has been used to estimate FDOM in marine waters (e.g,, [77]). DOM fluorescence in natural waters has been used as a proxy for the CDOM absorption coefficient since the early 1960s [76] and numerous investigators have observed a linear relationship between fluorescence and absorption coefficients [51, 68, 78–83]. Published empirical relationships between the absorption coefficient and fluorescence have been derived from results of field data collected in various coastal, marine, and estuarine environments on European, American Atlantic, and Pacific coasts (published coefficients of determination are usually higher than 0.9). DOM fluorescence is easy to measure with a research-grade spectrofluorometer and therefore can be used as the proxy of CDOM absorption. Green and Blough [84] reported FDOM apparent quantum yields in surface waters from various sites along the Atlantic coast and documented nearly two orders of magnitude variation in $a_{CDOM}(355)$, while variation in the calculated FDOM apparent quantum yield was only around 3-fold. However, within a given geographical area, variations in CDOM apparent quantum yield are much smaller, which may explain an overall very good correlation between CDOM absorption coefficient and fluorescence intensity. Stability of the FDOM apparent quantum yield enabled the successful deployment of single or multiple excitation wavelength fluorometers for in situ measurements of FDOM with high temporal and spatial frequency [16, 17, 85–87]. Figure 2 presents an example of the relationship between $a_{CDOM}(370)$ and fluorescence intensity measured in situ with use of a single excitation band fluorometer (TriOS GmbH, Germany) derived in the surface waters of the Baltic Sea.

The excitation–emission matrix (EEM) fluorescence spectrum [9] is obtained by acquiring emission spectra at a series of successively longer excitation wavelengths (Fig. 2b). Although slower to collect, EEM spectra provide a more complete picture of FDOM. The excitation–emission matrix spectra create a three-dimensional landscape, and specific peaks can be related to specific fluorophores associated with broad classes of dissolved organic compounds. The first EEM applications to field data analysis were conducted in the Atlantic Ocean [88], Black Sea [89] and in the Caribbean and Arabian Seas, and Gulf of Mexico [22, 56, 90]. It is also possible to use EEM spectra to follow changes in FDOM resulting from biological or physical processing of the material or trace CDOM from different sources ([9, 19]. It may be difficult to assess the dynamics of FDOM based on the EEM peak-picking technique because locations of individual peaks are sensitive to physical and chemical conditions [9]. However, the characterization of FDOM has improved with the application of a multivariate modeling approach called parallel factor analysis (PARAFAC) [14, 15], a powerful tool for interpreting the multidimensional nature of EEM data sets. The utility of using PARAFAC is that individual fluorescent components (Fig. 4; Table 1) can be identified and their relative concentrations quantified. This new approach has been successfully applied to study variability in FDOM composition and distribution in coastal and open ocean systems (e.g.,, [16, 17, 20, 91–93]).

3 Vertical Profiles and Relation with Water Masses

3.1 Vertical Distribution

Despite the fact that the definition of a water mass is generally associated with manifestations of fronts in temperature, salinity, and nutrients, several researchers have observed significant differences in the absorption and fluorescence properties of DOM associated with these oceanic fronts. If CDOM and FDOM are long-lived products of the degradation of DOM, it is logical to assume that the absorption and fluorescence properties can be very suitable for characterizing and differentiating water masses within the ocean interior. In this section, examples of absorption and fluorescence of DOM in the Atlantic, Pacific, and Arctic Oceans are presented.

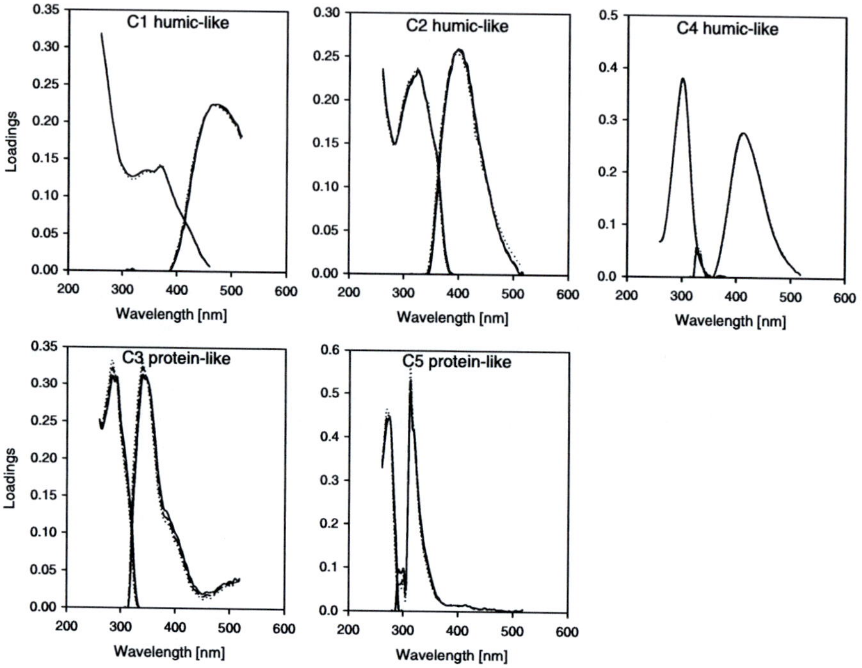

Fig. 4 Spectral characteristics of emission and excitation maxima of five components identified by PARAFAC modeling. C1–C5 are ordered by decreasing percent of explained variation. The *solid line* represents the loadings from the five-component model derived using the whole dataset whereas the *dotted* and *dashed lines* represent the loadings from two independent halves of the dataset (Adapted from [17])

3.1.1 Vertical Mixing and CDOM

The relationships of CDOM absorption properties with hydrography have been reported by numerous workers. These linkages usually exhibit substantial changes in absorption properties associated with the water mass formation. The basin-scale distribution of CDOM has been shown in the Atlantic Ocean along the Atlantic Meridional Transect in 1999–2000 (AMT 9–10) [54]. Extremely low $a_{\mathrm{CDOM}}(300)$ values were observed in the mixed layer of subtropical oligotrophic gyres in the northern and southern hemispheres separated by slightly elevated values close to equator. Nelson et al. [32], as part of the 2003 US CLIVAR/CO2 Repeat Hydrography survey, found that the deep layers showed relatively similar $a_{\mathrm{CDOM}}(325)$ and $S_{320\text{–}650}$ compared to the surface and intermediate layers where photobleaching and river or seasonal mixing significantly affect absorption properties. These results suggest that CDOM absorption can be used as a passive tracer within a deep ocean water mass.

Spectral slope S has also been successfully used to distinguish water masses in basin-scale sections in the Atlantic [32, 54, 94], Indian [94], Pacific [34, 94, 95], and

Arctic Oceans [96, 97]. For example, Kitidis et al. [54] found that the subsurface maxima of $a_{CDOM}(300)$ characterized by lowest spectral slope values, $S_{250-650}$, were associated with the deep chlorophyll-*a* maximum, confirming the role of plankton activity on CDOM production reported in experimental and field studies (e.g.,, [98–101]). The deep ocean CDOM in the world's ocean basins is characterized by low *S* values. In most deep layers in the North Atlantic, there were statistically significant relationships between $S_{320-650}$ and pCFC-12 age [32], indicating that low *S* values found in the ocean can result from the diagenesis of CDOM formed in situ, as well as from terrestrial-origin CDOM in surface layers.

The Arctic Basin receives 10% of the Global Ocean's freshwater inflow, although its volume is only 1% of the world's ocean, likely resulting in a greater load of CDOM in the Arctic than in any other oceans [96]. The increased load of DOC to the Arctic Basin will lead to serious implications in the basin's biogeochemistry, heat budget, and also global carbon cycles. While river runoff is a major source of CDOM to the Arctic Ocean [96, 102], contributions from sea-ice melt and formation and phytoplankton production are less well known. Recent studies of optical properties in Arctic seas have concluded that CDOM is a dominant light-absorbing water constituent, absorbing as much or more light than phytoplankton pigments in the surface waters [103, 104], thereby reducing the light available for primary productivity (e.g.,, [53, 105]). It has also recently been shown that CDOM contributes significantly to solar heating of Arctic surface water, with implications in the western Arctic that include accelerated sea-ice melt and increased stratification of the surface layer [104].

Vertical mixing is an efficient redistribution mechanism of CDOM from surface layers and coastal margins into deep ocean layers, serving to reduce its impact on surface heating. Matsuoka et al. [105] showed that $a_{CDOM}(440)$ was negatively correlated with salinity in the nutrient-rich Pacific-derived upper halocline formed in the Arctic Ocean. This relationship is primarily the result of brine rejection (as showed by $\delta^{18}O$; [102]) and lateral intrusion of Pacific summer waters into the layer. The negative trend in $a_{CDOM}(440)$ vs. salinity in the Pacific-derived upper and Atlantic-derived lower haloclines suggests that both water masses originated from Arctic coastal areas, namely, Chukchi and Barents Sea shelves, respectively [106–108]. Significant differences in absorption and spectral slopes *S* were also found between the polar mixed layer and the halocline layers in the Arctic Ocean [96] and between the polar waters and Atlantic layers in Fram Strait [109]. These results indicate that CDOM properties can assist in our understanding of the dynamics of CDOM, especially when used concurrently with detailed water mass tracers such as $\delta^{18}O$, dissolved oxygen, nutrients, and CFC [17, 32, 102, 109, 110].

3.1.2 Vertical Mixing and FDOM

Fluorescence measurements of DOM have been more common than absorption measurements in the open ocean, due primarily to their greater sensitivity and selectivity. Vertical profiles of FDOM humic fraction (Ex/Em 320/420 nm; humic-like) were lowest in the surface waters, increased with depth in mid-depth

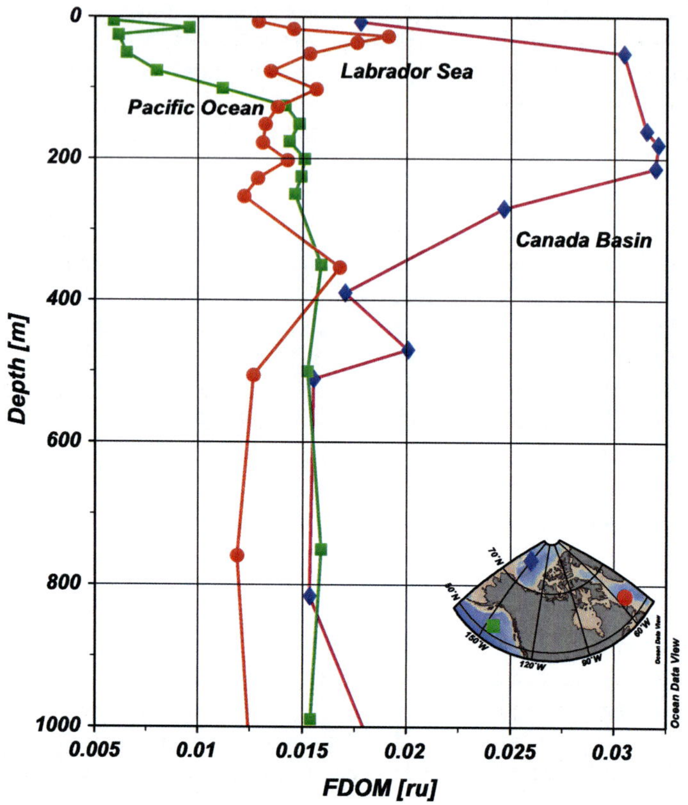

Fig. 5 Vertical profiles of humic-like FDOM in Labrador Sea, Canada Basin, and North Pacific Ocean (Guéguen, unpublished)

waters (~2,000 m), and were then found to occur in a relatively narrow range down to the bottom (Fig. 5). Such a characteristic profile of humic-like fluorescence intensity has been universally observed over oceanic environments, i.e., in the Southern Ocean [111, 112], the equatorial Pacific [113, 114], the North Pacific subtropical and subarctic [113, 115, 116], the Arabian Sea [90], the Sargasso Sea, North Atlantic [113, 117, 118], and the eastern Atlantic [119]. On the other hand, the vertical profiles of the protein-like fraction of FDOM (emission maximum < 400 nm) are reversed: the fluorescence intensity is highest in the surface waters and then gradually decreases with increased depth [120, 121].

Latitudinal and depth distribution of FDOM and CDOM were determined largely by major water masses separated by subtropical, tropical, and subpolar fronts and mixed layer thickness. Depth distribution patterns and optical properties have been used to identify frontal regions in the interhemispheric transect into the Atlantic Ocean, conducted within the 2010 Atlantic Meridional Transect program (AMT20) (Fig. 6). For example, CDOM/FDOM profiles taken between 50°N and

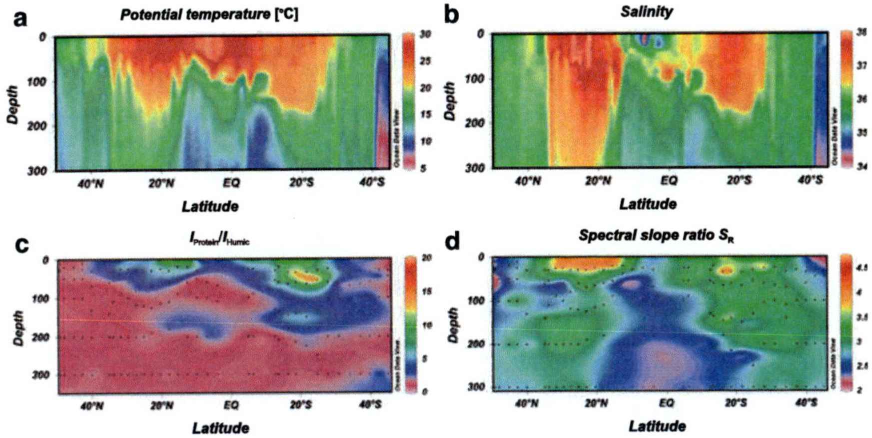

Fig. 6 Contour plots of (a) potential temperature, (b) salinity, (c) ratio between fluorescence intensity of protein-like and humic-like FDOM components $I_{Protein}/I_{Humic}$, and (c) CDOM absorption spectrum slope coefficient ratio S_R, according to Helms et al. [58], along the Atlantic Meridional Transect (AMT 20 cruise, on r/v James Cook, 13 October–21 November 2010). The salinity and temperature plots were provided by Rob Thomas from the British Oceanographic Data Centre. These contour plots were produced at the end of the AMT20 cruise using the ODV software [122]

45°S showed higher protein-like/humic-like fluorescence intensity ratio and S_R values in the top 200 m of water column, characteristics of warm and salty subtropical oligotrophic gyres. The dominance of low molecular weight (i.e., high S_R) and protein-like DOM likely results from the photodegradation of DOM produced during the spring algal bloom [98]. The distinct change in CDOM/ FDOM composition occurs below the mixed layer depth (MLD) which limits vertical mixing and thus exposure of DOM to UV solar radiation. On the other hand, the equatorial region is characterized by a very well marked advection of cool, fresher waters associated with low S_R and low protein-like/humic-like fluorescence intensity ratio values. The subpolar front south of 40°S is revealed by cold and freshwaters, low S_R, and low protein-like/humic-like fluorescence intensity ratio values. The DOM below MLD over the vast region of subtropical and tropical Atlantic and subpolar Northern and Southern Atlantic is characterized by low protein-like to humic-like fluorescence intensity ratio values and low S_R values, suggesting that DOM may consist of higher molecular weight, more aromatic humic-like compounds produced mostly by bacterial reworking of organic matter (in the deep ocean; [123, 124]) or be influenced by terrestrial input at continental margins (e.g.,, [16]). Similar vertical distribution of humic-like FDOM, with relation to molecular weight, has been observed in the Northern Pacific subtropical gyre [125, 126]. A lower fluorescence signal in the hydrophobic fraction and bulk DOM were observed in the surface layer of the gyre compared to deeper layers. This change in fluorescence intensity was also accompanied with a modification in the

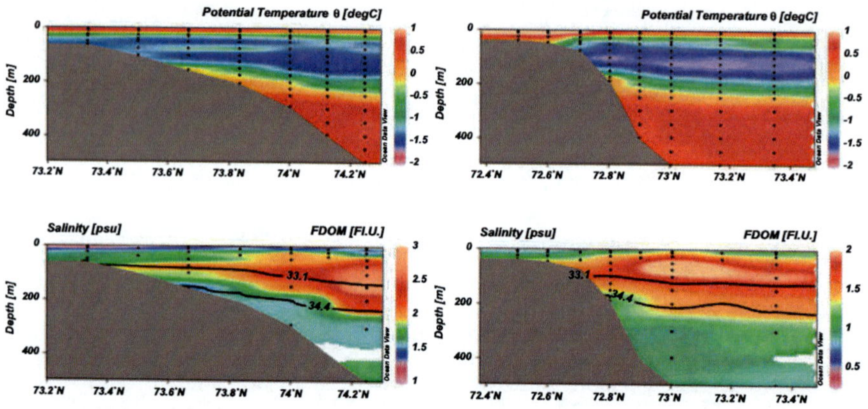

Fig. 7 Contour plots of (*top*) potential temperature and (*bottom*) FDOM (Ex/Em 370/457 nm, Fl.U.) along two transects across the shelf–basin interface in the western Arctic Ocean. Isohalines 33.1 and 34.4 ‰ are superimposed to highlight the cold halocline

vertical distribution of DOM molecular weights, with lower molecular weight substances in surface waters and higher molecular weight DOM in deeper layers (500- and 1,000-m depth). Omori et al. [126] have also conducted an irradiation experiment using samples collected in the deep waters of the Northern Pacific subtropical gyre, confirming photo degradation as the principle process responsible for the transformation of FDOM composition, loss of fluorescence intensity, and loss of molecular weight of DOM in surface waters.

This vertical profile of FDOM contrasts with that in the Arctic Ocean where a prominent maximum was found at depths between 40 and 200 m [17, 102], coinciding with low temperature, and centered around $S \sim 33.1$ PSU (Fig. 7). This prominent FDOM maximum was associated with the upper part of the halocline. Cooper et al. [127] also found a maximum FDOM fluorometer voltage in waters with salinity close to 33.1 PSU. This halocline overlays the more saline Atlantic waters ($S \sim 34.8$) where FDOM values are lower but relatively homogeneous (0.87 ± 0.24 ($1\ \sigma$) Fl.U., $n = 14$) and comparable to values found in previous studies. Differences in fluorescence intensity were observed between oceanic zones, suggesting FDOM distribution can be used to document the change in hydrography. In a recent comprehensive study using EEM–PARAFAC, a dramatic change in FDOM composition was found across frontal zones near the 180° meridian in the Western Arctic Ocean [17]. The humic-like fluorophores dominated FDOM in the westernmost region in the East Siberian Sea, whereas the contribution of protein-like fluorophores was predominant in Chukchi and Beaufort Seas. The significant difference in FDOM composition between East and West of the 180° meridian suggests the presence of a front that divides the study area into the Eastern Chukchi – Beaufort and East Siberian sides. The location of the hydrographic front is consistent with previous work using an in situ fluorometer [128].

3.2 Influence of Lateral Ventilation

Complex frontal systems can be characterized by different levels in dissolved oxygen which can in turn influence the optical properties of DOM. This is particularly significant in the dark ocean where correlations between CDOM/FDOM and AOU have been shown to vary between water masses [32, 34, 95, 123, 129].

In the dark ocean, DOM dynamics is influenced by microbial activities, respiration, and deep ocean circulation [32, 130, 131] which in turn can modify the absorption and fluorescence characteristics of DOM (e.g,, [100, 132]). The extent of respiration can be assessed using the Apparent Oxygen Utilization or AOU which is defined as

$$\mathrm{AOU} = [\mathrm{O_2}]_{\mathrm{saturation}} - [\mathrm{O_2}] \quad (4)$$

where $[O_2]_{saturation}$ is the saturation of dissolved oxygen concentration in equilibrium with the atmosphere at the temperature and salinity of the water and $[O_2]$ is the concentration of dissolved oxygen measured in the sample. Whereas temperature and salinity are conservative, the concentration of dissolved oxygen is not conservative and will decrease due to respiration while AOU will increase. The higher the AOU, the greater the amount of O_2 removed since the water mass was last seen at the surface.

Different AOU–FDOM relationships were found between the mesopelagic and bathypelagic layers in the Pacific Ocean [123, 124], indicative of change in FDOM composition and sources. The higher levels of terrigenous FDOM in the North Pacific intermediate water (NPIW) contrast with the bathypelagic layer which is supplied primarily from Circumpolar Deep Water. The North Atlantic deep layer differs from the other basins because of its relatively high fluorescence intensity compared to the AOU values. This pattern has also been observed in a global ocean study using EEM–PARAFAC [120]. The positive linear correlation between FDOM and AOU in the dark ocean suggests that net production of FDOM [94, 124] and in particular humic-like FDOM [120] is linked to microbial activities. This contrasts with protein-like FDOM distribution where no significant linear relation with AOU was observed, confirming that protein like is not linked to dark ocean microbial remineralization.

The absorbing fraction of DOM, $a_{\mathrm{CDOM}}(325)$ was found to be positively correlated with AOU in intermediate and deep water masses in the Atlantic, Pacific, and Indian Oceans [32, 34, 94]. A much weaker relationship was found for the Atlantic whereas nearly identical relationships were found for the deep Pacific and Indian Oceans [94]. The ventilation rates diagnosed using chlorofluorocarbon concentration in North and South Atlantic apparently exceeded the rate of in situ production of CDOM in the interior [32, 133]. This coupled with a comparatively greater input of terrestrial DOM masks the slow CDOM production at depth and thus the weaker correlation between CDOM and AOU. This work reveals the significant role of circulation and biogeochemical processes on CDOM distribution in the interior oceanic basin. The linear relationships between AOU and CDOM

[32, 34] and between AOU and FDOM [123, 124] suggest that both CDOM and FDOM are produced in situ in the deep ocean and can be used as tracers of deep water circulation.

3.3 Eddy and Upwelling

Regions influenced by upwelling events and strong fronts are characterized by cold water, high chlorophyll-*a*, and high nutrient content (as a result of decomposition of sinking organic matter from surface waters). Because the CDOM/FDOM composition in deep layers is significantly different from that in surface layers (Figs. 5, 6, and 7; Sect. 3.2), it suggests that these upwelling waters can be tracked using CDOM/FDOM measurements. Indeed, change in CDOM properties was reported as a result of water mass subduction in the North Atlantic, Equatorial Pacific, Caribbean, and South China Seas [32, 34, 134, 135]. Subduction of water masses associated with subtropical mode water has been revealed by a CDOM minimum in the North Atlantic and Caribbean Sea. The $S_{350–500}$ values also showed a clear tendency to decrease when going from hyperoligotrophic waters of the South Pacific Gyre to the eutrophic waters in the upwelling area off Chile [135]. The $a_{CDOM}(440)$ values were the lowest ($<0.01\ m^{-1}$) in the center of the South Pacific Gyre and increased by a factor of 4 in the upwelling area. Swan et al. [34] noticed a slight but significant offset of elevated CDOM values associated with the shoaling of isopycnal surfaces near the North American coast and in Equatorial Pacific between 10°N and 10°S, suggesting upwelling events. Similar observations of large humic-rich DOM typically associated with deep layers were reported in the top 300 m of water in Equatorial Atlantic (Fig. 6).

DOM fluorescence in upwelling systems [90, 134, 136–138] showed also significant differences in both intensity and composition relative to the surrounding waters. For example, the FDOM composition in the upwelling waters in the Arabian Sea displayed up to six fluorescence maxima (mainly humic-like DOM), whereas only two distinct fluorescence features were associated with the surrounding waters [90]. The humified material observed in upwelling waters is supported by the mineralization and respiration processes in bottom waters [136, 138]. Zhang et al. [134] documented the intrusion of old deep water masses with high FDOM signals into the shallow mesopelagic waters (25–450-m depth) in the South China Sea.

Together these results support the potential of CDOM/FDOM as tracers of ocean circulation processes for subducted water masses.

4 Coastal Frontal Zones

Estuaries constitute natural frontal zones as they encompass areas influenced by both terrestrial and marine forces, resulting in a strong gradient in DOM. Conditions are affected not only by seasonal changes and diurnal variability that influence marine waters but also by tides, multiple inflows, and fluctuating river flows. The behavior of CDOM/FDOM in estuaries has been studied by many researchers with respect to DOM export to the oceanic domain and biogeochemical alteration within the estuary. In combination with salinity, CDOM absorption coefficients have been shown to trace water mass mixing in coastal waters [102, 103, 139]. In the absence of local autochthonous production [57, 140], adsorption onto suspended particulate matter (SPM), and significant flocculation [22], a conservative mixing is observed between the CDOM-rich freshwater and CDOM-poor saline end-members. At low salinity, the freshwater end-member dominates, whereas at higher salinity, the mixing with low CDOM water mass is apparent. For example, Foden et al. [141] found that FDOM average values decreased by a factor of >20 between river and offshore waters in the southern North Sea. The impact of the river outflow on CDOM spatial distribution can be limited to the coastal domain [16] or detectable over long distances in the oceanic domain. For example, during high flow conditions, the influence of terrestrial CDOM in the Amazon River on the CDOM absorption coefficient at 440 nm, $a_{CDOM}(440)$, was detectable in the western tropical North Atlantic Ocean, over 1,000 km from the river mouth [142]. High CDOM absorption was observed more than 50 km offshore over the Mackenzie shelf by Guéguen et al. [23]. A conservative water mass mixing model based on the absorption coefficient and spectral slope of the three major end-members (i.e., Baltic outflow, German Bight, and the central North Sea) has revealed that 23% of Kattegat bottom water originated from the German Bight and the local CDOM inputs were very localized [66]. Hitchcock et al. [143] documented the extent of the Mississippi River plume in the Gulf of Mexico using surface drifters equipped with FDOM sensor. These drifters are designed to follow currents and make continuous measurements along their trajectories. Linear relationships between FDOM and salinity identified two subsurface sources of high-salinity water (salinity >35) underlying the surface plume in the Mississippi River plume. The CDOM data from the drifting observatory delineated the extent of the frontal zone in the Gulf of Mexico. Autonomous platforms equipped with DOM optical sensors were also used in real-time pilot projects to examine the factors controlling DOM distribution in highly dynamic systems such as the Gulf of Maine [144] and western English Channel [145]. These studies emphasize the great potential of using CDOM/FDOM to trace water mass mixing in combination with traditional conservative parameters (i.e., salinity and temperature).

Non-conservative mixing behavior of CDOM has also been reported in field studies. For instance, the 32% removal of $a_{CDOM}(350)$ observed in the humic-rich, turbid Tyne estuary (UK) was attributed to adsorption onto suspended particulate matter (SPM) and to a lesser degree flocculation processes [146]. Non-conservation behavior can be temporal and spatial as flocculation processes leading to loss of CDOM in estuarine waters require the presence of suitable particulate material [64]. Differences in particulate material composition between river systems throughout the year may be one of the causes behind spatial deviations from conservative mixing. The addition of CDOM/FDOM in the low turbidity area in the upper estuary suggests inputs from the river, sediment resuspension, and mangrove (e.g,, [140, 147]). Some studies have reported a marked deviation in CDOM/FDOM signal in the lower estuary (salinity >23–25), indicating that there are significant sinks of CDOM/FDOM within the estuary [16, 110]. Photochemical and microbial degradation of CDOM/FDOM along the estuary transect can also be evidenced by deviation from the conservative mixing curve [64, 148].

5 Benthic Boundary Layer

The nepheloid layers, i.e., layers with an increase in SPM, can also be viewed as frontal zones. Benthic fluxes of CDOM/FDOM from sediment pore waters [149–151] may contribute significantly to shallow water environments such as tidal flats [152] and coastal and estuarine environments [153]. Similarly, resuspension events of shelf sediments yield increases in CDOM/FDOM as water flows over these sediments. This process is largely involved in the CDOM/FDOM enrichment of the halocline layer in the Arctic Ocean [102, 150]. Indeed, results from shipboard incubation of DOM-rich Alaskan shelf sediments (up to 1.7% (w/w); [154]) indicated that sediment could be a net source of DOM to overlying waters [127]. Moreover, during ice formation, the brine that was generated formed high-salinity water that could be trapped along the shelf bottom. This dense water should flow over the surface of sediments, accumulating CDOM/FDOM in the process [149, 150]. This CDOM/FDOM accumulation may result from the leaching of sediments by brine waters [155]. Indeed, relatively higher ammonium content, protein-like fluorescence, and turbidity (Fig. 8) detected locally at the slope/break revealed the release of CDOM/FDOM from sediments, likely enhanced by benthic boundary layer processes, such as resuspension (e.g,, [156, 157]). The occurrence of a bottom nepheloid layer has been widely reported in other ocean margin regions (e.g., [43, 83]). Although shelf sediments [127, 150] and pore waters [149, 150] can be a source of DOM, the amount of CDOM/FDOM released into the water column has yet to be quantified.

6 Remote Sensing in Frontal Zones

Morel and Prieur [158] have classified water masses according to their bio-optical properties as Case 1 and Case 2. In the Case 1 waters, variability of IOP and AOP is primarily driven by dynamic impact of phytoplankton on absorption and the scattering of light. The Case 1 waters are mostly oligotrophic and mesotrophic open oceanic waters. Case 2 waters are represented by those regions where the bio-optical assumption on covariance of inherent and apparent optical properties of oceanic waters with phytoplankton biomass is not fulfilled, due to independent variability in optically significant water constituents. Case 2 waters are mainly

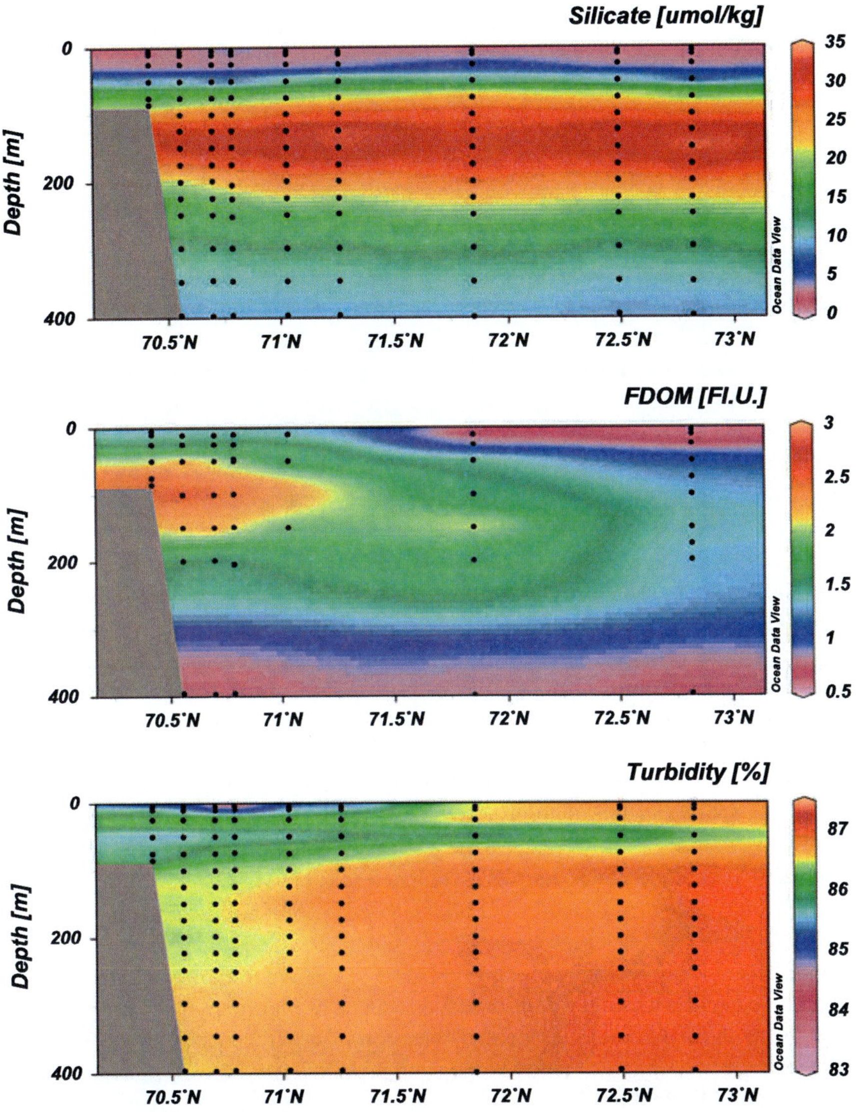

Fig. 8 Contour plots of (**a**) silicate, (**b**) humic-like FDOM, and (**c**) turbidity at the shelf–basin interface (Modified from [102])

influenced by riverine input of CDOM and suspended particles. The absorption of light by CDOM affects both the inherent and apparent optical properties of seawater. Therefore, optical methods, including remote sensing, may be applied to study the distribution of CDOM in the ocean. The theoretical approach by Morel and Prieur [158] to relate the spectral reflectance, which is a physical description of color of the sea, to the inherent optical properties in various marine waters resulted in the following equation:

$$R(\lambda) = f\frac{b_b(\lambda)}{a(\lambda)} = f\frac{b_{bw}(\lambda) + b_{bp}(\lambda)}{a_w(\lambda) + a_{ph}(\lambda) + a_{CDOM}(\lambda) + a_{NAP}(\lambda)}, \quad (5)$$

where the $b_w(\lambda)$ is the scattering coefficient of water molecules and $b_p(\lambda)$ is the scattering coefficient due to particles suspended in marine waters. $a_w(\lambda)$ is the absorption coefficient of water itself, $a_{CDOM}(\lambda)$ is the absorption coefficient of CDOM, $a_{ph}(\lambda)$ is the absorption coefficient of phytoplankton pigments, $a_{NAP}(\lambda)$ is the absorption coefficient of the non-algal particles, and f is the proportionality factor (0.33; [159]).

This equation is the foundation of ocean color satellite remote sensing, allowing continual estimations of phytoplankton biomass (expressed in chlorophyll-*a* concentration units) in global and regional scales since the launch of the first ocean color scanner, coastal zone color scanner in 1978. Earlier algorithms for retrieval of chlorophyll-a from coastal zone color scanner (CZCS) imagery were based on the assumption that the most important optically active components of marine waters were associated with phytoplankton abundance [160–162]. This assumption is valid in oligotrophic Case 1 waters [158], but in coastal waters and enclosed marine basins, which are influenced by suspended particles and CDOM of mainly terrigenous origin, the algorithms failed (Case 2 waters). Independent changes in CDOM and particulate absorption relative to chlorophyll-*a* were the main problems. The absorption spectra of chlorophyll-*a* and CDOM overlap in the blue region of the electromagnetic spectrum leading to overestimation of chlorophyll-*a* concentration, especially in coastal oceans and semi-enclosed seas [163].

Optical properties of CDOM have allowed ocean color remote sensing studies of organic carbon cycling on global and regional scales. Semi-analytical algorithms for estimating the inherent optical properties of seawaters from ocean color satellite imagery, e.g., GSM01 algorithm, produced values of chlorophyll-*a* concentrations together with CDOM and particulate absorption coefficients [164–167]. These algorithms were applied to produce maps on a basin scale, as well as of the global distribution of CDOM absorption and its seasonal variability [167]. The application of semi-analytical algorithms to observations of CDOM absorption by ocean color imagery has also improved chlorophyll-*a* retrievals resulting from the application of empirical algorithms in various oceanic basins [168] and hence a reassessment of the principal theoretical assumptions of ocean optics [169]. In coastal areas and semi-enclosed seas, which are optically complex Case 2 waters, empirical band ratio algorithms are often used to estimate CDOM absorption and its impact on light transmission through the water column in the UV and visible spectrum [71, 170–175]. Recently, neural network algorithms have been applied to detect CDOM absorption in Case 2 waters [176]. Figure 9 displays the distribution of the CDOM absorption coefficient, $a_{CDOM}(443)$, in North Atlantic estimated from the

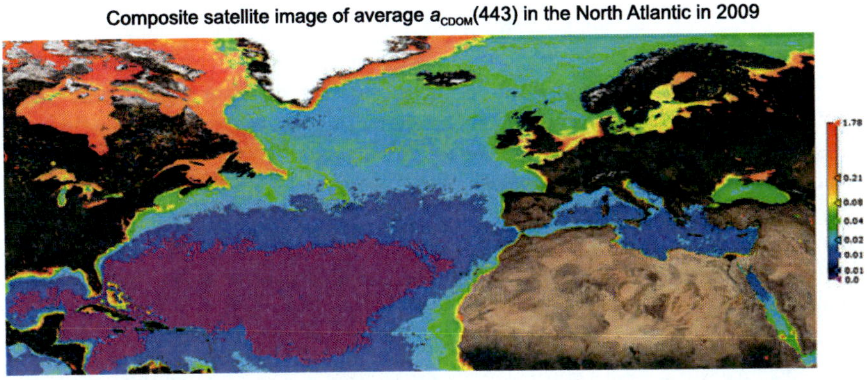

Fig. 9 A composite satellite image of average distribution of CDOM absorption coefficient, $a_{CDOM}(443)$, in the North Atlantic estimated from neural network algorithm [176] applied to all available MERIS images from year 2009. Image has been generated using BEAM-Case2R processor and Calvalus (Courtesy of Brockmann Consult)

neural network algorithm [176] applied to all available MERIS images from year 2009. This satellite imagery product clearly delineates the frontal system between major oceanic water masses that have distinctly different CDOM absorption properties. For example, a very strong gradient in $a_{CDOM}(443)$ could be observed along the South Eastern coast of the United States, where strong hydrological fronts separate the coastal waters influenced by riverine discharge of high CDOM concentrations from the Gulf Stream low absorbing oligotrophic waters [70, 93]. Further north, a distinct visible front between the Gulf of St. Lawrence estuarine waters [80] and the coastal jet of the Labrador Current that carries highly absorbing waters can also be observed. This contrasts with the subtropical gyre in the North Atlantic, where very low $a_{CDOM}(443)$ signals were found ($<0.01\ m^{-1}$). The oligotrophic gyre is fringed in the north by a front along the meandering boundary of the Gulf Stream that separates the productive temperate subpolar waters. Smaller frontal systems associated with riverine plumes of the Thames, Tyne Rhine, and Elbe Rivers in the North Sea, and with the Oder and Vistula Rivers in the Baltic Sea, could be also seen despite the turbid and highly absorbing waters in the eastern North Atlantic shelf marginal seas [177, 178]. Recent validation studies have evaluated the accuracy of neural network algorithms in the North Atlantic [179] and European Seas [180]. The neural network algorithms are capable of retrieving chlorophyll-*a* concentrations and the inherent optical properties of marine and oceanic waters and partitioning the total absorption coefficient into phytoplankton pigments, CDOM, and detritus absorption coefficients, with accuracy comparable to semi-analytical algorithms. In the Baltic Sea, which showcases optical properties dominated by CDOM absorption, the neural network algorithms retrieval accuracy of the $a_{CDOM}(412)$ from remote sensing reflectance was as good as 14% [180].

The influence of large fluxes of organic and inorganic species from rivers is not limited to the coastal regions, but can strongly influence biogeochemical properties in waters that are normally considered oligotrophic (Case 1 waters; [158]). Hessen et al. [181] documented how the Lena and Ob river waters are transported into the Kara Sea. The clear division of the marine and the river water masses seen from MODIS images can be used to monitor remotely the potential change in river water extent, as a result of river discharge increase (e.g,, [182]). Kowalczuk et al. [71, 183] showed the river plume dispersal in semi-enclosed Baltic Sea. Large amounts of terrigenous DOM delivered by the Mississippi River [22, 184], the Amazon River [142], and the Orinoco River [185] have been detected hundreds of kilometers into the normally oligotrophic offshore waters. An alternative to the commonly used GSM (Garver–Siegel–Maritorena) inversion method, in terms of retrieved CDOM concentrations and distribution patterns, is the CDOM index developed by Morel and Gentili [186]. This promising semi-analytical approach differentiates CDOM from the algal material by extricating their optical effects on the reflectance at four wavelengths (i.e., 412, 440, 490, 550 nm). After the global distribution of CDOM in the world's oceans [186], the location and extension of six subtropical oligotrophic gyres was determined using the CDOM index [4]. The clearest waters, corresponding to the lowest chlorophyll-*a* and CDOM concentrations, were found near Easter Island and near the Mariana Islands in the western part of the North Pacific Ocean. Recently, semi-analytical algorithms have been developed for the Mississippi and Atchafalaya regions [187] and Chukchi and Beaufort Seas [53, 188]. For example, Matsuoka et al. [53] reported the detection of Alaskan coastal water (ACW) using remote sensing derived $a_{CDOM}(440)$ of surface waters in Chukchi Sea. This study further supports the use of CDOM to track the surface circulation as a surrogate measure of the salinity and potential temperature, a traditional ACW tracer [189].

Another interesting application of satellite-derived CDOM is the remote ice-covered Southern Ocean. The Southern Ocean is one of the less studied oceanic domains, but is already experiencing climatic shifts. The Southern Ocean has a unique system of water masses where the transition of warm subtropical waters in the north to cold polar waters in the south occurs in a series of fronts including subtropical, subantarctic, and polar fronts [190–192]. This series of oceanic fronts result in the definition of different oceanographic zones where the development of endemic phytoplankton is favored. The presence of oceanic fronts is responsible for the important change in phytoplankton assemblages (e.g,, [193, 194]). Different phytoplankton groups have different light-absorbing pigment signatures and size classes, both of which have considerable impact on absorption properties (e.g,, [195, 196]). For example, the empirical relationships between chlorophyll-a and reflectance can generate a >10-fold range in retrieved chlorophyll, depending on the prevalence of phytoplankton species in the Southern Ocean [197]. Clemenston et al. [198] estimated that the change from underestimation to overestimation of the retrieved chlorophyll-a at 50°S coincided with the position of physical boundary of the subantarctic front (SAF). The most widely used empirical approach for empirically derived chlorophyll-a can be in error by a factor of 5 or more [197]. These substantial errors in satellite-derived chlorophyll-a should be acknowledged when inferring climate-related changes in marine biology.

Acknowledgements We thank the book editor, I. Belkin, for the invitation to prepare this chapter on CDOM in frontal zones. Céline Guéguen was supported by the Canada Research Chair program and by the Natural Sciences and Engineering Research Council of Canada. Piotr Kowalczuk was supported by the research grant no 546/N-AMT-CDOM/2009/0 entitled: "Sources and transformation of the Chromophoric Dissolved Organic Matter along the Atlantic Meridional Transect". The assessment with use of the measurements of the fluorescence Excitation–Emission Matrix spectra. AMT-CDOM. Partial support for PK was also provided by the project Satellite Monitoring of the Baltic Sea Environment – SatBałtyk, co-founded by the European Union through European Regional Development Fund contract No. POIG 01.01.02-22-011/09. Comments by three anonymous reviewers have greatly helped in improving the manuscript. We thank Paul Dainard and Tyler Jamieson for comments on earlier drafts of this chapter. This study is a contribution to the international IMBER project and was supported by the UK Natural Environment Research Council National Capability funding to Plymouth Marine Laboratory and the National Oceanography Centre, Southampton. This is contribution number 241 of the AMT program.

References

1. Simpson JH, Bowers D (1981) Models of stratification and frontal movement in shelf seas. Deep Sea Res I 28:727–738
2. O'Donnell J (1993) Surface fronts in estuaries: a review. Estuar Coast Shelf Sci 16:12–39
3. Gan J, Li L, Wang D, Guo X (2009) Interaction of a river plume with coastal upwelling in the northeastern South China Sea. Cont Shelf Res 29:728–740
4. Morel A, Claustre H, Gentili B (2010) The most oligotrophic subtropical zones of the global ocean: similarities and differences in terms of chlorophyll and yellow substance. Biogeosciences 7:3139–3151
5. de Baar HJW, de Jong JTM, Bakker DCE, Loscher BM, Veth C, Bathmann U, Smetacek V (1995) Importance of iron for plankton blooms and carbon dioxide drawdown in the Southern Ocean. Nature 373:412–415
6. Strass VH, Naveira Garabato ACN, Pollard RT, Fischer HI, Hense I, Allen JT, Read JF, Leach H, Smetacek V (2002) Deep Sea Res II 49:3735–3769
7. Bricaud A, Morel A, Prieur L (1981) Absorption by dissolved organic matter of the sea (yellow substance) in the UV and visible domains. Limnol Oceanogr 26:43–53
8. Müller-Karger FE, McClain CR, Fisher TR, Esaias WE, Varela R (1989) Pigment distribution in the Caribbean Sea: observation from space. Prog Oceanogr 23:23–64
9. Coble PG (1996) Characterization of marine and terrestrial DOM in seawater using excitation–emission matrix spectroscopy. Mar Chem 51:325–346
10. Blough NV, Green SA (1995) Spectroscopic characterization and remote sensing of non-living organic matter. In: Zepp RG, Sonntag C (eds) The role of non-living organic matter in the earth's carbon cycle. Wiley, New York, pp 23–45
11. Mopper K, Kieber DJ (2002) Photochemistry and the cycling of carbon, sulfur, nitrogen and phosphorus. In: Hansell D, Carlson C (eds) Biogeochemistry of marine organic matter. Academic, New York, pp 455–507
12. Chen Z, Li Y, Pan J (2004) Distributions of colored dissolved organic matter and dissolved organic carbon in the Pearl River Estuary, China. Cont Shelf Res 24:1845–1856
13. Tzortziou M, Neale PJ, Megonigal JP, Pow CL, Butterworth M (2011) Spatial gradients in dissolved carbon due to tidal marsh outwelling into a Chesapeake Bay estuary. Mar Ecol Prog Ser 2(426):41–56
14. Stedmon CA, Markager S, Bro R (2003) Tracing dissolved organic matter in aquatic environments using a new approach to fluorescence spectroscopy. Mar Chem 82:239–254
15. Stedmon CA, Bro R (2008) Characterizing dissolved organic matter fluorescence with parallel factor analysis: a tutorial. Limnol Oceanogr Method 6:572–579

16. Guéguen C, Granskog MA, McCullough G, Barber DG (2011) Characterization of colored dissolved organic matter in Hudson Bay and Hudson Strait using parallel factor analysis. J Mar Syst 88:423–433
17. Guéguen C, McLaughlin FA, Carmack EC, Itoh M, Narita H, Nishino S (2012) The nature of colored dissolved organic matter in the southern Canada Basin and East Siberian Sea. Deep Sea Res II 81–84:102–113
18. Bro R (1997) PARAFAC. Tutorial and applications. Chemometr Intell Lab 38:149–171
19. Parlanti E, Wörz K, Geoffroy L, Lamotte M (2000) Dissolved organic matter fluorescence spectroscopy as a tool to estimate biological activity in a coastal zone submitted to anthropogenic inputs. Org Geochem 31:1765–1781
20. Murphy KR, Stedmon CA, Waite TD, Ruiz GM (2008) Distinguishing between terrestrial and autochthonous organic matter sources in marine environments using fluorescence spectroscopy. Mar Chem 108:40–58
21. Walker SA, Amon RMW, Stedmon C, Duan S, Louchouarn P (2009) The use of PARAFAC modeling to trace terrestrial dissolved organic matter and fingerprint water masses in coastal Canadian Arctic surface waters. J Geophys Res 114:G00F06. doi:10.1029/2009JG000990
22. Del Castillo CE, Gilbes F, Coble PG, Müller-Karger FE (2000) On the dispersal of riverine colored dissolved organic matter over the West Florida Shelf. Limnol Oceanogr 45:1425–1432
23. Guéguen C, Guo L, Tanaka N (2005) Distributions and characteristics of colored dissolved organic matter in the western Arctic Ocean. Cont Shelf Res 25:1195–1207
24. Stedmon CA, Markager S (2001) The optics of chromophoric dissolved organic matter (CDOM) in the Greenland Sea: an algorithm for differentiation between marine and terrestrially derived organic matter. Limnol Oceanogr 46:2087–2093
25. Pope RM, Weidemann AD, Fry ES (2000) Integrating cavity absorption meter measurements of dissolved substances and suspended particles in ocean water. Dyn Atmos Oceans 31:307–320
26. Woźniak B, Dera J (2007) Light absorption in sea water. Springer, Dordrecht, 456 pp
27. Xiao YH, Sara-Aho T, Hartikainen H, Vähätalo AV (2013) Contribution of ferric iron to light absorption by chromophoric dissolved organic matter. Limnol Oceanogr 58:653–662
28. Jerlov NG (1976) Marine optics. Elsevier, New York, 231 pp
29. Kirk JTO (1994) Light and photosynthesis in aquatic ecosystems, 2nd edn. Cambridge University Press, New York
30. Mitchell G, Bricaud A, Carder K, Cleveland J, Ferrari G, Gould R, Kahru M, Kishino M, Maske H, Moisan T, Moore L, Nelson N, Phinney D, Reynolds R, Sosik H, Stramski D, Tassan S, Trees C, Weidemann A, Wieland J, Vodacek A (2000) Determination of spectral absorption coefficients of particles, dissolved material and phytoplankton for discrete water samples. In: Fargion GS, Mueller JL (eds) Ocean optics protocols for satellite ocean colour sensor validation, revision 2. NASA Technical Memorandum 2000 209966
31. D'Sa EJ, Steward RG, Vodacek A, Blough NV, Phinney D (1999) Optical absorption of seawater colored dissolved organic matter determined using a liquid capillary waveguide. Limnol Oceanogr 44:1142–1148
32. Nelson NB, Siegel DA, Carlson CA, Swan C, Smethie WM, Khatiwala S (2007) Hydrography of chromophoric dissolved organic matter in the North Atlantic. Deep Sea Res I 54:710–731
33. Nelson NB, Siegel DA (2002) Chromophoric DOM in the open ocean. In: Hansell DA, Carlson CA (eds) Biogeochemistry of marine dissolved organic matter. Academic, San Diego, pp 547–578
34. Swan CM, Siegel DA, Nelson NB, Carlson CA, Nasir E (2009) Biogeochemical and hydrographic controls on chromophoric dissolved organic matter distribution in the Pacific Ocean. Deep Sea Res I 56:2175–2192
35. Röttgers R, Schönfeld W, Kipp P-R, Doerffer R (2005) Practical test of a point-source integrating cavity absorption meter: the performance of different collector assemblies. Appl Optics 44:5549–5560

36. Röttgers R, Doerffer R (2007) Measurements of optical absorption by chromophoric dissolved organic matter using a point-source integrating-cavity absorption meter. Limnol Oceanogr Method 5:126–135
37. Pegau WS, Gray D, Zaneveld JRV (1997) Absorption and attenuation of visible and near-infrared light in water: dependence on temperature and salinity. Appl Optics 36:6035–6046
38. Moore CM, Zaneveld JRV, Kitchen JC (1992) Preliminary results from an in situ spectral absorption meter. Proc SPIE Ocean Opt XI 1750:330–337
39. Moore CM (1994) In situ, biochemical, oceanic, optical meters: spectral absorption, attenuation, fluorescence meters—a new window of opportunity for ocean scientists. Sea Technol 35:10–16
40. Twardowski MS, Sullivan JM, Donaghay PL, Zaneveld JRV (1999) Microscale quantification of the absorption by dissolved and particulate material in coastal waters with an ac-9. J Atmos Oceanic Tech 16:691–707
41. Twardowski MS, Boss E, Sullivan JM, Donaghay PL (2004) Modeling the spectral shape of absorption by chromophoric dissolved organic matter. Mar Chem 89:69–88
42. Loos EA, Costa M (2010) Inherent optical properties and optical mass classification of the waters of the Strait of Georgia, British Columbia, Canada. Prog Oceanogr 87:144–156
43. Boss E, Pegau WS, Zaneveld JRV, Barnard AH (2001) Spatial and temporal variability of absorption by dissolved material at a continental shelf. J Geophys Res 106(5):9499–9507
44. Twardowski MS, Donaghay PL (2001) Separating in situ and terrestrial sources of absorption by dissolved materials in coastal waters. J Geophys Res 106:2545–2560
45. Twardowski MS, Donaghay PL (2002) Photobleaching of aquatic dissolved materials: absorption removal, spectral alteration, and their interrelationship. J Geophys Res 107(C8). doi:10.1029/1999JC000281
46. Brown CA, Huot Y, Purcell MJ, Cullen JJ, Lewis MR (2004) Mapping coastal optical and biogeochemical variability using an autonomous underwater vehicle and a new bio optical inversion algorithm. Limnol Oceanogr Method 2:262–281
47. Nencioli F, Chang G, Twardowski M, Dickey TD (2010) Optical characterization of an eddy-induced diatom bloom west of the island of Hawaii. Biogeosciences 7:151–162
48. Antoine D, Siegel DA, Kostadinov T, Maritorena S, Nelson NB, Gentili B, Vellucci V, Guillocheau N (2011) Variability in optical particle backscaterring in contrasting bio-optical oceanic regimes. Limnol Oceanogr 56:955–973
49. Berthon JF, Zibordi G (2010) Optically black waters in the northern Baltic Sea. Geophys Res Lett 37. doi:10.1029/2010GL043227
50. Blough NV, Del Vecchio R (2002) Chromophoric DOM in the coastal environment. In: Hansell D, Carlson C (eds) Biogeochemistry of marine dissolved organic matter. Academic, New York, pp 509–546
51. Del Vecchio R, Blough NV (2004) Spatial and seasonal distribution of chromophoric dissolved organic matter and dissolved organic carbon in the Middle Atlantic Bight. Mar Chem 89:169–187
52. Stedmon CA, Markager S, Kaas H (2000) Optical properties and signatures of chromophoric dissolved organic matter (CDOM) in Danish coastal waters. Estuar Coast Shelf Sci 51:267–278
53. Matsuoka A, Hill V, Huot Y, Babin M, Bricaud A (2011) Seasonal variability in the light absorption properties of western Arctic waters: parameterization of the individual components of absorption for ocean color applications. J Geophys Res 116, C02007. doi:10.1029/2009JC005594
54. Kitidis V, Stubbins AP, Uher G, Upsill Goddard RC, Law CS, Woodward EMS (2006) Variability of chromophoric organic matter in surface waters of the Atlantic Ocean. Deep Sea Res II 53:1666–1684
55. Blough NV, Zafiriou OC, Bonilla J (1993) Optical absorption spectra of waters from the Orinoco river outflow: terrestrial input of coloured organic matter to the Caribbean. J Geophys Res 98:2271–2278

56. Del Castillo CE, Coble PG, Morell JM, Lopez JM, Corredor JE (1999) Analysis of the optical properties of the Orinoco River plume by absorption and fluorescence spectroscopy. Mar Chem 66:35–51
57. Kowalczuk P, Stedmon CA, Markager S (2006) Modeling absorption by CDOM in the Baltic Sea from season, salinity and chlorophyll. Mar Chem 101:1–11
58. Helms JR, Stubbins A, Ritchie JD, Minor EC, Kieber DJ, Mopper K (2008) Absorption spectral slopes and slope ratios as indicators of molecular weight, source, and photobleaching of chromophoric dissolved organic matter. Limnol Oceanogr 53:955–969
59. Guéguen C, Cuss CW (2011) Characterisation of aquatic dissolved organic matter by asymmetrical flow field-flow fractionation coupled to UV–Visible diode array and excitation emission matrix fluorescence. J Chromatogr A 1218:4188–4198
60. Carder KL, Steward RG, Harvey GR, Ortner PB (1989) Marine humic and fulvic acids. Their effect on remote sensing of ocean chlorophyll. Limnol Oceanogr 34:68–81
61. Moran AM, Sheldon WM, Zepp RG (2000) Carbon loss and optical changes during long-term photochemical and biological degradation of estuarine dissolved organic matter. Limnol Oceanogr 45:1254–1264
62. Grzybowski W (2000) Effect of short-term sunlight irradiation on absorbance spectra of chromophoric organic matter dissolved in coastal and riverine water. Chemosphere 40:1313–1318
63. Vähätalo AV, Wetzel RG (2008) Long-term photochemical and microbial decomposition of wetland-derived dissolved organic matter with alteration of 13C:12C mass ratio. Limnol Oceanogr 53(4):1387–1392
64. Stedmon CA, Markager S (2003) Behaviour of the optical properties of coloured dissolved organic matter under conservative mixing. Estuar Coast Shelf Sci 57:973–979
65. Xie H, Aubry C, Bélanger S, Song G (2012) The dynamics of absorption coefficients of CDOM and particles in the St. Lawrence estuarine system: biogeochemical and physical implications. Mar Chem 128–129:44–56
66. Stedmon CA, Osburn CL, Kragh T (2010) Tracing water mass mixing in the Baltic–North Sea transition zone using the optical properties of coloured dissolved organic matter. Estuar Coast Shelf Sci 87:156–162
67. Astoreca R, Rousseau V, Lancelot C (2009) Coloured dissolved organic matter (CDOM) in Southern North Sea waters: optical characterization and possible origin. Estuar Coast Shelf Sci 85:633–640
68. Vodacek A, Blough NV, DeGrandpre MD, Peltzer ET, Nelson RK (1997) Seasonal variation of CDOM and DOC in the Middle Atlantic Bight: terrestrial inputs and photooxidation. Limnol Oceanogr 42(2):674–686
69. Nelson JR, Guarda S (1995) Particulate and dissolved spectral absorption on the continental shelf of the southeastern United States. J Geophys Res 100(C5):8715–8732
70. Kowalczuk P, Cooper WJ, Whitehead RJ, Durako MJ, Sheldon W (2003) Characterization of CDOM in organic rich river and surrounding coastal ocean in the South Atlantic Bight. Aquat Sci 65:384–401
71. Kowalczuk P, Olszewski J, Darecki M, Kaczmarek S (2005) Empirical relationships between coloured dissolved organic matter (CDOM) absorption and apparent optical properties in Baltic Sea waters. Int J Remote Sens 26:345–370
72. Peuravouri J, Pihlaja K (1997) Molecular size distribution and spectroscopic properties of aquatic humic substances. Anal Chim Acta 337:133–149
73. Lawaetz AJ, Stedmon CA (2009) Fluorescence intensity calibration using the Raman scatter peak of water. Appl Spectrosc 63:936–940
74. Murphy KR, Butler KD, Spencer RGM, Stedmon CA, Boehme JR, Aiken GR (2010) The measurement of dissolved organic matter fluorescence in aquatic environments: an interlaboratory comparison. Environ Sci Tech 44:9405–9412
75. Kalle K (1949) Fluoreszenz und Gelbstoff im Botnischen und Finnischen Meerbusen. Dt Hydrogr Zeitung 2:117–124

76. Duursma EK (1965) The dissolved organic constituents of seawater. In: Riley JP, Skirrow G (eds) Chemical oceanography, vol 1. Academic, London, pp 433–475
77. Højerslev NK (1989) Bio-optical properties of Louisiana shelf waters off the Mississippi River Delta. Science Application International Corp, Raleigh, p 40
78. Ferrari GM, Tassan S (1991) On the accuracy of determining light absorption by "yellow substance" through measurements of induced fluorescence. Limnol Oceanogr 36(4):777–786
79. Hoge FE, Vodacek A, Blough NV (1993) Inherent optical properties of the ocean: retrieval of the absorption coefficient of chromophoric dissolved organic matter from fluorescence measurements. Limnol Oceanogr 38:1394–1402
80. Nieke B, Reuter R, Heuermann R, Wang H, Babin M, Therriault JC (1997) Light absorption and fluorescence of chromophoric dissolved organic matter (CDOM) in the St. Lawrence Estuary (Case 2 waters). Cont Shelf Res 17(3):235–252
81. Ferrari GM, Dowell MD (1998) CDOM absorption characteristics with relation to fluorescence and salinity in coastal areas of the southern Baltic Sea. Estuar Coast Shelf Sci 47:91–105
82. Ferrari GM (2000) The relationship between chromophoric dissolved organic matter and dissolved organic carbon in the European Atlantic coastal area and in the West Mediterranean Sea (Gulf of Lions). Mar Chem 70:339–357
83. Chen RF, Zhang Y, Vlahos P, Rudnick SM (2002) The fluorescence of dissolved organic matter in the Mid-Atlantic Bight. Deep Sea Res II 49:4439–4459
84. Green SA, Blough NV (1994) Optical absorption and fluorescence properties of chromophoric dissolved organic matter in natural waters. Limnol Oceanogr 39:1903–1916
85. Conmy RN, Coble PG, Del Castillo CE (2004) Calibration and performance of a new in situ multi-channel fluorometer for measurement of colored dissolved organic matter in the ocean. Cont Shelf Res 24:431–442
86. Belzile C, Roesler CS, Christensen JP, Shakhova N, Semiletov I (2006) Fluorescence measured using the WETStar DOM fluorometer as a proxy for dissolved matter absorption. Estuar Coast Shelf Sci 67:441–449
87. Kowalczuk P, Zabłocka M, Sagan S, Kuliński K (2010) Fluorescence measured in situ as a proxy of CDOM absorption and DOC concentration in the Baltic Sea. Oceanol 52:431–471
88. Traganza ED (1969) Fluorescence excitation and emission spectra of dissolved organic matter in sea water. B Mar Sci 19:897–904
89. Coble PG, Green S, Gagosian RB, Blough N (1990) Characterization of dissolved organic matter in the Black Sea by fluorescence spectroscopy. Nature 348:432–435
90. Coble PG, Del Castillo CE, Avril B (1998) Distribution and optical properties of CDOM in the Arabian Sea during the 1995 Southwest Monsoon. Deep Sea Res II 45:2195–2223
91. Stedmon CA, Markager SS (2005) Resolving the variability in dissolved organic matter fluorescence in a temperate estuary and its catchment using PARAFAC analysis. Limnol Oceanogr 50:686–697
92. Yamashita Y, Jaffé R, Maie N, Tanoue E (2008) Assessing the dynamics of dissolved organic matter (DOM) in coastal environments by excitation emission matrix fluorescence and parallel factor analysis (EEM-PARAFAC). Limnol Oceanogr 53:1900–1908
93. Kowalczuk P, Durako MJ, Young H, Kahn AE, Cooper WJ, Gonsior M (2009) Characterization of dissolved organic matter fluorescence in the South Atlantic Bight with use of PARAFAC model: interannual variability. Mar Chem 113:182–196
94. Nelson NB, Siegel DA, Carlson CA, Swan C (2010) Tracing global biogeochemical cycles and meridional overturning circulation using chromophoric dissolved organic matter. Geophys Res Lett 37, L03610. doi:10.1029/2009GL042325
95. Yamashita Y, Tanoue E (2009) Basin scale distribution of chromophoric dissolved organic matter in the Pacific Ocean. Limnol Oceanogr 54:598–609
96. Stedmon CA, Amon RMW, Rinehart AJ, Walker SA (2011) The supply and characteristics of colored dissolved organic matter (CDOM) in the Arctic Ocean: Pan Arctic trends and differences. Mar Chem 124:108–118

97. Fichot CG, Kaiser K, Hooker SB, Amon RMW, Babin M, Bélanger S, Walker SA, Benner R (2013) Pan-Arctic distributions of continental runoff in the Arctic Ocean. Sci Rep 3:1053. doi:10-1038-srep01053
98. Nelson NB, Siegel DA, Michaels AF (1998) Seasonal dynamics of colored dissolved material in the Sargasso Sea. Deep Sea Res I 45:931–957
99. Rochelle-Newall EJ, Fisher TR (2002) Production of chromophoric dissolved organic matter fluorescence in marine and estuarine environments: an investigation into the role of phytoplankton. Mar Chem 77:7–21
100. Romera-Castillo C, Sarmento H, Álvarez-Salgado AX, Gasol JM, Marrasé C (2010) Production of chromophoric dissolved organic matter by marine phytoplankton. Limnol Oceanogr 55:446–454
101. McIntyre AM, Guéguen C (2012) Binding interactions of algal-derived dissolved organic matter with metal ions. Chemosphere 90:620–626
102. Guéguen C, Guo L, Yamamoto-Kawai M, Tanaka N (2007) Colored dissolved organic matter dynamics across the shelf-basin interface in the western Arctic Ocean. J Geophys Res 112, C05038. doi:10.1029/2006JC003584
103. Granskog MA, Macdonald RW, Mundy CJ, Barber DG (2007) Distribution, characteristics and potential impacts of chromophoric dissolved organic matter (CDOM) in the Hudson Strait and Hudson Bay, Canada. Cont Shelf Res 27:2032–2050
104. Hill VJ (2008) Impacts of chromophoric dissolved organic material on surface ocean heating in the Chukchi Sea. J Geophys Res 113, C07024. doi:10.1029/2007JC004119
105. Matsuoka A, Bricaud A, Benner R, Para J, Sempéré R, Prieur L, Bélanger S, Babin M (2012) Tracing the transport of colored dissolved organic matter in water masses of the Southern Beaufort Sea: a relationship with hydrographic characteristics. Biogeosciences 9:929–940
106. Aagaard K, Carmack EC (1989) The role of sea ice and other freshwater in the Arctic circulation. J Geophys Res 94:14485–14498
107. Winsor P, Chapman DC (1989) Distribution and interannual variability of dense water production from coastal polynyas on the Chukchi Shelf. J Geophy Res 107:3079. doi:10.1029/2001JC000984, 2002
108. Shimada K, Itoh M, Nishino S, McLaughlin F, Carmack E, Proshutinsky A (2005) Halocline structure in the Canada Basin of the Arctic Ocean. Geophys Res Lett 32, L03605. doi:10.1029/2004GL021358
109. Granskog MA, Stedmon CA, Dodd PA, Amon RMW, Pavlov AK, de Steur L, Hansen E (2012) Characteristics of colored dissolved organic matter (CDOM) in the Arctic outflow in the Fram Strait: assessing the changes and fate of terrigenous CDOM in the Arctic Ocean. J Geophys Res 11, C12021. doi:10.1029/2012JC008075
110. Granskog MA, Macdonald RW, Kuzyk ZA, Senneville S, Mundy CJ, Barber DG, Stern GA, Saucier F (2009) Coastal conduit in southwestern Hudson Bay (Canada) in summer: rapid transit of freshwater and significant loss of colored dissolved organic matter. J Geophys Res 114, C08012. doi:10.1029/2009JC005270
111. Mopper K, Sarpal RS, Kieber DJ (1995) Protein and humic substance fluorescence of dissolved organic matter in Antarctic sea water. Antarct J 30:137–139
112. Wedborg M, Hoppema M, Skoog A (1998) On the relation between organic and inorganic carbon in the Weddell Sea. J Mar Syst 17:59–76
113. Chen RF, Bata JL (1992) The fluorescence of dissolved organic matter in seawater. Mar Chem 37:191–221
114. Hayase K, Shinozuoka N (1995) Vertical distribution of fluorescent organic matter along with AOU and nutrients in the equatorial Central Pacific. Mar Chem 48:283–290
115. Hayase K, Yamamoto M, Nakazawa I, Tsubota H (1987) Behavior of natural fluorescence in Sagami Bay and Tokyo Bay, Japan: vertical and lateral distribution. Mar Chem 48:283–290
116. Tani H, Nishioka J, Kuma K, Takata H, Yamashita Y, Tanoue E, Midorikawa T (2003) Iron (III) hydroxide solubility and humic-type fluorescent organic matter in the deep water column of the Okhotsk Sea and the northwestern North Pacific Ocean. Deep Sea Res I 50:1063–1078

117. Mopper K, Zhou XL, Kieber RJ, Kieber DJ, Sikorski RJ, Jones RD (1991) Photochemical degradation of dissolved organic carbon and its impact on the oceanic carbon cycle. Nature 353:60–62
118. Mopper K, Schultz CA (1993) Fluorescence as a possible tool for studying the nature and water column distribution of DOC components. Mar Chem 41:229–238
119. Determann S, Reuter R, Willkomm R (1996) Fluorescence matter in the eastern Atlantic Ocean: part 2. Vertical profiles and relation to water masses. Deep Sea Res 43:345–360
120. Jørgensen L, Stedmon CA, Kragh T, Markager S, Middelboe M, Søndergaard M (2011) Global trends in the fluorescence characteristics and distribution of marine dissolved organic matter. Mar Chem 126:139–148
121. Stedmon CA, Álvarez-Salgado XA (2011) Shedding light on a black box: UV–Visible spectroscopic characterization of marine dissolved organic matter. In: Jiao N, Azam F, Sanders S (eds) Microbial carbon pump in the ocean. American Association for the Advancement of Science, Washington, DC, pp 62–63
122. Schlitzer R (2012) Ocean data view. http://odv.awi.de
123. Yamashita Y, Tanoue E (2008) Production of bio-refractory fluorescent dissolved organic matter in the ocean interior. Nat Geosci 1:579–582
124. Yamashita Y, Cory RM, Nishioka J, Kuma K, Tanoue E, Jaffé R (2010) Fluorescence characteristics of dissolved organic matter in the deep waters of the Okhotsk Sea and the northwestern North Pacific Ocean. Deep Sea Res II 57:1478–1485
125. Omori YT, Hama MI, Saito S (2010) Relationship between the seasonal change in fluorescent dissolved organic matter and mixed layer depth in the subtropical western North Pacific. J Geophys Res 115, C06001. doi:10.1029/2009JC005526
126. Omori Y, Hama T, Ishii M, Saito S (2011) Vertical change in the composition of marine humic-like fluorescent dissolved organic matter in the subtropical western North Pacific and its relation to photoreactivity. Mar Chem 124:38–47
127. Cooper LW, Benner R, McClelland JW, Peterson BJ, Holmes RM, Raymond PA, Hansell DA, Grebmeier JM, Codispoti LA (2005) Linkages among runoff, dissolved organic carbon, and the stable oxygen isotope composition of seawater and other water mass indicators in the Arctic Ocean. J Geophys Res 110, G02013. doi:10.1029/2005JG000031
128. Guay CK, Klinkhammer GP, Falkner KK, Benner R, Coble PG, Whitledge TE, Black B, Bussell FJ, Wagner TA (1999) High-resolution measurements of dissolved organic carbon in the Arctic Ocean by *in situ* fiber-optic spectrometry. Geophys Res Lett 26(8):1007–1010
129. Yamashita Y, Tsukasaki A, Nishida T, Tanoue E (2007) Vertical and horizontal distribution of fluorescent dissolved organic matter in the Southern Ocean. Mar Chem 106:498–509
130. Hansell DA, Carlson CA (2002) Biogeochemistry of marine dissolved organic matter. Academic, New York, 774 pp
131. Santinelli C, Nannicini L, Seritti A (2010) DOC dynamics in the meso and bathypelagic layers of the Mediterranean Sea. Deep Sea Res II 57:1446–1459
132. Nieto-Cid M, Álvarez-Salgado XA, Pérez FF (2006) Microbial and photochemical reactivity of fluorescent dissolved organic matter in a coastal upwelling system. Limnol Oceanogr 51:1391–1400
133. Smethie WM, Fine RA (2001) Rates of North Atlantic deep water formation calculated from chlorofluorocarbon inventories. Deep Sea Res I 48:189–215
134. Zhang Y, Sintes E, Chen J, Zhang Y, Dai M, Jiao N, Herndl GJ (2009) Role of mesoscale cyclonic eddies in the distribution and activity of Archaea and Bacteria in the South China Sea. Aquat Microb Ecol 56:65–79
135. Bricaud A, Babin M, Claustre H, Ras J, Tièche F (2010) Light absorption properties and absorption budget of Southeast Pacific waters. J Geophys Res 115, C08009. doi:10.1029/2009JC005517
136. Nieto-Cid M, Alvarez-Salgado XA, Gago J, Perez FF (2005) DOM fluorescence, a tracer for biogeochemical processes in a coastal upwelling system (NW Iberian Peninsula). Mar Ecol Prog Ser 297:33–50

137. Baird ME, Suthers IM, Griffin DA, Hollings B, Pattiaratchi C, Everett JD, Roughan M, Oubelkheir K, Doblin M (2011) The effect of surface flooding on the physical-biogeochemical dynamics of a warm-core eddy off southeast Australia. Deep Sea Res II 58:592–605
138. Romera-Castillo C, Nieto-Cid M, Castro CG, Marrasé C, Largier J, Barton ED, Alvarez-Salgado XA (2011) Fluorescence: absorption coefficient ratio – tracing photochemical and microbial degradation processes affecting coloured dissolved organic matter in a coastal system. Mar Chem 125:26–38
139. Retamal L, Vincent WF, Martineau C, Osburn CL (2007) Comparison of the optical properties of dissolved organic matter in two river-influenced coastal regions of the Canadian Arctic. Estuar Coast Shelf Sci 72:261–272
140. Guo W, Stedmon CA, Han Y, Wu F, Yu X, Hu M (2007) The conservative and non-conservative behavior of chromophoric dissolved organic matter in Chinese estuarine waters. Mar Chem 107:357–366
141. Foden J, Sivyer DB, Mills DK, Devlin MJ (2008) Spatial and temporal distribution of chromophoric dissolved organic matter (CDOM) fluorescence and its contribution to light attenuation in UK waterbodies. Estuar Coast Shelf Sci 79:707–717
142. Del Vecchio R, Subramaniam A (2004) Influence of the Amazon River on the surface optical properties of the western tropical North Atlantic Ocean. J Geophys Res 109:C11. doi:10.1029/2004JC002503
143. Hitchcock GL, Chen RF, Gardner GB, Wiseman WJ Jr (2004) A Lagrangian view of fluorescent chromophoric dissolved organic matter distributions in the Mississippi River plume. Mar Chem 89:225–239
144. Pettigrew NR, Roesler CS, Neville F, Deese HE (2008) An operational real-time ocean sensor network in the Gulf of Maine. In: Nittel S, Labrinidis A, Stefanidis A (eds) Geosensor networks, vol 4540, Lecture notes in Computer Science. Springer, Berlin, pp 213–238
145. Smyth TJ, Fishwick JR, Gallienne CP, StephensJA BAJ (2010) Technology, design and operation of an autonomous buoy system in the western English Channel. J Atmos Ocean Tech 27:2056–2064
146. Uher G, Hughes C, Henry G, Upstill-Goddard RC (2001) Non-conservative mixing behaviour of colored dissolved organic matter in a humic-rich turbid estuary. Geophys Res Lett 28:3309–3312
147. Jaffé R, Boyer JN, Lu X, Maie N, Yang C, Scully NM, Mock S (2004) Source characterization of dissolved organic matter in a subtropical mangrove-dominated estuary by fluorescence analysis. Mar Chem 84:195–210
148. Bowers D, Evans DW, Thomas DN, Ellis K, Le PJ, Williams B (2004) Interpreting the colour of an estuary. Estuar Coast Shelf Sci 59:13–20
149. Chen RF, Bada JL, Suzuki Y (1993) The relationship between dissolved organic carbon (DOC) and fluorescence in anoxic porewaters: implications for estimating benthic DOC fluxes. Geochim Cosmochim Acta 57:2149–2153
150. Skoog A, Hall POJ, Hulth S, Paxeus N, Van Der Hoeff MR, Westerlund S (1996) Early diagenetic production and sediment-water exchange of fluorescent dissolved organic carbon in a coastal environment. Geochim Cosmochim Acta 60:3619–3629
151. Burdige DJ, Kline SW, Chen W (2004) Fluorescent dissolved organic matter in marine sediment pore waters. Mar Chem 89:289–311
152. Kim TH, Waska H, Kwon E, Suryaputa IGN, Kim G (2012) Production, degradation, and flux of dissolved organic matter in the subterranean estuary of a large tidal flat. Mar Chem 142–144:1–10
153. Sierra MMD, Donard OFX, Etcheber H, Soriano-Sierra EJ, Ewald M (2001) Fluorescence and DOC contents of pore waters from coastal and deep-sea sediments in the Gulf of Biscay. Org Geochem 32:1319–1328

154. Macdonald RW, Naidu AS, Yunker MB, Gobeil C (2004) The Beaufort Sea: distribution, sources, fluxes and burial of organic carbon. In: Stein R, Macdonald RW (eds) The organic carbon cycle in the Arctic Ocean. Springer, New York, pp 177–193
155. Moore RM, Lowings MG, Tan TC (1983) Geochemical profiles in the central Arctic Ocean: their relation to freezing and shallow circulation. J Geophys Res 88:2667–2674
156. Guo L, Santschi PH (2000) Sedimentary source of old high molecular weight dissolved organic carbon from the ocean margin benthic nepheloid layer. Geochim Cosmochim Acta 64:651–660
157. Hamilton P, Lugo-Fernandez A (2001) Observations of high speed deep currents in the northern Gulf of Mexico. Geophys Res Lett 28:2867–2870
158. Morel A, Prieur L (1977) Analysis of variations in ocean color. Limnol Oceanogr 22:709–722
159. Gordon HR, Brown OB, Jacobs MM (1975) Computed relationships between the inherent and apparent optical properties of a flat Homogeneous Ocean. Appl Optics 14(2):417–427
160. Gordon HR, Clark DK, Mueller JL, Hovis WA (1980) Phytoplankton pigments from Nimbus-7 coastal zone color scanner: comparison with surface measurements. Science 210:63–66
161. Gordon HR, Clark DK, Brown JW, Brown OB, Evans RH, Brinks WW (1983) Phytoplankton pigment concentration in the Middle Atlantic Bight; comparison of ship determination and CZCS estimates. Appl Optics 22:20–35
162. Smith RC, Baker KS (1982) Oceanic chlorophyll concentration as determined by satellite (Nimbus-7 coastal zone color scanner). Mar Biol 66:269–279
163. IOCCG (2000) Remote sensing of the ocean colour in the coastal and other optically complex waters. In: Sathyendranath S (ed) Reports of the international ocean-colour coordinating group, No. 3. IOCCG, Dartmouth, 140 pp
164. Garver SA, Siegel DA (1997) Inherent optical property inversion of ocean color spectra and its biogeochemical interpretation, 1. Time series from the Sargasso Sea. J Geophys Res 102:18607–18625
165. Carder KL, Chen FR, Lee ZP, Hawes SK, Kamykowski D (1999) Semianalytic moderate-resolution imaging spectrometer algorithms for chlorophyll *a* and absorption with bio-optical domains based on nitrate depletion temperatures. J Geophys Res 104:5403–5421
166. Maritorena S, Siegel DA, Peterson AR (2002) Optimization of a semi- analytical ocean color model for global-scale applications. Appl Optics 41:2705–2714
167. Siegel DA, Maritorena S, Nelson NB, Hansell DA, Lorenzi-Kayser M (2002) Global distribution and dynamics of colored dissolved and detrital organic materials. J Geophys Res 107:3228. doi:10.1029/2001JC000965
168. Siegel DA, Maritorena S, Nelson NB, Behrenfeld MJ, McClain CR (2005) Colored dissolved organic matter and its influence on the satellite-based characterization of the ocean biosphere. Geophys Res Lett 32, L20605. doi:10.1029/2005GL024310
169. Siegel DA, Maritorena S, Nelson NB, Behrenfeld MJ (2005) Independence and interdependencies among global ocean color properties: reassessing the bio-optical assumption. J Geophys Res 110, C07011. doi:10.1029/2004JC002527
170. Kahru M, Mitchell BG (1999) Empirical chlorophyll algorithm and preliminary SeaWiFS validation for the California Current. Int J Remote Sens 20:3421–3429
171. Kahru M, Mitchell BG (2001) Seasonal and non-seasonal variability of satellite derived chlorophyll and colored dissolved organic matter concentration in the California Current. J Geophys Res 106:2517–2529
172. D'Sa EJ, Miller RL (2003) Bio-optical properties in waters influenced by the Mississippi River during low flow conditions. Remote Sens Environ 84:538–549
173. Johannessen SC, Miller WL, Cullen JJ (2003) Calculation of UV attenuation and colored dissolved organic matter absorption spectra from measurements of ocean color. J Geophys Res 108:3301. doi:10.1029/2000JC000514
174. Fichot CG, Sathyendranath S, Miller WL (2008) SeaUV and SeaUV(C): algorithms for the retrieval of UV/visible diffuse attenuation coefficients from ocean color. Remote Sens Environ 112(4):1584–1602

175. Mannino A, Russ ME, Hooker SB (2008) Algorithm development and validation for satellite-derived distributions of DOC and CDOM in the U.S. Middle Atlantic Bight. J Geophys Res 113, C07051. doi:10.1029/2007JC004493
176. Doerffer R, Schiller H (2007) The MERIS case 2 water algorithm. Int J Remote Sens 28:517–535
177. Kowalczuk P (1999) Seasonal variability of yellow substance absorption in the surface layer of the Baltic Sea. J Geophys Res 104:30047–30058
178. Warnock RE, Gieskes WWC, van Laar S (1999) Regional and seasonal differences in light absorption by yellow substance in the Southern Bight of the North Sea. J Sea Res 42:169–178
179. Ioannou I, Gilerson A, Gross B, Moshary F, Ahmed S (2013) Deriving ocean color products using neural networks. Remote Sens Environ 134:78–91
180. D'Alimonte D, Zibordi G, Berthon J-F, Canuti E, Kajiyama T (2012) Performance and applicability of bio-optical algorithms in different European seas. Remote Sens Environ 124:402–412
181. Hessen DO, Carroll JL, Kjeldstad B, Korosov AA, Pettersson LH, Pozdnyakov D, Sorensen K (2010) Input of organic carbon as determinant of nutrient fluxes, light climate and productivity in the Ob and Yenisey estuaries. Estuar Coast Shelf Sci 88:53–62
182. Peterson BJ, Holmes RM, McClelland JW, Vorosmarty CJ, Shiklomanov IA, Shiklomanov AI, Lammers RB, Rahmstorf S (2002) Increasing river discharge to the Arctic Ocean. Science 298:2171–2173
183. Kowalczuk P, Darecki M, Zabłocka M, Górecka I (2010) Validation of empirical and semi-analytical remote sensing algorithms for estimating absorption by coloured dissolved organic matter in the Baltic Sea from SeaWiFS and MODIS imagery. Oceanol 52:171–196
184. D'Sa EJ, Miller RL, Del Castillo C (2006) Bio-optical properties and ocean color algorithms for coastal waters influenced by the Mississippi River during a cold front. Appl Optics 45(28):7410–7428
185. Odriozola AL, Varela R, Hu C, Astor Y, Lorenzoni L, Muller-Karger FE (2007) On the absorption of light in the Orinoco River plume. Cont Shelf Res 27:1447–1464
186. Morel A, Gentili B (2009) A simple band ratio technique to quantify the colored dissolved and detrital organic material from ocean color remotely sensed data. Remote Sens Environ 113:998–1011
187. Zhu W, Yu Q, Tian YQ, Chen RF, Gardner GB (2011) Estimation of chromophoric dissolved organic matter in the Mississippi and Atchafalaya river plume regions using above surface hyperspectral remote sensing. J Geophys Res 116, C02011. doi:10.1029/2010JC006523
188. Matsuoka A, Hooker SB, Bricaud A, Gentili B, Babin M (2013) Estimating absorption coefficients of colored dissolved organic matter (CDOM) using a semi-analytical algorithm for southern Beaufort Sea waters: application to deriving concentrations of dissolved organic carbon from space. Biogeosciences 10:917–927
189. Shimada K, Carmack EC, Hatakeyama K, Takizawa T (2001) Varieties of shallow temperature maximum waters in the western Canadian Basin of the Arctic Ocean. Geophys Res Lett 28:3441–3444. doi:10.1029/2001GL013168
190. Belkin IM, Gordon AL (1996) Southern Ocean fronts from the Greenwich meridian to Tasmania. J Geophys Res 101:3675–3696
191. Rintoul SR, Trull TW (2001) Seasonal evolution of the mixed layer in the Subantarctic Zone south of Australia. J Geophys Res 106:31447–31462
192. Sokolov S, Rintoul SR (2002) Structure of Southern Ocean fronts at 140° E. J Mar Syst 37:151–184
193. Arrigo KR, Robinson DH, Worthen DL, Schieber B, Lizotte MP (1998) Bio-optical properties of the southwestern Ross Sea. J Geophys Res 193:21683–21695
194. Balch WM, Gordon HR, Bowler BC, Drapeau DT, Booth ES (2005) Calcium carbonate measurements in the surface global ocean based on moderate-resolution imaging spectroradiometer data. J Geophys Res. doi:10.1029/2004JC-002560
195. Bricaud A, Morel A (1983) Optical efficiency factors of some phytoplankters. Limnol Oceanogr 28:816–832

196. Johnsen G, Samset O, Granskog L, Sakshaug E (1994) In vivo absorption characteristics in 10 classes of bloom forming phytoplankton Taxonomic characteristics and response to photoadaptation by means of discriminant and HPLC analysis. Mar Ecol Prog Ser 105:149–157
197. Diersen HM (2010) Perspectives on empirical approaches for ocean color remote sensing of chlorophyll in a changing climate. PNAS 107:17073–17078
198. Clementson LA, Parslow JS, Turnbull AR, McKenzie DC, Rathbone CE (2001) Optical properties of waters in the Australian sector of the Southern Ocean. J Geophys Res 106:31611–31625

Contamination by Persistent Organic Pollutants and Related Compounds in Deep-Sea Ecosystems Along Frontal Zones Around Japan

Shin Takahashi, Ramu Karri, and Shinsuke Tanabe

Abstract During the last few decades, contamination by anthropogenic chemicals such as persistent organic pollutants (POPs) has spread all over the world as evidenced by their detection in various environmental components and biota including those far from human activities. Particularly, research efforts on field observations and numerical models of global fate of POPs have revealed oceanic water bodies to be a global reservoir and final sink for these toxic contaminants that undergo transport from emission sources and partition between air and water and scavenge to deep-sea layers by various biogeochemical and geophysical processes. This chapter provides an overview of the contamination by POPs and related compounds in deep-sea ecosystems along frontal zones around Japan based on the results of the monitoring studies conducted by our laboratory during the last

S. Takahashi (✉)
Center of Advanced Technology for the Environment, Agricultural Faculty, Ehime University, Tarumi 3-5-7, Matsuyama 790-8566, Japan

Center for Marine Environmental Studies (CMES), Ehime University, Bunkyo-cho 2-5, Matsuyama 790-8577, Japan
e-mail: shint@agr.ehime-u.ac.jp

R. Karri
Integrated Coastal and Marine Area Management-Project Directorate, Ministry of Earth Sciences, NIOT Campus, Velacherry-Tambaram Main Road, Pallikkaranai, Chennai, 600100, India

Center for Marine Environmental Studies (CMES), Ehime University, Bunkyo-cho 2-5, Matsuyama 790-8577, Japan
e-mail: karriramureddy@gmail.com

S. Tanabe
Center for Marine Environmental Studies (CMES), Ehime University, Bunkyo-cho 2-5, Matsuyama 790-8577, Japan
e-mail: shinsuke@agr.ehime-u.ac.jp

Igor M. Belkin (ed.), *Chemical Oceanography of Frontal Zones*,
Hdb Env Chem (2022) 116: 319–354, DOI 10.1007/698_2013_252,

Published online: 17 May 2013

decade. In the chapter, we focus mainly on two regions, the western North Pacific (WNP), off-Tohoku, Japan, and the East China Sea (ECS). The WNP is a region influenced by various water masses and currents, making it one of the world's most biologically productive zones. The other region discussed in the chapter, the ECS, is an epicontinental sea with lots of continental inputs. Our studies in these regions were conducted with the objective of understanding the environmental transport and distribution and the specific accumulation characteristics of organohalogen and butyltin compounds in deep-sea organisms. Our results suggest the vertical transport of POPs and related compounds in high productive waters along oceanic fronts and the potential role of deep-sea bed as a final sink and reservoir for these persistent contaminants. Further, to implement and evaluate the effectiveness of international agreements to protect the marine environment from the deleterious effects of POPs, interdisciplinary approaches including studies on biogeochemical and geophysical processes in the ocean as well as field observations are required to delineate the global and regional fate of POPs.

Keywords Deep-sea ecosystem, East China Sea, Persistent organic pollutants, Western North Pacific

Contents

1 Introduction

Over the past few decades, large amounts of organic and inorganic contaminants have been released into the environment as a consequence of the worldwide urbanization and agricultural and industrial activities. Marine ecosystems are no exception and they have also been increasingly subject to anthropogenic chemical contamination, receiving contaminants from a variety of diffuse and point sources. The contaminants include industrial chemicals inadvertently released into the

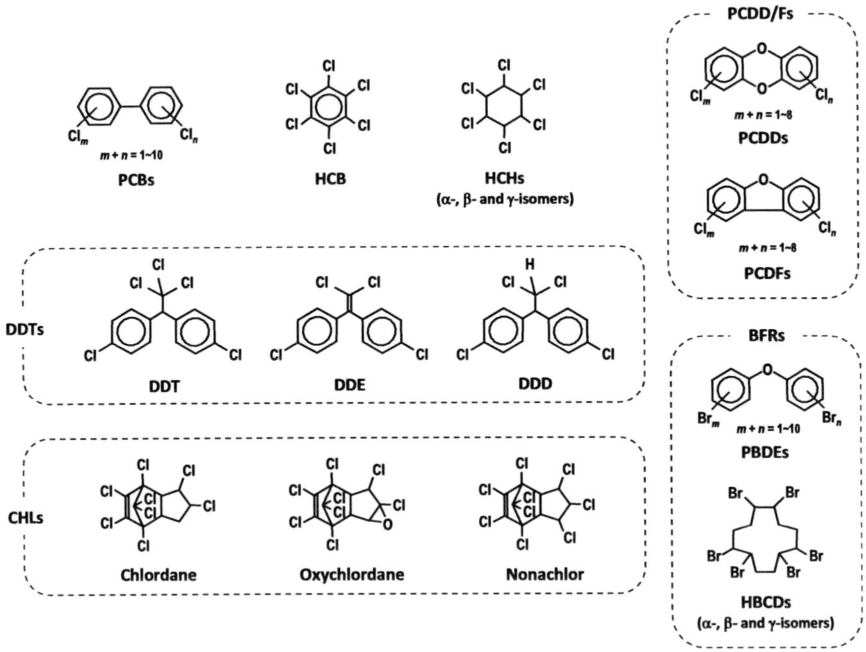

Fig. 1 Chemical structures of persistent organohalogen compounds (major POPs listed in the Stockholm Convention)

environment, as well as those derived from land use activities such as agricultural chemicals applied on crops. One such group of contaminants is the persistent organic pollutants (POPs), comprising various well-known organohalogen contaminants such as polychlorinated biphenyls (PCBs), polychlorinated dibenzo-*p*-dioxins and furans (PCDD/Fs), and organochlorine pesticides like dichlorodiphenyltrichloroethane (DDT) and its stable metabolites, DDE and DDD (DDTs), chlordane-related compounds (CHLs), hexachlorobenzene (HCB), and hexachlorocyclohexanes (HCHs) (Fig. 1). Besides, certain brominated organic compounds have been considered for inclusion in the list of POPs, i.e., polybrominated diphenyl ethers (PBDEs) and hexabromocyclododecanes (HBCDs) (Fig. 1). These compounds, which are added to electrical and electronic equipment, paints, textiles and building materials as brominated flame retardants (BFRs), are an emerging class of contaminants. In addition to the above contaminants, aquatic pollution by butyltins (BTs), particularly toxic tributyltin (TBT), used as a biocide in antifouling paints for boats and aquaculture nets, has been of concern due to the bioaccumulative potential and deleterious effects of BTs in organisms [1].

Both the emerging and legacy POPs are toxic, are chemically stable, and therefore do not easily degrade in the environment or in organisms. These chemicals tend to partition between various environmental media, such as air, water, soil, sediment, and biota, depending on the physicochemical properties.

Furthermore, being semi-volatile, these compounds are subject to co-distillation processes and can be transported globally through the atmosphere, which is well known as a concept of long-range atmospheric transport (LRAT) [2]. As a consequence, these substances have been detected in remote, relatively pristine locations, such as the polar regions and the deep seas. To protect human health and the environment, national and international control measures on the production and usage of POPs have been (or are being) introduced to reduce their emissions to the environment. For instance, the Stockholm Convention, negotiated under the support of the United Nations Environment Program (UNEP), established a globally binding framework initially targeting a group of POPs, informally called the "dirty dozen," for reduction and eventual elimination [3]. The dirty dozen include several organochlorine pesticides (aldrin, chlordane, dieldrin, endrin, heptachlor, hexachlorobenzene, mirex, toxaphene, and DDT), PCBs, and PCDD/Fs. Besides the pesticide use, HCB was also used for industrial processes. PCBs were widely applied for industrial fluids in electrical equipment. PCDD/Fs and some PCB congeners (i.e., dioxin-like PCBs or coplanar PCBs) are formed as unintentional by-products during incineration and other thermal processes. The Stockholm Convention also includes procedures for identifying and adding substances to the POPs list. In May 2009, nine other POPs, including several BDE congeners present in PBDE commercial mixtures, have been listed under the Stockholm Convention [4]. More recently, HBCD has also been listed to Annex A of the Stockholm Convention at the sixth Conference of Parties (COP 6) in 2013 [5].

Despite a limited number of studies, significant contamination by POPs in deep-sea fishes has been demonstrated in the early 1980s [6–9]. Besides, in recent years, POPs, including BFRs, have been found at appreciable concentrations in deep-sea organisms from various parts of the world [10–28]. Research efforts by various scientific groups in the last few decades have greatly increased our understanding on the global distribution of POPs. The world's oceans are thought to play an important role in the cycling and removal of POPs [29]. The oceans cover two-thirds of the Earth's surface and because of their large volume can contain a large inventory of POPs [30]. The sources of POPs to the marine environment are riverine transport, municipal and industrial discharges, continental runoff, and atmospheric deposition in open waters. Atmospheric deposition delivers a large proportion of the POPs present in the oceans, through various mechanisms such as diffusive air–water exchange of POPs and wet and dry deposition processes [31]. Once in the aquatic environment, POPs can be dissolved in the water phase or partition onto colloidal and suspended/settling particulate matter, incorporate into food webs, transfer to the deep waters with the sinking particles, and eventually deposit in bottom sediments. It should be noted as an important role in the global behavior of POPs that the bottom sediments of deep-sea can either act as a final sink or a reservoir for such persistent contaminants. Sediment-sorbed xenobiotics can be taken up by the epibenthic and infaunal biota as they feed. A limited desorption of POPs, independent from that mediated by the bottom currents at the superficial level of the sediment, also occurs in the interstitial water [32]. These features can lead to remobilization of sediment-bound POPs in the entire food web of the

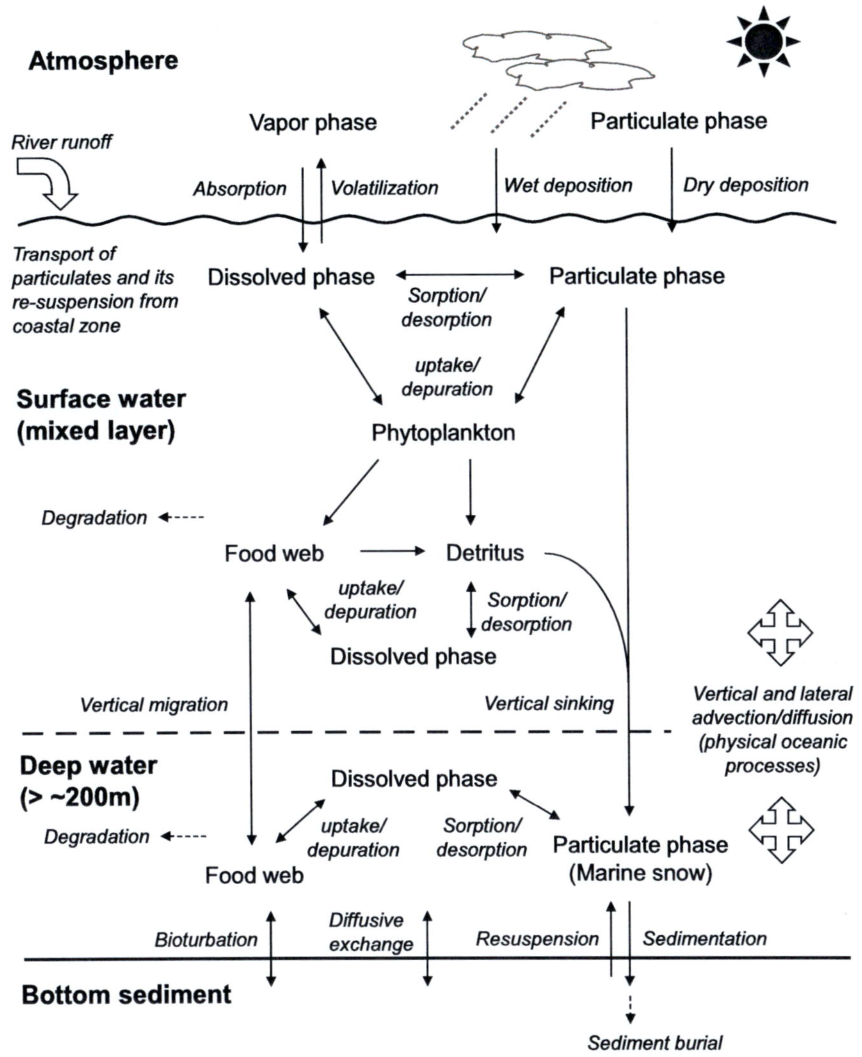

Fig. 2 Conceptual diagram showing the key processes affecting the transport of POPs between the atmosphere, water column, and bottom sediments

deep-sea benthos. Further, the fate of POPs are affected by the hydrodynamics of water masses such as turbulence and advection of water masses and the differential characteristics of coasts with respect to the open sea, i.e., enhanced stratification due to freshwater input from rivers, influence of tides etc. Figure 2 is a conceptual diagram showing the key processes affecting the transport of POPs between the atmosphere, water column, and bottom sediments. Depending on the physico-chemical properties of POPs (e.g., log K_{aw}), their global partition between environmental media and transport processes differ significantly among the compounds, congeners, or isomers [31, 33].

For the past four decades, our laboratory at the Center for Marine Environmental Studies (CMES), Ehime University, Japan, has been conducting research on the contamination status and spatial distribution, temporal trends, behavior, and fate of the above-listed contaminants in various matrices from different ecosystems (reviewed in Tanabe [34]; Tanabe et al. [35]; Tanabe and Ramu [36]). In the following chapter, we review some of our studies dealing with POPs in deep-sea organisms collected from various deep-sea environments. Although it is ideal to directly measure the levels of POPs in seawater to monitor the contamination status of POPs in marine ecosystems, there are limitations such as large volumes of seawater required for the analysis, sensitive analytical techniques crucial for detecting the low concentrations of POPs, and fluctuations of POP concentrations in seawater depending on the weather conditions. Therefore, in our studies, we often employ marine organisms like fish and shellfish as bioindicators to elucidate contamination status and spatial distribution of POPs in marine ecosystems. Aquatic organisms are very efficient in accumulating these contaminants, since, in addition to bioaccumulation through the diet, they are also subject to bioconcentration. Furthermore, data regarding levels and distributions of POPs in marine organisms, especially edible ones (and this includes an increasing number of deep-sea species), are important not only for assessing the state of the ecological environment but also from the human health perspective.

In the chapter, we focus mainly on two regions, the western North Pacific (WNP), off-Tohoku, Japan, and the East China Sea (ECS). The WNP is a region influenced by various water masses and currents making it one of the world's most biologically productive zones. The other region discussed in the chapter, the ECS, is an epicontinental sea with lots of continental inputs. Both the regions are characterized by the formation of numerous oceanic fronts. Belkin [37] defined oceanic fronts as a narrow zone of enhanced horizontal gradients of water properties (temperature, salinity, nutrients, etc.) that separates broader areas with different water masses or different vertical structure. These fronts are mostly characterized by strong mixing, stirring, enhanced bioproductivity, and ecotones. High productivities around the fronts may enhance flux of POPs into the oceans and their transport to deep waters because of phytoplankton uptake and the vertical flux of the particles play important roles in the biogeochemical cycles of POPs in the oceans [38, 39]. In general, land- and open-ocean-derived materials tend to converge at the frontal zones. The formation of coastal fronts has been suggested to be an important oceanic phenomenon determining the behavior and flux as well as the fate of persistent contaminants in the marine environment [40].

Detailed information on the sampling and analytical methods can be found in the cited papers [14, 41–44]. The reviewed studies may help improve our insights into the contamination status and the fate of legacy POPs and emerging contaminants such as BFRs in the deep-sea environments.

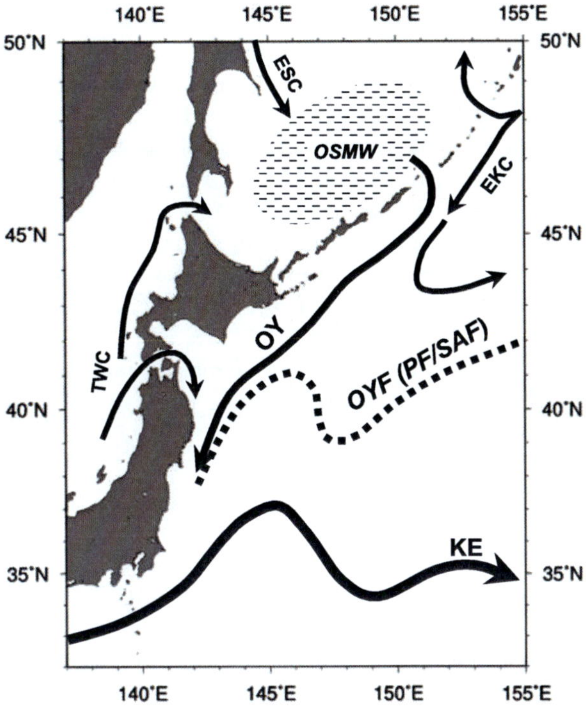

Fig. 3 A schematic circulation pattern of the major currents and frontal structures in the Kuroshio–Oyashio transition area, *EKC* East Kamchatka Current, *ESC* East Sakhalin Current, *OY* Oyashio, *TWC* Tsushima Warm Current, *KE* Kuroshio Extension, *OYF* Oyashio Front, *PF* Polar Front, *SAF* Subarctic Front, *OSMW* Okhotsk Sea Mode Water

2 Description of the Study Areas

2.1 The Western North Pacific, Off-Tohoku

The study area, the WNP, off-Tohoku, is characterized by a complex oceanographic structure represented by major ocean currents (Fig. 3). The Oyashio, a western boundary current of the Subarctic North Pacific, is a continuation of the East Kamchatka Current (EKC) and is fed by waters from the Western Subarctic Gyre and the Sea of Okhotsk [45]. The primary productivity of the Sea of Okhotsk is very high, especially on the continental shelf due to the relatively high insolation and the nutrient input from the Amur River and Pacific Ocean [46]. Recently, the dense shelf water flowing into Okhotsk Sea Mode Water (OSMW) has been considered as a potential source for exporting large amounts of organic matter and nutrients such as iron from the continental shelf to adjacent ocean interior along Oyashio [47, 48]. The eastward flowing Oyashio forms the Oyashio Front (OYF) which becomes Polar Front (PF) or Subarctic Front (SAF) to the east. The Kuroshio, a western boundary current from the southern tropic area, turns eastward from the eastern coast of Honshu, Japan. Then, warm and saline water is transported by this Kuroshio Extension [49]. The region between the Oyashio Front and the Kuroshio Extension Front is called the Kuroshio–Oyashio Transition Zone (KOTZ) or mixed water region, in which cold and warm waters mix and complex frontal structures are formed [49, 50].

Thus, the oceanography in the Tohoku area is complex and variable due to the confluence of various currents and water masses. Due to mixing of these water masses, the nutrient-rich subsurface water is brought into the euphotic zone promoting high phytoplankton production [51]. Marine biogeochemical activities including primary production, zooplankton grazing, microbial transformation, aggregation, and degradation processes of organic particles influence the flux of biogenic particles in the ocean. Vertical sinking of particle-associated pollutants such as PCBs may be enhanced by eutrophication since higher primary productivity leads to larger vertical fluxes of particles and organic carbon [38, 52]. This region known as Japan Trench is a depository of biogenic and lithogenic materials that are transported from the coastal shelf [53]. Thus, it can be assumed that a considerable proportion of anthropogenic contaminants may be transported into the deep water of North Pacific through the various biogeochemical processes.

2.2 *The East China Sea*

The ECS, located at midlatitudes between 25 and 35°N is an epicontinental sea surrounded by the Ryukyu archipelago, Japan, Korea, China, and Taiwan (Fig. 4). It has a broad continental shelf covering an area of 530×10^3 km^2 [54]. The major western boundary current, the Kuroshio, runs along the outer edge of the continental shelf, enters the ECS through the strait between Taiwan and the westernmost island of the Ryukyu Islands, flows northeastward along the shelf slope, and exits to the Philippine Sea after turning eastward near 30°N [55]. Two of the largest rivers in the world, the Yangtze River (Changjiang) and the Yellow River, discharge into the ECS. Thus, the cold, freshwater distributed on the continental shelf and the warm saline Kuroshio water that occupies the area around the shelf water lead to the formation of salinity front near the continental shelf break.

The Kuroshio, the Yangtze River runoff, and the East Asia monsoons are the dominant factors affecting the circulation in the ECS. The ECS continental shelf circulation pattern is characterized by the Kuroshio, Tsushima Current, Taiwan Warm Current, and other coastal waters and shelf fronts [56, 57] (Fig. 4). It has been demonstrated that the Kuroshio strongly influences not only the circulation in the ECS shelf but also its chemistry through water-mass exchanges [56–58].

Riverine runoff is an important mode to transport anthropogenic pollutants from terrestrial sources to adjacent oceans. The Yangtze River flows through densely populated areas with agriculture and industrial activities along both the banks, and Shanghai, the largest city in China, is situated at its mouth. Discharge of industrial wastes, application of fertilizers, pesticides, and herbicides in farming, as well as heavy metal pollution are said to make the Yangtze River one of the most polluted rivers in the world [59]. Li and Daler [60] reported that the environmental pollution of the Yangtze River basin greatly influences the state of the marine environment of the ECS. Consequently, the ECS has attracted much interest as a site for the study of the fate of terrestrial material in the marine environment [61]. The ECS shelf receives a rich supply of nutrients from the Yangtze River and the upwelled Kuroshio subsurface waters. As a result, the ECS shelf is one of the most productive

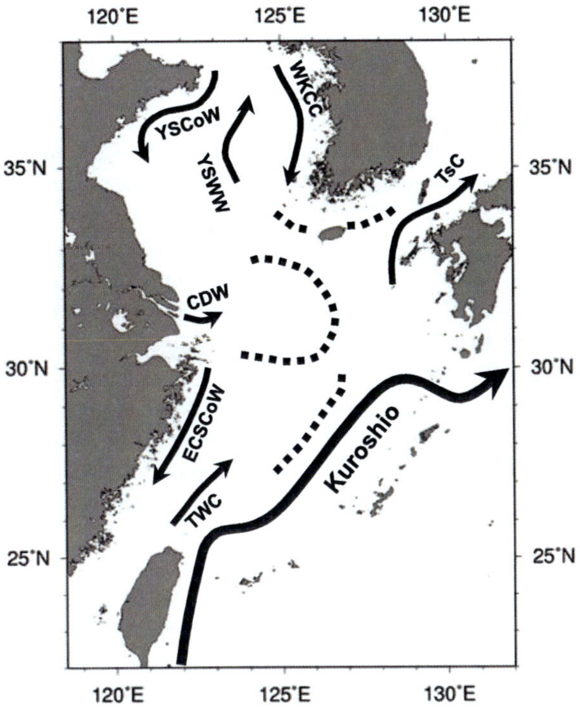

Fig. 4 A schematic circulation pattern of the major currents and frontal structures in the East China Sea. *TsC* Tsushima Current, *TWC* Taiwan Warm Current, *WKCC* Western Korea Cold Current, *CDW* Changjiang Diluted Water, *ECSCoW* East China Sea Coastal Water, *YSCoW* Yellow Sea Coastal Water, *YSWW* Yellow Sea Warm Water. *Dashed lines* indicate major shelf fronts in the East China Sea

marginal seas in the world [62]. Thus, it is obvious that the ECS receives enormous amounts of anthropogenic pollutants, suspended matter, and nutrients with the riverine runoff. Furthermore, the intensive exchange between the shelf water and the Kuroshio and the high primary productivity observed in this region may facilitate the flux of persistent contaminants to the bottom of the ECS.

3 POPs and Related Compounds in the Western North Pacific, Off-Tohoku

This section provides a synopsis of our studies conducted in the WNP, off-Tohoku [14, 41, 44], with the objective of understanding the distribution and the specific accumulation characteristics of organohalogen and butyltin compounds in deep-sea organisms in this region.

3.1 Contamination Status

Organochlorine compounds (OCs) were detected in all the deep-sea organisms (e.g., deep-sea eels, grenadiers, cods, eelpouts, sculpins, bikumins, flounders, myctophids, lantern sharks, ratfishes, squids, octopus, shrimps, crabs, gastropods) collected during three different time periods (see the above references for more details). The concentrations of PCBs and DDTs were the highest among the OCs analyzed, and the concentrations of other compounds were in the following order CHLs > HCHs ≥ HCB. The predominant accumulation of PCBs and DDTs in deep-sea organisms collected from the WNP, off-Tohoku, agreed with those reported in shallow- and deep-water fishes collected from other locations along the Pacific coast of Japan [12, 15] (Table 1). This reflects the higher bioaccumulative properties of PCBs and DDTs in marine food web as well as their significant usage.

Except for some higher trophic level organisms like snubnosed eels (*Simenchelys parasitica*), which contained some OCs at extremely high concentrations, the concentrations of OCs in deep-sea fish collected from the WNP, off-Tohoku, were generally lower than those in deep-sea organisms collected from other locations in the Atlantic Ocean, Mediterranean Sea, Ireland and Norwegian coasts, and the Arctic region [10, 11, 17–21, 25–27] (Table 2). The relatively low contamination by OCs in the present study area might be due to the smaller usage of these compounds in the WNP region compared to that of the North American and European countries. Besides, factors like variations in analytical methods and fish species (having different biological and ecological characters) may also influence the differences in the OC levels compared between the studies.

Comparing with studies carried out in adjoining areas on deep-sea fishes by our research group (Table 1), the concentrations of PCBs, DDTs, CHLs, and HCB in deep-sea fishes collected from the WNP, off-Tohoku, were comparable or lower than the data so far reported from Suruga and Tosa Bays [12, 15] and the ECS [42]. On the other hand, concentrations of HCHs in deep-sea fishes from this region were higher than those from other locations along the warm Kuroshio Current such as Tosa Bay and the ECS (Table 1). Higher concentrations of HCHs in cold waters along the Oyashio Current than in other offshore waters around Japan were also observed in a study using skipjack tuna (*Katsuwonus pelamis*) [63]. Due to the high vapor pressure, HCHs are known to rapidly evaporate and be transported from their pollution sources in the tropics and temperate regions to colder regions via the atmosphere [2, 30, 64]. The distribution patterns of HCHs found in the deep-sea organisms reflect the highly transportable nature of HCHs and its accumulation in the cold-water current of the WNP. No significant difference in the contamination status of HCHs between three research periods from 1994 to 2005 in this region (Table 1) is also suggestive of continuous flux of HCHs into the cold waters of the WNP region.

Among the three studies reviewed here, only the study by Takahashi et al. [44] reported the concentrations of PBDEs and HBCDs in deep-sea fishes collected from the western North Pacific, off-Tohoku, in 2005. The concentrations of PBDEs ranged from 1.3 to 8.5 ng/g with a mean of 3.6 ng/g lipid wt, while the concentrations of HBCDs ranged from 5.4 to 45 ng/g with a mean of 22 ng/g lipid wt (except for snubnosed eels). Despite the low levels compared to the other POPs, the detection of PBDEs and HBCDs in deep-sea fishes indicates the widespread presence of such "emerging POPs" even in deep oceans and their long-range transport. Similar to the

Table 1 Comparison of mean and range concentrations of organochlorines (ng/g lipid wt) between fishes from the western North Pacific and East China Sea

Location	Sampling year	PCBs	DDTs	CHLs	HCHs	HCB	References
Off-Tohoku myctophids	1994	160 (20–370)	110 (15–280)	22 (5–47)	13 (3.1–24)	4.0 (0.79–8.0)	[14]
Off-Tohoku deep-sea fish[a]	1995	420 (n.d. to 2,200)	230 (14–830)	110 (3.9–640)	30 (n.d. to 150)	16 (n.d. to 100)	[41]
Off-Tohoku deep-sea fish[a]	2005	150 (34–390)	110 (36–220)	54 (19–120)	25 (5–69)	21 (7–45)	[44]
Tosa Bay shallow-water fish	1997	310 (n.d. to 1,100)	220 (38–1,200)	46 (n.d. to 200)	5.3 (n.d. to 14)	5.7 (n.d. to 13)	[15]
Tosa Bay deep-sea fish	1997	340 (n.d. to 1,600)	290 (7.1–1,200)	44 (5.6–220)	11 (n.d. to 21)	11 (n.d. to 61)	[15]
Suruga Bay shallow-water fish	1993–1994	1,600 (540–2,600)	390 (80–1,700)	140 (46–320)	4.1 (n.d. to 8.0)	8.8 (3.7–24)	[12]
Suruga Bay deep-sea fish	1993–1994	1,000 (450–1,900)	390 (51–910)	260 (69–770)	25 (10–37)	20 (9.9–50)	[12]
East China Sea shallow-water fish	2001–2003	150 (20–830)	330 (110–1,200)	39 (5.9–180)	11 (2.2–62)	17 (n.d. to 70)	[42]
East China Sea deep-sea fish	2001–2003	230 (36–1,400)	720 (n.d. to 7,900)	40 (3.7–240)	11 (1.4–41)	18 (1.7–290)	[42]

Figures in parentheses indicate the range of concentrations

n.d. not detected (for calculation of mean, data less than detection limits was assumed to be 0)

[a]Data without snubnosed eel

Table 2 Comparison of mean concentrations of PCBs and DDTs in deep-sea fishes from various parts of the world

Species	Sampling year	Location	Depth (m)	PCBs (ng/g) wet wt	PCBs (ng/g) lipid wt	DDTs (ng/g) wet wt	DDTs (ng/g) lipid wt	References
Cartilaginous fish (2 sp.)	1992	Davis Strait (Greenland)	800–2,200	310	430	490	690	[10]
Bony fish (6 sp.)	1992	Davis Strait (Greenland)	200–2,100	300	790	250	670	[10]
Fish (6 sp.)	1993–1994	Suruga Bay, Japan	200–740	160	1,000	60	390	[12]
Cartilaginous fish (1sp.)	1992	Norway (Nordfjord)	400	1,800	2,400	4,600	6,000	[11]
Bony fish (3 sp.)	1992	Norway (Nordfjord)	400	3,900	8,700	8,800	20,000	[17, 18]
Armed grenadier	1992	North Atlantic	2,900	–	2,100	–	1,090	[17, 18]
Black halibut	1994–1998	South Atlantic	–	–	1,000	–	390	[17, 18]
Kingklip	1994–1998	South Atlantic	–	–	1,400	–	190	[17, 18]
Dover sole	1995	Monterey Bay Canyon	–	–	2,200	–	2,400	[17, 18]
Longspine thornyhead	1995	Monterey Bay Canyon	–	–	3,700	–	2,400	[17, 18]
Snubnosed eel	1995	Off-Tohoku, Japan	1,000	1,100	6,700	2,100	13,000	[41]
Fish (21 sp.)	1995	Off-Tohoku, Japan	150–1,300	81	400	30	220	[41]
Gadiform (*Mora moro*)	1996	Gulf of Lions	986–1,136	3,400	9,900	1,200	4,300	[19]
Fish (20 sp.)	1997–1998	Tosa Bay, Japan	150–400	39	350	31	290	[15]
Roundnose grenadier	1999	West of Ireland	1,000	450	770	730	1,300	[20]
Roundnose grenadier	1999	West of Ireland	2,000	800	1,900	1,000	2,400	[20]
Skate (3 sp.)	2000	Southern Adriatic Sea	–	310	890	–	–	[21]
Ghost shark	2000	Southern Adriatic Sea	–	280	390	–	–	[21]
Fish (37 sp.)	2001–2003	East China Sea	89–512	6.2	230	30	700	[42]
Fish (10 sp.)	2002	Sulu Sea	292–1,015	0.59	58	1.8	150	[22]
Hollowsnout grenadier	2003	Adriatic Sea	–	–	430	–	340	[25]

Roughtip grenadier	2003	Adriatic Sea	–	–	310	–	5,400	[25]
Hollowsnout grenadier	2006	Southern Adriatic Sea	–	–	1,200	–	760	[26]
Roughsnout grenadier	2006	Southern Adriatic Sea	–	–	12,000	–	5,400	[26]
Roundnose grenadier	2000–2002	Porcupine Seabight	1,000–1,900	–	2,400[a]	–	1,500[a]	[27]
Snubnosed eel	2005	Off-Tohoku, Japan	900	720	6,300	1,300	12,000	[44]
Fish (11 sp.)		Off-Tohoku, Japan	400–1,000	12	150	8.2	100	[44]

–: not available
[a]Median concentration
For calculation of mean, data less than detection limits was assumed to be 0
For comparison, all the data were rounded into two significant digits

result above, higher concentrations of HBCDs than PBDEs were observed in skipjack tuna collected from offshore waters around northern Japan [65, 66]. On the other hand, accumulation of PBDEs at higher concentrations than HBCDs was reported in fishes and cetaceans from East and South China Seas [35, 43]. Comparing to the concentration levels of PBDEs (5.1 and 19.9 ng/g lipid wt at mean concentrations) and HBCDs (<1.5 ng/g lipid wt) in two species of deep-sea grenadiers from the Mediterranean Sea [23], relatively high magnitude of contamination by HBCDs was noted in our data on deep-sea fish from the western North Pacific, off-Tohoku. Such differences among the studies and locations may reflect possible long-range transport of HBCDs to cold offshore waters, like HCHs, and/or significant usage of HBCDs (especially for thermal insulating materials) in northern Japan. Although HBCD is currently being proposed to be listed a POP in the Stockholm Convention, the contamination by this BFR in marine ecosystems around Japan may increase or continue in the meantime as its use or disposal, considering the increase in market demand for HBCD in Japan during late 1990s [67].

Concentrations of butyltin compounds (BTs) were reported by Takahashi et al. [14] and de Brito et al. [41]. Takahashi et al. [14] detected BTs in almost all myctophids analyzed (levels up to 46 ng/g wet wt for total BTs (∑BT: MBT + DBT + TBT)). Widespread contamination by BTs in deep-sea organisms from the western North Pacific, off-Tohoku (levels up to 530 ng/g wet wt in the liver of demon grenadier, *Coelorinchus gilberti*, for ∑BTs), was also reported by de Brito et al. [41]. Although the magnitude of contamination by BTs in open-water ecosystems of the western North Pacific seems to be relatively low, the results indicate the expansion of BT pollution in deep-sea ecosystems.

3.2 Species-Specific Accumulation and Trophic Magnification

Among various fish species analyzed, snubnosed eel (Fig. 5) accumulated hydrophobic organohalogens (with log K_{ow} ~ 6–8) such as PCBs, DDTs, CHLs, and PBDEs at extremely high concentrations (Table 2). This species can be considered as an organism at a high trophic level in the food web due to its strong carnivorous and unique feeding habit; this fish has been found to burrow into the body of large fish such as halibut and shark and feed on their intestine and blood [68]. Relatively higher concentrations of these compounds were also found in some large carnivore and/or bottom-dwelling fishes. These observations suggest that accumulation of the hydrophobic OC and PBDEs is influenced by trophic magnification and/or feeding habits of organisms and their long biological half-lives, i.e., higher uptake rate via food and slower equilibrium between ambient water and body lipids. In contrast, HCHs, HCB, and HBCDs, which have less lipophilicity than the other organohalogen compounds, showed smaller variation in concentrations among species of different trophic levels and feeding habits.

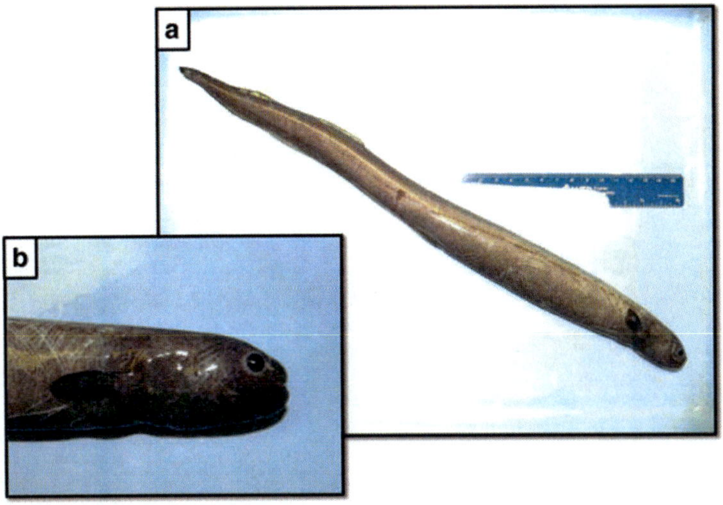

Fig. 5 Snubnosed eel (*Simenchelys parasitica*) (Photo by Dr. T. Kubodera, National Museum of Nature and Science, Tokyo, with permission). (**a**) Whole body; (**b**) head part

Isotopic signatures based on the ratios of different stable isotopes of elements have been broadly applied to investigate a variety of processes. In particular, stable carbon and nitrogen isotope ratios ($\delta^{13}C$ and $\delta^{15}N$) have been used in environmental studies for delineating carbon flows and trophic relationships in continental and marine ecosystems. The stable isotope ratios of nitrogen $\delta^{15}N$ reflect diets and represents time-integrated averages of assimilated foods [69]; thus, stable nitrogen isotope analysis has been used to assess the relative trophic level of aquatic biota and provide a quantitative continuous variable for studying the trophic magnification of contaminants within complex food webs [70, 71]. In contrast, the ratios of stable carbon isotopes $\delta^{13}C$, in biota, can help to elucidate trophic interactions by establishing the relative contribution of marine (or pelagic) versus coastal (or benthic) carbon sources [72]. Thus, the quantification of both stable carbon and nitrogen isotope ratios can provide valuable information on the feeding ecology of biota and its potential influence on the trophic enrichment of contaminants [70, 73].

To understand the trophic magnification of organohalogen compounds, $\delta^{15}N$ and $\delta^{13}C$ were determined in fishes collected in 2005. Their correlations to the concentrations of organohalogen compounds were discussed [44]. A wide range of $\delta^{13}C$ values among fishes was observed, implying the existence of multiple carbon sources in the food web of this region. Significant positive correlations were found between $\delta^{15}N$ (‰) and lipid normalized concentrations of PCBs, DDTs, and PBDEs ($p < 0.05$) showing their high trophic magnification potential in marine food webs (Fig. 6). No such relationship was found for HCB, HCHs, and HBCDs (Fig. 6). The rapid elimination through gills to surrounding water and by excretion via other means may prevent the trophic magnification of these compounds. In addition, less trophic magnification observed for some organohalogen

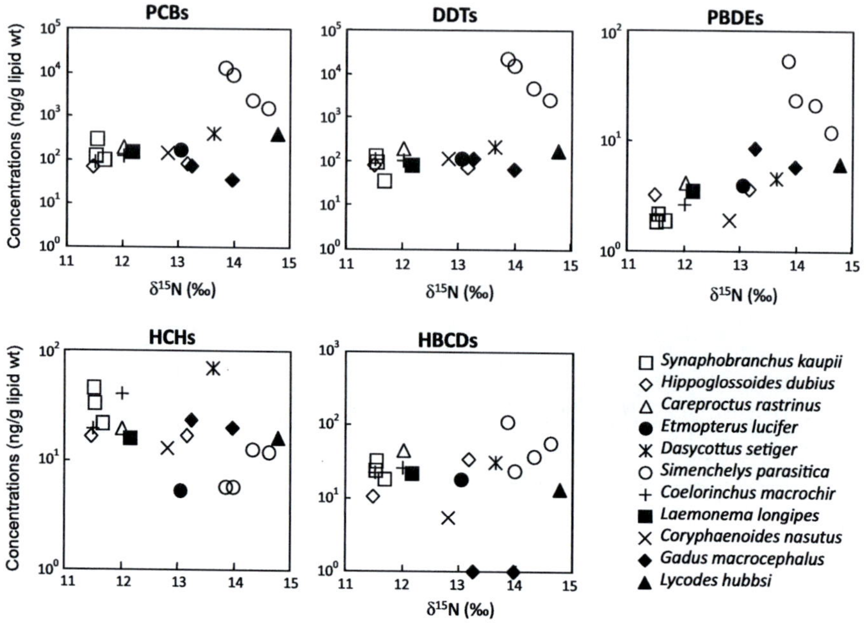

Fig. 6 Relationship between $\delta^{15}N$ (‰) and organohalogen compounds in deep-sea fishes from the western North Pacific, off-Tohoku, Japan

compounds in this study may be due to the smaller difference in trophic levels of fishes. In fact, there were no significant correlations between $\delta^{15}N$ values and the concentrations of almost all the compounds analyzed when the data of snubnosed eels was excluded (only a significant relationship for PBDEs with $p < 0.05$). Thus, the accumulation profiles of organohalogen compounds in deep-sea fishes reflect not only their trophic transfer but also the different sources of the contaminants in the food web and/or species-specific distribution of fishes. Especially, some bottom-dwelling fishes including snubnosed eels showed significant lower $\delta^{13}C$ values than the other fish species, reflecting influence of substances originated from sediments. Therefore, the data of high trophic bottom-dwelling fishes were excluded for further discussion on spatial and vertical distribution of contaminants.

A considerable variation was observed in the concentrations of BTs among tissues/organs and species. Among the tissues/organs analyzed, higher concentrations of BTs were found in the liver of fish (97 ± 212 ng ∑BT/g wet wt) and viscera of crabs (17 ± 17 ng ∑BT/g wet wt) [41]. Nevertheless, in the case of BTs, relatively high concentrations were observed not only in larger carnivore fishes (e.g., 9.7 ng ∑BT/g wet wt in the whole body of Pacific grenadier) but also in small fishes (up to 46 ng ∑BT/g wet wt in the whole body of myctophid fishes) and squids (up to 44 ng ∑BT/g wet wt in the whole body of Japanese hooked squid) [14, 41]. It is evident from previous studies that the bioaccumulation properties of BTs (i.e., organometallic compounds) were principally different from organohalogen compounds having hydrophobic and/or lipophilic nature. Less biomagnification of BTs was also documented in some food webs along Japanese coasts [74, 75].

3.3 Distribution in Relation to the Water-Mass Structure

Even among fishes showing similar trophic levels, considerable variation in the concentrations of POPs and related compounds was observed, which cannot be explained only by their feeding habits and trophic magnification. It has been documented that fishes can be used as suitable bioindicators to elucidate the environmental levels and distributions of persistent and lipophilic compounds in water bodies after careful consideration of their biological and ecological parameters [76].

Takahashi et al. [14] studied the accumulation profiles of OCs and BTs in relation to the sampling locations and diel vertical migration modes of myctophid species to delineate the spatial and vertical distribution of the contaminants in the WNP. While concentrations of OCs and BTs varied among species, no significant difference in the concentrations of fishes were observed between sampling locations (in the area: 38–41°N, 143–147°E). On the other hand, the accumulation patterns of contaminants showed a specific trend in accordance with the migration types (Fig. 7). Except for a migrant species *D. theta*, higher concentrations of PCBs, DDTs, and CHLs were found in non- or less-migratory species living in deeper waters. On the contrary, HCHs, HCB, and BTs accumulated at higher concentrations in migratory species as compared to nonmigratory fishes. A similar trend can also be suggested from mean concentrations of OCs and BTs in deep-sea fishes collected from the WNP, off-Tohoku [41] (Fig. 8). In general (except for snubnosed eels), higher concentrations of PCBs, DDTs, and CHLs were found in fishes from deeper waters (~1,000 m). On the contrary, HCHs, HCB, and BTs accumulated at higher concentrations in fishes from shallower waters.

These observations suggest the vertical distribution of these contaminants in the water column of this region. Tanabe and Tatsukawa [77] showed that the concentrations of HCHs significantly decreased with depth in the open-water column of the Pacific Ocean because of their volatility and high atmospheric mobility. Long-range atmospheric transport and cold condensation of volatile POPs, as noted in the above section, may enhance their flux to the surface waters in this WNP region. On the other hand, results of these studies are also suggestive of the vertical transport of hydrophobic OCs and their accumulation in benthic deep-sea organisms. OCs with lower water solubility and higher affinity for adsorption to particles would be transported to deeper waters and incorporated into the marine food web more readily than more water-soluble compounds. With regard to BTs, higher concentrations were found in shallow-water than deep-sea fishes. Recent input of BTs from the usage in surface waters (e.g., antifouling) could be well over the vertical transportation rate of these compounds, which results in higher levels in shallow waters.

In addition, the structure of the water mass around the WNP should be considered in evaluating the vertical profiles of contaminants. As noted in the description of the studied area, it has been suggested that the North Pacific Intermediate Water

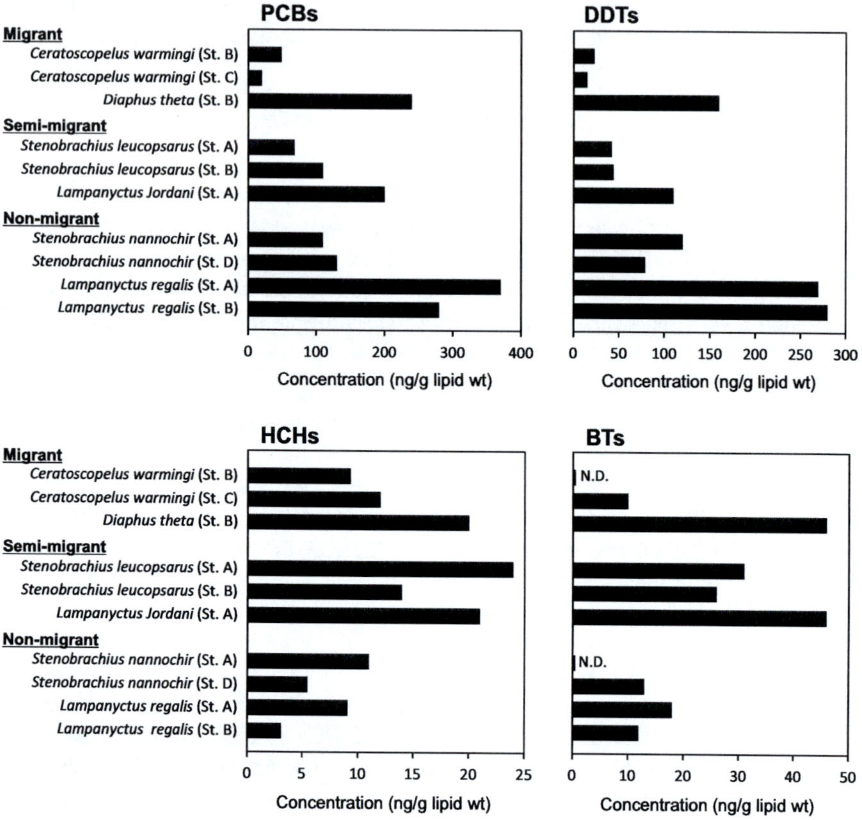

Fig. 7 Concentrations and residue patterns of PCBs, DDTs, HCHs, and BTs in myctophid fishes from the western North Pacific, off-Tohoku, with different migration types. Each *bar* indicates the concentration in a composite sample which was prepared from pooled and homogenized individual specimen belonging to the same species collected from the same sampling location [14]

(NPIW), which is found at depths of 300–800 m in the North Pacific subtropical gyre, originates from the water mass of OSMW, and new NPIW is formed along the Kuroshio–Oyashio interfrontal zone [50]. The dense shelf water flowing into the OSMW has been considered as a potential source for exporting large amounts of organic matter and nutrients such as iron from the continental shelf to adjacent ocean interior along Oyashio. This means that organic pollutants as well as the shelf-originated nutrients can be transported from the continental shelf to NPIW. Despite limited data available, elevated background levels of OCs such as DDTs and PCBs in the Okhotsk Sea and East Siberia regions have been suggested by several studies on air and water [78, 79] and wild animals such as seals and sea eagles [80, 81]. The sampling depths and locations in the study of Takahashi et al. [14] and de Brito et al. [41] overlapped with the area where an intrusion of water mass from the Okhotsk Sea into the mesopelagic zone and the formation of

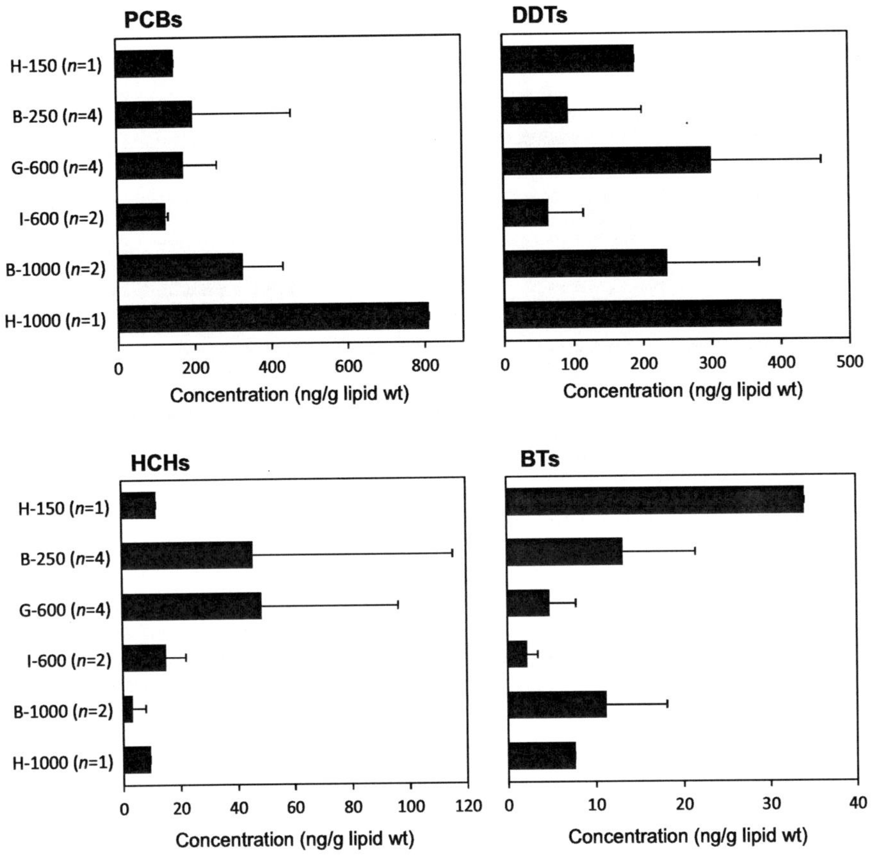

Fig. 8 Mean concentrations and residue patterns of PCBs, DDTs, HCHs, and BTs in deep-sea fishes (except for some high trophic benthic fishes) from the western North Pacific, off-Tohoku, with different sampling depth (*error bars* show standard deviations)

NPIW occur. In this context, the intermediate water in the study area can be expected to have higher concentrations of OCs such as PCBs and DDTs that originate from more contaminated Okhotsk Sea.

3.4 Compositions and Temporal Trends of Organohalogen Compounds

Influence of the water-mass structure and material transport around the Okhotsk Sea and Oyashio current on the vertical profiles of OCs can be also found in the compositions of DDTs in deep-sea fishes from the WNP, off-Tohoku. Among the DDT compounds, *p,p'*-DDE and *p,p'*-DDT were the predominant constituents in

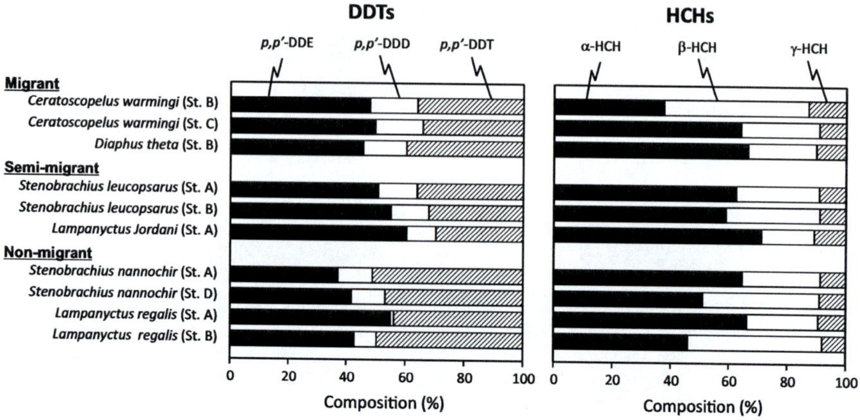

Fig. 9 Compositions of DDT compounds and HCH isomers in myctophid fishes from the western North Pacific, off-Tohoku. Each *bar* indicates the composition in a composite sample which was prepared from pooled and homogenized individual specimen belonging to the same species collected from the same sampling location [14]

almost all the deep-sea fishes (Figs. 9 and 10) [14, 44]. Similarly high proportions of *p,p'*-DDT (> 40%) were also found in Japanese common squid (*Todarodes pacificus*) from off-Tohoku, Japan, although the location is far away from any possible sources of DDT [82]. Such DDT composition (i.e., high proportions of a parent compound, *p,p'*-DDT) observed in fishes and squids from the WNP, off-Tohoku, is apparently different from those reported in previous studies from Suruga and Tosa Bays [12, 15] and in other recent researches from the Mediterranean Sea [25, 26] and North Atlantic Ocean [20, 27], where *p,p'*-DDE, a stable degradation compound, was dominant among DDTs in almost all the fishes analyzed. In this context, the intermediate water in the study area can be expected to have relatively fresh input of DDTs. Particularly in myctophids, higher proportions of *p,p'*-DDT were observed in nonmigratory fishes than in those of migratory ones ($p < 0.05$, Mann–Whitney U-test) (Fig. 9). This agrees with the concentration profile of DDTs as noted above.

Among the HCHs isomers, α-HCH was the predominant isomer in all the deep-sea fishes in the NWP, off-Tohoku (Figs. 9 and 10). It has been reported that the ratio of α-HCH to total HCH concentrations in seawater and fish has a tendency to increase with the increasing latitude [30, 63]. α-HCH is preferentially transported to northern colder regions due to higher vapor pressure among the HCH isomers.

Regarding to the temporal change in the composition of OCs, the percentages of *p,p'*-DDT in myctophids collected in 1994 (~40%) was apparently lower than those in myctophids collected from Yaizu, Suruga Bay, in 1976 (58%) [83]. This suggests the reduction of fresh input of DDTs in the WNP during the last few decades. The decreasing *p,p'*-DDT proportion in total DDT compounds has also been reported in the study on temporal trend of OCs in northern fur seals from the Pacific Coast of Japan [84]. In contrast to the composition of DDT compounds, no significant

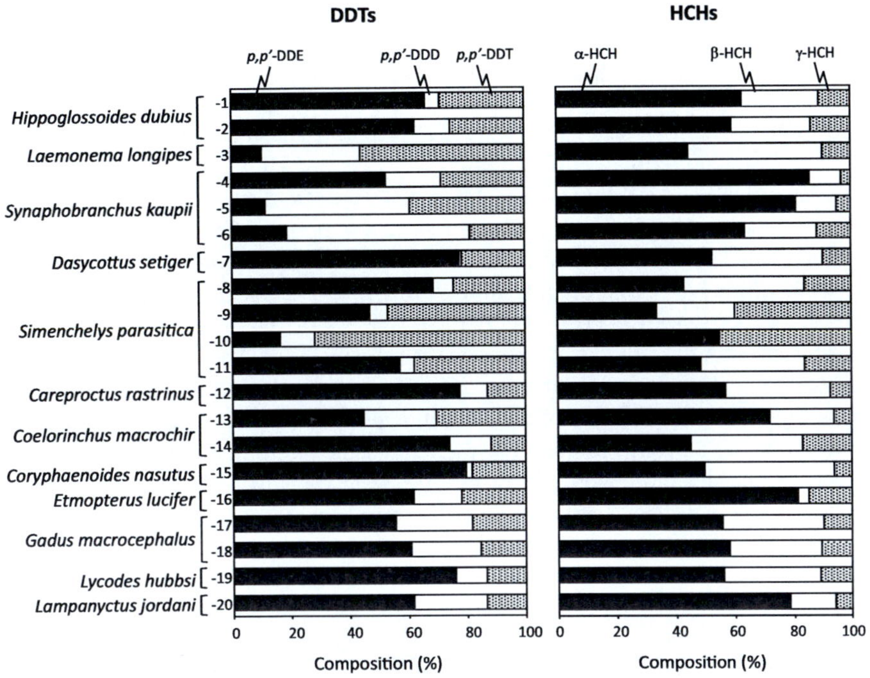

Fig. 10 Compositions of DDT compounds and HCH isomers in various deep-sea fishes from the western North Pacific, off-Tohoku

difference in the proportion of α-HCH (~60%) was observed between the myctophids collected in 1976 and 1994. Such a small variation in the temporal trend of HCH isomer composition has also been reported in the northern fur seals [84]. It may be attributable to the global transport of HCHs to high-latitude regions and/or the water-mass inflow from the Okhotsk Sea. Further monitoring along the Kuroshio–Oyashio interfrontal zone including marginal seas such as the Okhotsk Sea with geophysical and biogeochemical studies is required to delineate the distributions and fate of anthropogenic contaminants.

4 POPs and Related Compounds in the East China Sea

In this section our recent studies in the ECS [42, 43] have been reviewed for understanding the distribution and the specific accumulation characteristics of organohalogen and butyltin compounds in various deep-sea organisms (e.g., deep-sea eels, sea perches, hairtails, argentines, grenadiers, splitfins, flounders, gurnards, myctophids, lantern sharks, dogfish sharks, skates, squids, shrimps, prawns, lobsters, crabs, sea anemones; see the above references for more details).

4.1 Contamination Status

Among the organohalogen compounds analyzed in deep-sea organisms from the ECS, the concentrations of DDTs were the highest and those of the other compounds were approximately in the order of PCBs > CHLs > HCB > HBCDs > PBDEs = HCHs [42, 43] (Table 3). Deep-sea fishes from the ECS had the highest concentrations of DDTs among the data reported so far from our group for the WNP region (Table 1). Rapid industrial development and population growth in coastal areas of China have resulted in significant environmental pollution and damage to the aquatic ecosystems. From the 1950s to the 1980s, DDT was widely used in agriculture in China before it was legally banned in 1983 [85]. In the last two decades, many agricultural lands in China have been developed for commercial uses, thus, accelerating the remobilization of previously buried insecticides/pesticides. Yuan et al. [86] reported that the large-scale usage of DDT in agricultural practices and the subsequent runoff into the waterways have resulted in the high levels of DDTs in the freshwater, estuarine, and marine environment of China. In the case of PCBs, the concentrations in deep-sea fishes from ECS were comparable to that of deep-sea fishes from Tosa Bay [15] and off-Tohoku, Japan [14, 41, 44], but were significantly lower than that of deep-sea fishes from Suruga Bay [12] and other locations of the world (Table 2). On the other hand, the concentrations of HCHs and HCB in deep-sea organisms from the ECS were about one or two orders of magnitude lower than those of DDTs and PCBs. HCH isomers are less lipophilic when compared to other OCs, and, thus, they have lower biomagnification factors in aquatic ecosystems [87]. Significant fluxes (i.e., volatilization) of more volatile POPs, HCHs and HCB, to the atmosphere in the waters of low-latitude regions [2, 30, 64] may also be attributable to their less contamination in the ECS along the Kuroshio Current.

Unlike PCBs and DDT, which are largely a legacy of the past, BFRs such as PBDEs and HBCDs have been banned from usage since late 1990s–2000s or even currently used in various electronic devices, furniture and textiles. Being additive flame retardants they can leak out of the treated materials during the life cycle of the product, causing a continuous contamination of the environment even after the regulations on their production/application were implemented. Total concentrations of PBDEs and HBCDs in various deep-sea organisms from the ECS ranged from 0.31 to 57 ng/g lipid wt. and 0.15 to 210 ng/g lipid wt., respectively (Table 3). Reports on PBDEs and HBCDs in marine ecosystems and food webs, particularly for the deep-sea environment, are relatively scarce. The concentrations of PBDEs in deep-sea fishes from the ECS were higher than those from the WNP, off-Tohoku [44]. The detection of PBDEs and HBCDs in deep-sea organisms from the ECS indicates that these compounds are also transportable to the deep oceans, similar to that of other POPs. The sources of these BFRs in the ECS could be from the various manufacturing operations as well as e-waste recycling activities along the Chinese coast [88]. The recycling of e-wastes could mobilize these BFRs from the electronic components into the environment.

Table 3 Mean and range concentrations of organohalogen compounds (ng/g lipid wt) in deep-sea organisms from the East China Sea

Species	*n*	Fat (%)	PCBs	DDTs	CHLs	HCHs	HCB	PBDEs	HBCDs
Fishes	30	2.5 (0.48–9)	290 (36–1,400)	560 (<3.0–2,200)	51 (8.4–91)	11 (2.2–41)	23.0 (<1.5–290)	10.0 (0.31–40)	4.0 (0.25–20)
Cephalopods	3	2.1 (0.69–4.5)	120 (29–190)	220 (76–410)	39 (3.9–640)	5.7 (3.7–7.3)	8.5 (3.2–14)	11 (0.65–26)	71 (0.18–210)
Crustaceans	9	3.1 (0.47–12)	230 (25–850)	610 (21–2,500)	40 (7.5–140)	14 (4.5–28)	7.5 (0.55–17)	16 (1.1–57)	16 (0.15–73)
Mollusk	1	0.23	220	300	23	33	17	4.7	3.1
Echinoderm	1	0.041	590	72	1.9	3.4	<8.8	28	11

Figures in parentheses indicate the range of concentrations

n: number of species

For calculation of mean, data less than detection limits was assumed to be 0

The contamination status of organohalogen compounds in the ECS is probably affected by the various hydrological processes, notably the Yangtze River discharge and the Kuroshio Current. The Yangtze River flowing through big cities and important agricultural areas with an annual discharge of more than 9.0×10^{11} m^3 [89] is a significant source of anthropogenic pollutants from the terrestrial environment to the ECS. On the other hand, the Kuroshio, flowing to the northeast along the continental slope, is a unique source of oceanic materials of open-ocean origin to the shallow continental shelf [90].

Concentrations of butyltin compounds (BTs) and other organotin compounds including triphenyltin (TPT) were reported by Tanabe et al. [42]. BTs and TPT were detected in almost all deep-sea organisms analyzed (levels up to 62 ng/g wet wt in blackbelly lantern shark, *Etmopterus lucifer*, for ∑BT, and 40 ng/g wet wt in garden eel, *Conger myriaster*, for TPT). Although the magnitude of contamination by BTs in ECS seems to be relatively low comparing with that of the western North Pacific, off-Tohoku, the results indicate widespread contamination by organotin compounds including TPT in various deep-sea ecosystems around the East Asian region.

4.2 Species-Specific Accumulation and Composition of Organohalogen Compounds

The concentrations of most of the organohalogen compounds in deep-sea organisms from the ECS varied considerably between the various species analyzed. Factors like food web characteristics, habitat, and feeding strategies can contribute to the exposure of deep-sea organisms to organohalogen compounds. Higher concentrations of organohalogen compounds were found in fishes like garden eel (*Conger myriaster*), Japanese split fin (*Synagrops japonicus*), lantern shark (*Etmopterus lucifer*), dogfish shark (*Squalus japonicus*), sharpnose sevengill shark (*Heptranchias perlo*), and green eyes (*Chlorophthalmus acutifrons*). All these species are deep-sea species and feed at high trophic levels of the food web. Thus, the species-specific differences in the accumulation patterns of the organohalogen compounds observed are due to biomagnification and/or feeding habits of the organisms. In general, HCHs and HCB, which have lower molecular weights and are less lipophilic than the other organohalogen compounds, exhibited smaller variation in concentrations among species and trophic levels owing to their faster rates of attaining equilibrium with water concentrations [83, 91]. However, unicorn grenadier (*Caelorinchus productus*) showed very high levels of HCB (up to 290 ng/g lipid wt.). The other grenadier species from the ECS also showed elevated levels of HCB. High levels of HCB were also found in the same species from Tosa Bay, Japan [15], and off-Tohoku, Japan [41]. So far, it is not clear why grenadier species accumulate such high levels of HCB.

Among the DDT compounds, *p,p'*-DDE or *p,p'*-DDT were the predominant constituents in almost all the deep-sea organisms. The percentages of *p,p'*-DDT in shallow-water and deep-sea fishes, 37% and 40%, respectively, were higher than in

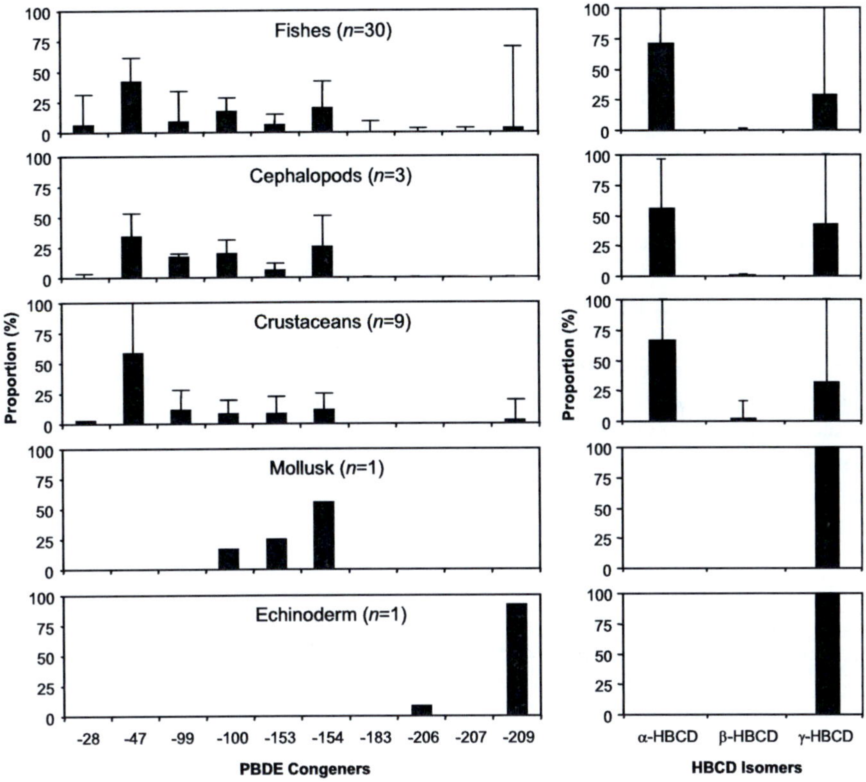

Fig. 11 Mean proportions of PBDEs and HBCDs in deep-sea organisms from the East China Sea (*error bars* show standard deviations)

the deep-sea fishes from Suruga Bay and Tosa Bay [12, 15] and other locations of the world such as the Mediterranean Sea [25, 26] and North Atlantic Ocean [20, 27], indicating that there is fresh input of DDT from China or other Southeast Asian countries to the ECS. Higher percentages of *p,p'*-DDT were also found in skipjack tuna (*Katsuwonus pelamis*) from the ECS [63]. In case of HCH composition, β-HCH was the dominant isomer in all the deep-sea organisms, and this may be because of its persistent nature. α-HCH which is the major isomer in technical HCH mixture was detected in most of the samples, but the percentage composition of α-HCH was far less when compared to deep-sea fishes from the WNP, off-Tohoku [14, 41, 44]. In spite of the usage of technical HCH mixture in China and nearby countries, the low levels of α-HCH may be due to the rapid volatilization of α-HCH in the lower latitudes [30].

Of the fourteen PBDE congeners analyzed, a total of ten congeners from tri- to deca-BDE were identified in the deep-sea organisms from the ECS. The congener profiles of PBDEs varied between the species and, in most cases, the congeners from tetra- to hexa-BDEs were predominant (Fig. 11). The higher brominated congeners from octa- to deca-BDE were detected in very few samples, suggesting the relatively low bioaccumulation potential as well as the low bioavailability of these congeners. Congener-specific biotransformation of PBDEs undoubtedly plays an important role in the bioaccumulation behavior of PBDEs. BDE 209 was occasionally present at concentrations above its detection limit. It is noteworthy that BDE 209 was the dominant congener in case of an echinoderm (*Holothuroidea* sp.) and in flatfish (*Chascanopsetta lugubris*), and this can be explained by their sediment-dwelling nature. BDE 209 is mainly associated with sediments and is a major congener in suspended particles and sediments and related environments. The low detection of BDE 209 in other deep-sea organisms could be due to either its large molecular size which lowers its uptake rates or its relatively rapid biotransformation [92]. However, the presence of BDE 209 in deep-sea organisms from the ECS indicates that this congener probably is capable of long-range transport and dispersal. Among the HBCD isomers, α- and γ-HBCDs were detected in all the samples, while β-isomer was below detection limits in most of the samples (Fig. 11). Similar to BDE 209, higher proportion of γ-HBCD was observed in sediment-dwelling organisms.

4.3 Trophic Magnification and Sources of Contaminants in the Food Web

As noted above in the section for the WNP, the quantification of both stable carbon and nitrogen isotope ratios can provide valuable information on the feeding ecology of biota and its potential influence on the trophic enrichment of contaminants. [43] examined $\delta^{15}N$ and $\delta^{13}C$ in deep-sea organisms from the ECS to assess trophic magnification of organohalogen contaminants in the food web of the ECS. The $\delta^{15}N$ values in the deep-sea organisms from the ECS ranged from 7.18‰ to 14.17‰. No clear relationships were observed between $\delta^{15}N$ (‰) and concentrations (both in wet and lipid wt) for all the contaminants. In the aquatic environment, the uptake of contaminants by biota usually includes direct partitioning from the surrounding environment and ingestion of contaminated dietary sources, whereas the loss processes include direct partitioning, biotransformation, growth dilution, maternal transfer to offspring, and egestion. Besides, intensive biological N_2 fixation by cyanobacterium, *Trichodesmium*, which plays an important role of primary production in the ECS, can affect variation of $\delta^{15}N$ in individual organisms and entire food web [93]. The large difference in nutrient and organic matter sources between the sampled locations and specimens can be plausible reasons for the lack of relationship between $\delta^{15}N$ (‰) and concentrations of organohalogen compounds in the ECS food web, as will be explained later.

Generally, the terrigenous organic matter is depleted in ^{13}C, while the marine organic matter is usually enriched in ^{13}C [94, 95]. Since the $\delta^{13}C$ of marine planktonic sources is different from those of most terrestrial organic material sources, the carbon isotopic composition can be used as a tracer to distinguish

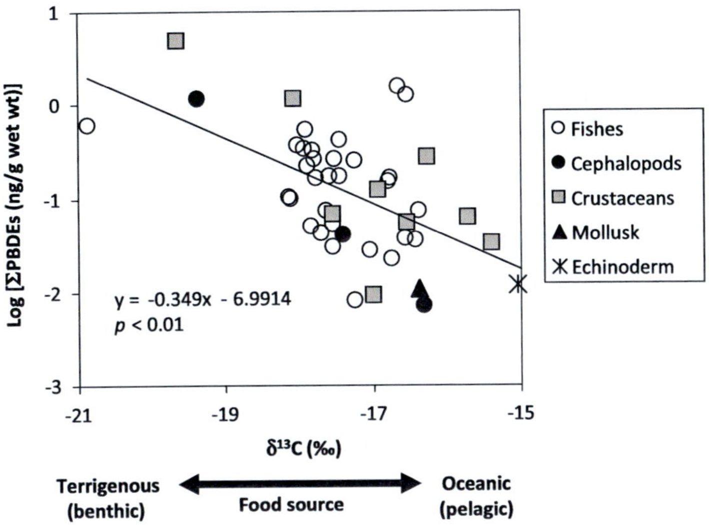

Fig. 12 Relationship between $\delta^{13}C$ (‰) and total PBDE concentrations in deep-sea organisms from the East China Sea

marine from terrigenous organics. In the present study, the $\delta^{13}C$ values in the deep-sea organisms from the ECS ranged from −15.54‰ to −20.87‰. Zhou et al. [96] reported the mean $\delta^{13}C$ values of suspended particulate matter (SPM) in the Changjiang Estuary to be −22.6‰. Therefore, the deep-sea organisms from the ECS with low $\delta^{13}C$ (−20.87‰) could indicate the input of terrigenous materials to the ECS. In the present study, strong negative relationships were observed for log-normalized lipid weight concentrations and $\delta^{13}C$ (‰) for all the analyzed organohalogen compounds (Fig. 12, data shown only for PBDEs). Thus, the negative relationships observed between the organohalogen concentrations and $\delta^{13}C$ (‰) indicate that organisms showing the higher concentrations than the other species accumulate these compounds from terrigenous (coastal benthic) sources rather than oceanic (pelagic) organic matter. Such specific pattern also reflects that steep slope in the concentrations of nutrients, terrigenous materials, and, possibly, contaminants from the coastal front zone is affected by Yangtze River discharge to oceanic waters along the Kuroshio Current.

4.4 Distribution and Transport of POPs into Deep Waters

Apart from the species-specific accumulation of organohalogen compounds, the variation in the concentrations of the organohalogen compounds among the organisms could be related to the distribution of these compounds in the marine ecosystem. Even among the fishes showing similar feeding habits, considerable variations

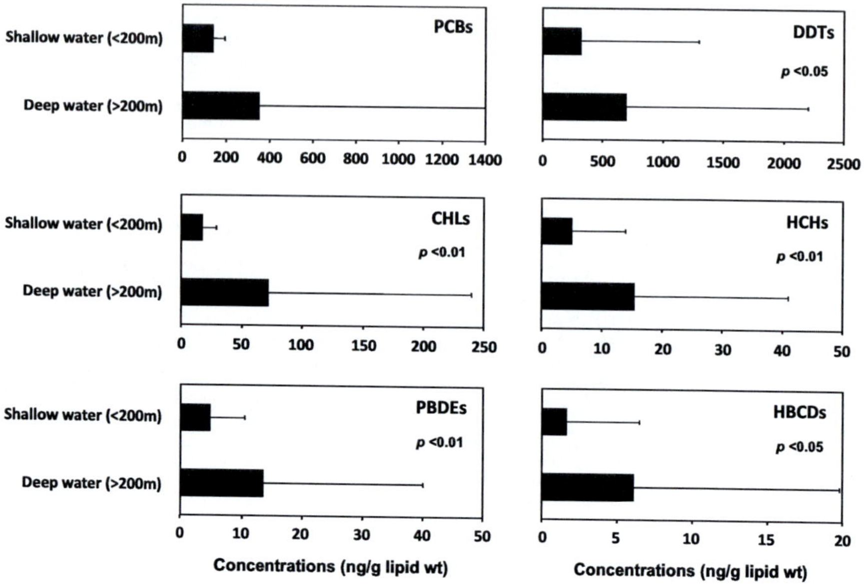

Fig. 13 Mean and range concentrations of organohalogen compounds in fishes collected from shallow (<200 m) and deep waters (>200 m) in the East China Sea (*error bars* show the maximum concentrations)

in the concentrations of organohalogen compounds were observed, which cannot be explained only by their feeding habits and food web magnification. To understand spatial and vertical distribution of the contaminants in the ECS, the fishes were grouped into two categories, shallow water (<200 m) and deep water (>200 m), based on their sampling depths. Interestingly, it was found that the concentrations of almost all the contaminants (except for PCBs) were significantly higher in fishes from greater depths than that from shallower depths ($p < 0.05$ for DDTs and HBCDs and $p < 0.05$ for CHLs, PBDEs, and HBCDs, Mann–Whitney U-test) (Fig. 13).

As it was mentioned before, the ECS shelf receives large amounts of freshwater, suspended matter, and nutrients with the Yangtze River discharge [62]. Furthermore, Gong et al. [97] reported that the ECS has a moderately high primary production (108–997 mg/m^2/d with a mean value of 425 mg/m^2/d). Thus, the high primary productivity and input of terrestrial materials with the Yangtze River discharge leads to an increase in the particulate matter. Kao et al. [61] tracked the export flux of organic matter from the shelf to the deep-sea by comparing carbon and nitrogen isotopic characteristics of sedimentary organic matter all over the ECS and the adjacent Okinawa Trough. The study found similar $\delta^{13}C$ and $\delta^{15}N$ values between sediments from the inner shelf near the China coast and from the Okinawa Trough, suggesting that the ECS shelf is a source of particulate matter to the deeper waters of the ECS.

In the marine environment, organohalogen compounds are either dissolved in the water phase or associated with suspended particulate matter or dissolved organic carbon depending on their hydrophobicity and organic carbon–water partition coefficients (K_{OC}). The particulate matter, on which POPs sorb by the process of flocculation and aggregation to be eventually deposited into the deep waters, could be of natural origin, like detritus, fecal pellets, and living planktonic organisms, or from an anthropogenic source. Studies have shown that eutrophication results in large vertical fluxes of particle-associated pollutants and organic carbon to the deep waters [38, 52]. Thus, all the examined organohalogen compounds being hydrophobic are expected to have high affinity to particulate matter and get transported to deep waters. Volatilization of semi-volatile compounds like HCHs and HCB to the atmosphere from the surface waters of ECS due to the high temperatures in the subtropical region could also contribute to the vertical gradient of their concentrations. On the other hand, biological and ecological factors which can influence the results obtained from various organisms should be considered for further discussions on behavior of POPs. It is necessary to develop a useful model including key biological and ecological parameters for further "normalization" of the contaminant concentrations in organisms (instead of simple lipid normalization) as well as to use better "bioindicators" comprising of same/similar species for future research.

5 Conclusions and Perspectives

It is noteworthy that higher concentrations of POPs and related compounds in deeper waters were indicated by using bioindicators such as fishes in the WNP, off-Tohoku, and ECS, which have been documented to be high productive frontal zones around East Asia. Expansion of contamination by emerging POPs including PBDEs into the deep-sea ecosystems was also suggested by our research. Our recent ocean survey on POPs in sediments from the ECS reported significantly higher concentrations of DDTs and PCBs in deep-sea sediment from the Okinawa Trough than those in the shallower continental shelf [98]. These results suggest the vertical transport of POPs and related compounds in the high productive waters along the oceanic fronts and the potential role of deep-sea bed as a final sink and reservoir for these persistent contaminants. The accumulation and distribution of the POPs and related compounds observed in the marine ecosystems reflects their physicochemical properties such as vapor pressure and hydrophobicity as well as biological and ecological factors of organisms. In addition, riverine discharges, advection by oceanic currents, and formation of oceanic fronts are some examples of the physical transport processes that influence the redistribution of POPs. Further interdisciplinary approach including studies on biogeochemical and geophysical processes in the ocean as well as field observations are required to delineate the global and regional fate of POPs.

Mathematical models are one of indispensable tools to describe the chemodynamics of POPs, including the mass budget, half-life, and long-range transport potential. Models have been used to study the role of atmospheric transport in the multi-compartmental global fate and distribution of persistent substances. However, high-resolution global models that predict nonsteady dynamics of POPs in oceans need to be developed to understand the potential impact of POPs on ecosystems, human health, and their ultimate fate. In this regard, our group has developed a model called the Finely Advanced Transboundary Environmental model (FATE) [99, 100]. The FATE is a global high-resolution multi-compartment model that computes nonsteady POPs biogeochemical cycles in and across the five environmental compartments (atmosphere, ocean, vegetation, soil, and cryosphere). This model enables to predict the three-dimensional distributions of POPs as a result of advective transports, and to quantify the amount of PCB exports to the deep ocean by several physical and biogeochemical processes including phytoplankton detritus settling. Furthermore, we have developed a three-dimensional high-resolution transport model for regional scale POPs fate in the ECS [101], which will be linked and evaluated by our POP observations in the ECS for three consecutive years from 2008 to 2010 [98]. Using such state-of-the-art models, we were able to predict global and regional sinks and reservoirs of POPs which could provide new insights into our current understanding of the fate and transport of POPs.

Recently, we have also started looking at the chemical aftermath of the Tohoku earthquake (the Great Eastern Japan Earthquake) and tsunami that took place on March 11th, 2011 [102]. Three nuclear reactors at Fukushima Daiichi Nuclear Power Station in the tsunami area suffered hydrogen explosions, and leaked radioactive substances including 29.3 PBq of ^{137}Cs were estimated to be released into the ocean [103]. These and other radioactive materials are now polluting the global environment, and it has been cautioned by many that these may accumulate in the biotic compartments and may ultimately reach human. This phenomenon needs the maximum attention of scientists working on the aftereffects of the Great Eastern Japan Earthquake. In addition, the tsunami had wiped out a strip of coast supporting a wide range of land uses and industries. Petro- and agrochemical plants, iron foundries, steel works, and automotive, electronics, food processing, paper, plastics, and pharmaceutical plants were among those that suffered damage. Numerous reports available indicate that various chemicals with potential environmental and health hazards may have been present at many facilities in heavily impacted locations and that old electrical equipment containing PCBs was washed away [104]. The complexity of the waste generated during the Great Eastern Japan Earthquake and tsunami may pose a serious threat of environmental pollution, which should also be monitored along with the effects of nuclear disaster. Our previous study [105] showed that the levels of ^{137}Cs was the lowest in the species of marine mammals obtained from off-Tohoku, Japan, when compared with the specimens caught from other parts of the world such as Lake Baikal, Black Sea, North Sea, and Northern Canadian waters. As reviewed in this chapter, the data on POPs and related compounds in various organisms in the WNP, off-Tohoku, Japan, was also based on the observations in the region where the Great Eastern Japan Earthquake occurred. Thus, we would like to reiterate here that work on POPs and other emerging contaminants (e.g., brominated flame retardants, perfluorochemicals, pharmaceutical and personal care products, toxic elements) as well as radionuclides on the specimens from this region now can give an insight into the most discussed environmental problem in the area. The papers reviewed here can provide the baseline data for comparison for studying the possibility of build-up of the chemical contaminants in the marine ecosystems along northeastern Japan. We hope that our outcomes and future trials on the marine pollution can contribute to avoid any possible catastrophe by such chemicals around our country and the world.

Acknowledgements Our studies reviewed here were conducted as part of a joint research project entitled "Study on Deep-Sea Fauna and Conservation of Deep-Sea Ecosystem" between the Department of Zoology, the National Museum of Nature and Science, and the Center for Marine Environmental Studies, Ehime University. We thank the late Prof. Kouichi Kawaguchi, Dr. Masatoshi Moku, and the crew and staff members of the R/V Tansei-maru, Ocean Research Institute, The University of Tokyo; Dr. Tsunemi Kubodera, National Museum of Nature and Science; and the crew and staff members of the R/V Wakataka-maru, Fisheries Research Agency, for their help on the collection of valuable samples and information of deep-sea organisms. Our studies were also supported by Global Center of Excellence Program of the Japanese Ministry of Education, Culture, Sports, Science and Technology and Grants-in-Aid for Scientific Research (S) (No. 20221003) and (B) (21310043) from the Japan Society for the Promotion of Science.

References

1. Fent K (1996) Ecotoxicology of organotin compounds. Crit Rev Toxicol 26:1–117
2. Wania F, Mackay D (1996) Tracking the distribution of persistent organic pollutants. Environ Sci Technol 30:390A–396A
3. UNEP (2005) Ridding the World of POPs: a guide to the Stockholm convention on persistent organic pollutants. http://www.pops.int
4. UNEP (2009) Report of the Conference of the Parties of the Stockholm Convention on Persistent Organic Pollutants on the work of its fourth meeting. United Nations Environment Programme, 2009 UNEP/POPs/COP.4/38
5. Sixth meeting of the conference of the parties to the stockholm convention. http://chm.pops.int/TheConvention/ConferenceoftheParties/Meetings/COP6/tabid/3074/mctl/ViewDetails/EventModID/870/EventID/396/xmid/10240/Default.aspx. Accessed 20 Nov 2013
6. Barber RT, Warlen SM (1979) Organochlorine insecticide residues in deep sea fish from 2500 m in the Atlantic Ocean. Environ Sci Technol 13:1146–1148
7. Krämer W, Buchert H, Reuter U, Biscoito M, Maul DG, Le Grand G, Ballschmiter K (1984) Global baseline pollution studies IX: C6-C12 organochlorine compounds in surface-water and deep-sea fish from the eastern North Atlantic. Chemosphere 13:1255–1267
8. Stegeman JJ, Kloepper-Sams PJ, Farrington JW (1986) Monooxygenase induction and chlorobiphenyls in the deep-sea fish *Coryphaenoides armatus*. Science 231:1287–1289
9. Melzian B, Zoffmann C, Spies RB (1987) Chlorinated hydrocarbons in lower continental slope fish collected near the Farallon Islands, California. Mar Pollut Bull 18:388–393
10. Berg V, Ugland KI, Hareide NR, Aspholm PE, Polder A, Skaare JU (1997) Organochlorine contamination in deep-sea fish from Davis Strait. Mar Environ Res 44:135–148
11. Berg V, Polder A, Skaare JU (1998) Organochlorines in deep-sea fish from the Nordfjord. Chemosphere 38:275–282
12. Lee JS, Tanabe S, Takemoto N, Kubodera T (1997) Organochlorine residues in deep-sea organisms from Suruga Bay, Japan. Mar Pollut Bull 34:250–258
13. Takahashi S, Lee JS, Tanabe S, Kubodera T (1998) Contamination and specific accumulation of organochlorine and butyltin compounds in deep-sea organisms collected from Suruga Bay, Japan. Sci Total Environ 214:49–64

14. Takahashi S, Tanabe S, Kawaguchi K (2000) Organochlorine and butyltin residues in mesopelagic myctophid fishes from the western North Pacific. Environ Sci Technol 34:5129–5136
15. Takahashi S, Hayashi S, Kasai R, Tanabe S, Kubodera T (2001) Contamination of deep-sea organisms from Tosa Bay, Japan by organochlorine and butyltin compounds. In: Fujita T, Saito H, Takeda M (eds) Deep-sea Fauna and Pollutants in Tosa Bay. National Science Museum Monographs, vol 20. National Science Museum, Tokyo, pp 363–380
16. de Boer J, Wester PG, Klamer JC, Lewis WE, Boon JP (1998) Do flame retardants threaten ocean life? Nature 394:28–29
17. Looser R, Froescheis O, Cailliet GM, Jarman WM, Ballschmiter K (2000) The deep-sea as a final sink of semi volatile persistent organic pollutants? Part 11: organochlorine pesticides in surface and deep-sea dwelling fish of the North and South Atlantic and the Monterey Bay Canyon (California). Chemosphere 40:661–670
18. Froescheis O, Looser R, Cailliet GM, Jarman WM, Ballschmiter K (2000) The deep-sea as a final global sink of semivolatile persistent organic pollutants? Part 1: PCBs in surface and deep-sea dwelling fish of the North and South Atlantic and the Monterey Bay Canyon (California). Chemosphere 40:651–660
19. Solé M, Porte C, Albaigés J (2001) Hydrocarbons, PCBs and DDT in the NW Mediterranean deep-sea fish *Mora moro*. Deep-Sea Res 48:495–513
20. Mormede S, Davies IM (2003) Horizontal and vertical distribution of organic contaminants in deep-sea fish species. Chemosphere 50:563–574
21. Storelli MM, Storelli A, Addabbo RD, Barone G, Marcotrigiano GO (2004) Polychlorinated biphenyl residues in deep-sea fish from the Mediterranean Sea. Environ Intern 30:343–349
22. Ramu K, Kajiwara N, Mochizuki H, Miyasaka H, Asante KA, Takahashi S, Nishida S, Tanabe S (2006) Occurrence of organochlorine pesticides, polychlorinated biphenyls and polybrominated diphenyl ethers in deep-sea fishes from the Sulu Sea. Mar Pollut Bull 52:1827–1832
23. Covaci A, Losada S, Roosens L, Vetter W, Santos FJ, Neels H, Storelli A, Storelli MM (2008) Anthropogenic and naturally occurring organobrominated compounds in two deep-sea fish species from the Mediterranean Sea. Environ Sci Technol 42:8654–8660
24. Unger MA, Harvey E, Vadas GG, Vecchione M (2008) Persistent pollutants in nine species of deep-sea cephalopods. Mar Pollut Bull 56:1498–1500
25. Storelli MM, Perrone VG, Marcotrigiano GO (2007) Organochlorine contamination (PCBs and DDTs) in deep-sea fish from the Mediterranean Sea. Mar Pollut Bull 54:1962–1989
26. Storelli MM, Losada S, Marcotrigiano GO, Roosens L, Barone G, Neels H, Covaci A (2009) Polychlorinated biphenyl and organochlorine pesticide contamination signatures in deep-sea fish from the Mediterranean Sea. Environ Res 109:851–856
27. Lemaire B, Priede IG, Collins MA, Bailey DM, Schtickzelle N, Thomé JP, Rees JF (2010) Effects of organochlorines on cytochrome P450 activity and antioxidant enzymes in liver of roundnose grenadier *Coryphaenoides rupestris*. Aquat Biol 8:161–168
28. Castro Jiménez J, Rotllant G, Ábalos M, Parera J, Dachs J, Company JB, Calafat A, Abad E (2013) Accumulation of dioxins in deep-sea crustaceans, fish and sediments from a submarine canyon (NW Mediterranean). Prog Oceanogr 118:260–272. http://www.sciencedirect.com/science/article/pii/S0079661113001304
29. Palm A, Cousins I, Gustafsson Ö, Axelman J, Grunder K, Broman D, Brorström-Lundén E (2004) Evaluation of sequentially-coupled POP fluxes estimated from simultaneous measurements in multiple compartments of an air-water-sediment system. Environ Pollut 128:85–97
30. Iwata H, Tanabe S, Sakai N, Tatsukawa R (1993) Distribution of persistent organochlorines in the oceanic air and surface seawater and the role of ocean on their global transport and fate. Environ Sci Technol 27:1080–1098
31. Wania F, Axelman J, Broman D (1998) A review of processes involved in the exchange of persistent organic pollutants across the air-sea interface. Environ Pollut 102:3–23

32. Knezovich JP, Harrison FL, Wilhelm RG (1987) The bioavailability of sediment-sorbed organic chemicals: a review. Water Air Soil Pollut 32:233–245
33. Lohmann R, Breivik K, Dachs J, Muir D (2007) Global fate of POPs: current and future research directions. Environ Pollut 150:150–165
34. Tanabe S (2007) Contamination by persistent toxic substances in the Asia-Pacific region. Dev Environ Sci 7:773–817
35. Tanabe S, Ramu K, Isobe T, Takahashi S (2008) Brominated flame retardants in the environment of Asia-Pacific: an overview of spatial and temporal trends. J Environ Monit 10:188–197
36. Tanabe S, Ramu K (2010) Environmental Specimen Bank (es-BANK) of Ehime University, Japan – current status and future perspectives. In: Isobe T, Nomiyama K, Subramanian A, Tanabe S (eds) Interdisciplinary studies on environmental chemistry. Environmental specimen bank: exploring possibility of setting-up esbs in developing countries, vol 4. TERRAPUB, Tokyo, pp 1–5
37. Belkin IM (2002) Front. In: Nybakken JW, Broenkow WW, Vallier TL (eds) Interdisciplinary encyclopedia of marine sciences, vol 1. Grolier, Danbury, pp 433–436
38. Dachs J, Eisenreich SJ, Hoff RM (2000) Influence of eutrophication on air-water exchange, vertical fluxes, and phytoplankton concentrations of persistent organic pollutants. Environ Sci Technol 34:1095–1102
39. Dachs J, Lohmann R, Ockenden WA, Mejanelle L, Eisenreich SJ, Jones KC (2002) Oceanic biogeochemical controls on global dynamics of persistent organic pollutants. Environ Sci Technol 36:4229–4237
40. Tanabe S, Nishimura A, Hanaoka S, Yanagi T, Takeoka H, Tatsukawa R (1991) Persistent organochlorines in coastal fronts. Mar Pollut Bull 22:344–351
41. de Brito APX, Takahashi S, Ueno D, Iwata H, Tanabe S, Kubodera T (2002) Organochlorine and butyltin residues in deep-sea organisms collected from the western North Pacific, off-Tohoku, Japan. Mar Pollut Bull 45:348–361
42. Tanabe S, Ramu K, Mochizuki H, Miyasaka H, Okuda N, Muraoka M, Kajiwara N, Takahashi S, Kubodera T (2005) Contamination and distribution of persistent organochlorine and organotin compounds in deep-sea organisms from East China Sea. In: Hasegawa K, Shinohara G, Takeda M (eds) Deep-sea Fauna and Pollutants in Nansei Islands. National Science Museum Monographs, vol 29, National Science Museum, Tokyo, pp 453–476
43. Ramu K, Kajiwara N, Isobe T, Takahashi S, Miyasaka H, Kubodera T, Tanabe S (2007) Organohalogen compounds in deep sea organisms from East China Sea: levels, profiles and fate. Organohalogen Compd 69:437–440
44. Takahashi S, Oshihoi T, Ramu K, Isobe T, Ohmori K, Kubodera T, Tanabe S (2010) Organohalogen compounds in deep-sea fishes from the western North Pacific, off-Tohoku, Japan: contamination status and bioaccumulation profiles. Mar Pollut Bull 60:187–196
45. Shimizu Y, Yasuda I, Ito SI (2001) Distribution and circulation of the coastal oyashio intrusion. J Phys Oceanogr 31:1561–1578
46. Sorokin YI, Sorokin PY (1999) Production in the sea of Okhotsk. J Plankton Res 21:201–230
47. Nakatsuka T, Toda M, Kawamura K, Wakatsuchi M (2004) Dissolved and particulate organic carbon in the Sea of Okhotsk: transport from continental shelf to ocean interior. J Geophys Res 109:C09S14. doi:10.1029/2003JC001909
48. Nishioka J, Ono T, Saito H, Nakatsuka T, Takeda S, Yoshimura T, Suzuki K, Kuma K, Nakabayashi S, Tsumune D, Mitsudera H, Johnson WK, Tsuda A (2007) Iron supply to the western subarctic Pacific: importance of iron export from the Sea of Okhotsk. J Geophys Res 112, C10012. doi:10.1029/2006JC004055
49. Yasuda I, Okuda K, Shimizu Y (1996) Distribution and modification of North Pacific intermediate water in the Kuroshio–Oyashio interfrontal zone. J Phys Oceanogr 26:448–465
50. Yasuda I (1997) The origin of the North Pacific intermediate water. J Geophys Res 102: 893–909

51. Honda MC, Imai K, Nojiri Y, Hoshi F, Sugawara T, Kusakabe M (2002) The biological pump in the northwestern North Pacific based on fluxes and major components of particulate matter obtained by sediment-trap experiments (1997–2000). Deep-Sea Res II 49:5595–5625
52. Jeff JD, Eisenreich SJ, Paterson MJ, Beaty KJ, Hecky R, Elser JJ (1999) Biogeochemical cycling of PCBs in lakes of variable trophic status: a paired-lake experiment. Limnol Oceanogr 44:889–902
53. Otosaka S, Noriki S (2000) REEs and Mn/Al ratio of settling particles: horizontal transport of particulate material in the northern Japan Trench. Mar Chem 72:329–342
54. Qin YS, Zhao YY, Chen LR, Zhao SL (eds) (1996) Geology of the East China Sea. Science Press, Beijing
55. Ichikawa H, Beardsley RC (2002) The current system in the Yellow and East China Seas. J Oceanogr 58:77–92
56. Hickox R, Belkin I, Cornillon P, Shan Z (2000) Climatology and seasonal variability of ocean fronts in the East China, Yellow and Bohai seas from satellite SST data. Geophys Res Lett 27:2945–2948
57. Chen CTA (2008) Chemical and physical fronts in the Bohai, Yellow and East China seas. J Mar Syst 78:394–410
58. Liu KK, Tang TY, Gong GC, Chen LY, Shiah FK (2000) Cross-shelf and along-shelf nutrient fluxes derived from flow fields and chemical hydrography observed in the southern East China Sea off northern Taiwan. Cont Shelf Res 20:493–523
59. Wong CM, Williams CE, Pittock J, Collier U, Schelle P (2007) World's top 10 rivers at risk. WWF International, Gland, Switzerland, http://assets.panda.org/downloads/worldstop10riversatriskfinalmarch13.pdf. Accessed 20 July 2013
60. Li D, Daler D (2004) Ocean pollution from land-based sources: East China sea, China. Ambio 33:107–113
61. Kao SJ, Linb FJ, Liu KK (2003) Organic carbon and nitrogen contents and their isotopic compositions in surficial sediments from the East China Sea shelf and the southern Okinawa Trough. Deep-Sea Res II 50:1203–1217
62. Hung JJ, Chen CH, Gong GC, Sheu DD, Shiah FK (2003) Distributions, stoichiometric patterns and cross-shelf exports of dissolved organic matter in the East China Sea. Deep-Sea Res II 50:1127–1145
63. Ueno D, Takahashi S, Tanaka H, Subramanian A, Fillman G, Nakata H, Lam PKS, Zheng J, Muchtar M, Prudente M, Chung KH, Tanabe S (2003) Global pollution monitoring of PCBs and organochlorine pesticides using skipjack tuna as a bioindicator. Arch Environ Contam Toxicol 45:378–389
64. Beyer A, Mackay D, Matthies M, Wania F, Webster E (2000) Assessing long-range transport potential of persistent organic pollutants. Environ Sci Technol 34:699–703
65. Ueno D, Kajiwara N, Tanaka H, Subramanian A, Fillmann G, Lam PKS, Zheng GJ, Muchtar M, Razak H, Prudente M, Chung KH, Tanabe S (2004) Global pollution monitoring of polybrominated diphenyl ethers using skipjack tuna as a bioindicator. Environ Sci Technol 38:2312–2316
66. Ueno D, Alaee M, Marvin C, Muir DCG, Macinnis G, Reiner E, Crozier P, Furdui VI, Subramanian A, Fillmann G, Lam PKS, Zheng GJ, Muchtar M, Razak H, Prudente M, Chung KH, Tanabe S (2006) Distribution and transportability of hexabromocyclododecane (HBCD) in the Asia-Pacific region using skipjack tuna as a bioindicator. Environ Pollut 144:238–247
67. Watanabe I, Sakai S (2003) Environmental release and behavior of brominated flame retardants. Environ Intern 29:665–682
68. Caira JN, Benz GW, Borucinska J, Kohler NE (1997) Pugnose eels *Simenchelys parasiticus* (Synaphobranchidae) from the heart of a shortfin mako *Isurus oxyrinchus* (Lamnidae). Environ Biol Fish 49:139–144
69. Hobson KA, Welch HE (1992) Determination of trophic relationships within a high arctic marine food web using $\delta^{13}C$ and $\delta^{15}N$ analysis. Mar Ecol Prog Ser 84:9–18

70. Fisk AT, Hobson KA, Norstrom RJ (2001) Influence of chemical and biological factors on trophic transfer of persistent organic pollutants in the Northwater Polynya food web. Environ Sci Technol 35:732–738
71. Ruus A, Ugland KI, Skaare JU (2002) Influence of trophic position on organochlorine concentrations and composition patterns in a marine food web. Environ Toxicol Chem 21:2356–2364
72. France RL, Peters RH (1997) Ecosystem differences in the trophic enrichment of 13C in aquatic food webs. Can J Fish Aquat Sci 54:1255–1258
73. Hobson KA, Fisk A, Karnovsky N, Holst M, Gagnon JM, Fortier M (2002) A stable isotope ($\delta^{13}C$, $\delta^{15}N$) model for the North Water Polynya food web: implications for evaluating trophodynamics and the flow of energy and contaminants. Deep-Sea Res II 49:5131–5150
74. Takahashi S, Tanabe S, Kubodera T (1997) Butyltin residues in deep-sea organisms collected from Suruga Bay, Japan. Environ Sci Technol 31:3103–3109
75. Takahashi S, Mukai H, Tanabe S, Sakayama K, Miyazaki T, Masuno H (1999) Butyltin residues in livers of humans and wild terrestrial mammals and in plastic products. Environ Pollut 106:213–218
76. Tanabe S, Subramanian A (2006) Bioindicator of POPs – monitoring in developing countries. Kyoto University Press/Trans Pacific Press, Kyoto
77. Tanabe S, Tatsukawa R (1983) Vertical transport and residence time of chlorinated hydrocarbons in the open ocean water column. J Oceanogr Soc Jpn 39:53–62
78. Iwata H, Tanabe S, Ouchi E, Tatsukawa R (1995) Persistent organochlorines in air and water from east Siberia. In: Abstracts, second SETAC (Society of Environmental Toxicology and Chemistry) World Congress. SETAC, Vancouver, p 77
79. McConnell LL, Kucklick JR, Bidleman TF, Ivanov GP, Chernyak SM (1996) Air–water gas exchange of organochlorine compounds in Lake Baikal, Russia. Environ Sci Technol 30:2975–2983
80. Nakata H, Tanabe S, Tatsukawa R, Koyama Y, Miyazaki N, Belikov S, Boltunov A (1998) Persistent organochlorine contaminants in ringed seals (*Phoca hispida*) from the Kara sea, Russian Arctic. Environ Toxicol Chem 17:1745–1755
81. Iwata H, Watanabe M, Kim EY, Gotoh R, Yasunaga G, Tanabe S, Masuda Y, Fujita S (2000) Contamination chlorinated hydrocarbons and lead in Steller's sea eagle and white-tailed sea eagle from Hokkaido, Japan. In: Ueta M, McGrady MJ (eds) First symposium on Steller's and white-tailed sea eagles in East Asia. Wild Bird Society of Japan, Tokyo, pp 91–106
82. Ueno D, Inoue S, Ikeda K, Tanaka H, Yamada H, Tanabe S (2003) Specific accumulation of polychlorinated biphenyls and organochlorine pesticides in Japanese common squid as a bioindicator. Environ Pollut 125:227–235
83. Tanabe S, Tanaka H, Tatsukawa R (1984) Polychlorobiphenyls, DDT, and hexachlorocyclohexane isomers in the western North Pacific ecosystem. Arch Environ Contam Toxicol 13:731–738
84. Tanabe S, Sung JK, Choi DY, Baba N, Kiyota M, Yoshida K, Tatsukawa R (1994) Persistent organochlorine residues in northern fur seal from the Pacific coast of Japan since 1971. Environ Pollut 85:305–314
85. Chen W, Zhang L, Xu L, Wang X, Hong L, Hong H (2002) Residue levels of HCHs, DDTs and PCBs in shellfish from coastal areas of east Xiamen Island and Minjiang estuary, China. Mar Pollut Bull 45:385–390
86. Yuan D, Yang D, Wade TL, Qian Y (2001) Status of persistent organic pollutants in the sediment from several estuaries in China. Environ Pollut 114:101–111
87. Colombo JC, Bilos C, Campanaro M, Rodriguez Presa MJ, Catoggio JA (1995) Bioaccumulation of polychlorinated biphenyls and chlorinated pesticides by the Asiatic clam *Corbicula fluminea*: its use as sentinel organism in the Rio de La Plata Estuary, Argentina. Environ Sci Technol 29:914–927

88. Ramu K, Kajiwara N, Sudaryanto A, Isobe T, Takahashi S, Subramanian A, Ueno D, Zheng GJ, Lam PKS, Takada H, Zakaria MP, Viet PH, Prudente M, Tana TS, Tanabe S (2007) Asian mussel watch program: contamination status of polybrominated diphenyl ethers and organochlorines in coastal waters of Asian countries. Environ Sci Technol 41:4580–4586
89. Jiang JH, Wang SM (2004) Primary analysis of water resource, disasters and environment in the Changjiang River catchment. Quat Sci 24:512–517 (in Chinese)
90. Lie HJ, Cho CH (2002) Recent advances in understanding the circulation and hydrography of the East China Sea. Fish Oceanogr 11:318–328
91. Ellgehausen H, Guth JA, Esser HO (1980) Factors determining the bioaccumulation potential of pesticides in the individual compartments of aquatic food chains. Ecotoxicol Environ Saf 4:137–157
92. de Wit CA (2002) An overview of brominated flame retardants in the environment. Chemosphere 46:583–624
93. Minagawa M, Wada E (1986) Nitrogen isotope ratios of red tide organisms in the East China Sea: a characterization of biological nitrogen fixation. Mar Chem 19:245–259
94. Wada E, Minagawa M, Mizutani H, Tsuji T, Imaizumi R, Karasawa K (1987) Biogeochemical studies on the transport of organic matter along the Otsuchi River watershed, Japan. Estuar Coast Shelf Sci 25:321–336
95. Middelburg JJ, Nieuwenhuize J (1998) Carbon and nitrogen isotopes in suspended matter and sediments from the Scheldt Estuary. Mar Chem 60:217–225
96. Zhou J, Wu Y, Zhang J, Kang Q, Liu Z (2006) Carbon and nitrogen composition and stable isotope as potential indicators of source and fate of organic matter in the salt marsh of the Changjiang Estuary, China. Chemosphere 65:310–317
97. Gong GC, Wen YH, Wang BW, Liu GJ (2003) Seasonal variation of chlorophyll a concentration, primary production and environmental conditions in the subtropical East China Sea. Deep-Sea Res II 50:1219–1236
98. Watanabe N, Yamano S, Shimizu J, Takahashi S (2011) Distribution of persistent organic pollutants (POPs) in the East China Sea. Rep Hydrographic Oceanogr Res 47:72–82 (in Japanese)
99. Kawai T, Handoh IC, Takahashi S (2009) The rise of the Finely-Advanced Transboundary Environmental model (FATE): a state-of-the-art model prediction of the global sink of persistent organic pollutants. Organohalogen Compd 71:1610–1615
100. Kawai T, Handoh IC, Suzuki N (2011) The roles of marine phytoplankton and ocean circulation in determining the global fate of polychlorinated biphenyls. In: Omori K, Guo X, Yoshie N, Fujii N, Handoh IC, Isobe A, Tanabe S (eds) Interdisciplinary studies on environmental chemistry—marine environmental modeling & analysis. TERRAPUB, Tokyo, pp 169–178
101. Ono J, Takahashi D, Guo X, Takahashi S, Takeoka H (2011) Process study on numerical simulation for persistent organic pollutants in the East China Sea. In: Omori K, Guo X, Yoshie N, Fujii N, Handoh IC, Isobe A, Tanabe S (eds) interdisciplinary studies on environmental chemistry—marine environmental modeling & analysis. TERRAPUB, Tokyo, Japan, pp 197–204
102. Christoudias T, Lelieveld J (2013) Modelling the global atmospheric transport and deposition of radionuclides from the Fukushima Dai-ichi nuclear accident. Atmos Chem Phys 13:1425–1438
103. Tanabe S (2011) Great eastern japan earthquake – possible marine environmental contamination by toxic pollutants. Mar Pollut Bull 62:883–884
104. Bird WA, Grossman E (2011) Chemical aftermath: contamination and cleanup following the Tohoku earthquake and tsunami. Environ Health Perspect 119:A290–A301
105. Yoshitome R, Kunito T, Ikemoto T, Tanabe S, Zenke H, Yamauchi M, Miyazaki N (2003) Global distribution of radionuclides (^{137}Cs and ^{40} K) in marine mammals. Environ Sci Technol 37:4597–4602

Dynamics of Transport, Accumulation, and Export of Plastics at Oceanic Fronts

G. Suaria, M. Berta, A. Griffa, A. Molcard, T. M. Özgökmen, E. Zambianchi, and S. Aliani

Contents

Abstract Every year, vast quantities of plastic debris arrive at the ocean surface. Nevertheless, our understanding of plastic movements is largely incomplete and many of the processes involved with the horizontal and vertical displacement of

G. Suaria (✉), M. Berta, A. Griffa, and S. Aliani
Institute of Marine Sciences (ISMAR), National Research Council (CNR), Lerici, Italy
e-mail: giuseppe.suaria@sp.ismar.cnr.it

A. Molcard
MIO, Université de Toulon, Aix-Marseille Univ., CNRS, IRD, Toulon, France

T. M. Özgökmen
Rosenstiel School of Marine and Atmospheric Science, University of Miami, Coral Gables, FL, USA

E. Zambianchi
Parthenope University and CoNISMa, Naples, Italy

Igor M. Belkin (ed.), *Chemical Oceanography of Frontal Zones*,
Hdb Env Chem (2022) 116: 355–406, DOI 10.1007/698_2021_814,

Published online: 19 October 2021

plastics in the ocean are still basically unknown. In this chapter we review the dynamics associated with the transport of plastics and other pollutants at oceanic fronts. Fronts had been historically defined as simple barriers to exchange, but here we show that the role of these structures in influencing the transport of plastics is more complex. The tools used to investigate the occurrence of frontal structures at various spatial scales are reviewed in detail, with a particular focus on their potential applications to the study of plastic pollution. Three selected case studies are presented to better describe the role of fronts in favoring or preventing plastic exchanges: the large-scale Antarctic Circumpolar Current, a Mediterranean mesoscale front, and the submesoscale fronts in the Gulf of Mexico. Lastly, some aspects related to the vertical subduction of plastic particles at oceanic fronts are discussed as one of the most promising frontiers for future research. The accumulation of floating debris at the sea surface is mainly affected by the horizontal components of frontal dynamics. At the same time, vertical components can be relevant for the export of neutrally buoyant particles from the surface into the deep sea. Based on these evidences, we propose that submesoscale processes can provide a fast and efficient route of plastic transport within the mixed layer, while mesoscale instabilities and associated vertical velocities might be the dominant mechanism to penetrate the deeper ocean on slower but broader scales. We conclude that given the ubiquitous presence of fronts in the world's ocean, their contribution to the global plastic cycle is probably not negligible and the role of these processes in vertically displacing neutrally buoyant microplastics should be investigated in more detail.

Keywords Contaminant dispersal, Frontal zones, Marine debris, Marine litter, Microplastics, Plastic pollution, Sinking

1 Introduction

Since plastic production began, humankind comprehensively produced around 8,300 million tons (Mt) of synthetic polymers [1], and every year, about 200 Mt. of municipal plastic waste are generated and disposed of around the world [2]. Of this amount, between 4.8 and 12.7 Mt of mismanaged plastic items are estimated to enter the oceans every year through various sources [3], an estimate which is predicted to grow up to 53 Mt per year by 2030, if significant global reduction measures for environmental plastic emissions will not be implemented [4]. As a result, plastic has been accumulating in marine ecosystems for decades, and synthetic polymers of various sizes, shapes, and typologies are now widespread, from the highly urbanized Mediterranean Sea to the most remote polar waters, although in varying concentrations [5–7].

Much progress has been made since the first reports of plastic pollution appeared in the scientific literature, and we now have a much better understanding of what are the main sources and impacts of many synthetic polymers commonly found in the marine environment [5, 8]. However, a clear understanding of the global plastic cycle is still far from being achieved. According to the most recent global estimates, the total amount of plastic floating at sea (<0.3 Mt, [9–11]) represents only a small portion of the total estimated inputs in the marine environment (~8 Mt $year^{-1}$, [3]), although this theory has been recently challenged [12]. Recent studies suggested that backshores and coastal margins [13, 14], the water column [15, 16], or deep-sea sediments [17, 18] can all be accounting for this missing fraction [19]. In spite of this growing body of information, this remarkable mismatch demonstrates that our understanding of plastic fluxes and pathways between different compartments is largely incomplete, and some fundamental aspects of plastic transport and distribution in the ocean are still basically unknown.

Plastic concentration is very inhomogeneous in the ocean. Research has shown that high-concentration zones occur not only at the large scale in oceanic garbage patches, but they are distributed in all regions at various spatial scales and intermittent in time [20]. This inhomogeneity can lead to an important sampling bias, with obvious consequences for global plastic estimates. At the same time ocean plastics occur in a wide range of size classes, further complicating our understanding of the main factors associated with their cycling through the marine environment. A comprehensive description of the main physical processes driving and influencing horizontal and vertical debris transport at different scales of ocean dynamics has been recently made by van Sebille et al. [21]. The large-scale dynamics responsible for the well-known "garbage patches," i.e., the floating litter accumulation zones found at midlatitude in all the world's oceans, has been studied by many authors (e.g., [9, 10, 22]; other references in [21]), but in general, scant information is available about the small-scale transport of debris driven by submesoscale, 3D turbulence, and microscale processes. The transport of other pollutants by mesoscale eddies, geostrophic currents, and internal waves is better reported [23–27]. At the same time, the continuous findings of massive debris accumulations on the seafloor (including low-density polymers) suggest that one of the ultimate sinks for plastic debris is the deep ocean, and that marine sediments can be likely considered as a final plastic repository [28–30], although the role of resuspension mechanisms, especially in the presence of intense benthic nepheloid dynamics still needs to be clarified (c.f. [31]). In the upper ocean, therefore, there must be processes capable of removing plastic from the sea surface and transporting it to the deep sea.

In principle, once in the marine environment, plastic particles (here defined as synthetic particles smaller than a few millimeters), as any other small particles in the ocean, move together with the surrounding water, and with a good approximation the movement of plastic particles follows the movement of water parcels [21]. In an ideal uniform flow, the relative distances of particles in the flow, i.e. its dilution or accumulation, would mainly depend on the particle's properties (e.g., size, buoyancy, density) rather than on the fluid's properties. But in the actual ocean, the flow is highly complex and the physical properties of ocean currents play a very important role in shaping and defining plastic distribution. In addition, many plastic items such as expanded polystyrene and synthetic foams also contain trapped air, which further increases their buoyancy, subsequently aiding their wind-mediated dispersal. So, if we want to understand plastic movements in the oceans, we first need to understand ocean dynamics and their spatio-temporal variability. From this point of view, a better understanding of the physical processes governing the transport of solid pollutants in the marine environment is a fundamental prerequisite to finally close the plastic budget, ultimately improving our capacity to address and mitigate the negative impacts of this global emerging challenge.

Oceanic fronts are physical features that have not been investigated thoroughly with respect to plastic litter. These frontal systems are often characterized by the presence of convergence zones that accumulate floating debris at the ocean surface together with planktonic organisms and higher trophic levels organisms attracted by enhanced productivity [32, 33], and downwelling areas that can potentially export neutrally buoyant debris into the deep sea over short distances possibly contributing to their final sinking. Convergences at frontal zones have the potential to boost the interactions between litter and marine life, when high densities of plastics and frontal processes coexist. Ingestion, entanglement or colonization of litter surfaces increases [32] with many consequences. The ingestion likelihood of plastics depends on the probability for the species to encounter debris, therefore planktonic organisms passively accumulated in the convergence zones are more susceptible to plastic ingestion, but at the same time, also ocean-going predators can be attracted by an increase in local preys' abundance and by the shade created by large floating debris that attract both small fish and large predators around them [34]. Interactions of biota with plastics at frontal convergence zones are also relevant as the growth of biofouling can noticeably alter the buoyancy of the plastic particles [35–37], thus favoring and/or accelerating their sinking velocities [29, 38]. Nevertheless, the available literature on these processes is still very limited, even at the ocean surface, where sampling tends to be easier. The ubiquitous presence of fronts in the world's ocean at any spatial scale suggests that the global contribution of these structures to the plastic cycle is probably not negligible and the roles of these processes in creating surface concentration zones and in removing debris from the sea surface are not fully understood yet.

In this chapter we focus on the transport, accumulation, and export dynamics of plastic debris and other pollutants at oceanic fronts (Sects. 2 and 3). The tools used to investigate the occurrence of frontal structures at various spatial scales are reviewed in detail, with a particular focus on their potential applications to the study of plastic pollution (Sect. 4). Three selected case studies are presented to better describe the role of fronts in favoring or preventing plastic exchanges: the large-scale Antarctic Circumpolar Current (Sect. 5.1), a Mediterranean mesoscale front (Sect. 5.2), and the submesoscale fronts in the Gulf of Mexico (Sect. 5.3). Lastly, some innovative aspects related to the vertical subduction of plastic particles at oceanic fronts are presented as one of the most promising frontiers for future research (Sect. 6).

2 Fronts as Boundaries for Plastic Exchanges

Fronts are defined from the geometrical point of view as areas characterized by a strong gradient in one direction (cross-front), and a much weaker gradient in the perpendicular direction (along front). They can be characterized by various physical and biogeochemical properties and are commonly observed at the ocean surface with scales that can vary from kilometers to meters. The process of frontal formation and sharpening is called frontogenesis [39] and is especially prominent and fast acting in the case of active fronts based on density gradients, involving the development of vertical secondary circulation in the cross-front direction [40]. The frontogenetic processes lead to enhanced gradients of physical properties such as temperature and salinity, that determine density, but also of the chemical and biological properties that characterize the water masses. Indeed, it is often observed that fronts are not typical of just one property but are usually reflected in many properties [41, 42], with separation sometimes clearly visible by macroscopic water properties e.g., color, particle loads, foams, and flotsams including anthropogenic litter (Fig. 1). While fronts appear mostly as horizontal boundaries, various instability processes occur at different scales that can facilitate property mixing. They include mesoscale instabilities, generating meanders and eddies, and submesoscale instabilities, generating smaller features and active convective cells. So, the interaction between water masses involves a set of complex dynamics at different spatial scales and not a simple separation.

Oceanic fronts occur on a variety of scales, from a few meters up to many thousand kilometers [39]. Some of them are short-lived, but most of the large- and mesoscale fronts are quasi-stationary and seasonally persistent: they emerge and disappear at the same locations during the same season, year after year. The most prominent fronts are present year-around. Small submesoscale fronts, on the other hand, associated with mixed layer inhomogeneities or water mass filaments detaching from mesoscale eddies or jets typically have very short time scales, of the order of days, and lead to mixed layer restratification. There is no definitive classification of fronts, but a partial listing of them would include tidal fronts, upwelling fronts, estuarine fronts, shelf-break fronts, river plume fronts, fronts associated with the convergence or divergence of water masses in the open ocean, frontal eddies, and fronts associated with geomorphologic features such as headlands, islands, and canyons [43]. All of them have the potential to concentrate flotsams and contaminants, although with different effectiveness, but so far, the actual presence of debris at frontal systems has not been studied as deeply as other pollutants [44, 45].

Alternating zones of downwelling and upwelling flow are usually produced in correspondence of frontal systems. These phenomena can induce vertical flows enhancing, in turn, peculiar physical and biological features. Converging surface water necessarily sinks at the front line [46], separating light and heavy waters that form a floating lens of light water around the convergence and a downwelling plume of dense water, respectively. Zones of convergence are a common pattern in oceanic

Fig. 1 Airborne images of surface frontal systems: in the top panel oil slicks aligned into a front in the Gulf of Mexico. In the bottom panels: fronts associated with the Mississippi river input into the Gulf of Mexico. The water masses have different optical properties that can be spotted in the visible field from airborne images (Photo credit: Maristella Berta)

fronts. Similar to water parcels, debris is transported toward the convergence where the heavy portion sinks and lighter floating material (natural or anthropogenic) accumulates at the surface where it is visible by naked eye and at least in theory, by satellites too [47–50]. Preliminary data and modeling experiments suggest that in frontal convergences, surface accumulation processes can generate densities of floating litter (and other pollutants) orders of magnitude higher than in the surrounding waters [51–55], but dedicated experiments are necessary for an accurate description and quantification of these structures' concentration properties.

Even though fronts can form at any depth, it should be emphasized that in the following sections we will mainly focus on surface fronts for the purpose of studying their interactions with marine litter. Although some data on the vertical distribution of microplastics along the water column are starting to appear in the scientific literature (see Liu et al. [56] for a recent review), virtually no data exist about microplastics in the water column at deep oceanic frontal systems. As a matter of fact, the vertical distribution of floating plastic depends not only on the particle's buoyancy, but also on the dynamical environment induced by vertical movements of ocean water (see Sect. 6). So, downwelling at frontal systems has the potential to sequester floating material from the ocean's surface and can be considered as potential debris sinking zones from the surface to the ocean interior [21]. The transport of material at the ocean surface, however, is expected to play a major role for most buoyant and nearly buoyant plastics (i.e., the majority of all synthetic polymers currently produced worldwide). In addition, surface fronts are indeed the most active ones from the physical point of view, not only because of the direct interaction with atmospheric forcing, but mostly because the absence of vertical velocities at the interface between ocean and atmosphere allows for enhanced stirring within mesoscale eddies, causing gradient sharpening and therefore further enhancing frontal properties [40, 57, 58].

3 Ocean Currents and the Transport of Plastics: A Problem of Scale

Focusing on surface processes, over the past few decades, there has been a tremendous amount of progress in understanding and predicting ocean currents and their associated transport of anthropogenic pollutants. Perhaps a suitable beginning for such rapid progress can be attributed to MODE, i.e. the Mid-Ocean Dynamics Experiment [59] during which "ocean weather" consisting of mesoscale eddies on spatial scales of hundreds of kilometers, evolving on time scales of months was discovered. This view of the ocean containing long-lasting coherent structures is different from the earlier perspective based on a large-scale mean flow that governs the advection of material for years and decades. Early numerical modeling using quasi-geostrophic equations (arising from a balance of forces between the Earth's rotation and pressure gradient) showed that barotropic and baroclinic instabilities lead to a ubiquitous emergence of mesoscale eddies in the ocean circulation [60].

The second major step forward was taken with the advent of satellite oceanography, in particular the Topex/Poseidon mission, which allowed estimation of sea surface height anomaly over the range of scales from 100 km to several thousand kilometers [61]. Using the geostrophic approximation, only a few satellite altimeters allowed estimation of much of the ocean's near-surface velocity field uninterrupted by cloud coverage [62]. The availability of horizontal velocity data from satellite measurements allowed calculation of many quantities of interest. One such quantity is the turbulent kinetic energy spectrum. This is important not only for parameterizing and modeling turbulence, but also for improving the accuracy of climate forecasts.

The zeroth-order expectation, on the basis of high aspect ratio (lateral vs vertical scale) and strong influence of rotation, would be that the ocean exhibits 2-dimensional (2D) turbulence characteristics at geostrophic scales, with the wavenumber power spectrum of kinetic energy obeying a power law. However, estimates based on satellite altimeter measurements consistently showed a slope that is flatter than expected [63]. This result indicates that the ocean's surface is more energetic than can be explained on the basis of 2D turbulence alone. The mechanism causing this discrepancy in slope was not immediately clear, until the realization of the relevance of the so-called submesoscale flows. The existence of small-scale convergent circulations in the ocean was known long before [64]. More recently, it has been demonstrated that the slope can be also altered by barotropic tides and their interaction with mesoscale dynamics [65, 66]. Submesoscale dynamics, however, are currently surmised to act as a main conduit between the nearly-2D geostrophic mesoscales and classical 3D turbulence [67, 68].

Submesoscales are generically defined as flows on the margin of loss of geostrophic balance, with Rossby number (i.e., ratio of relative and planetary vorticity) on the $O(1)$, with horizontal space scales of 0.1–10 km and evolution time scales of hours to days. Visible observations from the Space Shuttle of spiral eddies with radii of 5 km [69] and satellite observations of chlorophyll supported the notion that such flows exist in the ocean. The reader is referred to reviews by McWilliams [70] and Klein et al. [71] for full references on submesoscale flows. For the purposes of this chapter, it is important to note that the first order effect of submesoscale flows is a change in the character of the flow from being mostly rotational to mostly convergent at the surface. As such, submesoscale flows are intimately related to the transport of all surface material [72].

Overall, then the emerging picture of transport in the ocean points toward a complex time integrated function of processes ranging from large scales (>100 km and order 1 month), to mesoscales (in the range 100–10 km and month-days), reaching submesoscales (in the range 10–0.1 km and 1 day), and high frequency processes such as waves and 3D turbulence. All these scales and processes interact and, as we will see below, their relative importance depends on the specific problem and application considered.

4 Tools, Prediction, and Validation Methodologies

Several tools are currently available to study and improve our knowledge about the processes discussed above. Indeed, the occurrence of floating macro and microplastics in the marine environment has been traditionally studied by net sampling (for microplastics) and visual surveys (for macroplastics). A detailed description of the methods used for sampling and measuring ocean plastics is out of the scope of this chapter, but comprehensive reviews can be found in Hidalgo-Ruz et al. [73] and Mai et al. [74], among the others. For the purposes of this section, it is important to keep in mind that these techniques have important limitations, as discrete, point samples are often too sparse in space and time to provide enough coverage for accurately describing dynamic events such as accumulation and/or dispersal of plastics at oceanic fronts. In addition, ship-based visual transects and/or surface net trawls often integrate information over scales of a few kilometers or hundreds of meters, which are too coarse to accurately describe accumulation dynamics happening at much smaller spatial scales. Therefore, due to the inadequacy of the most commonly used sampling methods and to the previously mentioned high spatial heterogeneity in plastic distribution over the ocean surface [20], traditional sampling tools should always be integrated with other observational methods that help locate the sampling results within the current velocity field, in order to fully understand the effects of physical processes on plastic transport and accumulation dynamics.

It is worth mentioning that the majority of techniques described in this section – such as HF radar, Lagrangian drifters, satellite and airborne data – are intrinsically devoted to the study of the ocean surface, as this is where most of the work has been done so far. Surface measurements can provide useful synoptic information and are very valuable for identifying the surface expression of oceanographic fronts at various spatial and temporal scales. This is often the first, important step, to guide more in-depth observations at fronts, using research vessels or gliders [75].

From the ocean dynamics point of view it is well known that water column dynamics plays a key role in changing ocean stratification and the mixed layer structure through vertical fluxes of mass, energy, and tracers [58, 76]. Indeed, surface information alone is not able to fully resolve the complex 3D pathways that are expected to influence plastic movements in the ocean. Therefore, besides allowing to characterize surface transport in greater detail, the joint use of surface measurements and 3D numerical models is very promising in investigating and improving the accuracy of plastic transport pathways predictions at sea and for this reason 3D ocean dynamics represents a very active field of research which is already showing promising results [75, 77–79]. At the same time there is mounting evidence about the increasing presence of submerged plastic debris, either on the seafloor [17, 28, 80] and in the water column [15, 81], even though subsurface transport processes and plastic sinking behavior are still far from being thoroughly understood. Vertical dynamics plays also a major role in oil spill events where multiple phases fluxes (liquid, gas, and heat components) interact with ocean turbulence and mixing processes [82–84]. So, as we further suggest in Sect. 6, direct measurements of both the physical ocean dynamics and the tracer's component remain critical for a comprehensive understanding of 3D current structure and its role in the fate and vertical displacement of oil, plastic debris, and other contaminants [21, 85].

4.1 Lagrangian Drifters

Transport properties of oceanic and coastal flows have been widely investigated both at the regional scale and at global level through the deployment of drifters since the 1980s. Modern drifters are floating devices equipped with Global Navigation Satellite Systems (GNSS) receivers designed to drift with currents at different depths according to the shape and depth of the underwater drogue (Fig. 2). The most common and widespread types are the "Coastal Ocean Dynamics Experiment" (CODE) drifter [86] sampling the first meter of water, and the "Surface Velocity Program" (SVP) drifter [87] sampling currents at 15 m depth. Some of these devices may also carry other sensors, for example measuring surface temperature, or atmospheric pressure, used for cross-validation of remote sensing observations and for monitoring climate variability [88]. In the last years, new designs of drifters have been proposed, and particular attention has been devoted to minimize the environmental impact of deploying expendable devices at sea. For instance, the surface (60 cm-depth) drifter developed by the Consortium for Advanced Research of Transport of Hydrocarbon in the Environment (CARTHE) is designed to be 85% biodegradable [89]. The CARTHE drifter has been widely used for massive deployments in the Gulf of Mexico to unravel surface transport pathways in areas deeply exploited by oil and gas companies, to describe the fate of oil released into the environment, and to inform and guide mitigation efforts [90].

GPS drifter-based connectivity studies allow the investigation of passive transport processes through the analysis of drifters' trajectories. These studies find application in many scientific fields, since ocean currents have a crucial role in mixing ocean properties as well as in carrying and dispersing floating material and substances of anthropogenic and natural origin [91]. The investigation of predominant transport pathways driven by currents through drifters, together with modeling

Fig. 2 CARTHE (top) and CODE (bottom) drifters before (left panel) and after (right panel) deployment at sea (Photo credit: Maristella Berta)

and genetic studies [92–95], puts in evidence the connectivity among neighboring sea areas, that is used to study sources and fate of biological material influencing population distribution and dynamics [96]. These studies introduce the important concept of oceanographic distance, for which two sea areas may be geographically next to each other but their exchange of properties and substances is enhanced (or inhibited) by the currents encompassing them [97]. Therefore, a specific area may preferentially exchange water properties with an area not necessarily spatially adjacent, because prevailing sea currents put them in connection [98].

The importance of oceanographic connectivity among remote areas, and of the time scales involved, has been evidenced as well in studies investigating the complex pathways of floating marine debris both at global and basin scales. For example, Maximenko et al. [99] analyzed global drifter trajectories to identify five major oceanic debris accumulation areas centered in the subtropical gyres; van Sebille et al. [100] investigated the origin and evolution of the global garbage patches from observational data of the Global Drifter Program in the open ocean and coastal regions on interannual to centennial timescales; and Zambianchi et al. [101] used the historical Lagrangian dataset collected in the Mediterranean Sea since the 1980s to estimate the probability of debris particles reaching different sub-basins.

The distribution of plastic debris and biological material is also strongly affected by processes taking place at the meso- and submesoscales, and in particular along fronts, where the interaction of different water masses drives the transport and accumulation of substances. For instance, in the framework of the 2016 CARTHE experiment in the Gulf of Mexico, devoted to the investigation of oil spills' dispersion properties, D'Asaro et al. [102] characterized the effectiveness of submesoscale frontal activity to organize surface material by analyzing surface drifter trajectories launched across a frontal zone and converging along the frontal line. Customized surface drifters have also been used by Meyerjürgens et al. [103] to investigate marine litter pathways in the German Bight, characterized by high concentration of floating debris coming from the English Channel, and rich of submesoscale fronts in the coastal and estuarine area, shaping surface accumulation patterns of floating debris. Similarly, an interesting citizen science-based wooden drifter launch and recapture approach was used by Schöneich-Argent and Freund [104], to investigate the dispersal and accumulation of floating litter from coastal, riverine, and offshore sources in the German Bight.

The GPS drifter release campaigns provide valuable and direct information on surface transport trajectories. Dispersion properties can also be drawn from single or multiple particle statistics (Taylor [105] and Ottino [106], respectively; see the review by LaCasce [107], and the recent assessment of different dispersion regimes in the ocean from drifter data by Corrado et al. [108]), but generally require a large number of drifters to be released. The analysis of the dispersion properties gives a picture of the regimes characterizing the circulation, and may help to evidence regions of marine debris accumulation or transport barriers. However, it is difficult and rare to repeat these campaigns with real drifters or to make massive releases such as those performed by the CARTHE experiments. Most observational experiments generally use ~$O(10)$ drifters, whereas virtual particle-tracking experiments can release ~O(10000) or more particles, thus overcoming this limitation. Since numerical models and HF radars provide Eulerian fields, larger statistics could be easily obtained on synthetic Lagrangian trajectories based on the velocity field for advection, plus, in case, a parameterization of subgrid-scale, diffusive processes.

4.2 Eulerian Velocity Fields

As a first approximation plastic particles could be considered as passively advected water parcels. Therefore, knowledge of surface currents is essential to study the distribution of plastics at the ocean surface. There are currently only two ways to obtain temporal evolution of total (geostrophic and ageostrophic components) synoptic surface currents over large horizontal areas: numerical models and HF radars, or a combination of both via data assimilation techniques. Satellite remote sensing currently only provides fairly coarse information on sea level from which only the geostrophic component can be deduced.

Numerical models applied to the ocean are now acknowledged for their validity and performance thanks to the increasing data assimilation and computer capabilities. The Navier-Stokes equations, more or less approximated (hydrostatic, Boussinesq, LES) are discretized and numerically resolved on a mesh, which defines the spatial resolution. Subgrid-scale processes are parametrized with empirical formulae based on theoretical, in-situ and laboratory experiments. While being aware of the limitations of the numerical approach, it is still the only way to have a three-dimensional, synoptic view of hydrology and ocean dynamics, and to make realistic forecasts [109]. The age of machine learning and artificial intelligence may change this principle in the coming years. Most of the research that has been done in the last decade on marine debris was based on these numerical simulations, to predict, and later confirm the existence of the great Pacific garbage patch or other large-scale convergence zones, as well as to highlight the main driving forces behind large-scale plastic distribution patterns (e.g., Lebreton et al. [110]; Maximenko et al. [99], and many others). More recently, higher-resolution numerical models have been also applied to study the dynamics of litter movements at oceanic fronts and at smaller spatial scales [26, 52].

On the observational side, HF radars can produce coastal surface current maps at high spatial and temporal resolution over a relatively wide range (up to 200–250 km). They are land-based remote sensing instruments that gained popularity in the last few decades [111–113]. Their functioning is based on the coherent Bragg scattering from the "lattice" represented by surface gravity wave trains propagating at the surface of coastal basins. This happens when their typical wavelength approaches half the wavelength of the electromagnetic radiation emitted by the radar. A first Doppler shift in the frequency of the backscattered signal occurs if surface waves approach or recede from the transmitting antennas. A second Doppler effect is associated with the presence of currents underlying the gravity waves, from which it is possible to compute the surface velocity field. Since this procedure provides the radial current with respect to the antenna location, it is necessary to have at least two stations to reconstruct the surface velocity vector field. HF radars are characterized by very high spatial and temporal resolution; the former is determined by the operating frequency of the radar, which in turn dictates its range and coverage (see Table I in Rubio et al. [113]). Their characteristics allow to get a repeated in time, synoptic view of surface currents which would be otherwise unobtainable from in-situ or space-based remote sensing techniques. More recently, also ship-based X-band marine radars emerged as useful tools for diagnosing frontal features in offshore areas [114] as well as floating patches of plastic debris in coastal areas [115].

It is clear from above that such systems, as well as the numerical models, do not provide direct measurements of plastics or other pollutants floating at the ocean surface. However, synoptic observations of the surface current fields can be used to reconstruct the transport and fate of buoyant substances, which can then be validated or ground-truthed by field sampling. This is best done through Lagrangian, i.e., water following methods, as discussed in the following sections. However, in the framework of fronts' identification, the approach of using high-resolution numerical models is very powerful, as dynamical velocity-based fronts can be associated with strong gradients in the water mass characteristics. The time evolution of the 3D front shape can be followed [116, 117], as well as the interactions with the atmosphere (wind-stress, air-sea fluxes) and the surrounding environment (gyres, coast, etc.). The required resolution necessary to simulate surface fronts is the mesoscale to submesoscale (depending on the Rossby Radius of deformation). Eventually, the non-hydrostatic versions of a numerical model may improve the resolution of the strong vertical velocities associated with fronts. HF radars could evidence sharp velocity, vorticity, and divergence gradients that may reflect the presence of a coastal front often associated with a boundary current system. By associating information about surface currents obtained with HF radars and numerical models with particle advection, one can obtain a good representation of plastic distribution and accumulation pathways.

4.3 Virtual Trajectories and Lagrangian Coherent Structures

The Eulerian velocity fields described in Sect. 4.2 can be used as a basis to compute virtual particle trajectories, i.e., to provide the so-called Lagrangian view of the flow. The Lagrangian description is especially useful to describe transport, and many Lagrangian metrics and tools have been used to provide indications on passive transport of pollutants such as plastics. Many of the historical metrics used to describe passive particle behavior are based on the concepts of dispersion. Single particle or absolute dispersion describes the average distance covered by a random particle from its initial condition over a given time, while two-particle dispersion provides the average distance between particle pairs [105, 106]. These statistics characterize the main spreading properties of a passive tracer and can be used also to characterize different fluid dynamic regimes (see, for instance, reviews by Corrado et al. [108] and LaCasce [107]). Dispersion properties have been computed in the ocean using in-situ GPS drifters (see Sect. 4.1) but their number is always necessarily limited. For this reason, using virtual particle in numerical models is considered a very useful approach to complement and guide direct in-situ observations.

Different Lagrangian particle-tracking models coupled to ocean circulation models have been widely used to evaluate and predict distribution and beaching of marine debris, assuming a 2D approach. Neglecting the complex 3D physics or the biological processes experienced by different plastic items during their journey does not preclude interesting results. At the global oceanic scale, Lebreton et al. [110] simulated 30 years of plastic debris distribution with specific continuous input (rivers or cities) resulting in accumulation zones in the large oceanic gyres. Martinez et al. [118] or Lebreton et al. [22] included additional transport mechanisms (Stokes drift and windage) and analyzed long-term marine debris pathways. Equivalent studies have been carried out at basin and regional scales [26, 119–121], yielding interesting results on the main regional drivers and patterns of floating litter distribution.

Broadening the perspective to HF radar Lagrangian studies for the very purpose of pollutant transport studies, the stage for specific HF radar application to oil pollution events was set back in 1998: Heron et al. [122] looked at the spreading of a patch of Lagrangian particles advected by the surface current field observed by a HF radar in Port Phillip Bay, Australia. Uttieri et al. [123] studied the onset of a major pollution outbreak in the Gulf of Naples. Mantovanelli and Heron [124] used Lagrangian reconstructions obtained on the basis of HF radar-derived surface velocities in the Australian Great Barrier Reef area to investigate the fate of oil spilled by a coal carrier, as well as the spreading of estuarine and inshore waters probably responsible for the spreading of a fish disease. Transport studies, focused on oil spill applications, based on HF radar data were also performed in the framework of the TOSCA project [125], also using non-conventional blending techniques [126] of radar observations with other sources of information (drifters). More recently, Sciascia et al. [96] and Cianelli et al. [127] looked at biological transport in two coastal regions through forward and backward Lagrangian reconstructions, respectively; the latter paper, in particular, succeeded in discriminating between physical and biological mechanisms in species succession. While some work has been recently done using X-band radars [115], to the best of our knowledge, HF radars have not yet been used to study the movements and dynamics of plastic debris, and this will surely represent a future promising application for this useful technology.

An additional powerful tool that could help to easily identify accumulation or divergence areas in the velocity field that could impact plastic transport is the Lagrangian Coherent Structure (LCS) method, coming from the dynamical systems theory and recently applied to geophysical fluid dynamics [128–132]. LCS are lines or surfaces that can act as barriers between regions where tracers exhibit different behaviors, and can be easily mapped with the largest Finite-Time (FTLE) or Finite-Size (FSLE) Lyapunov Exponents. Where the FTLE measure the exponential rate of separation of trajectories initially close, FSLE measure the time required for particles to separate. These methods require a large number of particles which motivates the use of virtual trajectories and have been recently used to characterize regimes of dispersion and mixing [133, 134], to effectively correlate LCS barriers with frontal regions [135], as well as to successfully identify sources, pathways, barriers and transport dynamics of anthropogenic debris in subtropical embayments [136, 137]. Other derived Lagrangian metrics (Finite Domain Lagrangian divergence or Lagrangian Eddy Kinetic Energy) have also been recently used to relate frontal dynamics and small-scale processes to phytoplankton distribution [138, 139].

4.4 Remote Sensing

Since the launch of TOPEX/Poseidon and ERS-1 missions in the early 1990s, the growing constellation of altimetric satellites observations allowed us to characterize large-scale ocean circulation variability down to mesoscales [140–142]. Presently global sea surface topography is inferred from multi-satellite altimeters combination (Jason-2, Criosat-2, Altika), being able to resolve features of about 50 km at weekly time scales [143]. Several projects are now arising to increase the capability of satellite altimetry to resolve even smaller space scales [144], i.e., submesoscales (in the order of 10 km or less), such as the Surface Water and Ocean Topography (SWOT) mission and the Winds and Currents Mission (WaCM). Monitoring and understanding mesoscale and submesoscale dynamics is essential to interpret the variability of many other physical and biological ocean properties [145–148], as well as to investigate the pathways of transport and dispersion of pollutants at sea, such as oil and plastics. Ongoing efforts are currently underway to develop algorithms able to identify oceanic fronts [149, 150], and some active or incoming satellite missions already provide observations that can be used for marine debris monitoring based on sensors originally meant for monitoring other physical and chemical processes at the ocean surface, with promising results (e.g., [49, 50]). Nevertheless, considering the observational limitations of each sensor (related to spectral resolution and range, sensitivity, spatio-temporal resolution, and coverage), the integration of complementary and targeted sensors on the same observing platform represents a significant technological challenge to enhance the performances of marine debris monitoring from space [151]. The combination of concurrent satellite sensors resolving both the physical dynamics (in the near future down to submesoscales) and the distribution and composition of pollutants at the sea surface will provide crucial information on the role of fronts and currents in influencing the transport and dispersion of floating debris.

Remote sensing is also essential in case of oil spill accidents, during which a timely and accurate assessment of the distribution, thickness, and type of the oil patches and slicks is fundamental to plan efficient response and mitigation strategies [152]. Typically, observations of oil spills are provided by visible, infrared, and hyperspectral sensors on board of satellite missions or aerial surveys. Despite the regular availability and global coverage of satellite missions, the identification and monitoring of a specific spill event can be limited by the time of satellite overpasses and, in case of sensors based on the visible and infrared light spectrum detection, by daylight conditions, cloud cover, or sun glint [153].

Complementary monitoring comes from airborne observations, which can provide targeted surveys of areas affected by oil spills with higher resolution in space and time [154] as well as large-scale surveys of floating macro-litter [155–157]. As for satellite platforms, specific sensors on aircrafts, such as the thermal and imagery components, have also proven their effectiveness in the investigation of small-scale circulation processes with surface signature but related to the dynamics of the upper ocean layer. Rascle et al. [158] and [159] investigated the dynamics of a very intense and sharp (30–50 m) front in the Gulf of Mexico through aerial sun glint images used to retrieve surface roughness and therefore surface current gradients. Their characterization of the frontal activity was confirmed by concurrent remote and in-situ observations, such as satellite SAR and radiometers, X-band radar, and drifters. Another example of airborne observations of the small-scale dynamics is based on infrared images of a very sharp (about 100 m) river plume front surface signature in the Gulf of Mexico (Fig. 3), whose dynamics have been further investigated through in-situ samplings of the water column properties (thermohaline and currents), together with drifter trajectory observations in the proximity of the front [160].

Remote observations of small scale (10–100 m) ocean fronts and of the induced convergence lines trapping floating material can be also retrieved by optically tracking the shape evolution of floater clusters through automated aerial imaging systems mounted on ship-tethered aerostats [161]. In particular, targeted experiments clearly evidenced the influence of sustained wind forcing at the ocean surface, which drives energetic Langmuir circulation. Under such conditions, the floaters dense patch starts aligning in windrows after some minutes from deployment, and in the next hours the windrows spacing grows through windrows merging (Fig. 4). The

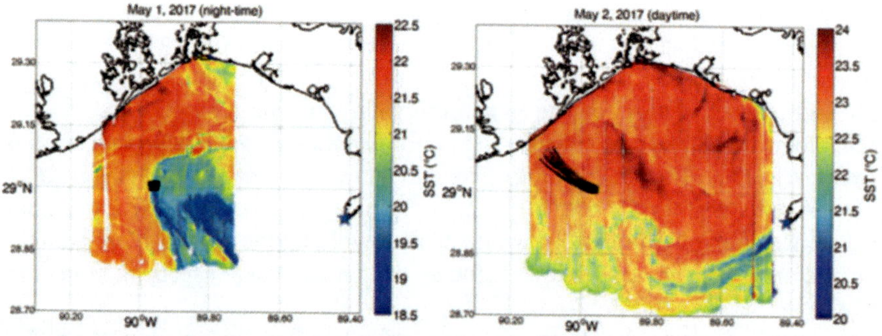

Fig. 3 Left panel: nighttime identification of a plume front (Mississippi river) from airborne infrared images and the contextual deployment of a cluster of drifters along the front (black dots). Right panel: daytime evolution of the same front (advection toward North-West and warm up by sunlight). Black lines represent the trajectories of drifters trapped in the frontal margin. The blue star denotes the position of the river mouth. Color bar limits of the panels are different to enhance details visibility despite daytime temperature excursion (Modified after Solodoch et al. [160])

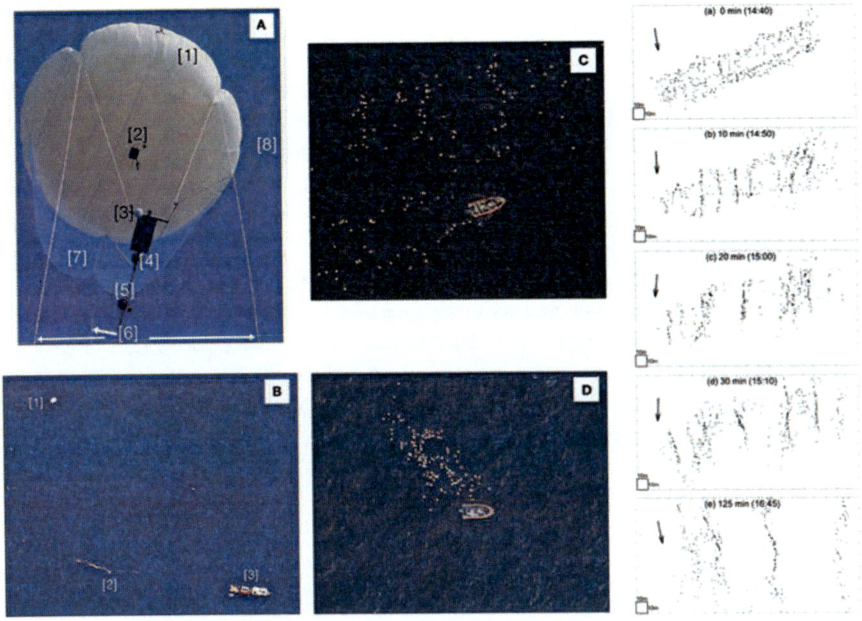

Fig. 4 The study of the formation and evolution over time of Langmuir windrows at the sea surface. Left and middle images (**A**–**D**): ship-tethered aerostat system with camera providing aerial pictures of bamboo plates cluster deployment and evolution by surface currents. Right panels (**a**–**e**): rectified plates position and alignment at different times after deployment (Images reproduced from Carlson et al. [161] and Chang et al. [162]. See these papers for a more accurate description of these images)

evolution of the streaks, aligned with the wind, allowed to characterize crosswind and downwind surface dispersion and the spatio-temporal scales involved [162]. These encouraging applications of aerial observations of contaminant dispersion as well as small-scale physical processes further motivate the development of coordinated observations of both aspects; moreover, the use of drones [163] is also very promising to enhance the comprehension of the role of ocean currents in influencing the distribution and transport of plastics and other pollutants at the sea surface [164–166].

5 Large-Scale, Mesoscale, and Submesoscale Frontal Systems: Selected Case Studies

In this section we discuss some specific cases of frontal systems and their potential impact on plastic distribution in the world ocean, going from the large-scale Antarctic Circumpolar Current (ACC), to the mesoscale, describing a NW Mediterranean coastal front, and lastly, to the submesoscale, with a riverine frontal system in the Northern Gulf of Mexico (GoM). While the scale distinction is very useful to differentiate the main frontal properties, different spatial scales actually interact with each other (as shown in the following sections). Large fronts can be seen as the coalescence of several mesoscale fronts, and mesoscale frontal systems can generate filaments leading to submesoscale fronts and instabilities.

The three case studies differ not only in terms of main scales, but also in terms of basic scientific questions and basic results presented here. The AAC provides a textbook example of how a large-scale frontal system can act as both a barrier and a mixer, because of the great number of processes involved. The nature of the "leakage" through the front and its consequences in terms of transport and tracer distribution is discussed. In the NW Mediterranean case, we concentrate mostly on the specific mechanism of interaction between a mesoscale coastal front and the winds that appear to play a key role in the strengthening and disruption of the front, as well as in the distribution of tracers and plastics. The potential impact of submesoscale features, observed by HF radar within the larger mesoscale frontal system, is also discussed. Finally, in the Northern GoM case, we focus on the influence of submesoscale frontal processes generated by high riverine gradients on the properties of dispersion and trapping of surface tracers. Results from intense in-situ surveys including hundreds of drifters and modeling studies performed in the area during the last decade are discussed, leading to a new perspective on submesoscale frontal dynamics.

5.1 The Antarctic Circumpolar Current: An Imperfect Barrier?

The competition between impermeability and cross-frontal exchange in a jet current has been subject to investigation for decades: a good example of this is provided by studies on the Gulf Stream in this perspective. Back in 1985 Amy Bower and colleagues [167] tried to determine the mutual importance of the two contrasting mechanisms of the current, as "a barrier or a blender," which is still under debate. Different scales of motion appear to play very different roles in such a framework (see, e.g., the very recent paper by [168], who suggest that submesoscale processes may be at the root of an important fraction of the cross-frontal transport). The same ambivalence holds true for one of the most prominent large-scale frontal systems: the Antarctic Circumpolar Current (ACC), where both barring and blending mechanisms seem to coexist. The ACC flows in a latitudinal band of the Southern Ocean dominated by the westerlies, which represent its major forcing. Together with the absence of meridional boundaries (due to emerged land, unique to this latitude range), this zonal pattern of the atmospheric forcing, mediated by the Coriolis effect, allows for the development of a massive circumpolar current connecting the three major ocean basins through the Southern Ocean. This powerful system has in turn a wide influence on the global ocean circulation, the world's climate and large-scale biogeochemical transport.

As first recognized by Deacon [169], the ACC structure is not that of a simple swift zonal current marking the transition from warm, light subtropical water in the north to cold, dense Antarctic water in the south. The ACC is characterized by a very complex structure of multiple jets and fronts. In their classical definition [170], these fronts develop zonally around the entire Antarctic continent; they are the Southern ACC Front (SACCF), the Polar Front (PF), and the Subantarctic Front (SAF), (see Fig. 5) all encompassed by the Southern Tropical Front, or Northern Boundary (NB) in the North and the so-called Southern Boundary (SB) in the South. They can

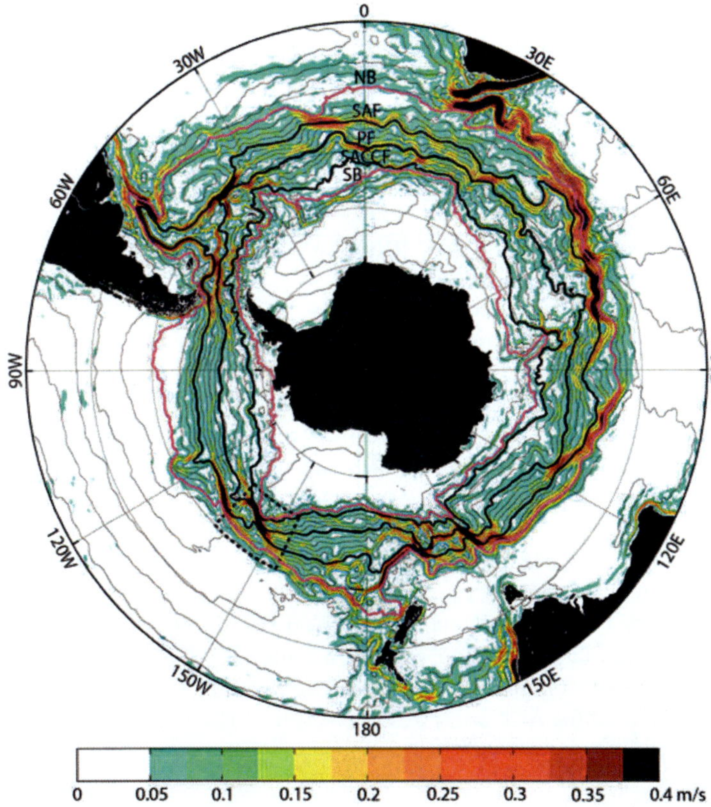

Fig. 5 Mean Dynamic Topography (light black lines every 0.1 m) of the Southern Ocean from the Centre National d'Etudes Spatiales-Collect Localisation Satellites 2018 Mean Dynamic Topography data set. Thick black lines stand for three major Antarctic Circumpolar Current fronts, from the north: SAF = Subantarctic Front; PF = Polar Front; SACCF = Southern Antarctic Circumpolar Current Front. The NB = northern boundary and SB = southern boundary encompassing the Antarctic Circumpolar Current are indicated by thick magenta lines. The color bar refers to the intensity of surface geostrophic currents (Reproduced from Park et al. [171])

be identified in meridional transects as locations of strong property gradients in the horizontal and of steep isopycnal slopes [170].

In the velocity field the latter characteristic is mirrored by maxima of the zonal current [172–174]; this determines the multiple jet pattern which characterizes the ACC. These fronts separate "zones" [175] relatively homogeneous in terms of hydrological and biogeochemical properties and even biological populations, differentiated from, and with little exchange with, one another [173]: the Subantarctic Zone, the Polar Frontal Zone, and the Antarctic Zone [176]. In terms of the three-dimensional velocity field, this is the surface expression of a meridional sequence of convergences and divergences (see the classical association of the Antarctic convergence with the Polar Front by Wyrtky [177]). The barrier effect of such a sharp, banded structure is further enhanced at the surface by the northward Ekman transport associated with the westerlies, in particular in correspondence of the Subantarctic and Polar Fronts [178].

In principle, all the above makes both coastal and offshore Antarctic waters the most isolated region of the global ocean, and for this reason, the Antarctic region has been traditionally considered relatively unaffected by plastic pollution, even though crossing of the Antarctic Polar Front by driftwood and fishing-related materials was already reported in both directions since the early 1960s [179, 180]. In the 1980s, the arrival of microplastics in Antarctic waters was inferred based on the presence of ingested plastics in Antarctic seabird species which remain south of the Polar Front year-round [181, 182], and as a matter of fact, plastic debris has been washing up on sub-Antarctic islands for decades [183–188].

More recently, our view of the circumpolar fronts as impermeable barriers has been further challenged by the evidence that storm-driven surface waves and ocean eddies facilitate crossing of the polar fronts, resulting in more frequent north-south dispersal of drifting kelp and other materials than previously thought [189, 190]. It is thus becoming increasingly clear that circumpolar fronts are not impenetrable, and this finding is further supported by the increasing evidence that the Southern Ocean is not completely exempt from the arrival of plastic pollution from lower latitudes [191–193], suggesting that the Antarctic continent is not as isolated from the rest of the world as previously thought.

The permeability of the ACC occurs most likely where the zonal frontal structure of the current is not sharply defined: splitting and merging of the multiple fronts have been broadly observed in different sectors of the Southern Ocean, giving rise to an intricate structure of the ACC fronts [173, 174, 176, 194–198]. The eddy field has a peculiar importance for the ACC system, because it is key to its response to the atmospheric forcing [172, 199, 200], showing patterns clearly dependent on the bottom topography. In particular, available potential energy due to an increased wind stress yields a growth of available potential energy by increasing the steepness of isopycnals, that is eventually released through baroclinic instability (this is the so-called eddy saturation regime, see [201–203], and references therein). Such instabilities develop mainly in the regions downstream of major topographic features, as nicely illustrated, e.g., by Rintoul [204], which are characterized by meandering motion, where front merging or splitting takes place (Fig. 6).

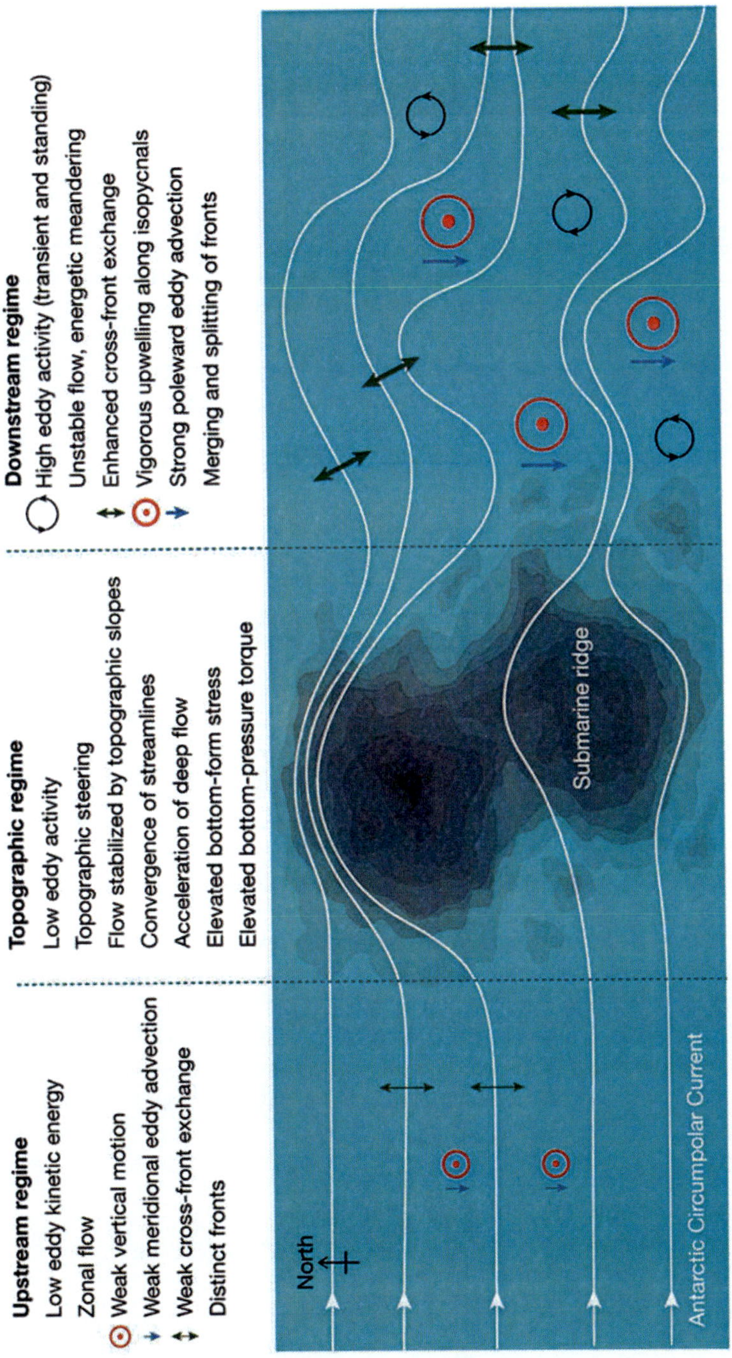

Fig. 6 Processes developing when the ACC interacts with major topographic features (reproduced from Rintoul [204])

What is more important to our context is that these are the regions where enhanced mesoscale eddy activity is responsible for cross-frontal exchange at different depths [205, 206], therefore these instabilities have been defined as "hotspots" of transport, or "leaky jet" segments ([204, 207, 208]; and many references therein).

All the above suggest that the main cross-frontal exchange mechanism, which makes the southernmost portion of the Southern Ocean accessible to floating pollutants, and in particular to plastics, is most likely represented by mesoscale eddy transport. However, the role of these processes in favoring cross-frontal transport of plastics and other buoyant material clearly deserves to be investigated in more detail.

Lastly, it is worth adding that very little is known about submesoscale motions in correspondence of the ACC, stemming from mesoscale features [209, 210]: studies have been mainly focused on their associated vertical transport [211–214], which leaves the question to their possible contribution to cross-frontal transport enhancement open. With this regard, a recent modeling study by Wichmann et al. [215] nicely demonstrated that the role of the ACC in preventing transported matter from entering the Southern Ocean weakens significantly with depth. Moreover, wind- and wave-driven mixing and Langmuir turbulence can greatly enhance the submersion of buoyant plastic debris [216, 217]. Therefore, it can be hypothesized that in the highly dynamic Southern Ocean, microplastics can be more prone to be transported southward by subsurface currents, hence explaining the low concentrations of plastics usually found in Antarctic surface waters [218], especially when compared to subtropical waters north of the STF [192].

As an additional process, Stokes' drift may act as a supplementary cross-frontal, meridional transport mechanism, considering in particular that the Southern Ocean is subject to a very strong wind regime (home to the roaring 40s, furious 50s, and screaming 60s, see https://oceanservice.noaa.gov/facts/roaring-forties.html). Stokes' drift is a second-order effect which causes a weak Lagrangian transport in the direction of propagation of surface waves which is maximal at the surface and decays with depth [219]. Recent modeling studies [190] suggest that the combination of eddy transport and Stokes' drift, in cases of storms, may be at the origin of a limited, occasional meridional "permeability" of the ACC frontal system to biological particle (kelp fragments) southward motion, eventually accessing the periantarctic ocean. As underlined by Onink et al. [24], the incorporation of Stokes drift in Lagrangian simulations can lead to a remarkable increase in the arrival of particles in the Southern Ocean, but it remains unclear whether this is also the case for microplastics and other pollutants, and further investigations on this topic are clearly needed.

5.2 *A Mesoscale Front in the NW Mediterranean Sea*

The Mediterranean Sea is recognized as one of the world's hot-spot for plastics concentration [6, 11, 156, 220] and in this basin, the concentration of plastics and other pollutants often exceeds by orders of magnitude those reported from the main

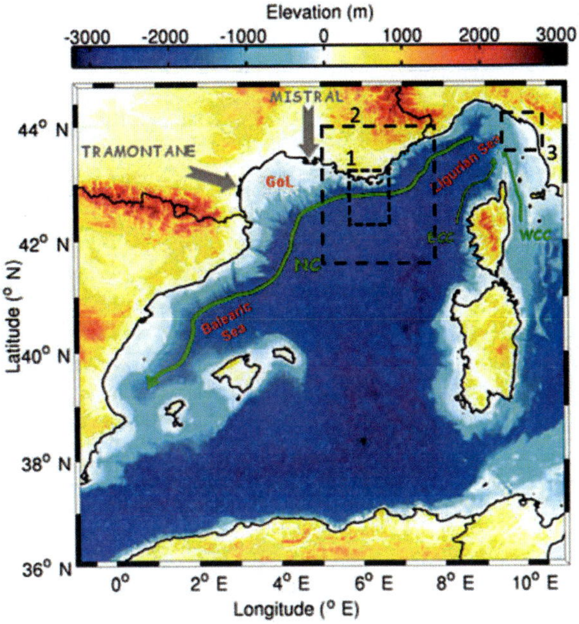

Fig. 7 Schematic of the Western Mediterranean circulation: colors represent topography/bathymetry, green arrows show the main circulation branches to the North known as Eastern Corsica Current (ECC), Western Corsica Current (WCC), and Northern Current (NC). The Gulf of Lion (GoL) is an important dense water formation location due to the interaction with Tramontane and Mistral wind (gray arrows). The three boxes show the location of the measurements reviewed in Sect. 5.2 (Modified from Fig. 1 of Berta et al. [77] with permission of the authors)

global accumulation zones (i.e., the subtropical gyres). The lack of permanent accumulation zones in the Mediterranean Sea has been already reported by multiple authors [101, 120, 121, 221, 222], and at the same time, a relatively high variability is generally observed in both macro and microplastics distribution surveys [6, 223, 224]. This spatial heterogeneity is likely due to a number of different factors: inhomogeneous distribution of plastic sources, meteorological conditions, sampling designs, methodological considerations, or the effect of surface currents [20, 225, 226]. Although a clear understanding of the respective contribution of all these factors is currently lacking, a better discernment of the spatio-temporal variability in plastic concentrations is important not only because it has obvious implications for the validity of basin-scale estimates, but also because it enables a better targeting of future and ongoing mitigation efforts. In this section, we concentrate on mesoscale processes and their intrinsic variability, by providing some key examples related to a major mesoscale front in the North-Western (NW) Mediterranean Sea.

The NW Mediterranean basin is characterized by a broad cyclonic circulation (Fig. 7) with a well-defined boundary current, the Northern Current (NC), or Liguro-Provencal current. The NC is characterized by a frontal zone, separating lighter coastal waters of Atlantic origin [227, 228] from the denser interior water masses, and it has been intensively studied in the last decades from both the experimental and modeling points of view. It forms from the confluence of two northward branches flowing on both sides of the Corsica Island and flows westward in the upper 200–300 m along the continental slope of Italy, France, and Spain. It is strongly seasonally modulated [229–232], with an extended width of approximately 40 km and relatively weak flow (<50 cm/s) during summer, while during winter the flow

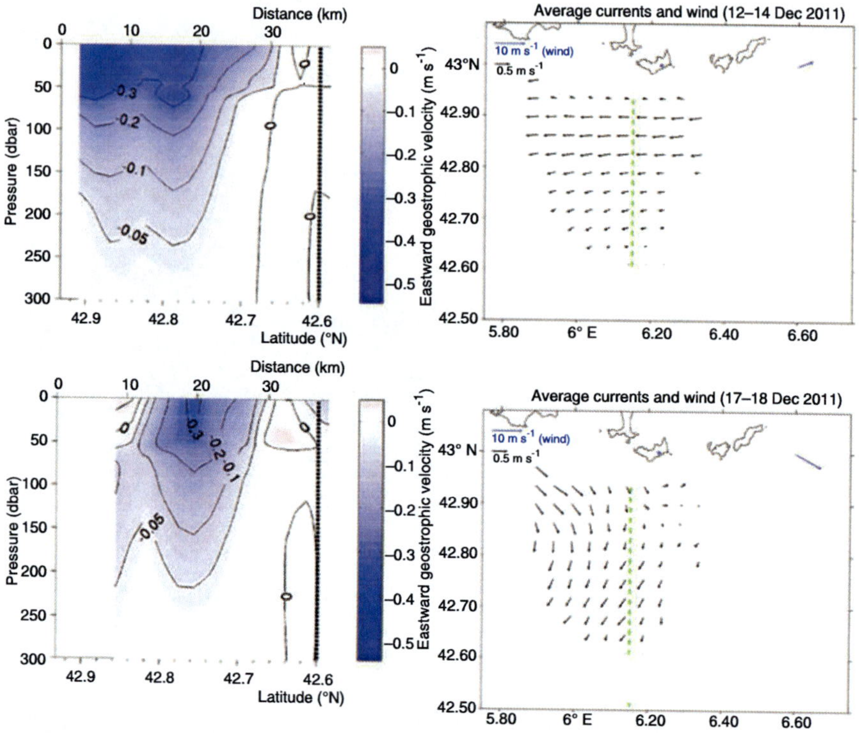

Fig. 8 Examples of Northern Current velocity measured by HF radar (right panels) and glider (left panels) during a calm wind period (upper panels), and a period dominated by westerly winds (lower panels). The black arrows in the right panels are the surface currents from HFR averaged during the period in which the glider transect (green dashed line) falls inside the HFR coverage (Box 1 of Fig. 7). The left panels show the geostrophic velocity along the transect as computed from the hydrographic glider data (modified from Figs. 5 and 6 of Berta et al. [77] with permission of the authors)

accelerates up to 100 cm/s, moves closer to the coast becoming narrower and deeper and the transport increases up to 2 Sv. The flow is also strongly modulated by meso and submesoscale instabilities [227, 233], especially during winter, as well as by meteorological forcing responses [74, 234].

From the dynamical point of view, the NC is a mesoscale current delimited by a frontal area which is primarily characterized by geostrophic balance. Figure 8 (upper right panel) shows a typical velocity pattern at the sea surface, characterized by an along-slope maximum, as measured by HF radars in the area of Toulon in a period of calm winds in December (see Box 1 in Fig. 7). The corresponding depth-dependent geostrophic velocity within the frontal region is shown in Fig. 8 (upper left panel) as estimated by hydrographic data collected along a glider section. The interesting point is that the surface values of the geostrophic velocity exactly coincide with the values measured by HF radars along the section [77], confirming the predominant geostrophic nature of the current. Such balance, though, can be strongly altered by the presence of instabilities and even more so by meteorological forcing. An example is shown in Fig. 8 (lower panels), in presence of a westerly wind lasting more than 3 days, that causes upwelling in the water column, disrupting the front and causing offshore surface transport as captured by the radar. Indeed, in this case the geostrophic velocity strongly differs from the surface radar velocity and the interior current transport decreases to approximately half of its values reaching 0.5 Sv.

As shown in Fig. 8, wind effects play a fundamental role in inducing ageostrophic processes in the frontal NC. Prevailing winds in the area are along the same axis as the current, but they can be either in the same direction as the current (easterlies) or opposite to it (westerlies). Westerly winds, blowing opposite to the current direction, cause upwelling and tend to move offshore the current core. When they are sufficiently strong and long lasting, the NC front can actually be temporarily disrupted and arrested. Easterly winds, on the other hand, blow in the same direction as the current and are downwelling prone and tend to sharpen the front and move the core of the current toward the coast [234].

Modifications to the frontal zone and processes of frontogenesis and frontolysis can also occur because of dynamical instabilities of the system. Mesoscale instabilities in the NC are prevalent [227, 235] and are characterized by the formation of meanders and the generation of secondary circulation with vertical velocities close to the front, as diagnosed in the framework of quasi-geostrophic dynamics [236]. Enhanced vertical motion is expected to occur in presence of submesoscale processes when the geostrophic balance breaks [39, 70], especially within frontal boundaries and filaments. Indications of submesoscale instabilities in the NC are shown also by glider sections [237]. Even though vertical velocities have not been directly measured in the NC front, interior physical and biological patterns suggest that vertical processes are likely to be very relevant [228, 238–240].

Overall, the processes summarized above suggest that the NC frontal zone can impact the distribution and transport of tracers, and in particular of plastics, in several ways:

1. The basic structure of the mesoscale frontal zone and associated currents are expected to induce enhanced horizontal shear dispersion within the current, while acting as a barrier toward the rest of the basin.
2. The barrier effect is expected to be disrupted in presence of flow instabilities and even more so in case of strong winds. Westerly winds are expected to move surface plastics offshore, while easterly winds can generate an onshore stranding effect.
3. Instabilities and wind forcing in the frontal zone can lead to frontogenetic effects, and induce accumulation of buoyant materials at the front. Vertical velocities, with expected subduction from the dense side of the front can influence the vertical distribution of microplastics in a way that depends on their properties and distribution.

Observations of smaller microplastics in the NW Mediterranean area [221, 241, 242], on the other hand, do not show accumulation patterns in the frontal zone. It is not clear at this point whether this is due to the sparse sampling that does not allow to

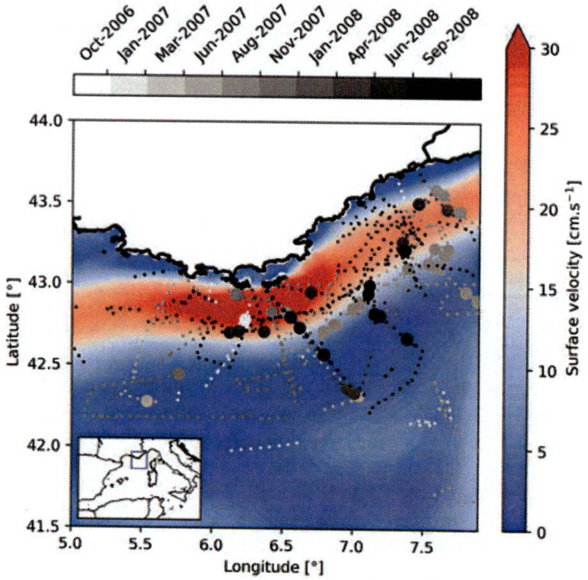

Fig. 9 Floating marine litter distribution observed along the French Riviera (Box 2, Fig. 7) during the EcoOcean cruises (October 2006–October 2008) superimposed to the mean surface velocity. Large and small circles mark the positions and dates of observed high (>10 items/km^2) and low (<10 items/km^2) concentrations of floating debris, respectively (Reproduced from Ourmières et al. [243])

effectively resolve the front, or whether indeed the response of microplastics is different from the response of buoyant macroplastics. This could be due to the complex interplay between the physical processes, including vertical cells, and the nature and sources of microplastics, its buoyancy as a function of materials and fragmentation history. Further studies are needed to sort these points out.

Direct measurements of plastics in the NC frontal zone are scarce, but a first synthesis of results obtained from repeated surveys of floating marine debris has been provided by Ourmières et al. [243] in the French Riviera (Box 2 in Fig. 7). Visual surveys of floating litter were performed as part of the Marine Strategy Framework Directive along 67 transects in the period 2006–2008 [244, 245]. The results are summarized in Fig. 9 where measured floating litter concentrations are shown, superimposed to the NC path as depicted from a numerical model in terms of surface velocity field averaged over the transects period [243]. In most surveys, higher debris concentrations were reported in the NC core and frontal zone, despite the potential importance of coastal sources. Only in 18% of the cases, high litter concentrations were observed offshore with respect to the front, and in most of these cases, a westerly upwelling prone wind event occurred in the previous weeks. From repeated beach survey observations of stranded litter carried out in three beaches in 2010, episodic enhancement appears to be connected not only to rainfall effects on the local river, but also to the occurrence of easterly, downwelling prone winds, in agreement also with previous observations of jellyfish stranding on beaches [234]. Overall, these results are in good agreement with what can be expected from the frontal dynamics effects summarized above.

Another interesting aspect that should be further investigated by future research is the potential role of submesoscale features and coastal eddies on plastic distribution.

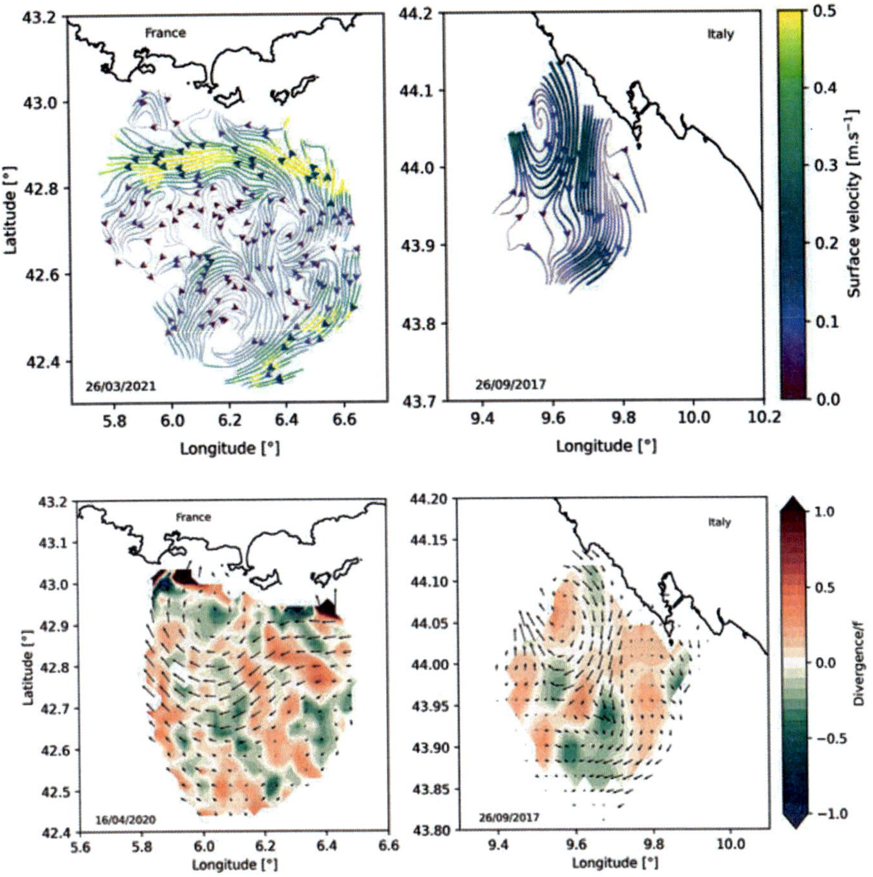

Fig. 10 HF radar surface currents measured on the 26th of April 2021 in the area of Toulon, France (left panels, corresponding to Box 1 of Fig. 7), and in the area of La Spezia, Italy (right panels, corresponding to Box 3 of Fig. 7) measured on the 26th of September 2017. Vorticity/f (upper) and Divergence/f (lower panels) colored background are superimposed to the velocity field

While the NC has a well-defined structure (as shown in Fig. 8), in most areas where the continental shelf is narrow, there are other areas along its path where the presence of an extended shelf allows for the formation of submesoscale vortices, especially during the summer period when the current is weaker. This is the case, for instance, in the Eastern Ligurian Sea (Box 3 in Fig. 7) and in the area of Toulon (Box 1 Fig. 7) as evidenced by HF radar observations. Figure 10, shows an example of small-scale features (order of 5–7 km) depicted from HF radar at both locations at different dates. These features are characterized by high values of divergence and vorticity reaching order f (Coriolis parameter; Berta et al. [246]). As shown by Goldstein et al. [20] and Gove et al. [32], the presence of submesoscale features with high divergence values can lead to a high spatio-temporal inhomogeneity in plastic concentration, with this having obvious implications when trying to extrapolate general features in plastic distribution based on sparse microplastic samples and discrete macro-litter observations.

5.3 Submesoscale Fronts in the Northern Gulf of Mexico

The Gulf of Mexico (GoM) is highly affected by anthropic activities with several consequences, going from frequent red tides episodes [247], massive oil spills due to platform failures or explosions [248], and high macro- and microplastic concentrations [249, 250]. The large-scale circulation in the GoM is dominated by the Loop Current (LC) entering from the Yucatan Channel and flowing anti-cyclonically eastward, while occasionally shedding large LC eddies and displaying a wide variety of oscillations [251]. In the Northern GoM, the circulation is due to an interplay between occasional eddy intrusion, wind forcing, and the outflow of the Mississippi River, entering the Gulf through the typical bird foot delta. The river influx causes a web of freshwater filaments, creating extended regions characterized by submesoscale fronts with extensive submesoscale instabilities [252]. The Deepwater Horizon (DwH) oil spill occurred in the Northern Gulf in 2010 and it is considered to be the largest marine oil spill in the history of the petroleum industry. This event was a dramatic demonstration of the effects of such circulation on transport, since oil is a good tracer and millions of barrels of oil were released into the ocean for almost 3 months, making the impact of convergence regions on the distribution of surface material visibly evident compared to many past oceanographic experiments. However, images taken during this disaster did not provide adequate scientific quantification of the effect of fronts on ocean surface transport.

With the objective of understanding processes that may influence transport of oil near the surface of the ocean, as well as evaluating the accuracy of current-generation ocean models, the CARTHE Consortium (http://carthe.org/) carried out several field campaigns in the northern Gulf of Mexico in the last few years. Due to the high information content and high spatio-temporal variability of ocean fronts, a large number of biodegradable drifters [89] were selected as the observation main tool (Fig. 2). In addition to GPS drifters, a variety of new instruments were created for this program to achieve unprecedented levels of dense and overlapping datasets that span a wide range of spatial and temporal scales (Fig. 11). The reader is referred to Lund et al. [114], Carlson et al. [161], Chang et al. [162], and Lodise et al. [253] for a detailed description of the results obtained from the spectrum of tools shown in Fig. 11.

Overall, on the basis of Gulf of Mexico experiments, the influence of oceanic fronts on the transport of flotsam can be summarized as follows:

1. *Modified dispersion*: On the basis of two-point dispersion metrics, data from 300 drifters released during the Grand LAgrangian Deployment (GLAD) clearly show local or scale dependent transport across all scales measured (summarized in [252] and references therein). A dispersion deficit in the submesoscale/convergent regime can occur when these processes are not resolved adequately in ocean general circulation models. This local dispersion hypothesis contrasts with the alternative hypothesis being that mesoscales can govern dispersion at all scales, including smaller ones (see Fig. 2 in [254]). Once the dispersion rates are measured, it is possible to parameterize this Lagrangian drift in ocean models,

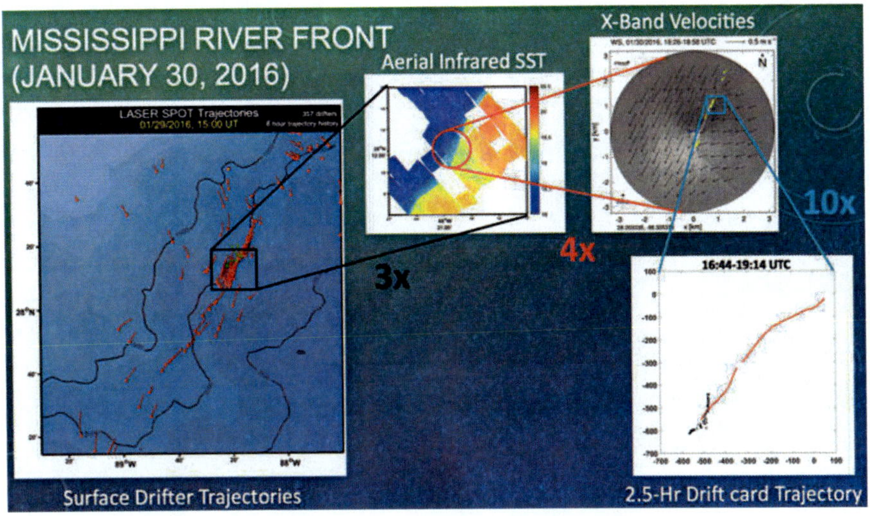

Fig. 11 Collection of overlapping observations from an experiment performed in the Gulf of Mexico, in which different instruments were used to progressively zoom in by a factor of 120x (Image courtesy of Dr. Henry Chang). The reader is referred to Lund et al. [114], Carlson et al. [161], Chang et al. [162], and Lodise et al. [253] for a detailed description of the results obtained from the spectrum of tools shown here

using Lagrangian subgrid models presented in Haza et al. [255]. Poje et al. [252] also showed that geostrophic currents derived from altimeters have large differences, further strengthening the point that ageostrophic (submesoscale) processes are responsible for this difference. By using an ocean model as well as real drifters' trajectories, Pearson et al. [256] showed that over scales <10 km, second-order structure functions of surface drifters consistently have shallower slopes than Eulerian statistics, suggesting that surface drifter structure functions differ systematically. The main reason for this discrepancy is the anisotropy when drifters encounter fronts at scales below 10 km. As shown in a modeling study by Haza et al. [257], enhanced dispersion by submesoscales is not limited to northern Gulf of Mexico where freshwater fronts are in abundance. Mixed layer instabilities taking place along the rim of mesoscale eddies can also create enhanced dispersion by facilitating the flotsam to leak through the mesoscale transport barriers.

2. *Trapping and blocking of surface material*: Huguenard et al. [258], Roth et al. [259], Rascle et al. [158], Androulidakis et al. [260], and Solodoch et al. [160] document that upper ocean fronts created by coastal fresh water outflows act as barriers to transport, exerting a strong influence on the pathways of flotsam coming from the ocean (several examples are shown in Fig. 12). River plume dynamics influence material transport at the surface, its landfall locations and timing. In the coastal experiment described by Solodoch et al. [160] all drifters deployed along the targeted Mississippi plume front that were not recovered by

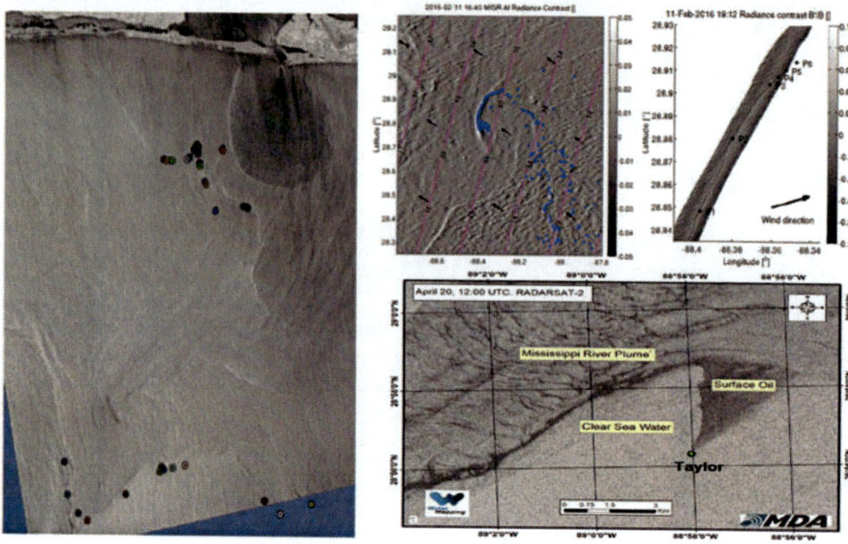

Fig. 12 Images from various expeditions in the Gulf of Mexico showing the impact of upper ocean fronts on blocking the pathways of flotsam. (Left panel) Ocean drifters released in the vicinity of a tidal freshwater outflow near Destin FL during a CARTHE experiment, superimposed on SAR (synthetic aperture radar) image (From Huguenard et al. [258]). (Right upper panels) Drifters (marked in blue and as P1 to P6) released in northern Gulf of Mexico superimposed on airplane-based visible image of sea surface roughness (from Rascle et al. [158]). (Right lower panel) SAR image of Taylor oil spill from northern Gulf of Mexico propagating along the boundary of Mississippi river outflow (from Androulidakis et al. [260]). In all cases, observational data indicate that fronts control transport of upper ocean material

the end of the in-situ survey continued the shoreward propagation and made landfall a few hours later. Trapping of material was known for a long time, but quite dramatic examples of complete blocking and diverting the transport to locations that are hundreds of kilometers from the source of the surface material are documented in the above-mentioned studies. Furthermore, all ocean fronts observed in these Gulf of Mexico experiments had a lateral scale on the $O(1\text{ m})$, and evolved on time scales of several hours, much smaller and faster than known before.

3. *Vertical transport to depth*: Densely-launched drifter data sets also contain a large number of "ocean sink holes." A very dramatic example is described by D'Asaro et al. [102], where 300 drifters initially deployed by two ships on a frontal region over 30 km by 30 km (the size of a large city) collapsed into 30 m by 30 m (the size of a large classroom), corresponding to a contraction in area by one million times in a few days. Vertical velocities of several cm/s were measured by deploying a Lagrangian float on the dense side of the rolling front. The implication of this observation, made possible only by the availability of large enough number of drifters, is that the creation of cyclonic submesoscale eddies by frontal instabilities (anticyclonic ones are prohibited by inertial instability; Wang and

Özgökmen [261]) causes enough vertical velocity to provide a pathway for material transport from surface to the depth of the oceans (see following section).

6 The Vertical Challenge

There is never enough data in the ocean, and this is particularly true for plastics, which are typically sampled and observed in a non-systematic way, and whose distribution may be highly variable in space and time. However, it has been recognized that what we measure at the ocean surface is a very tiny percentage of the global yearly input of plastics into the ocean (definitely $<5\%$, with estimates as low as 1%, as in Koelmans et al. [262]; but see also Cózar et al. [9] or Kaandorp et al. [263]). These figures have been recently downscaled by Weiss et al. [12], however a substantial fraction of the missing plastic budget is probably found at the bottom of the ocean, even reaching the possible deepest locations [264], part of it is probably trapped in coastal areas [13, 14], ingested by marine biota [265], or fragmented below our current detection limit [266]. Still, another large share of the global plastic budget should be probably sought in the water column [16], that is the largest habitat on Earth, where measurements are challenging, and very scarce.

A (small) number of possible mechanisms have been hypothesized as responsible for plastic sinking in the ocean, such as entrainment into organic and inorganic matter, buoyancy loss due to biofouling, incorporation into fecal pellets or transport through the food web flows [21, 36, 37, 263, 267, 268]. Recent numerical simulations showed that particle properties combined with the global spatio-temporal variability of physical and biological properties can result in different microplastic sinking timescales in different regions of the ocean [29, 269]. Also, flash-flood generated hyperpycnal flows [30] and other extreme events such as tropical cyclones and tsunamis [270–273] have all been suggested to play an important role for the sinking of plastic debris in coastal areas, while much less is known about the offshore sinking processes of the most abundant micrometer-sized particles. We also know that mixing is taking place in the upper ocean when winds blow over the surface with enough energy to activate wave-generated mixing and that microplastics can be advected downwards by these processes. When estimating microplastic densities based upon manta skimming of the upper surface for instance, it has been demonstrated that wind- and wave-driven vertical mixing can noticeably affect the estimation of microplastics concentrations, and correction factors shall be applied to compensate for these subduction mechanisms [216, 217, 274].

In general terms, we expect that plastic sinking is due to several interacting processes, involving both direct gravitational sinking and buoyancy transformation on the one side, and vertical advection by marine currents on the other. Both of these broad mechanisms, and their potential interaction, are today still very far from being satisfactorily understood and the discussion is still open. In a recent paper, Egger et al. [15] suggested that the presence of plastics in the water column is mainly the result of a direct “fallout” of small plastic fragments from surface waters, based on a positive correlation between the number of plastic debris at the sea surface and the depth-integrated concentrations of plastic fragments in the water column of the Great Pacific Garbage Patch. It should be noted, though, that the sinking mechanisms of marine plastics are still largely unknown, differently from the sinking mechanisms of oil [275], and vertical fluxes of organic matter and other substances that have been relatively well studied. Plastics sedimentation rates greatly depend on the particle size, shape, density, and fouling-mediated drag. Microplastics are an extremely heterogeneous class of solid contaminants, mostly composed by irregularly-shaped fibers and fragments of varying sizes and materials, and experimental research has shown that under steady and dynamic flow conditions, particle shape can noticeably affect microplastic settling rates [276–278].

The possible role of vertical velocities on plastic fluxes is still largely unknown. Being extremely variable in terms of polymer composition and physicochemical characteristics, plastic particles are not the ideal passive tracer, but they can be treated as such at least in first approximation, since many particles, especially those with densities similar to seawater, behave in a comparable way. Identifying 3D transport pathways of passive tracers in the ocean, including the vertical component, is very challenging since vertical velocities are typically orders of magnitude smaller than horizontal velocities, because of the combined effects of stratification and earth rotation. 3D Lagrangian simulations also require a much greater computational efforts if compared to 2D simulations and are thus still limited in number [29, 53, 269, 276, 279]. Yet, there are specific processes that can enhance vertical downward and upward velocities, through the formation of convergent and divergent areas. An important role in this direction is played by the wind. As already discussed in the previous sections, wind induced Ekman transport at large- and mesoscale can cause significant downwelling and upwelling, inducing vertical velocities that can reach, in specific episodes, orders of magnitude of mm/s [53, 280]. Also, at much smaller scales (order 100 m–1 km) the effects of wind and waves at the ocean surface can induce phenomena such as Langmuir cells and windrows with significant vertical velocities capable of causing smaller plastic particles subduction and mixing [51]. This phenomenon is potentially very important because it is ubiquitous in the world ocean, but is expected to be limited to the upper ocean, i.e. the first 10–20 m. This is suggested also by recent results from drifter clusters [281], showing high convergence values at those scales commonly measured at the surface in several areas of the world ocean, while at 15 m depth the values are significantly lower.

Another key element to understand vertical advection is given by frontal dynamics. At the mesoscale, frontal instabilities are expected to generate vertical velocities of the same order as the Ekman induced velocities, i.e. of the order of mm/s [280], acting over significant time and space scales (order of tens of km and weeks). As a matter of fact, Stukel et al. [282] demonstrated that the flux of sinking particles is twice as high at a frontal system, relative to surrounding waters or to non-frontal conditions and that mesoscale fronts may be responsible for over a quarter of total organic carbon sequestration in the California Current. Although the focus of their work was on organic carbon, it can be assumed that the role of these processes for the global plastic cycle is also likely to be significant.

Frontal dynamics are also intensified and further accelerated at the surface through the formation of submesoscale instabilities [39] that can generate significant vertical velocities, of the order of cm/s [102]. Lapeyre and Klein [283] have shown that thin submesoscale filaments (with a width of 5–10 km) formed by frontogenic eddy interactions are an important source of tracer injection and can lead to large vertical velocities within these structures, concluding that on a global scale, the tracer flux associated with these filaments is as significant as that associated with eddies. Numerical simulations have also shown that buoyant tracers (including microplastics) accumulate at the surface along submesoscale fronts where they are subducted downward into the water column [210].

A possible hypothesis that needs further testing is that submesoscale processes, that are especially prominent in frontal structures, provide a fast and efficient route of transport within the mixed layer until its base (that can be of the order of 50–100 m, depending on ocean region and season), while mesoscale instabilities and associated vertical velocities might be the dominant mechanism to penetrate the deeper ocean on slower but broader scales. This is suggested, for instance, by recent glider observations collected by Mahadevan et al. [75] in a frontal system in the Alboran Sea, showing deep chlorophyll penetration (up to 120–140 m) originated from a mesoscale instability (Fig. 13).

Despite all the progress made in observing and modeling fronts at the surface of the ocean, several knowledge gaps still need to be addressed in order to understand the relevance of submesoscale frontal processes on a global scale. The first is that since submesoscale fronts cover only a small fraction of the ocean's surface, and evolve rapidly, they are hard to find and track using ocean vessels alone. Aerial platforms have limited range and duration, while satellite sensors have spatio-temporal resolution limitations. So, there are clearly observational challenges. Second, the broad probability distributions of these fronts are not known, given the limited observations available. If the frequency deviate from a Gaussian/normal distribution, then this could imply that rare, extreme events, such as the ocean sink hole identified by almost pure coincidence and sheer data density in D'Asaro et al. [102] might in fact govern how plastics and other pollutants are transported into the ocean, similarly to the storm-driven cross-frontal transport observed in the Southern Ocean by Fraser et al. [190].

Finally, just because oceanic fronts are surface signatures of strong vertical downwelling surfaces, this does not mean that they are effective conduits for transporting and sequestering surface material into the ocean's depths. For instance, a Lagrangian float deployed by D'Asaro et al. [102] surfaced later through a broad upwelling zone, without crossing the boundary between the mixed layer base and stratified layers below (Fig. 14). In the case of plastic transport, a possible hypothesis, that would need verification, is that the combination of vertical transport and buoyancy property changes occurring at fronts due to increased biological activity (i.e., fouling) could lead to enhanced and more effective sinking [38]. On the other hand, for neutral or slightly buoyant particles, the death of fouling organism in the

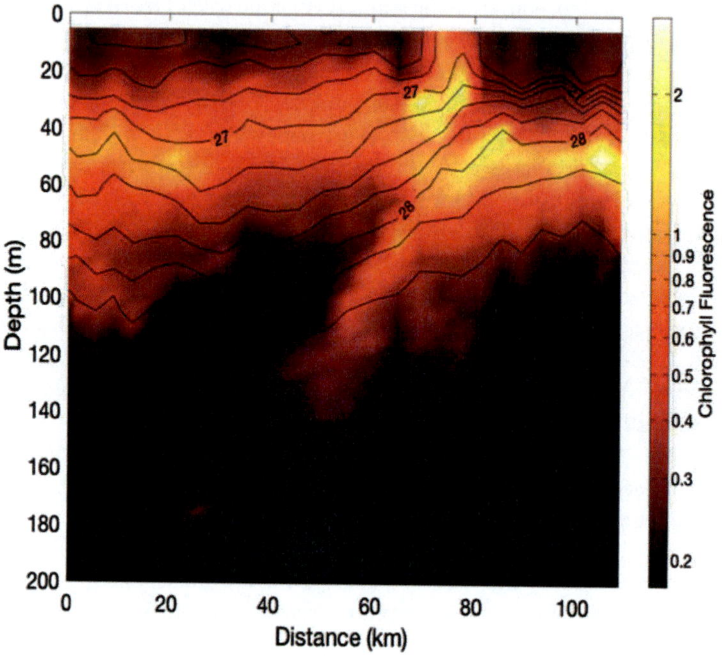

Fig. 13 Chlorophyll (color, log-scale) and density (contours) measured by a glider crossing a gyre front in the Alboran Sea in June 2018, showing the downward advection of phytoplankton from about 50 m to 120 m depth at the front (Reproduced from Fig. 3 of Mahadevan et al. [75] with permission of the authors)

deep would probably lead to a re-surfacing of the particles after a certain amount of time as originally proposed by Ye and Andrady [284], with the so-called yo-yo effect, leading to additional particle redistribution.

Overall, our understanding about the connection between the surface mixed layer, where many plastics and pollutants are first released, and the deeper ocean remains limited, making observation and understanding the vertical transport of material in the ocean a grand challenge. Progress has been made recently in better understanding the so-called Mixed Layer Pump, which appears to contribute significantly to particulate organic carbon export from the mixed layer to the mesopelagic zone especially at high latitudes, where seasonal variations of the mixed layer depth are large [285]. However, whether this process is also relevant for plastic pollution is still unknown. Future research should be aimed at revealing if and how plastic and other contaminants sink at frontal zones, and how relevant these processes are on a global scale, ultimately adding a fundamental piece of knowledge to the plastic puzzle, and helping to finally quantify intercompartmental fluxes between plastic reservoirs (c.f. [286]).

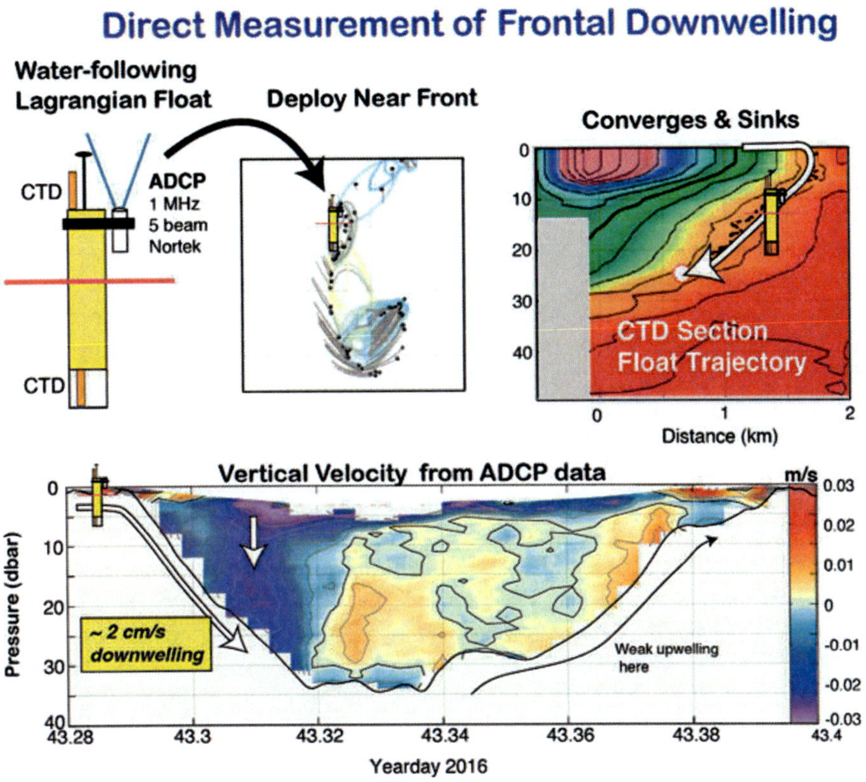

Fig. 14 Behavior of a Lagrangian float deployed along the downwelling frontal boundary of the sink hole as described in D'Asaro et al. [102]

7 Conclusions and Outlook

The global mass of plastic produced by mankind is greater than the overall mass of all terrestrial and marine animals combined [287] and every year, a remarkable portion of mismanaged plastic debris arrives at the ocean surface where it has now become a constant and ubiquitous presence. Nevertheless, our understanding of plastic fluxes, pathways, and fate in oceanic environments is largely incomplete and many of the biological, physical, and geochemical processes that influence the horizontal and vertical transport of plastic at sea (e.g., degradation, sinking, fragmentation, ingestion, beaching) still need to be investigated in detail.

In this chapter we focused our attention on the role of oceanic fronts in influencing the accumulation of plastics and other pollutants, shaping their distribution over the ocean surface, and possibly contributing to their final sinking. We showed that in correspondence with frontal systems, convergence zones potentially accumulate large amounts of floating debris, while at the same time they can favor or prevent the spread of floating material between compartments. We also discussed the still challenging questions related to the potential role of fronts in the vertical transport of microplastics. Direct measurements of vertical microplastic distribution at fronts are not available yet, and also the investigation of vertical velocity mechanisms is a still open and active area of research, indicating that downwelling frontal areas can potentially export large amounts of debris into the deep sea over relatively short spatio-temporal scales. Oceanic fronts are difficult to locate, but given their ubiquitous presence in the world's ocean, their contribution to the global plastic cycle is probably not negligible and their role in creating surface concentration zones and in removing debris from the sea surface clearly deserves further investigation.

The interaction of mesoscale and submesoscale processes with the general oceanic circulation can potentially create tremendously complex patterns in the distribution of plastic debris. Even an ocean at rest could accommodate a significant inhomogeneity through geographical variation in plastic sources and environmental variables, time-dependent forcings, and different polymer behaviors. Understanding the mechanisms responsible for such a high observed variability in plastic distributions is thus an extremely difficult task. Recent advances in measurement technologies will likely play a pivotal role in advancing future research capabilities. At the same time, the linkage of such measurements with more advanced numerical models (e.g., based on artificial intelligence, data assimilation and machine learning techniques) will likely yield important insights into the mechanisms controlling plastic patchiness in the ocean.

Future research should be focused on better understanding plastic movements in the ocean, in order to more efficiently target intervention and mitigation measures. Being punctiform in space and time, it is becoming increasingly clear that traditional plastic sampling alone is not comprehensive enough to describe complex evolving dynamics such as those involved with frontal systems and oceanic circulation. Improving remote sensing capabilities, developing automated and/or autonomous plastic sensors, and most of all, implementing process-oriented sampling designs are urgent priorities in marine litter research. Estimating the global relevance of all the processes reviewed by this chapter is not an easy task. Future studies should ideally make use of an interdisciplinary array of innovative tools and integrated earth observational instruments that will ultimately enable us to predict and locate more precisely fronts, plastic accumulation areas, concentration hotspots and at last, more effective remediation targets.

Acknowledgments The authors of this chapter received funding through the following research projects: Gulf of Mexico Research Initiative (GoMRI); IPSODES "Investigating the predictability of the Southern Ocean dynamics through ensemble simulation hindcasts" (PNRA18_00199-C); PRIN EMME "Exploring the fate of Mediterranean microplastics: from distribution pathways to biological effects" (2017WERYZP); JPI-Oceans FACTS "Fluxes and Fate of Microplastics in Northern European Waters"; ONR award N00014-18-1-2782 within the CALYPSO DRI; JERICO-S3 (grant agreement No. 871153) and NAUTILOS "New Approach To Underwater Technologies For Innovative, Low-Cost Ocean Observation (grant agreement No. 101000825) projects that received funding from the European Union's Horizon 2020 research and innovation programme; and by the Discovery Element of the European Space Agency's Basic Activities project WASP "Mapping Windrows as Proxy for Marine Litter Monitoring from Space" (ESA Contract. 4000130627/20/NL/GLC).

References

1. Geyer R, Jambeck JR, Law KL (2017) Production, use, and fate of all plastics ever made. Sci Adv 3:e1700782. https://doi.org/10.1126/sciadv.1700782
2. Lebreton L, Andrady A (2019) Future scenarios of global plastic waste generation and disposal. Palgrave Commun 5:6. https://doi.org/10.1057/s41599-018-0212-7
3. Jambeck JR, Geyer R, Wilcox C et al (2015) Plastic waste inputs from land into the ocean. Science 347:768–771. https://doi.org/10.1126/science.1260352
4. Borrelle SB, Ringma J, Law KL et al (2020) Predicted growth in plastic waste exceeds efforts to mitigate plastic pollution. Science 369:1515–1518. https://doi.org/10.1126/science.aba3656
5. Law KL (2017) Plastics in the marine environment. Annu Rev Mar Sci 9:205–229. https://doi.org/10.1146/annurev-marine-010816-060409
6. Suaria G, Avio CG, Mineo A et al (2016) The Mediterranean plastic soup: synthetic polymers in Mediterranean surface waters. Sci Rep 6:37551. https://doi.org/10.1038/srep37551
7. Tirelli V, Suaria G, Lusher AL (2020) Microplastics in polar samples. In: Rocha-Santos T, Costa M, Mouneyrac C (eds) Handbook of microplastics in the environment. Springer International Publishing, Cham, pp 1–42
8. Gong J, Xie P (2020) Research progress in sources, analytical methods, eco-environmental effects, and control measures of microplastics. Chemosphere 254:126790. https://doi.org/10.1016/j.chemosphere.2020.126790
9. Cozar A, Echevarria F, Gonzalez-Gordillo JI et al (2014) Plastic debris in the open ocean. Proc Natl Acad Sci 111:10239–10244. https://doi.org/10.1073/pnas.1314705111
10. Eriksen M, Lebreton LCM, Carson HS et al (2014) Plastic pollution in the world's oceans: more than 5 trillion plastic pieces weighing over 250,000 tons afloat at sea. PLoS One 9: e111913. https://doi.org/10.1371/journal.pone.0111913
11. van Sebille E, Wilcox C, Lebreton L et al (2015) A global inventory of small floating plastic debris. Environ Res Lett 10:124006. https://doi.org/10.1088/1748-9326/10/12/124006
12. Weiss L, Ludwig W, Heussner S et al (2021) The missing ocean plastic sink: gone with the rivers. Science 373:107–111. https://doi.org/10.1126/science.abe0290
13. Olivelli A, Hardesty BD, Wilcox C (2020) Coastal margins and backshores represent a major sink for marine debris: insights from a continental-scale analysis. Environ Res Lett 15:074037. https://doi.org/10.1088/1748-9326/ab7836
14. Lebreton L, Egger M, Slat B (2019) A global mass budget for positively buoyant macroplastic debris in the ocean. Sci Rep 9:12922. https://doi.org/10.1038/s41598-019-49413-5
15. Egger M, Sulu-Gambari F, Lebreton L (2020) First evidence of plastic fallout from the North Pacific garbage patch. Sci Rep 10:7495. https://doi.org/10.1038/s41598-020-64465-8
16. Pabortsava K, Lampitt RS (2020) High concentrations of plastic hidden beneath the surface of the Atlantic Ocean. Nat Commun 11:4073. https://doi.org/10.1038/s41467-020-17932-9
17. Woodall LC, Sanchez-Vidal A, Canals M et al (2014) The deep sea is a major sink for microplastic debris. R Soc Open Sci 1:140317. https://doi.org/10.1098/rsos.140317
18. Barrett J, Chase Z, Zhang J, Holl MMB, Willis K, Williams A, Hardesty BD, Wilcox C (2020) Microplastic pollution in deep-sea sediments from the Great Australian Bight. Front Mar Sci 7:576170. https://doi.org/10.3389/fmars.2020.576170
19. Thompson RC, Olsen Y, Mitchell RP et al (2004) Lost at sea: where is all the plastic? Science 304:838–838. https://doi.org/10.1126/science.1094559

20. Goldstein MC, Titmus AJ, Ford M (2013) Scales of spatial heterogeneity of plastic marine debris in the Northeast Pacific Ocean. PLoS One 8:e80020. https://doi.org/10.1371/journal.pone.0080020
21. van Sebille E, Aliani S, Law KL et al (2020) The physical oceanography of the transport of floating marine debris. Environ Res Lett 15:023003. https://doi.org/10.1088/1748-9326/ab6d7d
22. Lebreton L, Slat B, Ferrari F et al (2018) Evidence that the great pacific garbage patch is rapidly accumulating plastic. Sci Rep 8:4666. https://doi.org/10.1038/s41598-018-22939-w
23. Brach L, Deixonne P, Bernard M-F et al (2018) Anticyclonic eddies increase accumulation of microplastic in the North Atlantic subtropical gyre. Mar Pollut Bull 126:191–196. https://doi.org/10.1016/j.marpolbul.2017.10.077
24. Onink V, Wichmann D, Delandmeter P, Sebille E (2019) The role of Ekman currents, geostrophy, and stokes drift in the accumulation of floating microplastic. J Geophys Res Oceans 124:1474–1490. https://doi.org/10.1029/2018JC014547
25. Shanks AL (1987) The onshore transport of an oil spill by internal waves. Science 235:1198–1200. https://doi.org/10.1126/science.235.4793.1198
26. Stanev EV, Ricker M (2019) The fate of marine litter in semi-enclosed seas: a case study of the Black Sea. Front Mar Sci 6:660. https://doi.org/10.3389/fmars.2019.00660
27. Zhang Z, Wang W, Qiu B (2014) Oceanic mass transport by mesoscale eddies. Science 345:322–324. https://doi.org/10.1126/science.1252418
28. Canals M, Pham CK, Bergmann M et al (2020) The quest for seafloor macrolitter: a critical review of background knowledge, current methods and future prospects. Environ Res Lett. https://doi.org/10.1088/1748-9326/abc6d4
29. Lobelle D, Kooi M, Koelmans AA et al (2021) Global modeled sinking characteristics of biofouled microplastic. J Geophys Res Oceans. https://doi.org/10.1029/2020JC017098
30. Pierdomenico M, Casalbore D, Chiocci FL (2019) Massive benthic litter funnelled to deep sea by flash-flood generated hyperpycnal flows. Sci Rep 9:5330. https://doi.org/10.1038/s41598-019-41816-8
31. Gardner WD, Richardson MJ, Mishonov AV (2018) Global assessment of benthic nepheloid layers and linkage with Upper Ocean dynamics. Earth Planet Sci Lett 482:126–134. https://doi.org/10.1016/j.epsl.2017.11.008
32. Gove JM, Whitney JL, McManus MA et al (2019) Prey-size plastics are invading larval fish nurseries. PNAS 116:24143–24149. https://doi.org/10.1073/pnas.1907496116
33. González Carman V, Acha EM, Maxwell SM et al (2014) Young green turtles, Chelonia mydas, exposed to plastic in a frontal area of the SW Atlantic. Mar Pollut Bull 78:56–62. https://doi.org/10.1016/j.marpolbul.2013.11.012
34. Josse E, Dagorn L, Bertrand A (2000) Typology and behaviour of tuna aggregations around fish aggregating devices from acoustic surveys in French Polynesia. Aquat Living Resour 13:183–192. https://doi.org/10.1016/S0990-7440(00)00051-6
35. Fazey FMC, Ryan PG (2016) Biofouling on buoyant marine plastics: an experimental study into the effect of size on surface longevity. Environ Pollut 210:354–360. https://doi.org/10.1016/j.envpol.2016.01.026
36. Kooi M, van Nes EH, Scheffer M, Koelmans AA (2017) Ups and downs in the ocean: effects of biofouling on vertical transport of microplastics. Environ Sci Technol 51:7963–7971. https://doi.org/10.1021/acs.est.6b04702
37. Kowalski N, Reichardt AM, Waniek JJ (2016) Sinking rates of microplastics and potential implications of their alteration by physical, biological, and chemical factors. Mar Pollut Bull 109:310–319. https://doi.org/10.1016/j.marpolbul.2016.05.064
38. Kaiser D, Kowalski N, Waniek JJ (2017) Effects of biofouling on the sinking behavior of microplastics. Environ Res Lett 12:124003. https://doi.org/10.1088/1748-9326/aa8e8b
39. McWilliams JC (2021) Oceanic frontogenesis. Annu Rev Mar Sci 13:227–253. https://doi.org/10.1146/annurev-marine-032320-120725

40. Lapeyre G, Klein P, Hua BL (2006) Oceanic restratification forced by surface frontogenesis. J Phys Oceanogr 36:1577–1590. https://doi.org/10.1175/JPO2923.1
41. Acha EM, Piola A, Iribarne O, Mianzan H (2015) Ecological processes at marine fronts: oases in the ocean. Springer, Cham
42. Belkin IM, Cornillon PC, Sherman K (2009) Fronts in large marine ecosystems. Prog Oceanogr 81:223–236. https://doi.org/10.1016/j.pocean.2009.04.015
43. Mann KH, Lazier JRN (2006) Dynamics of marine ecosystems: biological-physical interactions in the oceans.3rd edn. Blackwell Pub, Malden
44. Klemas V, Polis DF (1977) A study of density fronts and their effects on coastal pollutants. Remote Sens Environ 6:95–126. https://doi.org/10.1016/0034-4257(77)90009-8
45. Klemas V (1980) Remote sensing of coastal fronts and their effects on oil dispersion. Int J Remote Sens 1:11–28. https://doi.org/10.1080/01431168008948208
46. Spall MA (1995) Frontogenesis, subduction, and cross-front exchange at upper ocean fronts. J Geophys Res 100:2543. https://doi.org/10.1029/94JC02860
47. Arias M, Sumerot R, Delaney J et al (2021) Mapping windrows as proxies for marine litter monitoring from space (WASP). EGU general assembly 2021, online, 19–30 Apr 2021, EGU21-15275. https://doi.org/10.5194/egusphere-egu21-15275
48. Martínez-Vicente V, Clark JR, Corradi P et al (2019) Measuring marine plastic debris from space: initial assessment of observation requirements. Remote Sens 11:2443. https://doi.org/10.3390/rs11202443
49. Topouzelis K, Papageorgiou D, Suaria G, Aliani S (2021) Floating marine litter detection algorithms and techniques using optical remote sensing data: a review. Mar Pollut Bull 170:112675. https://doi.org/10.1016/j.marpolbul.2021.112675
50. Biermann L, Clewley D, Martinez-Vicente V et al (2020) Finding plastic patches in coastal waters using optical satellite data. Sci Rep 10:5364. https://doi.org/10.1038/s41598-020-62298-z
51. Cózar A, Aliani S, Basurko OC et al (2021) Marine litter windrows: a strategic target to understand and manage the ocean plastic pollution. Front Mar Sci 8:571796. https://doi.org/10.3389/fmars.2021.571796
52. Gutow L, Ricker M, Holstein JM et al (2018) Distribution and trajectories of floating and benthic marine macrolitter in the south-eastern North Sea. Mar Pollut Bull 131:763–772. https://doi.org/10.1016/j.marpolbul.2018.05.003
53. Ricker M, Stanev EV (2020) Circulation of the European northwest shelf: a Lagrangian perspective. Ocean Sci 16:637–655. https://doi.org/10.5194/os-16-637-2020
54. Ruiz I, Basurko OC, Rubio A et al (2020) Litter windrows in the south-east coast of the Bay of Biscay: an ocean process enabling effective active fishing for litter. Front Mar Sci 7:308. https://doi.org/10.3389/fmars.2020.00308
55. Shanks AL (2021) Observational evidence and open questions on the role of internal tidal waves on the concentration and transport of floating plastic debris. Front Mar Sci 8:621062. https://doi.org/10.3389/fmars.2021.621062
56. Liu K, Courtene-Jones W, Wang X et al (2020) Elucidating the vertical transport of microplastics in the water column: a review of sampling methodologies and distributions. Water Res 186:116403. https://doi.org/10.1016/j.watres.2020.116403
57. Mensa JA, Garraffo Z, Griffa A et al (2013) Seasonality of the submesoscale dynamics in the Gulf Stream region. Ocean Dyn 63:923–941. https://doi.org/10.1007/s10236-013-0633-1
58. McWilliams JC (2019) A survey of submesoscale currents. Geosci Lett 6:3. https://doi.org/10.1186/s40562-019-0133-3
59. The MODE Group (1978) The mid-ocean dynamics experiment. Deep-Sea Res 25:859–910. https://doi.org/10.1016/0146-6291(78)90632-X
60. Holland W (1978) The role of mesoscale eddies in the general circulation of the ocean numerical experiments using a wind-driven quasi-geostrophic model. J Phys Oceanogr 8:363–392

61. Fu L, Smith R (1996) Global ocean circulation from satellite altimetry and high-resolution computer simulation. Bull Am Meteorol Soc 77:2625–2636. https://doi.org/10.1175/1520-0477(1996)077%3C2625:GOCFSA%3E2.0.CO;2
62. Ducet N, Le Traon PY, Reverdin G (2000) Global high-resolution mapping of ocean circulation from TOPEX/Poseidon and ERS-1 and -2. J Geophys Res 105:19477–19498. https://doi.org/10.1029/2000JC900063
63. Le Traon PY, Klein P, Hua BL, Dibarboure G (2008) Do altimeter wavenumber spectra agree with the interior or surface quasigeostrophic theory? J Phys Oceanogr 38:1137–1142. https://doi.org/10.1175/2007JPO3806.1
64. Langmuir I (1938) Surface motion of water induced by wind. Science 87:119–123. https://doi.org/10.1126/science.87.2250.119
65. Stanev EV, Ricker M (2020) Interactions between barotropic tides and mesoscale processes in deep ocean and shelf regions. Ocean Dyn 70:713–728. https://doi.org/10.1007/s10236-020-01348-6
66. Tchilibou M, Gourdeau L, Morrow R et al (2018) Spectral signatures of the tropical Pacific dynamics from model and altimetry: a focus on the meso−/submesoscale range. Ocean Sci 14:1283–1301. https://doi.org/10.5194/os-14-1283-2018
67. McWilliams JC (2008) Fluid dynamics at the margin of rotational control. Environ Fluid Mech 8:441–449. https://doi.org/10.1007/s10652-008-9081-8
68. Müller P, McWilliams J, Molemaker M (2005) Routes to dissipation in the ocean: the two-dimensional/three-dimensional turbulence conundrum. In: Baumert HZ, Simpson J, Sündermann J (eds) Marine turbulence: theories, observations, and models. Results of the CARTUM project. Cambridge University Press
69. Munk W, Armi L, Fischer K, Zachariasen F (2000) Spirals on the sea. Proc R Soc Lond A 456:1217–1280. https://doi.org/10.1098/rspa.2000.0560
70. McWilliams JC (2016) Submesoscale currents in the ocean. Proc R Soc A 472:20160117. https://doi.org/10.1098/rspa.2016.0117
71. Klein P, Lapeyre G, Siegelman L et al (2019) Ocean-scale interactions from space. Earth Space Sci 6:795–817. https://doi.org/10.1029/2018EA000492
72. Lévy M, Ferrari R, Franks PJS et al (2012) Bringing physics to life at the submesoscale. Geophys Res Lett 39. https://doi.org/10.1029/2012GL052756
73. Hidalgo-Ruz V, Gutow L, Thompson RC, Thiel M (2012) Microplastics in the marine environment: a review of the methods used for identification and quantification. Environ Sci Technol 46:3060–3075. https://doi.org/10.1021/es2031505
74. Mai L, Bao L-J, Shi L et al (2018) A review of methods for measuring microplastics in aquatic environments. Environ Sci Pollut Res 25:11319–11332. https://doi.org/10.1007/s11356-018-1692-0
75. Mahadevan A, Pascual A, Rudnick DL et al (2020) Coherent pathways for vertical transport from the surface ocean to interior. Bull Am Meteorol Soc 101:E1996–E2004. https://doi.org/10.1175/BAMS-D-19-0305.1
76. Thomas LN, Tandon A, Mahadevan A (2008) Submesoscale processes and dynamics. In: Hecht MW, Hasumi H (eds) Geophysical monograph series. American Geophysical Union, Washington, pp 17–38
77. Berta M, Bellomo L, Griffa A et al (2018) Wind-induced variability in the northern current (northwestern Mediterranean Sea) as depicted by a multi-platform observing system. Ocean Sci 14:689–710. https://doi.org/10.5194/os-14-689-2018
78. Manso-Narvarte I, Fredj E, Jordà G et al (2020) 3D reconstruction of ocean velocity from high-frequency radar and acoustic Doppler current profiler: a model-based assessment study. Ocean Sci 16:575–591. https://doi.org/10.5194/os-16-575-2020
79. Manso-Narvarte I, Rubio A, Jordà G et al (2021) Three-dimensional characterization of a coastal mode-water eddy from multiplatform observations and a data reconstruction method. Remote Sens 13:674. https://doi.org/10.3390/rs13040674

80. Chiba S, Saito H, Fletcher R et al (2018) Human footprint in the abyss: 30 year records of deep-sea plastic debris. Mar Policy 96:204–212. https://doi.org/10.1016/j.marpol.2018.03.022
81. Choy CA, Robison BH, Gagne TO et al (2019) The vertical distribution and biological transport of marine microplastics across the epipelagic and mesopelagic water column. Sci Rep 9:7843. https://doi.org/10.1038/s41598-019-44117-2
82. Boufadel M, Liu R, Zhao L et al (2020) Transport of oil droplets in the upper ocean: impact of the eddy diffusivity. J Geophys Res Oceans 125. https://doi.org/10.1029/2019JC015727
83. Fabregat Tomàs A, Poje AC, Özgökmen TM, Dewar WK (2016) Dynamics of multiphase turbulent plumes with hybrid buoyancy sources in stratified environments. Phys Fluids 28:095109. https://doi.org/10.1063/1.4963313
84. Soloviev AV, Haus BK, McGauley MG et al (2016) Surface dynamics of crude and weathered oil in the presence of dispersants: laboratory experiment and numerical simulation. J Geophys Res Oceans 121:3502–3516. https://doi.org/10.1002/2015JC011533
85. Röhrs J, Dagestad K-F, Asbjørnsen H et al (2018) The effect of vertical mixing on the horizontal drift of oil spills. Ocean Sci 14:1581–1601. https://doi.org/10.5194/os-14-1581-2018
86. Davis RE (1985) Drifter observations of coastal surface currents during CODE: the method and descriptive view. J Geophys Res 90:4741. https://doi.org/10.1029/JC090iC03p04741
87. Niiler PP, Sybrandy AS, Bi K et al (1995) Measurements of the water-following capability of holey-sock and TRISTAR drifters. Deep-Sea Res I Oceanogr Res Pap 42:1951–1964. https://doi.org/10.1016/0967-0637(95)00076-3
88. Centurioni LR, Turton J, Lumpkin R et al (2019) Global in situ observations of essential climate and ocean variables at the air–sea interface. Front Mar Sci 6:419. https://doi.org/10.3389/fmars.2019.00419
89. Novelli G, Guigand CM, Cousin C et al (2017) A biodegradable surface drifter for ocean sampling on a massive scale. J Atmos Ocean Technol 34:2509–2532. https://doi.org/10.1175/JTECH-D-17-0055.1
90. D'Asaro EA, Carlson DF, Chamecki M et al (2020) Advances in observing and understanding small-scale open ocean circulation during the Gulf of Mexico research initiative era. Front Mar Sci 7:349. https://doi.org/10.3389/fmars.2020.00349
91. van Sebille E, Griffies SM, Abernathey R et al (2018) Lagrangian Ocean analysis: fundamentals and practices. Ocean Model 121:49–75. https://doi.org/10.1016/j.ocemod.2017.11.008
92. Alberto F, Raimondi PT, Reed DC et al (2011) Isolation by oceanographic distance explains genetic structure for Macrocystis pyrifera in the Santa Barbara Channel. Mol Ecol 20:2543–2554. https://doi.org/10.1111/j.1365-294X.2011.05117.x
93. Berline L, Rammou A-M, Doglioli A et al (2014) A connectivity-based eco-regionalization method of the Mediterranean Sea. PLoS One 9:e111978. https://doi.org/10.1371/journal.pone.0111978
94. Carlson DF, Griffa A, Zambianchi E et al (2016) Observed and modeled surface Lagrangian transport between coastal regions in the Adriatic Sea with implications for marine protected areas. Cont Shelf Res 118:23–48. https://doi.org/10.1016/j.csr.2016.02.012
95. White C, Selkoe KA, Watson J et al (2010) Ocean currents help explain population genetic structure. Proc R Soc B 277:1685–1694. https://doi.org/10.1098/rspb.2009.2214
96. Sciascia R, Berta M, Carlson DF et al (2018) Linking sardine recruitment in coastal areas to ocean currents using surface drifters and HF radar: a case study in the Gulf of Manfredonia, Adriatic Sea. In situ observations/current field/surface/Mediterranean Sea. Ocean Sci 14 (6):1461–1482
97. Celentano P, Falco P, Zambianchi E (2020) Surface connection between the Ionian Sea and different areas of the Mediterranean derived from drifter data. Deep-Sea Res I Oceanogr Res Pap 166:103431. https://doi.org/10.1016/j.dsr.2020.103431
98. Jönsson BF, Watson JR (2016) The timescales of global surface-ocean connectivity. Nat Commun 7:11239. https://doi.org/10.1038/ncomms11239

99. Maximenko N, Hafner J, Niiler P (2012) Pathways of marine debris derived from trajectories of Lagrangian drifters. Mar Pollut Bull 65:51–62. https://doi.org/10.1016/j.marpolbul.2011.04.016
100. van Sebille E, England MH, Froyland G (2012) Origin, dynamics and evolution of ocean garbage patches from observed surface drifters. Environ Res Lett 7:044040. https://doi.org/10.1088/1748-9326/7/4/044040
101. Zambianchi E, Trani M, Falco P (2017) Lagrangian transport of marine litter in the Mediterranean Sea. Front Environ Sci 5. https://doi.org/10.3389/fenvs.2017.00005
102. D'Asaro EA, Shcherbina AY, Klymak JM et al (2018) Ocean convergence and the dispersion of flotsam. PNAS 115:1162–1167. https://doi.org/10.1073/pnas.1718453115
103. Meyerjürgens J, Badewien TH, Garaba SP et al (2019) A state-of-the-art compact surface drifter reveals pathways of floating marine litter in the German bight. Front Mar Sci 6:58. https://doi.org/10.3389/fmars.2019.00058
104. Schöneich-Argent RI, Freund H (2020) Trashing our own "backyard" – investigating dispersal and accumulation of floating litter from coastal, riverine, and offshore sources in the German Bight using a citizen science-based wooden drifter recapture approach. Mar Environ Res 162:105115. https://doi.org/10.1016/j.marenvres.2020.105115
105. Taylor GI (1922) Diffusion by continuous movements. Proc Lond Math Soc s2-20:196–212. https://doi.org/10.1112/plms/s2-20.1.196
106. Ottino JM (1989) The kinematics of mixing: stretching, chaos, and transport. Cambridge University Press, Cambridge
107. LaCasce JH (2008) Statistics from Lagrangian observations. Prog Oceanogr 77:1–29. https://doi.org/10.1016/j.pocean.2008.02.002
108. Corrado R, Lacorata G, Palatella L et al (2017) General characteristics of relative dispersion in the ocean. Sci Rep 7:46291. https://doi.org/10.1038/srep46291
109. Jacobs GA, D'Addezio JM, Bartels B, Spence PL (2021) Constrained scales in ocean forecasting. Adv Space Res 68:746–761. https://doi.org/10.1016/j.asr.2019.09.018
110. Lebreton LC-M, Greer SD, Borrero JC (2012) Numerical modelling of floating debris in the world's oceans. Mar Pollut Bull 64:653–661. https://doi.org/10.1016/j.marpolbul.2011.10.027
111. Paduan JD, Washburn L (2013) High-frequency radar observations of ocean surface currents. Annu Rev Mar Sci 5:115–136. https://doi.org/10.1146/annurev-marine-121211-172315
112. Roarty H, Cook T, Hazard L et al (2019) The global high frequency radar network. Front Mar Sci 6:164. https://doi.org/10.3389/fmars.2019.00164
113. Rubio A, Mader J, Corgnati L et al (2017) HF radar activity in European coastal seas: next steps toward a pan-European HF radar network. Front Mar Sci 4. https://doi.org/10.3389/fmars.2017.00008
114. Lund B, Haus BK, Horstmann J et al (2018) Near-surface current mapping by shipboard marine X-band radar: a validation. J Atmos Ocean Technol 35:1077–1090. https://doi.org/10.1175/JTECH-D-17-0154.1
115. Serafino F, Bianco A (2021) Use of X-band radars to monitor small garbage islands. Remote Sens 13:3558. https://doi.org/10.3390/rs13183558
116. Holt J, Umlauf L (2008) Modelling the tidal mixing fronts and seasonal stratification of the northwest European continental shelf. Cont Shelf Res 28:887–903. https://doi.org/10.1016/j.csr.2008.01.012
117. Mahadevan A (2006) Modeling vertical motion at ocean fronts: are nonhydrostatic effects relevant at submesoscales? Ocean Model 14:222–240. https://doi.org/10.1016/j.ocemod.2006.05.005
118. Martinez E, Maamaatuaiahutapu K, Taillandier V (2009) Floating marine debris surface drift: convergence and accumulation toward the South Pacific subtropical gyre. Mar Pollut Bull 58:1347–1355. https://doi.org/10.1016/j.marpolbul.2009.04.022
119. Carlson DF, Suaria G, Aliani S et al (2017) Combining litter observations with a regional ocean model to identify sources and sinks of floating debris in a semi-Enclosed Basin: the Adriatic Sea. Front Mar Sci 4. https://doi.org/10.3389/fmars.2017.00078

120. Liubartseva S, Coppini G, Lecci R, Clementi E (2018) Tracking plastics in the Mediterranean: 2D Lagrangian model. Mar Pollut Bull 129:151–162. https://doi.org/10.1016/j.marpolbul.2018.02.019
121. Mansui J, Darmon G, Ballerini T et al (2020) Predicting marine litter accumulation patterns in the Mediterranean basin: Spatio-temporal variability and comparison with empirical data. Prog Oceanogr 182:102268. https://doi.org/10.1016/j.pocean.2020.102268
122. Heron M, Prytz A, Kalangi P (1998) Monitoring dispersion of buoyant pollutants using HF ocean surface radar. J Adv Mar Sci Technol Soc 4:231–241
123. Uttieri M, Cianelli D, Nardelli BB et al (2011) Multiplatform observation of the surface circulation in the Gulf of Naples (southern Tyrrhenian Sea). Ocean Dyn 61:779–796. https://doi.org/10.1007/s10236-011-0401-z
124. Mantovanelli A, Heron M (2012) Radar-based tracking of pollutants/larvae in the Coral Sea. In: Proceedings of the 12th international coral reef symposium, Cairns, Australia
125. Bellomo L, Griffa A, Cosoli S et al (2015) Toward an integrated HF radar network in the Mediterranean Sea to improve search and rescue and oil spill response: the TOSCA project experience. J Oper Oceanogr 8:95–107. https://doi.org/10.1080/1755876X.2015.1087184
126. Berta M, Bellomo L, Magaldi MG et al (2014) Estimating Lagrangian transport blending drifters with HF radar data and models: results from the TOSCA experiment in the Ligurian Current (North Western Mediterranean Sea). Prog Oceanogr 128:15–29. https://doi.org/10.1016/j.pocean.2014.08.004
127. Cianelli D, D'Alelio D, Uttieri M et al (2017) Disentangling physical and biological drivers of phytoplankton dynamics in a coastal system. Sci Rep 7:15868. https://doi.org/10.1038/s41598-017-15880-x
128. Boffetta G, Lacorata G, Redaelli G, Vulpiani A (2001) Detecting barriers to transport: a review of different techniques. Phys D Nonlinear Phenom 159:58–70. https://doi.org/10.1016/S0167-2789(01)00330-X
129. Haller G, Yuan G (2000) Lagrangian coherent structures and mixing in two-dimensional turbulence. Phys D Nonlinear Phenom 147:352–370. https://doi.org/10.1016/S0167-2789(00)00142-1
130. Haller G (2015) Lagrangian coherent structures. Annu Rev Fluid Mech 47:137–162. https://doi.org/10.1146/annurev-fluid-010313-141322
131. Nencioli F, d'Ovidio F, Doglioli AM, Petrenko AA (2011) Surface coastal circulation patterns by in-situ detection of Lagrangian coherent structures. Geophys Res Lett 38. https://doi.org/10.1029/2011GL048815
132. Olascoaga MJ, Haller G (2012) Forecasting sudden changes in environmental pollution patterns. Proc Natl Acad Sci 109:4738–4743. https://doi.org/10.1073/pnas.1118574109
133. Haza AC, Özgökmen TM, Griffa A et al (2010) Transport properties in small-scale coastal flows: relative dispersion from VHF radar measurements in the Gulf of La Spezia. Ocean Dyn 60:861–882. https://doi.org/10.1007/s10236-010-0301-7
134. Schroeder K, Chiggiato J, Haza AC et al (2012) Targeted Lagrangian sampling of submesoscale dispersion at a coastal frontal zone: LIDEX10-REP10. Geophys Res Lett 39. https://doi.org/10.1029/2012GL051879
135. d'Ovidio F, Pascual A, Wang J et al (2019) Frontiers in fine-scale in situ studies: opportunities during the SWOT fast sampling phase. Front Mar Sci 6:168. https://doi.org/10.3389/fmars.2019.00168
136. Ghosh A, Suara K, McCue SW et al (2021) Persistency of debris accumulation in tidal estuaries using Lagrangian coherent structures. Sci Total Environ 781:146808. https://doi.org/10.1016/j.scitotenv.2021.146808
137. Suara K, Khanarmuei M, Ghosh A et al (2020) Material and debris transport patterns in Moreton Bay, Australia: the influence of Lagrangian coherent structures. Sci Total Environ 721:137715. https://doi.org/10.1016/j.scitotenv.2020.137715

138. Hernández-Carrasco I, Alou-Font E, Dumont PA, Cabornero A, Allen J, Orfila A (2020) Lagrangian flow effects on phytoplankton abundance and composition along filament-like structures. Prog Oceanogr 189. https://doi.org/10.1016/j.pocean.2020.102469
139. Hernández-Carrasco I, Orfila A, Rossi V et al (2018) Effect of small scale transport processes on phytoplankton distribution in coastal seas. Sci Rep 8:8613. https://doi.org/10.1038/s41598-018-26857-9
140. Fu L-L, Chelton D, Le Traon P-Y, Morrow R (2010) Eddy dynamics from satellite altimetry. Oceanography 23:14–25. https://doi.org/10.5670/oceanog.2010.02
141. Lee T, Hakkinen S, Kelly K et al (2010) Satellite observations of ocean circulation changes associated with climate variability. Oceanography 23:70–81. https://doi.org/10.5670/oceanog.2010.06
142. Lee Z, Shang S, Hu C et al (2010) Time series of bio-optical properties in a subtropical gyre: implications for the evaluation of interannual trends of biogeochemical properties. J Geophys Res 115:C09012. https://doi.org/10.1029/2009JC005865
143. Chelton DB, Schlax MG, Samelson RM (2011) Global observations of nonlinear mesoscale eddies. Prog Oceanogr 91:167–216. https://doi.org/10.1016/j.pocean.2011.01.002
144. Chelton DB, Schlax MG, Samelson RM et al (2019) Prospects for future satellite estimation of small-scale variability of ocean surface velocity and vorticity. Prog Oceanogr 173:256–350. https://doi.org/10.1016/j.pocean.2018.10.012
145. Capet X, McWilliams JC, Molemaker MJ, Shchepetkin AF (2008) Mesoscale to submesoscale transition in the California current system. Part I: flow structure, Eddy flux, and observational tests. J Phys Oceanogr 38:29–43. https://doi.org/10.1175/2007JPO3671.1
146. Klein P, Lapeyre G (2009) The oceanic vertical pump induced by mesoscale and submesoscale turbulence. Annu Rev Mar Sci 1:351–375. https://doi.org/10.1146/annurev.marine.010908.163704
147. Lévy M, Klein P, Treguier A-M (2001) Impact of sub-mesoscale physics on production and subduction of phytoplankton in an oligotrophic regime. J Mar Res 59:535–565. https://doi.org/10.1357/002224001762842181
148. Mahadevan A, Thomas LN, Tandon A (2008) Comment on "Eddy/wind interactions stimulate extraordinary Mid-Ocean plankton blooms". Science 320:448b. https://doi.org/10.1126/science.1152111
149. Li Q, Zhong G, Xie C, Hedjam R (2021) Weak edge identification network for ocean front detection. IEEE Geosci Remote Sens Lett. https://doi.org/10.1109/LGRS.2021.3051203
150. Lima E, Sun X, Yang Y, Dong J (2017) Application of deep convolutional neural networks for ocean front recognition. J Appl Remote Sens 11:1. https://doi.org/10.1117/1.JRS.11.042610
151. Maximenko N, Corradi P, Law KL et al (2019) Toward the integrated marine debris observing system. Front Mar Sci 6:447. https://doi.org/10.3389/fmars.2019.00447
152. Fingas M, Brown C (2017) A review of oil spill remote sensing. Sensors 18:91. https://doi.org/10.3390/s18010091
153. Leifer I, Lehr WJ, Simecek-Beatty D et al (2012) State of the art satellite and airborne marine oil spill remote sensing: application to the BP deepwater horizon oil spill. Remote Sens Environ 124:185–209. https://doi.org/10.1016/j.rse.2012.03.024
154. Zielinski O, Hengstermann T, Robbe N (2006) Detection of oil spills by airborne sensors. In: Gade M, Hühnerfuss H, Korenowski GM (eds) Marine surface films. Springer, Berlin, pp 255–271
155. Garaba SP, Aitken J, Slat B et al (2018) Sensing ocean plastics with an airborne hyperspectral shortwave infrared imager. Environ Sci Technol 52:11699–11707. https://doi.org/10.1021/acs.est.8b02855
156. Lambert C, Authier M, Dorémus G et al (2020) Setting the scene for Mediterranean litterscape management: the first basin-scale quantification and mapping of floating marine debris. Environ Pollut 263:114430. https://doi.org/10.1016/j.envpol.2020.114430

157. Unger B, Herr H, Viquerat S et al (2021) Opportunistically collected data from aerial surveys reveal spatio-temporal distribution patterns of marine debris in German waters. Environ Sci Pollut Res 28:2893–2903. https://doi.org/10.1007/s11356-020-10610-9
158. Rascle N, Molemaker J, Marié L et al (2017) Intense deformation field at oceanic front inferred from directional sea surface roughness observations. Geophys Res Lett 44:5599–5608. https://doi.org/10.1002/2017GL073473
159. Rascle N, Chapron B, Molemaker J et al (2020) Monitoring intense oceanic fronts using sea surface roughness: satellite, airplane, and in situ comparison. J Geophys Res Oceans 125. https://doi.org/10.1029/2019JC015704
160. Solodoch A, Molemaker JM, Srinivasan K et al (2020) Observations of shoaling density current regime changes in internal wave interactions. J Phys Oceanogr 50:1733–1751. https://doi.org/10.1175/JPO-D-19-0176.1
161. Carlson DF, Özgökmen T, Novelli G et al (2018) Surface ocean dispersion observations from the ship-tethered aerostat remote sensing system. Front Mar Sci 5:479. https://doi.org/10.3389/fmars.2018.00479
162. Chang H, Huntley HS, Kirwan AD et al (2019) Small-scale dispersion in the presence of Langmuir circulation. J Phys Oceanogr 49:3069–3085. https://doi.org/10.1175/JPO-D-19-0107.1
163. Johnston DW (2019) Unoccupied aircraft systems in marine science and conservation. Annu Rev Mar Sci 11:439–463. https://doi.org/10.1146/annurev-marine-010318-095323
164. Garcia-Garin O, Borrell A, Aguilar A et al (2020) Floating marine macro-litter in the north western Mediterranean Sea: results from a combined monitoring approach. Mar Pollut Bull 159:111467. https://doi.org/10.1016/j.marpolbul.2020.111467
165. Streßer M, Carrasco R, Horstmann J (2017) Video-based estimation of surface currents using a low-cost quadcopter. IEEE Geosci Remote Sens Lett 14:2027–2031. https://doi.org/10.1109/LGRS.2017.2749120
166. Topouzelis K, Papakonstantinou A, Garaba SP (2019) Detection of floating plastics from satellite and unmanned aerial systems (plastic litter project 2018). Int J Appl Earth Obs Geoinf 79:175–183. https://doi.org/10.1016/j.jag.2019.03.011
167. Bower AS, Rossby HT, Lillibridge JL (1985) The Gulf Stream - barrier or blender? J Phys Oceanogr 15:24–32. https://doi.org/10.1175/1520-0485(1985)015%3C0024:TGSOB%3E2.0.CO;2
168. Wenegrat JO, Thomas LN, Sundermeyer MA et al (2020) Enhanced mixing across the gyre boundary at the Gulf Stream front. Proc Natl Acad Sci U S A 117:17607–17614. https://doi.org/10.1073/pnas.2005558117
169. Deacon G (1937) The hydrology of the Southern Ocean. Discov Rep 15:3–122
170. Orsi AH, Whitworth T, Nowlin WD (1995) On the meridional extent and fronts of the Antarctic circumpolar current. Deep-Sea Res I Oceanogr Res Pap 42:641–673. https://doi.org/10.1016/0967-0637(95)00021-W
171. Park Y-H, Park T, Kim TW, Lee SH, Hong CS, Lee JH, Rio M-H, Pujol MI, Ballarotta M, Durand I, Provost C (2019) Observations of the Antarctic circumpolar current over the Udintsev fracture zone, the narrowest choke point in the Southern Ocean. J Geophys Res Oceans 124(7):4511–4528
172. Falco P, Zambianchi E (2011) Near-surface structure of the Antarctic circumpolar current derived from world ocean circulation experiment drifter data. J Geophys Res 116:C05003. https://doi.org/10.1029/2010JC006349
173. Sokolov S, Rintoul SR (2009) Circumpolar structure and distribution of the Antarctic circumpolar current fronts: 1. Mean circumpolar paths. J Geophys Res 114:C11018. https://doi.org/10.1029/2008JC005108
174. Sokolov S, Rintoul SR (2009) Circumpolar structure and distribution of the Antarctic circumpolar current fronts: 2. Variability and relationship to sea surface height. J Geophys Res 114:C11019. https://doi.org/10.1029/2008JC005248

175. Whitworth T (1980) Zonation and geostrophic flow of the Antarctic circumpolar current at Drake Passage. Deep Sea Res Part A Oceanogr Res Pap 27:497–507. https://doi.org/10.1016/0198-0149(80)90036-9
176. Belkin IM, Gordon AL (1996) Southern Ocean fronts from the Greenwich meridian to Tasmania. J Geophys Res 101:3675–3696. https://doi.org/10.1029/95JC02750
177. Wyrtki K (1960) The Antarctic circumpolar current and the Antarctic polar front. Dtsch Hydrogr Z 13:153–174. https://doi.org/10.1007/BF02226197
178. Talley LD, Pickard GL, Emery WJ (eds) (2011) Descriptive physical oceanography: an introduction.6th edn. Academic Press, Amsterdam
179. Barber HN, Dadswell HE, Ingle HD (1959) Transport of driftwood from South America to Tasmania and Macquarie Island. Nature 184:203–204. https://doi.org/10.1038/184203a0
180. Coombs DS, Landis CA (1966) Pumice from the south sandwich eruption of march 1962 reaches New Zealand. Nature 209:289–290. https://doi.org/10.1038/209289b0
181. Ryan PG (1987) The origin and fate of artefacts stranded on islands in the African sector of the Southern Ocean. Envir Conserv 14:341–346. https://doi.org/10.1017/S0376892900016854
182. van Franeker JA, Bell PJ (1988) Plastic ingestion by petrels breeding in Antarctica. Mar Pollut Bull 19:672–674. https://doi.org/10.1016/0025-326X(88)90388-8
183. Convey P, Barnes D, Morton A (2002) Debris accumulation on oceanic island shores of the Scotia Arc, Antarctica. Polar Biol 25:612–617. https://doi.org/10.1007/s00300-002-0391-x
184. Eriksson C, Burton H, Fitch S et al (2013) Daily accumulation rates of marine debris on sub-Antarctic island beaches. Mar Pollut Bull 66:199–208. https://doi.org/10.1016/j.marpolbul.2012.08.026
185. Gregory M, Kirk R, Mabin M (1984) Pelagic tar, oil, plastics and other litter in surface waters of the NZ sector of the Southern Ocean, and on Ross dependency shores. N Z Antarct Rec 6:12–28
186. Ryan PG (1987) The incidence and characteristics of plastic particles ingested by seabirds. Mar Environ Res 23:175–206. https://doi.org/10.1016/0141-1136(87)90028-6
187. Slip DJ, Burton HR (1991) Accumulation of fishing debris, plastic litter, and other artefacts, on heard and Macquarie Islands in the Southern Ocean. Environ Conserv 18:249–254. https://doi.org/10.1017/S0376892900022177
188. Walker TR, Reid K, Arnould JPY, Croxall JP (1997) Marine debris surveys at Bird Island, South Georgia 1990–1995. Mar Pollut Bull 34:61–65. https://doi.org/10.1016/S0025-326X(96)00053-7
189. Fraser CI, Kay GM, du Plessis M, Ryan PG (2017) Breaking down the barrier: dispersal across the Antarctic polar front. Ecography 40:235–237. https://doi.org/10.1111/ecog.02449
190. Fraser CI, Morrison AK, Hogg AM et al (2018) Antarctica's ecological isolation will be broken by storm-driven dispersal and warming. Nat Clim Change 8:704–708. https://doi.org/10.1038/s41558-018-0209-7
191. Ivar do Sul JA, Barnes DKA, Costa MF et al (2011) Plastics in the Antarctic environment: are we looking only at the tip of the iceberg? Oecol Austr 15:150–170. https://doi.org/10.4257/oeco.2011.1501.11
192. Suaria G, Perold V, Lee JR et al (2020) Floating macro- and microplastics around the Southern Ocean: results from the Antarctic circumnavigation expedition. Environ Int 136:105494. https://doi.org/10.1016/j.envint.2020.105494
193. Waller CL, Griffiths H, Waluda CM et al (2018) Microplastics in the Southern Ocean. Antarct Environ Portal. https://doi.org/10.18124/D4JK8V
194. Anilkumar N, George JV, Chacko R et al (2015) Variability of fronts, fresh water input and chlorophyll in the Indian Ocean sector of the Southern Ocean. N Z J Mar Freshw Res 49:20–40. https://doi.org/10.1080/00288330.2014.924972
195. Billany W, Swart S, Hermes J, Reason CJC (2010) Variability of the Southern Ocean fronts at the Greenwich Meridian. J Mar Syst 82:304–310. https://doi.org/10.1016/j.jmarsys.2010.06.005

196. Kim YS, Orsi AH (2014) On the variability of Antarctic circumpolar current fronts inferred from 1992–2011 altimetry. J Phys Oceanogr 44:3054–3071. https://doi.org/10.1175/JPO-D-13-0217.1
197. Park Y-H, Lee J-H, Durand I, Hong C-S (2014) Validation of Thorpe-scale-derived vertical diffusivities against microstructure measurements in the Kerguelen region. Biogeosciences 11:6927–6937. https://doi.org/10.5194/bg-11-6927-2014
198. Sparrow MD, Heywood KJ, Brown J, Stevens DP (1996) Current structure of the south Indian Ocean. J Geophys Res 101:6377–6391. https://doi.org/10.1029/95JC03750
199. Hughes CW, Ash ER (2001) Eddy forcing of the mean flow in the Southern Ocean. J Geophys Res 106:2713–2722. https://doi.org/10.1029/2000JC900332
200. Morrow R, Coleman R, Church J, Chelton D (1994) Surface eddy momentum flux and velocity variances in the Southern Ocean from Geosat altimetry. J Phys Oceanogr 24:2050–2071. https://doi.org/10.1175/1520-0485(1994)024%3C2050:SEMFAV%3E2.0.CO;2
201. Hogg AMC, Meredith MP, Blundell JR, Wilson C (2008) Eddy heat flux in the Southern Ocean: response to variable wind forcing. J Clim 21:608–620. https://doi.org/10.1175/2007JCLI1925.1
202. Menna M, Cotroneo Y, Falco P et al (2020) Response of the Pacific sector of the Southern Ocean to wind stress variability from 1995 to 2017. J Geophys Res Oceans 125. https://doi.org/10.1029/2019JC015696
203. Thompson AF, Naveira Garabato AC (2014) Equilibration of the Antarctic circumpolar current by standing meanders. J Phys Oceanogr 44:1811–1828. https://doi.org/10.1175/JPO-D-13-0163.1
204. Rintoul SR (2018) The global influence of localized dynamics in the Southern Ocean. Nature 558:209–218. https://doi.org/10.1038/s41586-018-0182-3
205. Naveira Garabato AC, Ferrari R, Polzin KL (2011) Eddy stirring in the Southern Ocean. J Geophys Res 116:C09019. https://doi.org/10.1029/2010JC006818
206. Thompson AF, Sallée J-B (2012) Jets and topography: jet transitions and the impact on transport in the Antarctic circumpolar current. J Phys Oceanogr 42:956–972. https://doi.org/10.1175/JPO-D-11-0135.1
207. Dufour CO, Griffies SM, de Souza GF et al (2015) Role of mesoscale eddies in cross-frontal transport of heat and biogeochemical tracers in the Southern Ocean. J Phys Oceanogr 45:3057–3081. https://doi.org/10.1175/JPO-D-14-0240.1
208. Patel RS, Phillips HE, Strutton PG et al (2019) Meridional heat and salt transport across the subantarctic front by cold-core eddies. J Geophys Res Oceans 124:981–1004. https://doi.org/10.1029/2018JC014655
209. Bachman SD, Taylor JR, Adams KA, Hosegood PJ (2017) Mesoscale and submesoscale effects on mixed layer depth in the Southern Ocean. J Phys Oceanogr 47:2173–2188. https://doi.org/10.1175/JPO-D-17-0034.1
210. Taylor JR (2018) Accumulation and subduction of buoyant material at submesoscale fronts. J Phys Oceanogr 48:1233–1241. https://doi.org/10.1175/JPO-D-17-0269.1
211. Rosso I, Hogg AMC, Kiss AE, Gayen B (2015) Topographic influence on submesoscale dynamics in the Southern Ocean. Geophys Res Lett 42:1139–1147. https://doi.org/10.1002/2014GL062720
212. Rosso I, Hogg AMC, Matear R, Strutton PG (2016) Quantifying the influence of sub-mesoscale dynamics on the supply of iron to Southern Ocean phytoplankton blooms. Deep-Sea Res I Oceanogr Res Pap 115:199–209. https://doi.org/10.1016/j.dsr.2016.06.009
213. Rosso I, Hogg AMC, Strutton PG et al (2014) Vertical transport in the ocean due to sub-mesoscale structures: impacts in the Kerguelen region. Ocean Model 80:10–23. https://doi.org/10.1016/j.ocemod.2014.05.001
214. Siegelman L, Klein P, Rivière P et al (2020) Enhanced upward heat transport at deep submesoscale ocean fronts. Nat Geosci 13:50–55. https://doi.org/10.1038/s41561-019-0489-1

215. Wichmann D, Delandmeter P, van Sebille E (2019) Influence of near-surface currents on the global dispersal of marine microplastic. J Geophys Res Oceans 124:6086–6096. https://doi.org/10.1029/2019JC015328
216. Brunner K, Kukulka T, Proskurowski G, Law KL (2015) Passive buoyant tracers in the ocean surface boundary layer: 2. Observations and simulations of microplastic marine debris. J Geophys Res Oceans 120:7559–7573. https://doi.org/10.1002/2015JC010840
217. Kukulka T, Proskurowski G, Morét-Ferguson S et al (2012) The effect of wind mixing on the vertical distribution of buoyant plastic debris. Geophys Res Lett 39:L07601. https://doi.org/10.1029/2012GL051116
218. Lacerda ALDF, Rodrigues LDS, Van Sebille E et al (2019) Plastics in sea surface waters around the Antarctic peninsula. Sci Rep 9:3977. https://doi.org/10.1038/s41598-019-40311-4
219. Kundu PK (1990) Fluid mechanics. Academic, San Diego
220. Everaert G, De Rijcke M, Lonneville B et al (2020) Risks of floating microplastic in the global ocean. Environ Pollut 267:115499. https://doi.org/10.1016/j.envpol.2020.115499
221. Cózar A, Sanz-Martín M, Martí E et al (2015) Plastic accumulation in the Mediterranean Sea. PLoS One 10:e0121762. https://doi.org/10.1371/journal.pone.0121762
222. Zambianchi E, Iermano I, Suaria G, Aliani S (2014) Marine litter in the Mediterranean Sea: an oceanographic perspective. In: Briand F (ed) Marine litter in the Mediterranean and black seas. CIESM workshop monograph n° 46. CIESM Publisher, Monaco, 180 pp
223. de Haan WP, Sanchez-Vidal A, Canals M (2019) Floating microplastics and aggregate formation in the Western Mediterranean Sea. Mar Pollut Bull 140:523–535. https://doi.org/10.1016/j.marpolbul.2019.01.053
224. Suaria G, Aliani S (2014) Floating debris in the Mediterranean Sea. Mar Pollut Bull 86:494–504. https://doi.org/10.1016/j.marpolbul.2014.06.025
225. Aliani S, Griffa A, Molcard A (2003) Floating debris in the Ligurian Sea, North-Western Mediterranean. Mar Pollut Bull 46:1142–1149. https://doi.org/10.1016/S0025-326X(03)00192-9
226. Aliani S, Molcard A (2003) Hitch-hiking on floating marine debris: macrobenthic species in the Western Mediterranean Sea. Hydrobiologia 503:59–67. https://doi.org/10.1023/B:HYDR.0000008480.95045.26
227. Millot C (1999) Circulation in the western Mediterranean Sea. J Mar Syst 20:423–442. https://doi.org/10.1016/S0924-7963(98)00078-5
228. Stemmann L, Prieur L, Legendre L et al (2008) Effects of frontal processes on marine aggregate dynamics and fluxes: an interannual study in a permanent geostrophic front (NW Mediterranean). J Mar Syst 70:1–20. https://doi.org/10.1016/j.jmarsys.2007.02.014
229. Alberola C, Millot C, Font J (1995) On the seasonal and mesoscale variabilities of the northern current during the PRIMO-0 experiment in the western Mediterranean-Sea. Oceanol Acta 18:163–192
230. Astraldi M, Gasparini GP (1992) The seasonal characteristics of the circulation in the North Mediterranean basin and their relationship with the atmospheric-climatic conditions. J Geophys Res 97:9531. https://doi.org/10.1029/92JC00114
231. Ourmières Y, Zakardjian B, Béranger K, Langlais C (2011) Assessment of a NEMO-based downscaling experiment for the north-western Mediterranean region: impacts on the northern current and comparison with ADCP data and altimetry products. Ocean Model 39:386–404. https://doi.org/10.1016/j.ocemod.2011.06.002
232. Taupier-Letage I, Millot C (1986) General hydrodynamical features in the Ligurian Sea inferred from the dyome experiment. Oceanol Acta 9:119–131
233. Barrier N, Petrenko AA, Ourmières Y (2016) Strong intrusions of the northern Mediterranean current on the eastern Gulf of Lion: insights from in-situ observations and high resolution numerical modelling. Ocean Dyn 66:313–327. https://doi.org/10.1007/s10236-016-0921-7
234. Berline L, Zakardjian B, Molcard A et al (2013) Modeling jellyfish Pelagia noctiluca transport and stranding in the Ligurian Sea. Mar Pollut Bull 70:90–99. https://doi.org/10.1016/j.marpolbul.2013.02.016

235. Sammari C, Millot C, Prieur L (1995) Aspects of the seasonal and mesoscale variabilities of the northern current in the western Mediterranean Sea inferred from the PROLIG-2 and PROS-6 experiments. Deep-Sea Res I Oceanogr Res Pap 42:893–917. https://doi.org/10.1016/0967-0637(95)00031-Z
236. Allen JT, Brown L, Sanders R et al (2005) Diatom carbon export enhanced by silicate upwelling in the northeast Atlantic. Nature 437:728–732. https://doi.org/10.1038/nature03948
237. Niewiadomska K, Claustre H, Prieur L, d'Ortenzio F (2008) Submesoscale physical-biogeochemical coupling across the Ligurian current (northwestern Mediterranean) using a bio-optical glider. Limnol Oceanogr 53:2210–2225. https://doi.org/10.4319/lo.2008.53.5_part_2.2210
238. Boucher J, Ibanez F, Prieur L (1987) Daily and seasonal variations in the spatial distribution of zooplankton populations in relation to the physical structure in the Ligurian Sea Front. J Mar Res 45:133–173. https://doi.org/10.1357/002224087788400891
239. Goffart A, Hecq J, Prieur L (1995) Control of the phytoplankton of the Ligurian Basin by the Liguro-Provencal Front (Corsican sector). Oceanol Acta 18:329–342
240. Sournia A, Brylinski J, Dallot S et al (1990) Fronts hydrologiques au large des côtes françaises : Les sites-ateliers de programme Frontal. Oceanol Acta 13:413–438
241. de Lucia GA, Caliani I, Marra S et al (2014) Amount and distribution of neustonic micro-plastic off the western Sardinian coast (Central-Western Mediterranean Sea). Mar Environ Res 100:10–16. https://doi.org/10.1016/j.marenvres.2014.03.017
242. Pedrotti ML, Petit S, Elineau A et al (2016) Changes in the floating plastic pollution of the Mediterranean Sea in relation to the distance to land. PLoS One 11:e0161581. https://doi.org/10.1371/journal.pone.0161581
243. Ourmières Y, Mansui J, Molcard A et al (2018) The boundary current role on the transport and stranding of floating marine litter: the French Riviera case. Cont Shelf Res 155:11–20. https://doi.org/10.1016/j.csr.2018.01.010
244. Di-Méglio N, Campana I (2017) Floating macro-litter along the Mediterranean French coast: composition, density, distribution and overlap with cetacean range. Mar Pollut Bull 118:155–166. https://doi.org/10.1016/j.marpolbul.2017.02.026
245. Gerigny O, Henry M, Tomasino C, Galgani F (2011) Déchets en mer et sur le fond. in rapport de l'évalution initiale, Plan d'action pour le milieu marin - Mediterranée Occidentale
246. Berta M, Griffa A, Corgnati L et al (2020) Submesoscales variability from surface drifter and HF radar measurements: scale and wind dependence of kinematic properties. In: EGU General Assembly 2020, 4–8 May 2020. EGU2020-18620, Online
247. Sklar FH, Browder JA (1998) Coastal environmental impacts brought about by alterations to freshwater flow in the Gulf of Mexico. Environ Manag 22:547–562. https://doi.org/10.1007/s002679900127
248. Buskey EJ, White HK, Esbaugh AJ (2016) Impact of oil spills on marine life in the Gulf of Mexico: effects on plankton, nekton and deep-sea benthos. Oceanography 29:174–181
249. Di Mauro R, Kupchik MJ, Benfield MC (2017) Abundant plankton-sized microplastic particles in shelf waters of the northern Gulf of Mexico. Environ Pollut 230:798–809. https://doi.org/10.1016/j.envpol.2017.07.030
250. Lecke-Mitchell KM, Mullin K (1992) Distribution and abundance of large floating plastic in the north-central Gulf of Mexico. Mar Pollut Bull 24:598–601. https://doi.org/10.1016/0025-326X(92)90279-F
251. Oey L-Y, Ezer T, Lee H-C (2005) Loop current, rings and related circulation in the Gulf of Mexico: a review of numerical models and future challenges. In: Sturges W, Lugo-Fernandez A (eds) Geophysical monograph series. American Geophysical Union, Washington, pp 31–56
252. Poje AC, Ozgokmen TM, Lipphardt BL et al (2014) Submesoscale dispersion in the vicinity of the deepwater horizon spill. Proc Natl Acad Sci 111:12693–12698. https://doi.org/10.1073/pnas.1402452111

253. Lodise J, Özgökmen T, Gonçalves R et al (2020) Investigating the formation of submesoscale structures along mesoscale fronts and estimating kinematic quantities using Lagrangian drifters. Fluids 5(3):159
254. Özgökmen TM, Poje AC, Fischer PF et al (2012) On multi-scale dispersion under the influence of surface mixed layer instabilities and deep flows. Ocean Model 56:16–30. https://doi.org/10.1016/j.ocemod.2012.07.004
255. Haza AC, Özgökmen TM, Griffa A et al (2012) Parameterization of particle transport at submesoscales in the Gulf Stream region using Lagrangian subgridscale models. Ocean Model 42:31–49. https://doi.org/10.1016/j.ocemod.2011.11.005
256. Pearson J, Fox-Kemper B, Barkan R et al (2019) Impacts of convergence on structure functions from surface drifters in the Gulf of Mexico. J Phys Oceanogr 49:675–690. https://doi.org/10.1175/JPO-D-18-0029.1
257. Haza AC, Özgökmen TM, Hogan P (2016) Impact of submesoscales on surface material distribution in a gulf of Mexico mesoscale eddy. Ocean Model 107:28–47. https://doi.org/10.1016/j.ocemod.2016.10.002
258. Huguenard KD, Bogucki DJ, Ortiz-Suslow DG et al (2016) On the nature of the frontal zone of the Choctawhatchee Bay plume in the Gulf of Mexico. J Geophys Res Oceans 121:1322–1345. https://doi.org/10.1002/2015JC010988
259. Roth MK, MacMahan J, Reniers A et al (2017) Observations of inner shelf cross-shore surface material transport adjacent to a coastal inlet in the northern Gulf of Mexico. Cont Shelf Res 137:142–153. https://doi.org/10.1016/j.csr.2016.12.017
260. Androulidakis Y, Kourafalou V, Özgökmen T et al (2018) Influence of river-induced fronts on hydrocarbon transport: a multiplatform observational study. J Geophys Res Oceans 123:3259–3285. https://doi.org/10.1029/2017JC013514
261. Wang P, Özgökmen TM (2015) How do hydrodynamic instabilities affect 3D transport in geophysical vortices? Ocean Model 87:48–66. https://doi.org/10.1016/j.ocemod.2015.01.002
262. Koelmans AA, Kooi M, Law KL, van Sebille E (2017) All is not lost: deriving a top-down mass budget of plastic at sea. Environ Res Lett 12:114028. https://doi.org/10.1088/1748-9326/aa9500
263. Kaandorp MLA, Dijkstra HA, van Sebille E (2020) Closing the Mediterranean marine floating plastic mass budget: inverse modeling of sources and sinks. Environ Sci Technol 54:11980–11989. https://doi.org/10.1021/acs.est.0c01984
264. Jamieson AJ, Brooks LSR, Reid WDK et al (2019) Microplastics and synthetic particles ingested by deep-sea amphipods in six of the deepest marine ecosystems on earth. R Soc Open Sci 6:180667. https://doi.org/10.1098/rsos.180667
265. López-Martínez S, Morales-Caselles C, Kadar J, Rivas ML (2021) Overview of global status of plastic presence in marine vertebrates. Glob Change Biol 27:728–737. https://doi.org/10.1111/gcb.15416
266. Hale RC, Seeley ME, La Guardia MJ et al (2020) A global perspective on microplastics. J Geophys Res Oceans 125. https://doi.org/10.1029/2018JC014719
267. de la Fuente R, Drótos G, Hernández-García E et al (2021) Sinking microplastics in the water column: simulations in the Mediterranean Sea. Ocean Sci 17:431–453. https://doi.org/10.5194/os-17-431-2021
268. Loughlin C, Marques Mendes AR, Morrison L, Morley A (2021) The role of oceanographic processes and sedimentological settings on the deposition of microplastics in marine sediment: Icelandic waters. Mar Pollut Bull 164:111976. https://doi.org/10.1016/j.marpolbul.2021.111976
269. Mountford AS, Morales Maqueda MA (2019) Eulerian modeling of the three-dimensional distribution of seven popular microplastic types in the global ocean. J Geophys Res Oceans 124:8558–8573. https://doi.org/10.1029/2019JC015050
270. Gündoğdu S, Çevik C, Ayat B et al (2018) How microplastics quantities increase with flood events? An example from Mersin Bay NE Levantine coast of Turkey. Environ Pollut 239:342–350. https://doi.org/10.1016/j.envpol.2018.04.042

271. Lo H-S, Lee Y-K, Po BH-K et al (2020) Impacts of Typhoon Mangkhut in 2018 on the deposition of marine debris and microplastics on beaches in Hong Kong. Sci Total Environ 716:137172. https://doi.org/10.1016/j.scitotenv.2020.137172
272. Murray CC, Maximenko N, Lippiatt S (2018) The influx of marine debris from the Great Japan Tsunami of 2011 to North American shorelines. Mar Pollut Bull 132:26–32. https://doi.org/10.1016/j.marpolbul.2018.01.004
273. Sulistiawati D, Safir M, Putra AE et al (2021) The amount and type of plastics in the Baiya Beach after the Palu Bay tsunami. J Phys Conf Ser 1763:012072. https://doi.org/10.1088/1742-6596/1763/1/012072
274. Kukulka T, Brunner K (2015) Passive buoyant tracers in the ocean surface boundary layer: 1. Influence of equilibrium wind-waves on vertical distributions. J Geophys Res Oceans 120:3837–3858. https://doi.org/10.1002/2014JC010487
275. Jacketti M, Beegle-Krause CJ, Englehardt JD (2020) A review on the sinking mechanisms for oil and successful response technologies. Mar Pollut Bull 160:111626. https://doi.org/10.1016/j.marpolbul.2020.111626
276. Bagaev A, Mizyuk A, Khatmullina L et al (2017) Anthropogenic fibres in the Baltic Sea water column: field data, laboratory and numerical testing of their motion. Sci Total Environ 599–600:560–571. https://doi.org/10.1016/j.scitotenv.2017.04.185
277. Khatmullina L, Isachenko I (2017) Settling velocity of microplastic particles of regular shapes. Mar Pollut Bull 114:871–880. https://doi.org/10.1016/j.marpolbul.2016.11.024
278. Wang Z, Dou M, Ren P, et al (2021) Sedimentation of irregular shaped microplastics under steady and dynamic flow conditions (in Review)
279. Iwasaki S, Isobe A, Kako S et al (2017) Fate of microplastics and mesoplastics carried by surface currents and wind waves: a numerical model approach in the Sea of Japan. Mar Pollut Bull 121:85–96. https://doi.org/10.1016/j.marpolbul.2017.05.057
280. Vélez-Belchí P, Tintoré J (2001) Vertical velocities at an ocean front. Sci Mar 65:291–300. https://doi.org/10.3989/scimar.2001.65s1291
281. Esposito G, Berta M, Centurioni L et al (2021) Submesoscale vorticity and divergence in the Alboran Sea: scale and depth dependence. Front Mar Sci. https://doi.org/10.3389/fmars.2021.678304
282. Stukel MR, Aluwihare LI, Barbeau KA et al (2017) Mesoscale ocean fronts enhance carbon export due to gravitational sinking and subduction. Proc Natl Acad Sci U S A 114:1252–1257. https://doi.org/10.1073/pnas.1609435114
283. Lapeyre G, Klein P (2006) Impact of the small-scale elongated filaments on the oceanic vertical pump. J Mar Res 64:835–851. https://doi.org/10.1357/002224006779698369
284. Ye S, Andrady AL (1991) Fouling of floating plastic debris under Biscayne Bay exposure conditions. Mar Pollut Bull 22:608–613. https://doi.org/10.1016/0025-326X(91)90249-R
285. Lacour L, Briggs N, Claustre H et al (2019) The intraseasonal dynamics of the mixed layer pump in the subpolar North Atlantic Ocean: a biogeochemical-Argo float approach. Global Biogeochem Cycles 33:266–281. https://doi.org/10.1029/2018GB005997
286. Hoellein TJ, Rochman CM (2021) The "plastic cycle": a watershed-scale model of plastic pools and fluxes. Front Ecol Environ. https://doi.org/10.1002/fee.2294
287. Elhacham E, Ben-Uri L, Grozovski J et al (2020) Global human-made mass exceeds all living biomass. Nature 588:442–444. https://doi.org/10.1038/s41586-020-3010-5

Lagrangian Methods for Visualizing and Assessing Frontal Dynamics of Floating Marine Litter with a Focus on Tidal Basins

Marcel Ricker, Jens Meyerjürgens, Thomas H. Badewien, and Emil V. Stanev

Contents

Abstract Lagrangian methods are a common tool in physical oceanography. Due to the quasi-Lagrangian characteristics of floating marine litter (FML) and the chemical

M. Ricker (✉)
Institute of Coastal Systems – Analysis and Modeling, Helmholtz-Zentrum Hereon, Geesthacht, Germany

Institute for Chemistry and Biology of the Marine Environment, University of Oldenburg, Oldenburg, Germany
e-mail: marcel.ricker@hereon.de

J. Meyerjürgens and T. H. Badewien
Institute for Chemistry and Biology of the Marine Environment, University of Oldenburg, Oldenburg, Germany

E. V. Stanev
Institute of Coastal Systems – Analysis and Modeling, Helmholtz-Zentrum Hereon, Geesthacht, Germany

Department of Meteorology and Geophysics, Faculty of Physics, Sofia University "St. Kliment Ohridski", Sofia, Bulgaria

Igor M. Belkin (ed.), *Chemical Oceanography of Frontal Zones*,
Hdb Env Chem (2022) 116: 407–442, DOI 10.1007/698_2021_812,

Published online: 20 October 2021

substances released from it, Lagrangian methods can be used to study this environmental threat. Most of the existing investigations of this topic have been carried out in the deep ocean, where baroclinic dynamics dominate. In contrast, studies of tidally dominated, shallow regions are much fewer in number. Compared to the deep ocean, shallow shelves are more strongly influenced by freshwater inputs, bottom stress, complex coastlines, and wind, which imply higher diffusion rates, especially in the presence of tides. Furthermore, they steer the transport of FML from rivers to the deep ocean with fronts as an important driver. The present chapter reviews Lagrangian methods for visualizing and assessing frontal dynamics in tidal basins with data obtained from numerical modeling and satellite-tracked drifters. The specific requirements for the two data sources are described and discussed. Some of these methods are applied in the North Sea, located on the European northwest shelf, where tidal mixing fronts and fronts due to freshwater runoff exist. It is demonstrated how surface convergence and gradients in temperature, salinity, and density are connected with the accumulation of virtual and satellite-tracked drifters. The effect of tides on the propagation of Lagrangian particles is shown to be significant and demonstrates the importance of tidal forces and vertical dynamics in Lagrangian simulations in tidal basins. The chapter ends with the future outlook, illuminating the numerous knowledge gaps remaining and proposing areas for future research.

Keywords Litter accumulation, Microplastics, Multiple particle statistics, North Sea, Particle tracking, Satellite-tracked drifters, Shelf seas, Surface convergence, Tidal coastal area, Vertical dynamics

Abbreviations

ADR	Average dilation rate
AUV	Autonomous underwater vehicle
CM	Complexity method
CODE	Coastal Dynamics Experiment
CS	Cluster strength
DKPs	Differential kinematic properties
EKE	Eddy kinetic energy
ENWS	European northwest shelf
EX1	Experiment 1
EX2	Experiment 2
FDLD	Finite-domain Lagrangian divergence
FML	Floating marine litter
FSLE	Finite-size Lyapunov exponent
FTLE	Finite-time Lyapunov exponent
GPS	Global Positioning System
HF radar	High frequency radar
LCS	Lagrangian coherent structure

LD	Lagrangian divergence
NCPD	Normalized cumulative particle density
NEMO	Nucleus for European Modeling of the Ocean
REF	Reference experiment
ROFI	Region of freshwater influence
SAR	Search and rescue
SPM	Suspended particulate matter
SST	Sea surface temperature

1 Introduction

Tidal basins are often surrounded by highly populated coastal regions, which impose a great deal of stress on marine environments such as the North Sea. One of these stressors is floating marine litter (FML), which has received much attention during the last decade due to the raising awareness for its damage to the ecology and economy. In particular, very small pieces emerging from the fragmentation of plastic litter (from macro- to micro- to nanoplastics) or from primary sources, such as sewage treatment plants, can be hazardous to both animals and humans. The negative effects of macro- and microplastics are the subject of many studies, and several different kinds of impacts have been reported, such as ingestion, entanglement, transport of alien species, aesthetic problems, or problems with fisheries [1–8]. Less well understood are the impacts of nanoplastics, which were found to be able to cross selective membranes, such as the placenta or the blood-brain barrier, and have effects on the metabolism of cells. In addition, FML releases chemical substances, such as bisphenol A or phthalates, which are known to exert hormonal effects on living organisms [8–11]. The effects of plastics on living beings are expected to be far-reaching considering the large amounts of plastics in many marine habitats across the world's oceans [12–16].

In shelf seas, geostrophic and baroclinic dynamics are less pronounced than in the deep ocean, and in turn, the underlying physics of fronts in both regimes differ. Shelf sea fronts are influenced by freshwater runoff and tides and are known to be dynamically important for the levels of substances such as chlorophyll, and subsequently, for primary production [17–21] or may act as convergence zones for flotsam [20, 22, 23]. Frontal dynamics are well understood from theoretical and experimental approaches, and a variety of methods have been developed to study them in the global ocean (e.g., [24, 25]) and in tidal basins [26–28]. These studies are often based on satellite measurements but also on shipborne, or autonomous underwater vehicle (AUV) data. However, satellite data can only provide temporal means, which can be problematic in tidal basins. Ship-based measurements are expensive, and only single fronts can be studied. Furthermore, most of these approaches do not allow for dynamical investigations of the fronts and related effects on concentrations of floating objects, such as FML. The most intuitive approach for analyzing the dynamics of floating objects at fronts are Lagrangian methods, which motivates this chapter.

In the case of numerical approaches, velocity data are typically provided by ocean models, high-frequency (HF) radars, or satellite-based velocity products. In the following, virtual Lagrangian tracers are named "particles" to make them distinguishable from Eulerian tracers and satellite-tracked drifters ("drifters"). Moreover, the term "particle concentrations" is related to Lagrangian particles and not to Eulerian concentrations. Eulerian concentrations, computed for each grid cell, have also been used to describe FML accumulation patterns [29, 30] but are not part of this chapter.

The present chapter reviews and provides Lagrangian methods suitable for visualizing and studying frontal dynamics. After an introduction to the physical oceanography of fronts (Sect. 1.1) and related FML dynamics in tidal basins (Sect. 1.2), methods for numerical modeling (Sect. 2.1) and for drifters (Sect. 2.2) are reviewed and applied to the region of the North Sea. The influence of tides on particle dynamics is presented in Sect. 2.3. The general hydrodynamic conditions in the North Sea are not part of this chapter and are provided in relevant reviews [31–33]. The chapter ends with a discussion and future outlook in Sect. 3.

1.1 *Physical Oceanography of Fronts in Tidal Basins*

Fronts in tidal basins of shelf seas form due to several different physical processes [34, 35]. The freshwater discharge from rivers can form low-salinity lenses spreading with the coast on their right (in the Northern Hemisphere). Along the lens margin, parallel to the coast, a front can form. The slope of isopycnals depends on the strength of vertical mixing and the amount of freshwater discharge (Fig. 1). The distance of the front to the coast is approximately the internal Rossby radius if the isopycnals are not vertical. However, in the presence of strong vertical mixing

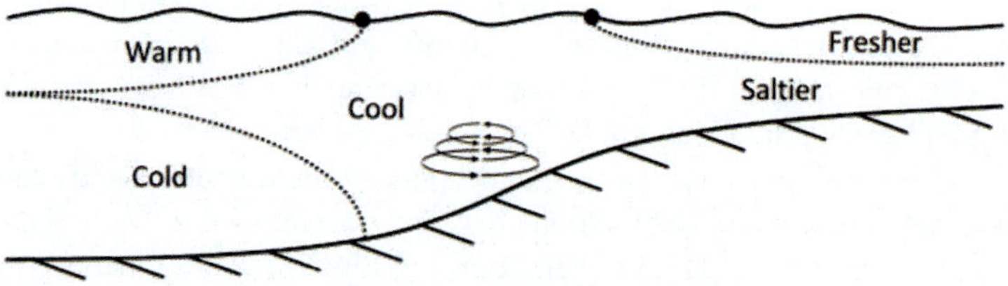

Fig. 1 Sketch of two different fronts (dashed lines) on the continental shelf. On the right, a freshwater front is depicted, which evolves due to a freshwater lens spreading over saltier shelf water. The slope of the isopycnal depends on the strength of vertical mixing and the amount of freshwater input. On the left, a tidal mixing front in summer is shown. The surface heating prevents vertical mixing and a thermocline develops. Strong tidally induced mixing (the spiral) can cause a vertically mixed water column in shallow regions. The positions of surface fronts are indicated by black dots

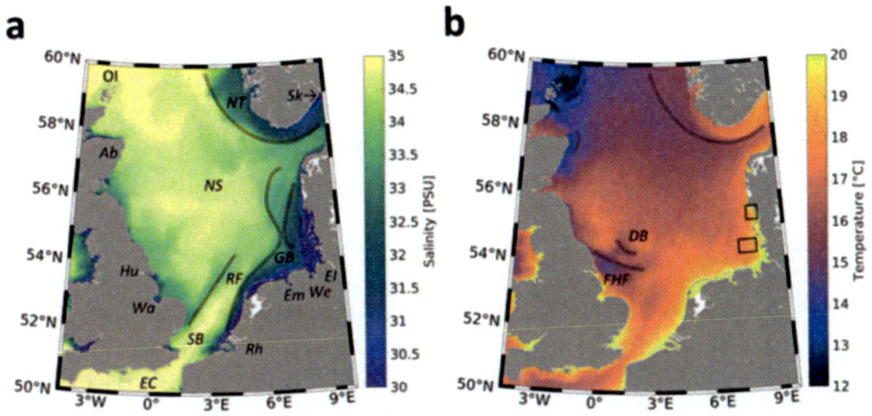

Fig. 2 Surface (**a**) salinity and (**b**) temperature averaged over the period July 12 to August 14, 2019. The abbreviations are: *Ab* Aberdeen, *DB* Dogger Bank, *EC* English Channel, *El* Elbe River, *Em* Ems River, *FHF* Flamborough Head Front, *GB* German Bight, *Hu* Humber Estuary, *NS* North Sea, *NT* Norwegian Trench, *OI* Orkney Islands, *RF* Rhine Front, *Rh* Rhine River, *SB* Southern Bight, *Sk* Skagerrak, *Wa* The Wash Estuary, and *We* Weser River. Fronts described in the text are marked with a double line. The black squares in (**b**) are the subsets shown in Fig. 6. The model data are freely available under https://resources.marine.copernicus.eu/?option=com_csw&view=details&product_id=NORTHWESTSHELF_ANALYSIS_FORECAST_PHY_004_013

(induced by tides and wind), the front can reach the seafloor, and the distance to the coast increases. Data from the operational ocean model AMM15 of the European northwest shelf (ENWS) [36] show that vast coastal regions of the North Sea are dominated by freshwater runoff. Most notably, the Rhine and Elbe outflows dilute the southeastern basin considerably (Fig. 2a). Rhine water spreads along the Dutch coast, where a sharp salinity gradient forms, enhanced by saline Atlantic water entering the North Sea through the English Channel. Along the coasts of the British Isles, another area of strong thermohaline gradients appears due to the waters from the Humber and Wash estuaries. When the Rhine water enters the German Bight, it is separated by a front from the inner North Sea water, and the highest salinity gradients stretch to the northeast, heading toward the Skagerrak. The Elbe River outflow additionally adds freshwater input into the German Bight, and a second gradient is present pointing from the Elbe mouth to the northwest. The Weser and Ems rivers also contribute to an overall salinity decrease in the German Bight. Some of these fronts are shown in Fig. 2a, which is based on the freely available operational AMM15 model data (see its caption for details).

The second type of front is tidal mixing fronts. Away from regions of freshwater influence (ROFIs), tides can induce sufficient vertical turbulence to mix the whole water column vertically (Fig. 1). Typically, this happens in the summer when shelf seas are stratified. During this time, warm water is located above cold water (if not vertically mixed). The transition zone from stratified to vertically mixed water is quite sharp and has lateral gradients of approximately 1 °C km^{-1} [35, 37]. Cool water regions and high temperature gradients can be well observed in satellite images. The effect of tides on frontogenesis has been known since the 1970s. It was found that the balance of stabilizing surface heating and the counteracting tidally induced turbulence at the seafloor can be predicted with a certain threshold of $\log_{10}\left(h\,\overline{u}^{-3}\right)$, which is also known as the Simpson-Hunter parameter [38]. In this equation, h is the water depth, and $\overline{u}$ is the amplitude of the tidal currents. Below (above) ~2.7, the water column is well mixed (stratified), whereas contours of this value often coincide with tidal mixing fronts. Although this simple equation can predict the fronts' position quite well, it has been constantly improved in the following years by adding more physically reasoned relations [35]. The energy needed to vertically mix the water column can be expressed as a fraction of available tidal stirring power ($\sim 3 \times 10^{-3}$) and is found to be an intrinsic property in this process [39]. Tidal mixing fronts induce a density-driven, secondary circulation. At the surface, downwelling can be observed along the front, leading to surface convergence, which is observable as distinct lines of flotsam [22, 40–43]. On the stratified side, an along-front geostrophic jet is established. For continuity reasons, on the mixed side, upwelling is observed, bringing cold water to the surface. Model results and observations have shown typical jet velocities of 10–15 cm s^{-1} [42]. In the North Sea, several tidal mixing fronts have been observed [44] and modeled [45]. The surface temperature of the ENWS operational model shows that in the North Sea, four regions are prone to tidal mixing fronts: the Flamborough Head front north of the Humber Estuary, around the Dogger Bank, north of Aberdeen, and around the Orkney Isles (Fig. 2b). They are easily identifiable from the temperature decrease on their shallow side, and the positions of the thermal fronts coincide with frontal positions obtained from satellite data [24, 27]; they typically appear in summer. Tidal fronts or, at least, a transition from nonstratified to stratified waters in the German Bight were found in model and in situ data [43, 45–48], and their positions can be considerably altered by wind [47, 49] or river runoff [47]. Additionally, an influence of waves on fronts has been shown [50].

The present chapter will focus on Lagrangian dynamics at the ROFIs and tidal mixing fronts in the North Sea since they are the most pronounced ones in this area with regard to surface dynamics. However, other kinds of fronts are also present on the ENWS, namely, shelf break fronts [51, 52], deep cold water pool fronts [53–56], local upwelling [47, 57], and fronts in the Norwegian Trench/Skagerrak induced by Baltic Sea water inflow (Fig. 2a), which will not be discussed. Further theoretical considerations related to fronts are given by, e.g., McWilliams [58].

1.2 Floating Marine Litter at Shelf Sea Fronts

The dynamics of FML have been investigated in various ways, and its pathways on the global scale are well understood [59–62]. In contrast, much less is known about the dynamics in marginal seas. As transition zones of riverine litter on their way to the open ocean, they play an important role in the FML transport chain. In the region of the North Sea, modeling attempts of FML in rivers and estuaries are rare, so the dynamics there are not well understood from the modeling perspective. It is known that surface particles can form axial accumulation patterns due to secondary circulation [63–65] but are also affected by turbidity maximum zones [66] and other dynamics [67]. The axial strips can even reach open waters in the North Sea and thereby form localized input areas. Once the litter reaches the North Sea, it can become trapped within the river plume [68]. Litter released within rivers will probably also beach in their waterways [65, 69]. Despite that, a considerable amount of litter enters the North Sea via rivers, especially at its southern coast [70], where the main litter discharge originates from the rivers Elbe, Ems, and Weser in Germany [71], from the Rhine in the Netherlands [72], the Thames in Great Britain, and the Seine (English Channel) in France [73].

Shelf sea dynamics can be very complex due to the strong currents induced by tides, winding coastlines, increased turbulence, and other processes. This variety of processes makes the analysis of FML in shelf seas challenging. Attempts to study specific oceanic regions have been made to investigate their specific processes and dynamics [49, 74–77]; an extensive overview of FML modeling studies is given by Ricker [65] (Table 1). It has long been known that flotsam can accumulate along coastal fronts in general [78], tidal mixing fronts [22], or bottom salinity fronts [66], but such reports are still limited. Windrows, defined as submesoscale accumulation of FML, are sometimes also related to fronts [79, 80]. Even fewer studies report the modeling of pollutants at shelf sea fronts. Tracer experiments revealed a strong correlation between salinity and tracer concentrations at a river plume front [68], providing a first hint for possible frontal impacts on FML. These tracer experiments also showed that cross-frontal diffusion is two orders of magnitude lower than along-front diffusion, demonstrating the dynamic influence of fronts on the spread of river plumes [81]. The accumulation and transport of FML at fronts has been hypothesized for the Russian Arctic Seas [82]. The accumulation of modeled particles along river plumes in the Black Sea has been reported and was supported by aerial FML images [83]. The authors pointed out that the necessary criterion for FML accumulation along river plume fronts is a stagnation point; otherwise, the surface convergence would be too variable to induce accumulation. It was also shown that tidal mixing fronts and ROFI fronts act as barriers and accumulation zones for surface released particles [49, 84], suspended particulate matter (SPM) [84, 85], or chlorophyll [21, 35]. Extensive and area-wide (Lagrangian) modeling attempts to study the impacts of shelf sea fronts on surface FML dynamics are still missing and further motivate this chapter.

2 Lagrangian Methods

In this section, methods for studying and visualizing fronts in tidal basins are presented. These methods are either rather suitable for numerical Lagrangian models (Sect. 2.1), as they often require the release of numerous drifters, or they are also suitable for GPS drifter experiments (Sect. 2.2), for which usually fewer drifters are available. Section 2.3 demonstrates the importance of tides for particle dynamics in tidal basins by providing some additional methods for numerical models. To complete the first step in the direction of the understanding of the specific characteristics of each method, this section aims to provide an inventory of available methods and categorizes them. It is also discussed how the methods can reveal frontal processes. A summary of the presented methods together with their strengths and weaknesses is given in Table 1 in the Appendix.

2.1 Methods for Numerical Modeling

Lagrangian particle tracking allows us to determine the trajectories of virtual particles that are released within the ocean model domain. The purposes of particle tracking can be very diverse, as the large number of examples demonstrates. Typical applications track the pathways of biological quantities, oil slicks, and FML. An introduction to particle dynamics in the coastal ocean is given by Lynch et al. [86], and a more specified summary for microplastics is given by Hardesty et al. [87]. Depending on the temporal integration direction, particles can be tracked backward or forward in time to determine sources and sinks, respectively. For more realistic results, particles can have specific physical characteristics, such as buoyancy, beaching probabilities, or active movement. In addition, subgrid diffusion can be added in the case of forward integration. Often, instantaneous snapshots of particle positions or trajectories are shown. Although these methods provide new insights [83, 88], their ability to reliably illustrate particle accumulation regions can be limited [49]. For example, instantaneous plots of the results can underrepresent possible accumulation areas. In addition, temporal snapshots are not necessarily connected to the mean state of a certain period of time. Thus, different methods are required that are able to depict the frontal divergence and attraction zones in a Lagrangian manner. A common approach is to generate the so-called pseudo-Eulerian plots, which display Lagrangian results on a specific grid, e.g., the ocean model grid. In contrast to mapping particle positions, this approach is able to take into account the possible overlapping of particles. An overlapping of particles could lead to a visual underestimation of particle accumulation, so particle distributions should only be used for qualitative analyses. Trajectory plots also suffer from this fact (see Sect. 2.3 for examples of trajectory analyses). In addition, it is often useful to apply a uniform initial release of the particles to enable a comprehensive and area-wide representation of the particle dynamics. To overcome such graphical issues, generalized visualization guidelines have been prepared and should be taken into account [89].

The need for nonstationary velocity fields is also noteworthy, which per se excludes some approaches to advect the particles, such as transition matrices [61, 62, 74, 90–93] or satellite data [29, 94, 95]. The effect of temporal changes in surface velocities on particle accumulation has been demonstrated by Stanev and Ricker [77] in the case of the Black Sea.

2.1.1 Particle Concentrations

To detect particle accumulation and dispersal, it is straightforward and intuitive to count particles on a predefined Eulerian grid (also called binning). Subsequently, (averaged) particle concentrations, also called particle densities, can be computed by taking into account the particle trajectory. These concentrations can be absolute numbers, percentages, or other relative numbers, such as probability maps when the number of particles in a bin is normalized by the total amount of particles. Such quantities have often been applied in the field of FML, and the advantages are easy and fast computation as well as simple application to a variety of purposes. Applications to marginal seas have been made to the ENWS [49], Mediterranean Sea [96–100], Adriatic Sea [101], Aegean Sea [76], Baltic Sea [102, 103], Sea of Japan [75], Barents Sea [104], and Black Sea [77]. What makes this method challenging is (1) to correctly setup the seeding, which has to be adapted to the processes of interest; (2) to correctly choose the properties of the density, such as the spatial and temporal averaging, to be able to interpret the results adequately with respect to (1); and (3) to ensure statistical relevance. In particular, the temporal averaging should be treated with care because a mixture of time scales could lead to misinterpretations due to the depiction of processes taking place on different time scales. Considering the numerous examples of this method focusing on different processes affecting the distribution of FML, almost no studies have investigated the effect of fronts on particle concentration, especially in shelf seas [49].

To demonstrate the application of some of the methods discussed in this chapter, particles were released uniformly (one particle per grid cell) in the ENWS operational AMM15 model on the surface in the North Sea. The grid cells of the model had an approximate size of 2.6 × 1.5 km (longitude × latitude), which resulted in a total of ~200 k particles. The particles were advected by the freely available open-source offline model OpenDrift [105], and neither vertical velocities nor wind and diffusion were taken into account. These properties of the particles represent FML (or released chemical substances), which floats at the surface but is fully submerged. Although this approach only represents a specific kind of FML, it is additionally able to capture the pure Lagrangian properties of the ocean. After 34 days of integration, pronounced features in particle distribution occur (Fig. 3a). Notable is the particle dispersal along coasts in the southern North Sea, the English Channel, and the Norwegian Trench. Furthermore, particle accumulation takes place along a thin strip from the Southern Bight to the Skagerrak and along the southern rim of the Norwegian Trench. However, it is difficult to reliably evaluate the result by analyzing the particle patterns due to the possible overlapping of particles. The instantaneous particle density is sometimes used to overcome this problem, but surface particles are very sensitive to wind, which can alter the particle density considerably, as stressed by Ricker and Stanev [49]. It is also noted that instantaneous particle

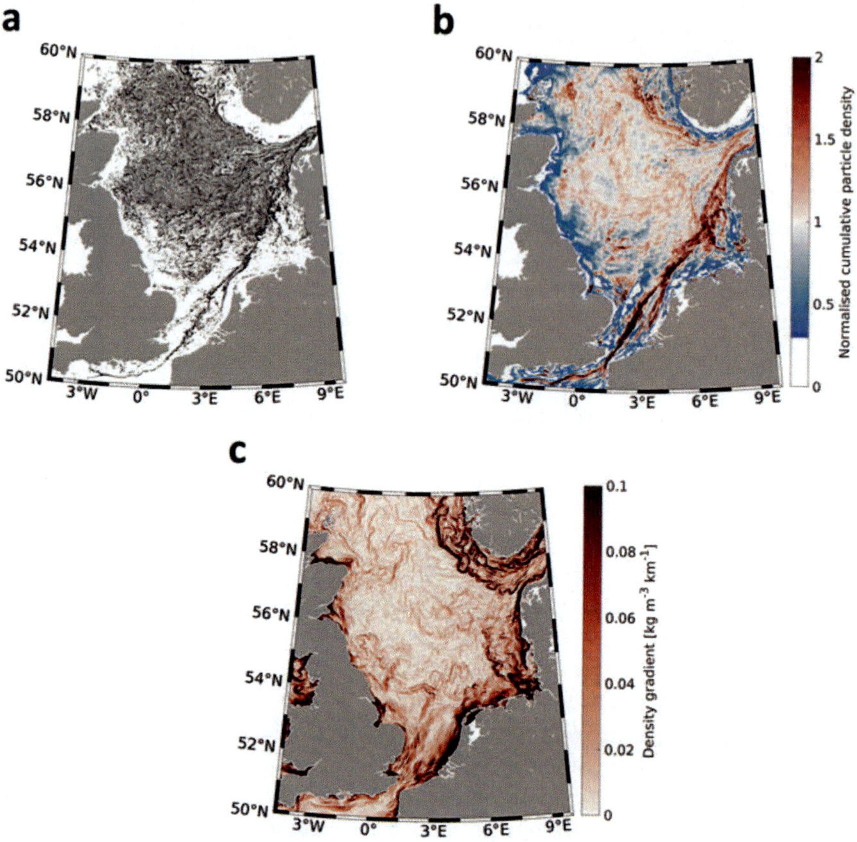

Fig. 3 (**a**) Particle positions on August 14, 2019. The particles (in total ~200 k) were released uniformly in every model grid on the sea surface on July 12, 2019. The velocities for particle advection are obtained from the dataset given in Fig. 2. Vertical velocities are not taken into account. (**b**) Normalized cumulative particle density (NCPD) computed from the hourly particle positions of (**a**). Particle accumulation/dispersal is associated with NCPD larger/smaller than 1. Values smaller than 0.3 are supposed to have no statistical significance and were removed. (**c**) Surface density gradient magnitude averaged over the same period using the Sobel operator as proposed by, e.g., Belkin and O'Reilly [25]

concentrations need to be gridded coarsely so that spatially sufficient particles can be taken into account. Otherwise, the result would be very noisy.

In contrast, averaged accumulation takes into account short-term alterations in particle positions that is a rather appropriate approach in tidal basins. For example, the normalized cumulative particle density (NCPD) is computed with $NCPD(x, y, t_n) = \frac{\sum_{i=0}^{n} N_{u\neq0}(X, Y, t_i)}{\sum_{i=0}^{n} N_{u=0}(X, Y, t_i)}$, where (X, Y) are the coordinates of an arbitrary grid cell in the longitudinal and latitudinal directions, respectively, with grid dimensions (dX, dY) (here equal to the ocean model grid size); n is the number of time steps from t_0 to t_n; $\boldsymbol{u}$ denotes the velocity field, and N is the number of particles at time step i in grid (X, Y) [49]. In the case of averaged concentrations, (dX, dY) can be smaller than in the case of instantaneous concentration maps, as stressed above. The main accumulation area (NCPD>1) in the North Sea appears along a distinct strip from the Southern Bight to the Skagerrak (Fig. 3b), which has been shown to be also present in FML situ data [88]. In the Southern Bight, this strip is split into southern and northern parts. In addition, the southern rim of the Norwegian Trench is shaped by particle accumulation and some minor accumulation areas are of interest, specifically thin strips in the German Bight with a north-south orientation. Most of these NCPD features can be associated with thermohaline fronts, as they approximately align with the surface density gradient maxima (Fig. 3c). Similar results were obtained by Korshenko et al. [83] for river plumes in the Black Sea. In particular, the transition from low to high gradients seems to be favorable for particle accumulation. However, tidal mixing fronts, which are mainly thermal fronts, do not occur in either the NCPD or density gradient. This means that their contribution to density changes is rather minor, as is their effect on particle dynamics. Only the western margin of the Dogger Bank is an exception, where the surface temperature is reduced (Fig. 2b). NCPD is also similar to the frontal positions of SPM [28] and is another hint of the accuracy of the accumulation simulations. Nonetheless, SPM has spatially inhomogeneous sources and is thus not fully suitable for comparisons with NCPD. Mansui et al. [96, 97] used another approach to obtain averaged particle densities by averaging over a large number of model runs.

2.1.2 Lyapunov Exponents

Lyapunov exponents and their related quantities are commonly used for the detection of particle attraction and repulsion. In the field of modeling, it is possible to release a large number of particles initially located close to each other. From these particles, a flow map $F_{t_0}^t(\boldsymbol{x}_0)$ at time t can be obtained, where $\boldsymbol{x}_0$ are the initial particle positions. In turn, the flow map enables the computation of the Cauchy Green Strain Tensor $C(\boldsymbol{x}_0) = \left|\nabla F_{t_0}^t(\boldsymbol{x}_0)\right|^T \nabla F_{t_0}^t(\boldsymbol{x}_0)$. C provides the magnitude of particle separation growth of the small perturbations of the particle release positions and is specified by its eigenvalues λ and the related eigenvectors. The exponential growth rate of a finite-time interval is also known as the finite-time Lyapunov exponent (FTLE) and is given by $FTLE(\boldsymbol{x}_0, t_0, \tau) = \frac{1}{\tau} \ln \sqrt{\lambda_{max}}$, where $\tau = t - t_0$, and λ_{max} is the maximum eigenvalue of C. The ridges of local FTLE maxima are called Lagrangian coherent structures (LCS) and correspond with repellent LCS (unstable manifolds) at t_0. The FTLE field can also be computed backward in time, which reveals attracting LCS (stable manifolds) and has been shown to coincide with temperature gradients [106]. In contrast to FTLE, which represents the separation of particles in a specific period, the finite-size Lyapunov exponent (FSLE) gives the time needed for a specified particle separation. FSLE ridges can also be used to obtain LCS, and their stability is even given in the presence of noise and low horizontal resolution of the model [107]. Nevertheless, although FSLE is expected to have some shortcomings in terms of LCS detection [108], it can be used to detect cross-stream transport barriers [109]. More detailed derivations and other methods to compute LCS can be found in several reviews [110–118]. The Lyapunov exponent and related quantities have been frequently used, e.g., for the visualization of oceanic transport barriers [119], the advection properties in an archipelago [120], the transport of algal blooms [121], the relation of fishery conditions and fronts [122], and to explain the influence of boundary currents on the beaching rates of FML [123]. However, knowledge about the interconnection of thermohaline fronts and LCS in shelf seas is currently restricted compared to the open ocean [124] and identifying such connections can be challenging due to highly variable dynamics there.

2.1.3 Other Methods

Cluster strength (CS), defined as $CS(\boldsymbol{x}^*) = \sum_{n=1}^{N} e^{-(d_n/L)^2}$ with $d_n = |\boldsymbol{x}_n - \boldsymbol{x}^*|$, where L is a scaling factor and d_n is the distance from the nth particle $\boldsymbol{x}_n$ to the position $\boldsymbol{x}^*$ at a specified time step, has been developed to display regions of particle accumulation and is comparable to particle concentrations [125, 126]. Although Eulerian divergence $\nabla \cdot \boldsymbol{u}$ drives particle accumulation, it was shown that Eulerian divergence is not able to fully explain the particle distribution in both deep and coastal regions [125, 126]. A more suitable approach is the computation of the Lagrangian divergence $LD = \nabla \cdot \boldsymbol{u}[\boldsymbol{x}(\tau), \tau]$, which is the divergence along a particle's trajectory. Its temporal integral is called the average dilation rate (ADR) or finite-domain Lagrangian divergence (FDLD) and is given by $\Delta = \frac{1}{t}\int_0^t \nabla \cdot \boldsymbol{u}[\boldsymbol{x}(\tau), \tau] d\tau$. This quantity showed a good correlation with particle accumulation ($\Delta < 0$) and dispersal ($\Delta > 0$) [125, 126] and temperature gradients [106]. Thus, it is computed for the North Sea based on the operational ENWS model and the particle experiment described in Sect. 2.1.1. In the North Sea, the NCPD is in good accordance with density gradients and FDLD (Fig. 4a–c), whereas the Eulerian divergence (Fig. 4d) is highly variable in space and does not show a qualitative correlation with the NCPD. However, compared to NCPD (1 month), a shorter integration period for FDLD (1 week) had to be chosen to obtain usable results. Thus, weekly-based analyses are presented in Fig. 4. The shorter integration time is probably due to the high tide-induced diffusion in the North Sea. Diffusion could force particles to leave frontal regions and decrease the correlation between FDLD and NCPD. In addition, the accumulation strips observable in NCPD (red areas) are broader than the strips in the FDLD field. On the other hand, this reveals an advantage of FDLD compared with NCPD: shorter periods can reveal clearer accumulation areas compared to the longer period needed to compute NCPD (compare Figs. 3 and 4 for monthly and weekly results). FTLE can also be computed with $FTLE = {}^{(\Delta+\Sigma)}\!/_{2}$, where Σ is the stretch rate [125, 126]. Δ is related to area changes spun up by particles at the sea surface, whereas Σ is related to area deformations without area changes. Compared to FTLE and stretch rate, FDLD shows a higher correlation with

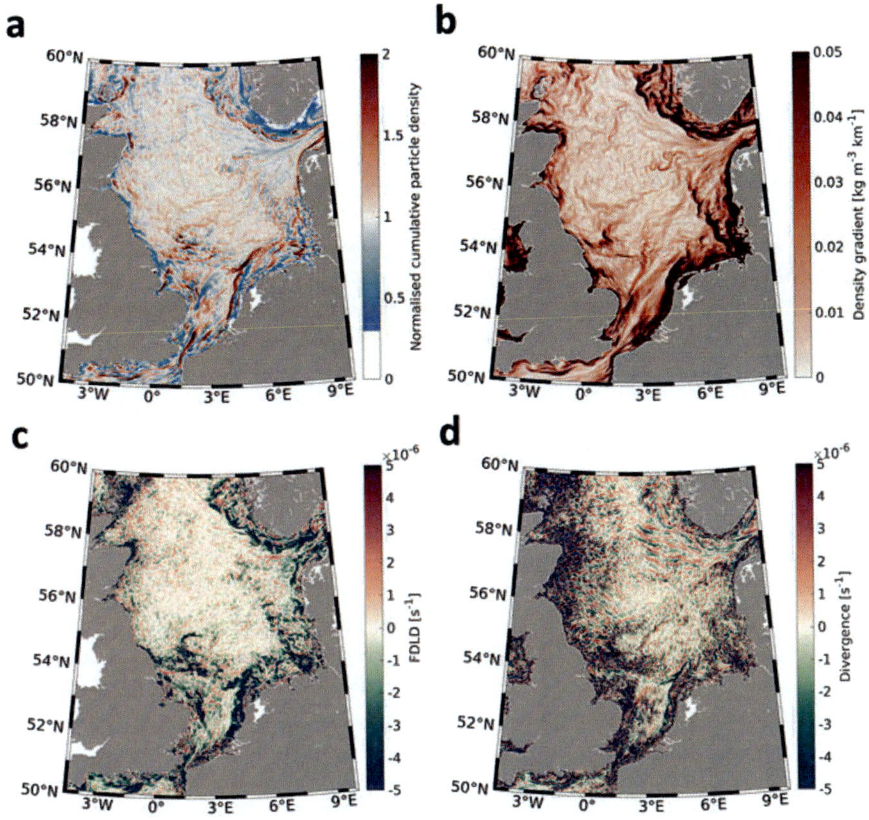

Fig. 4 Surface (**a**) NCPD, (**b**) density gradient magnitude, (**c**) forward in time finite-domain Lagrangian divergence (FDLD), and (**d**) Eulerian divergence of the period August 2 to 9, 2019. Negative values in (**c**) and (**d**) are associated with convergence and thus with particle accumulation. These results are based on the same dataset and particle experiments as described in Figs. 2 and 3. However, only 1 week is analyzed, which enables a better visualization of FDLD. This week represents a period with wind directions mainly from the south-west comparable to long-term means

CS and is thus more suitable to display particle accumulation, as demonstrated with chlorophyll-a concentrations [127]. However, both CS and FDLD are computationally costly because a large number of particles is needed and the divergence has to be integrated along the particle trajectories.

Similar approaches were developed by Kalda et al. [128] by introducing finite-time compressibility and by Duran et al. [129], who computed an along-path area change. The former method measures the area changes of particle triads, and the latter can show whether convergence (negative divergence) or confluence (divergence-free attraction) is responsible for particle accumulation. The complexity method (CM), which includes quantities such as the correlation dimension of a trajectory or the ergodicity effect, can detect coherent structures [130]. Dynamical systems theory has been applied to detect hyperbolic trajectories and their stable and unstable manifolds [131, 132]. It has been shown that these manifolds can visualize the location of fronts, at least in the deep ocean [133]. Particle dispersion is often applied to drifters (see also Sect. 2.2.2) but could also be used for simulated particles [106, 134]. Approaches measuring particle travel times [135, 136]; Lagrangian residence times [137]; or tracking temperature, salinity, and other properties [134, 138, 139] are expected to provide insight into water mass dynamics and so into frontal regions; connectivity studies could also have such a potential [140–143]. However, most of these methods have not yet been applied to frontal dynamics, yet. A promising approach is the so-called Lagrangian flow network and its ability to detect dynamic barriers such as oceanic fronts [144, 145].

2.2 *Methods Applied to Drifter Data*

Lagrangian observations with drifters are a direct method for identifying the dynamics of the turbulent flow field at the ocean surface in time and space. Drifter sampling has been used to study the pathways of biological substances [146], oil spills [147], and climate-relevant tracers [148], for search and rescue (SAR) applications [149, 150] and for the tracking of FML [61, 151]. The representation of FML is similar to the surface particles of the previous section but with a slightly higher direct wind influence. Lagrangian observations greatly improved the understanding of tracer advection and accumulation, and provided validation data for the parameterization of particle tracking models [62, 152]. In recent decades, several drifter designs have been developed to mimic the water-following characteristics of various tracers. For example, the Coastal Dynamics Experiment (CODE) drifter was developed with the goal of reducing direct wind and Stokes drift to describe surface currents accurately and, for example, the spread of oil spills on the water surface [153]. Other drifter designs aimed to simulate the pathways and dispersal of FML, taking into account wind-induced motions [151, 154]. To study convergence and divergence along and across oceanic fronts with drifters, drifter pairs or clusters are often released in frontal regions. The spreading of a cloud of drifters around its center of mass can provide information to identify areas of convergence or divergence. In the following section, common methods for drifter pairs and drifter clusters applicable to tidal basins are presented. An exemplary analysis of drifter convergence in the area of a front in the German Bight is given in Sect. 2.2.1.

2.2.1 Pairwise Drifter Statistics

Several experiments with pairwise deployed or clustered drifters have been conducted to study dispersion scaling laws in the oceans [153, 155–157]. Various studies have investigated dispersion characteristics and eddy diffusivities at the submesoscale, mesoscale, and large scale in various ocean basins [147, 158]. For this purpose, the separation of drifters deployed in pairs or clusters over time is considered, defined as $R(t) = \frac{1}{N}\sum_{i,j}\left|\boldsymbol{x}_i(t) - \boldsymbol{x}_j(t)\right|$, where i and j refer to each drifter of a pair in a cluster of N drifter pairs, and $\boldsymbol{x}(t)$ corresponds to the position of the drifters at time t. A common metric in two-particle pair statistics involving drifters is the relative dispersion, which is the second-order moment of $R(t)$, defined as $\left\langle R(t)^2\right\rangle = \frac{1}{N}\sum_{i,j}\left|\boldsymbol{x}_i(t) - \boldsymbol{x}_j(t)\right|^2$. Decreasing dispersion and separation rates have been observed in the vicinity of fronts, which is an indicator of convergent zones in these areas. Estimates of drifter separation rates in the Gulf of Mexico have shown that fronts can contribute to massive drifter convergence into an area 400 times smaller than the initial release area [159].

A drifter cluster consisting of five drifters tracked with the Global Positioning System (GPS) deployed in the area of the front in the southeastern German Bight shows two scenarios of cluster convergence during a wind-induced cross-front movement (Fig. 5). The dataset is freely available [160], and a technical description of the deployed drifters is given by Meyerjürgens et al. [151]. The average drifter separations for the first scenario (Fig. 5, blue period) decreased from approximately 3,500 to 1,600 m as they crossed the front (Fig. 5a), while the satellite sea surface temperature (SST) dropped from approximately 19 to 17.5°C (Figs. 5b and 6a). The satellite chlorophyll concentration also showed a gradient during this period (Figs. 5c and 6b). For a detailed description of the satellite data, the reader is referred to the website given in the caption of Fig. 5. A second scenario of drifter convergence (Fig. 5, red period) was observed, where the average drifter separation decreased from approximately 2,500 to 1,600 m while crossing a front with a temperature drop of 0.7°C (Figs. 5b and 6c) and an increase in the chlorophyll concentration of approximately 2 mg m^{-3} (Figs. 5c and 6d).

A drifter separation decrease occurred simultaneously with the highest gradients in SST and chlorophyll, revealing the frontal ability to converge drifters (Fig. 5). Along the front, where SST and chlorophyll were almost constant (August 1–5, 2019), no convergent trend was observed in the average drifter separation. These results are consistent with drifter observations on the southern flank of Georges Bank, where cross-front convergence is generated by a tidal mixing front [161, 162]. Analyses of cluster separations in the Gulf of Mexico also show the potential for submesoscale filaments and fronts to concentrate drifters and other floating materials, while larger mesoscale structures contribute to the separation between different clusters [159, 163].

2.2.2 Multiple Drifter Cluster Statistics and Velocity Gradients

Several studies have derived kinematic properties such as divergence and vorticity from drifters to quantify convergent features at oceanic fronts and eddies [159, 164, 165]. Drifter triplets can be used to calculate the area rates of change of the parcel by $\frac{1}{A}\frac{dA}{dt} = \left(\frac{\partial u}{\partial x} + \frac{\partial v}{\partial y}\right)$, where A is the area spanned by the three drifters; u and v are the

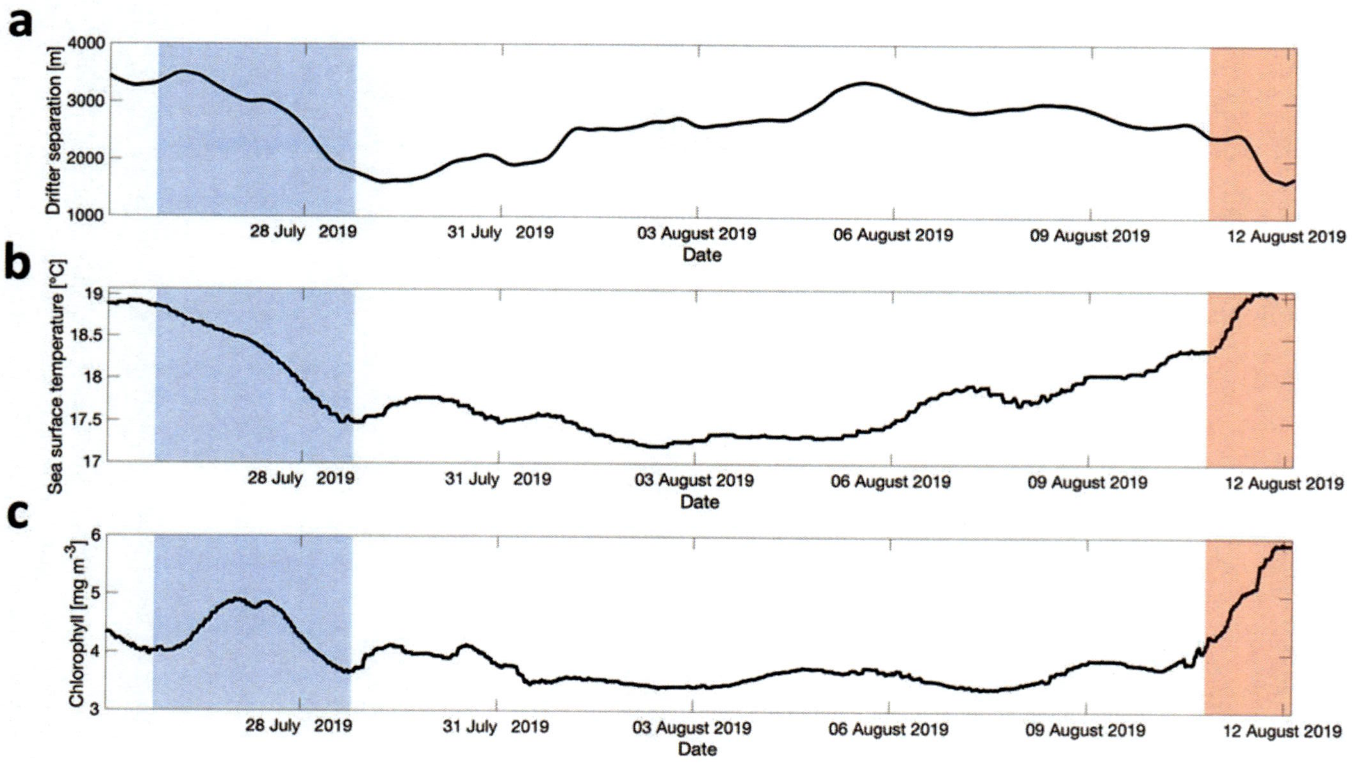

Fig. 5 (**a**) Average drifter separation R of five satellite-tracked drifters. Colored patches mark periods of drifter convergence. The trajectories have been low-pass filtered by a 25-h Butterworth filter in order to remove the dominating M_2 tide. (**b**) Daily sea surface temperature (SST) from satellite data with a resolution of ~2.5 km interpolated to the average positions of the filtered drifter trajectories. (**c**) Daily chlorophyll concentration from satellite data with a resolution of ~1 km interpolated to the average positions of the filtered drifter trajectories. The corresponding drifter trajectories of the blue and red periods are shown in Fig. 6. The SST and chlorophyll datasets are freely available under https://resources.marine.copernicus.eu/?option=com_csw&view=details&product_id=SST_ATL_SST_L4_NRT_OBSERVATIONS_010_025 and https://resources.marine.copernicus.eu/?option=com_csw&view=details&product_id=OCEANCOLOUR_ATL_CHL_L4_NRT_OBSERVATIONS_009_037, respectively

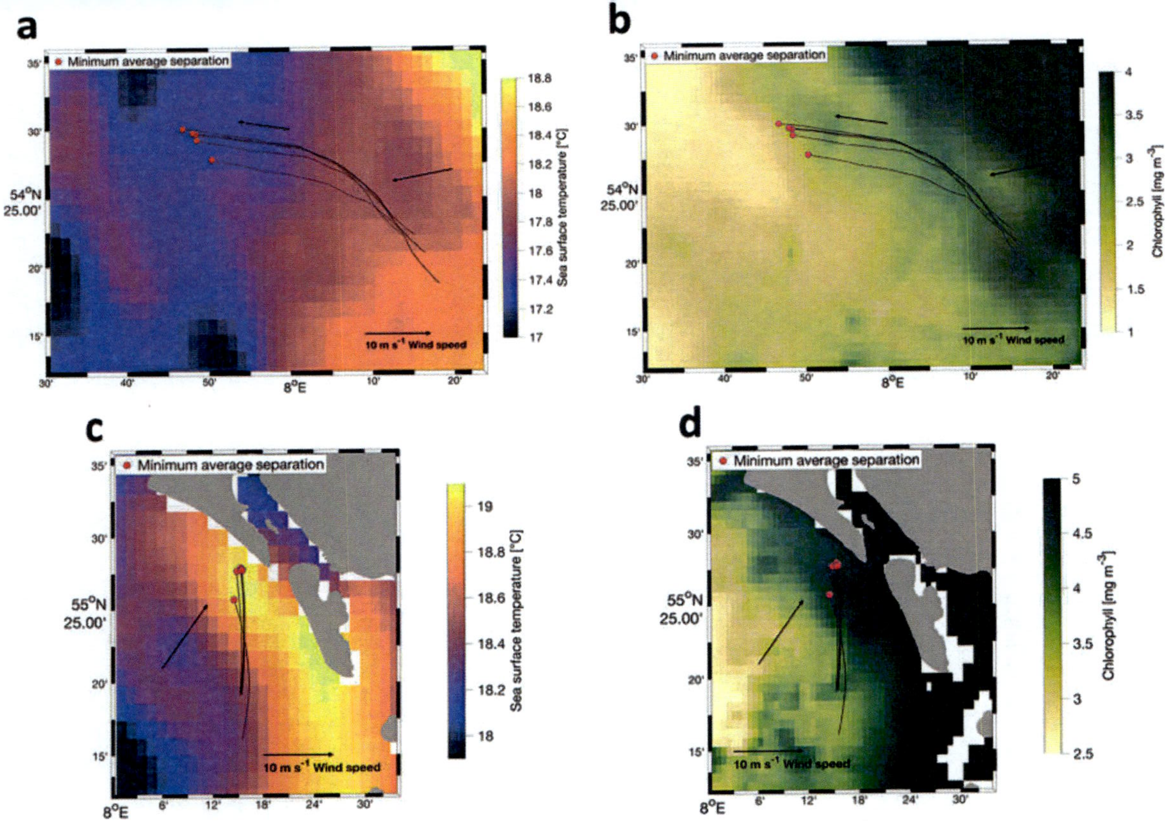

Fig. 6 Filtered drifter trajectories from (**a**, **b**) July 25 to 28, 2019 and (**c**, **d**) August 11 to 12, 2019 during periods of convergence across a front (blue and red periods in Fig. 5, respectively). Red circles mark the location of minimum average drifter separation as well as the final positions of the drifters. Average model wind speeds are indicated by black arrows and are divided into two 1.5-day periods in case of the first period. Background colors show the (**a**, **c**) average SST and (**b**, **d**) average chlorophyll concentrations corresponding to the shown period of the trajectories. The SST and chlorophyll datasets are the same as used in Fig. 5. See Fig. 2b for the location of the maps

zonal and meridional velocity components of the center of mass of the triplet, and x and y are the spatial coordinates, respectively. This method can calculate the surface divergence by estimating the change in the area of the drifter cluster (see also Sect. 2.1.3). This method was presented by Reed [166] and first applied to drifter clusters by Molinari and Kirwan [167]. The latter study also calculated vorticity, shearing, and stretching deformation rates, which can also be valuable indicators for drifter convergence at oceanic fronts [159, 164]. These differential kinematic properties (DKPs) can be calculated using differential characteristic velocities, such as the relative vorticity $\zeta = \left(\frac{\partial v}{\partial x} - \frac{\partial u}{\partial y}\right)$, shearing deformation $S = \left(\frac{\partial v}{\partial x} + \frac{\partial u}{\partial y}\right)$, and stretching deformation $N = \left(\frac{\partial u}{\partial x} - \frac{\partial v}{\partial y}\right)$. These kinematic properties are directly linked to the deformation of drifter clusters and can provide information about convergent and divergent dynamics. To cover the spatial scale of an oceanic front and increase the statistical significance of the DKPs, high quantities of drifter triplets along the front are necessary [164]. Targeted drifter deployments covering large spatial and temporal scales can be used to project Lagrangian trajectories to Eulerian velocity fields to compute DKPs on frontal scales. Eulerian velocity fields have been reconstructed with drifter observations along a mesoscale front in the Gulf of Mexico with convergence magnitudes of 20 times the planetary vorticity f [168]. Rascle et al. [169] combined DKPs from drifter clusters with remote sensing data and found that current gradients were mainly caused by cross-front convergence with cyclonic vorticities, while along-front shear was observed with anticyclonic vorticities.

2.2.3 Drifter Trajectory Analysis with Lagrangian Coherent Structures

As described in Sect. 2.1.2, the positions of LCSs are often associated with oceanic fronts and form transport barriers in these areas [170–172]. The combination of ship-based measurements, LCS derived from altimetry data, and Lagrangian observations can provide valuable insights into the transport processes of drifters along fronts. Several experiments demonstrated that drifters tend to accumulate at the LCS and are transported along with these structures. Experiments in the western Mediterranean Sea revealed that LCS can block connectivity and clearly separate different water masses [173]. Horizontal convergence of drifters can be inferred from LCS, as indicated by negative FTLE values [174]. Drifter trajectories typically provide sparse trajectory information on spatial and temporal scales and thereby prevent the reconstruction of an Eulerian velocity field to compute LCS. Allshouse and Peacock [115] presented two methods to estimate LCS from sparse trajectory information. Cluster-based analysis has already been applied to a global ocean drifter dataset [175] and braid-based analysis is also expected to be applicable to drifter data. In addition, strong tidal oscillations and inertial motions can lead to the distortion of the FTLE field [176, 177], suggesting that LCS are probably not optimal for Lagrangian analysis in the North Sea.

2.3 *The Impact of Tides*

One of the main processes driving shelf sea dynamics is tidal fluctuations. Examples in relation to Lagrangian dynamics are, e.g., that (over) tides are important for matter transport [178, 179] and that Lagrangian mean velocities are needed to adequately take into account tidal dynamics for the computation of residual currents [180]. Recently, a scale dependency of tidal contribution to diffusivity has been shown [155]. Simulations of floating microplastic accumulation revealed that on global and multiannual scales, barotropic tides have no significant impact on accumulation patterns, at least for linearly added tidal currents [181]. In contrast, the implications of barotropic tides on Lagrangian particle dynamics on intra-annual scales have been demonstrated for the deep ocean and continental slope [182]. A reason for this impact is the interactions of tides with mesoscale processes manifesting in, e.g., eddy kinetic energy (EKE), work induced by wind and M_4 wave trains. Particles released at 150 m depth and advected by three-dimensional model velocities show that not only did the lengths of the trajectories change considerably but the ratio of vertical and horizontal scales of motion also changed.

To assess the influence of tides on particle dynamics in the North Sea, two experiments were conducted. These experiments were performed with an ocean model setup of the Nucleus for European Modeling of the Ocean (NEMO) with a horizontal resolution of 9.7 × 7.4 km (longitude × latitude) and is an extension of the experiments of Stanev and Ricker [182]. Model validations can be found in Ricker and Stanev [49] and Stanev and Ricker [182]. In the first experiment (EX1), barotropic tides were switched off at the beginning of the simulation (January 1, 2014), and the model was run until January 10, 2015. It is expected that, during this period, the hydrodynamic changes due to the missing tides reached a steady state. In the second experiment (EX2), the model was also initialized on January 1, 2014, but the barotropic tides were switched off on January 1, 2015. This run also ended on January 10, 2015, and was made to obtain a transition period from a steady state including tides to a steady state without tides. In these two experiments, neutrally buoyant particles were released uniformly (one particle per grid cell) at the surface on January 1, 2015, which resulted in a total of ~11 k particles. Particle advection was performed with the NEMO online tool to take into account high-frequency and vertical dynamics. Such particle properties could, e.g., represent neutrally buoyant FML, chemical substances, or very small objects, such as microplastics. Trajectory integration was stopped on January 10, 2015. For comparison, a reference run (REF) in which barotropic tides were not switched off was prepared for the same period (January 1, 2014 to January 10, 2015).

A drastic decrease in surface trajectory lengths in the absence of tides is observed in the North Sea (Fig. 7a, red areas). Horizontal displacement is correlated with the dominating M_2 tidal amplitude (see, e.g., Stanev and Ricker [182] Fig. 3a for a tidal chart), whereas reduced vertical displacements (Fig. 7b, red areas) are correlated with (tidally induced) surface deformation, which is given by $|D| = \sqrt{S^2 + N^2}$ (see Ricker and Stanev [49] Fig. 2g, h). Another interesting correlation becomes apparent

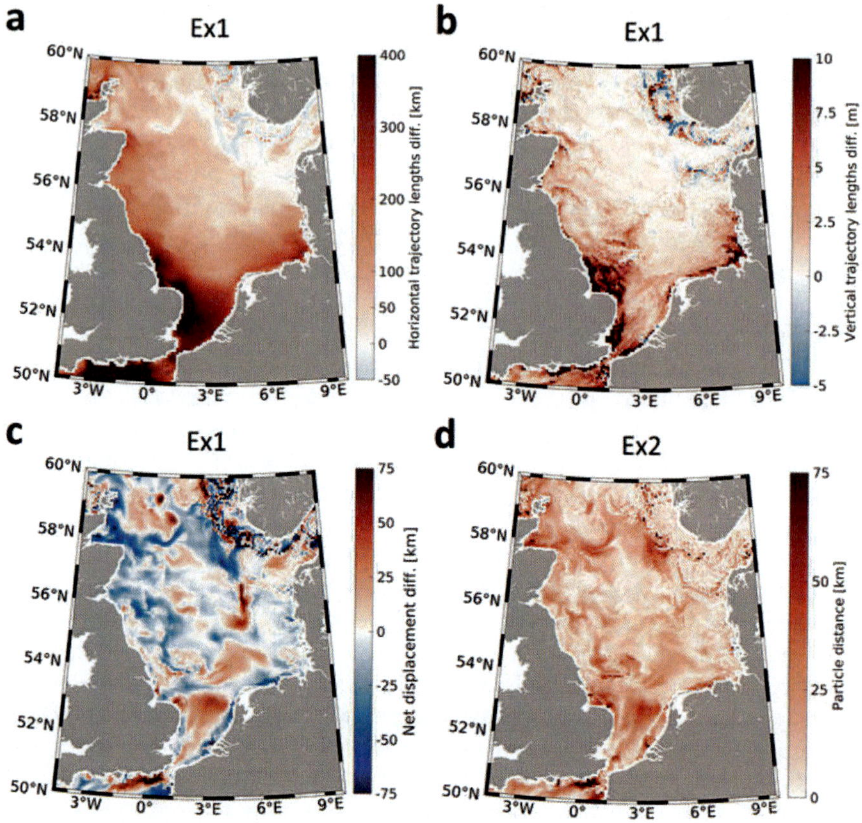

Fig. 7 The impact of tides on Lagrangian particles in the North Sea. An ocean model was used in which two sensitivity experiments (EX1 and EX2) were conducted (see the text for details). Particles were released uniformly at the surface (1 particle per grid cell; in total ~11 k) and were tracked for 10 days. Vertical velocities were taken into account. The presented differences were computed by subtracting the results of EX1 and EX2 from the results of the reference run (REF). Differences of the trajectory lengths (REF minus EX1) were computed from (**a**) the horizontal and (**b**) the vertical motions. (**c**) Net displacement difference computed from the distance from initial to final particle positions (REF minus EX1). (**d**) Particle distances of their final positions in REF and EX2

when comparing vertical displacements and thermohaline gradients (see Ricker and Stanev [49] Fig. 2c–f for consistent ocean model results): high gradients of vertical displacement differences coincide with high thermohaline gradients, of which the latter are associated with particle accumulation. Moreover, tides affect the net displacements of particles in both tidally dominated regions with strong tidal currents and tidally less dominated regions with weak tidal currents, such as the inner North Sea (Fig. 7c). Frontal positions partly manifest as strips without changes in displacements and further exhibit the ability of fronts to act as dynamic barriers for Lagrangian particles. Differences in advection directions can be estimated from the different patterns of net displacements (Fig. 7c) and particle final position differences (Fig. 7d). However, particle accumulation is impacted differently. In the absence of tides, shifts in ROFI fronts and the disappearance of tidal mixing fronts were reported. Consequently, particle accumulation patterns change accordingly in terms of shifts and disappearances of particle accumulation strips [49]. With respect to drifter observations, Callies et al. [183] found that tides could induce short-term convergence of surface drifters, which are probably induced by bathymetry and not by fronts. This nonfrontal convergence should be accounted for when analyzing short time scales of high-resolution simulations to avoid a mix-up with frontal-induced convergence.

3 Discussion and Outlook

A variety of Lagrangian methods, categorized according to their typical field of application, has been described. Several of the methods for numerical models have already been used to study frontal dynamics. However, these applications are often restricted to the deep ocean (e.g. [129]), and their application to shelf seas and tidal basins mostly focuses on the general transport of flotsam (see the examples given in Sect. 2.1.1). Consequently, the specific characteristics of each method are often not known, but it has been demonstrated that the results of each method depend on the analyzed processes, the method's configuration, and the analyzed period. This requires an adaption of the methods to the specific process or application of interest. In addition, drifter data are usually limited in shelf seas, which implies a knowledge gap regarding the Lagrangian dynamics of fronts in shelf seas concerning not only physics. Frontal dynamics are also considered crucial for other disciplines, as they affect biogeochemical substances as well as pollutants, such as plastics and chemicals. Nonetheless, many methods developed for other regions or applications seem to be capable of studying frontal dynamics in tidal basins. This is demonstrated with some selected methods applied to the North Sea. The presented results illustrate the promising potential of Lagrangian methods in tidal basins for future studies. Developments in LCS techniques have already been proven to be promising tools for frontal analyses [184–186].

More specifically, for the field of numerical modeling, some more specific questions have not yet been addressed. The attraction of particles by fronts has been shown, but it is less clear how single physical processes, such as tides, wind, and diffusion, affect fronts and, in turn, particle dynamics. A start in this direction was made by Ricker and Stanev [49], and it was demonstrated that a deeper analysis of interactions of physical processes contributes to a better understanding of Lagrangian dynamics [182, 187]. For example, wind variability and river runoff induce strong frontal variability in the North Sea and German Bight [47, 49]. Another driver of particle advection is tides, which have been shown to impact single trajectories, as well as particle accumulation. In the difference patterns of particle experiments with and without tides, frontal structures are often present and reveal their role as dynamic barriers (Fig. 7). In addition, it could be possible that tides weaken the linear dependency of FDLD and CS [125, 126] due to the induced shear stresses. Although tides are a dynamically important process in shelf seas, their role in particle dynamics has yet to be determined in more detail because the existing literature and suitable methods are sparse. Other dependencies between parameters, such as NCPD and FDLD, could also exist and could have specific properties when strong tides are present. In contrast to pseudo-Eulerian quantities, pure Lagrangian techniques are also expected to reveal new insights, as they can visualize frontal dynamics in much more detail. However, generalized methods have not been proposed, yet. As shown by Jacobs et al. [126], particle clustering appears on different temporal scales, of which the smaller scales evolve faster than the larger scales. This can also be observed in the North Sea by comparing Fig. 3b with 4a in terms of NCPD. This emphasizes the need for awareness when averaging particle concentrations over time and the issue of mixing different time scales and the related clustering processes. It is suggested to take into account this fact when analyzing the physical processes of particle accumulation. Interestingly, an increase in spatial resolution in ocean models not only induces spectral energy to higher but also to lower frequencies of FDLD, which reveals a transfer of clustering potential from smaller to larger scales [126]. This will also affect shelf sea models as their spatial resolution is constantly increasing. By increasing the resolution of basin-scale models, frontal dynamics will be better resolved. In particular, the cross-frontal dynamics require a high model resolution due to their typical scales between O (1 km) and O(10 m) on shelves. Therefore, it could be possible to investigate their full three-dimensional dynamics and the related attraction properties in the future. The use of HF radar velocity data to advect particles is also expected to improve the results. An important contribution to the analysis of submesoscale dynamics at an ocean front was recently made by Verma and Sarkar [134], who combined several of the methods presented herein. In addition, turbulence parameterizations in particle tracking have to be assessed in frontal regions and, if needed, to be optimized [188]. Another need for the future will be to add realistic characteristics to the particles depending on the object or chemical substance of interest. The present chapter only considers passive particles to determine the pure Lagrangian effects of fronts. However, changing the particle properties is expected to alter the results; therefore, problem-oriented simulations are necessary. This, in turn, will require more field and laboratory experiments to determine the respective physical properties as well as the distribution in the environment.

Lagrangian observations with drifters provide valuable insight into the physical processes affecting the dynamics of floating materials at oceanic fronts. Targeted experiments with drifter pairs and clusters at fronts can improve the understanding of the dispersion and convergence of cross- and along-front dynamics. High-resolution in situ measurements of hydrodynamic parameters and remote sensing data are necessary for observing temporal and spatial patterns of frontogenesis. Combining the latter data with Lagrangian observations can shed light on complex submesoscale and mesoscale frontal dynamics. However, satellite remote sensing data often cannot provide the temporal and spatial resolution necessary to identify submesoscale processes accurately. In contrast, ship-based measurements can supply high-resolution in situ data but entail the risk of biasing submesoscale processes, particularly at the water surface. To overcome this obstacle, surface drifters that record sea surface temperature, conductivity, gas exchange, and heat flux are needed to investigate Lagrangian dynamics at fronts with noninvasive collection of data on water mass properties. These data could be compared to ocean model data in which the temperature and salinity of the particles are tracked. Experiments with these types of drifters must be conducted over spatial scales that cover the processes of frontogenesis. Furthermore, airplane-based FML detection methods are promising [189] and could complement ship- and drifter-based measurements. Combining different kinds of in situ data is expected to further advance the understanding of the accumulation and dispersal of floating substances at fronts. In addition, the combination of in situ with ocean model data would be beneficial for both disciplines. A good basis for such a combination of model and in situ data is given in Sect. 2.2.3 but is currently restricted to LCS. Other numerical methods presented in this chapter would require numerous, area-wide drifter data. For example, the Gulf of Mexico [190], the Norwegian Sea [191], and the Mediterranean Sea [192] show high spatial drifter coverage, and some of these methods could be suitable or are already being applied, such as the calculation of convergence and divergence from large numbers of drifter clusters [163]. In contrast, other regions, such as the North Sea, have very sparse drifter data coverage, and thus, the number of applicable methods is limited. Hence, an increase in drifter coverage would be beneficial for future analyses, as area-wide simulations of Lagrangian dynamics still need to be confirmed by in situ data. Conversely, the model results can provide information about potentially interesting areas for drifter deployment and other in situ experiments.

Acknowledgements We thank Axel Braun, Andreas Sommer, Michael Butter, Elisa Schöpe, and Lars Meyer-Hagg for the support in constructing the drifters. The master and crew aboard the R/V Alkor (cruise AL523-3) supported the drifter deployments and financial support was received by the project "Macroplastics Pollution in the Southern North Sea – Sources, Pathways and Abatement Strategies" (grant no. ZN3176). It is underlined that this work would not have been possible without freely available data and numerical models. The authors thank the four anonymous reviewers for their critical and constructive reviews.

Appendix

Table 1 Summary of Lagrangian methods for visualizing and assessing frontal dynamics in tidal basins as presented in this chapter. Note that most of the methods can be applied in two and in three dimensions. Empty fields mean that the literature is too limited to give details and further investigations are needed. This table is not supposed to be complete and strengths and weaknesses can vary among applications. It is rather supposed to serve as a first orientation

Data source	Category (Sect.)	Method	Strengths	Weaknesses	Frontal dynamics
Numerical Lagrangian model	Particle positions (Sect. 2.1)	Particle positions	Intuitive; easy to interpret	Temporal snapshot; visually misleading	Direct visualization of Lagrangian ocean dynamics
		Particle trajectories	Easy to compute and interpret; takes into account the particle's past	Visually misleading	
	Particle concentrations (Sect. 2.1.1)	Instantaneous concentrations	Fast computation; easy to interpret	Misleading in regions with highly variable dynamics; many particles needed	Convergent and divergent regions
		Averaged concentrations	Less particles needed compared to instantaneous concentrations; robust against variable dynamics	The particle seeding and averaging strategy have to be carefully determined and adapted to the specific needs	Convergent and divergent regions as temporal average
	Lyapunov exponents (Sect. 2.1.2)	Finite-time/size Lyapunov exponent (FTLE/FSLE)	Suitable for short periods	Computationally costly; stretching rate included; analyses of long periods difficult	Attracting or repelling regions including stretching rate
		Lagrangian coherent structures (LCS)			
	Other methods (Sect. 2.1.3)	Cluster strength (CS)	Takes into account neighbored particles; high resolution for temporal snapshots	Computationally costly	Convergent and divergent regions
		Lagrangian divergence (LD)/finite-domain Lagrangian divergence (FDLD)/average dilation rate (ADR)	Probably the most physical representation of Lagrangian divergence; shorter periods possible compared to average concentrations	Computationally costly	Convergent and divergent regions

		Finite-time compressibility	Has a direct physical relation to area changes (stretching rate)	Convergence and divergence cannot be separated	Area change (convergence or divergence)
		Along-path area change	Detection of divergence-free attraction possible		Area change (convergence and divergence)
		Complexity method (CM)	See also FTLE/LCS; applicable to sparse data sets; less noisy compared to FTLE	Theoretical knowledge beneficial	Attracting or repelling regions
		Dynamical systems theory	See FTLE and LCS		
		Particle dispersion	Easy to compute and interpret		Convergence or divergence
		Temporal measures	Temporal characterization of processes		Time scales
		Physical properties transported by Lagrangian particles (temperature, salinity, etc.)	Allow for thermohaline and more comprehensive analyses		Comprehensive water mass dynamics
		Connectivity studies/Lagrangian flow network	Determination of individual hydrodynamics regions		Detection of transport barriers
		Trajectory lengths/particle displacement (Sect. 2.3)	Easy to compute and analyze	Total trajectory length computationally costly	Water mass transport; detection of transport barriers
GPS drifters	Pairwise drifter statistics (Sect. 2.2.1)	Drifter separation/relative dispersion	Temporal and spatial in situ data of surface dispersion	High number of drifters required	Convergence
	Velocity gradients (Sect. 2.2.2)	Differential kinematic properties (DKPs)	Direct measure to identify convergent and divergent regions	High number of drifters required	Convergence and divergence
	Lyapunov exponents (Sect. 2.2.3)	Lagrangian coherent structures (LCS)	Direct measure to identify attracting and repelling regions	Specific methods for sparse data sets needed	Attracting or repelling regions

References

1. Barnes DKA, Galgani F, Thompson RC, Barlaz M (2009) Accumulation and fragmentation of plastic debris in global environments. Philos Trans R Soc B 364:1985–1998. https://doi.org/10.1098/rstb.2008.0205
2. Thompson RC, Moore CJ, vom Saal FS, Swan SH (2009) Plastics, the environment and human health: current consensus and future trends. Philos Trans R Soc B 364:2153–2166. https://doi.org/10.1098/rstb.2009.0053
3. Andrady AL (2011) Microplastics in the marine environment. Mar Pollut Bull 62:1596–1605. https://doi.org/10.1016/j.marpolbul.2011.05.030
4. Cole M, Lindeque P, Halsband C, Galloway TS (2011) Microplastics as contaminants in the marine environment: a review. Mar Pollut Bull 62:2588–2597. https://doi.org/10.1016/j.marpolbul.2011.09.025
5. Wright SL, Thompson RC, Galloway TS (2013) The physical impacts of microplastics on marine organisms: a review. Environ Pollut 178:483–492. https://doi.org/10.1016/j.envpol.2013.02.031
6. Ivar do Sul JA, Costa MF (2014) The present and future of microplastic pollution in the marine environment. Environ Pollut 185:352–364. https://doi.org/10.1016/j.envpol.2013.10.036
7. Kühn S, Bravo Rebolledo EL, van Franeker JA (2015) Deleterious effects of litter on marine life. In: Bergmann M, Gutow L, Klages M (eds) Marine anthropogenic litter. Springer, Cham, pp 75–116. https://doi.org/10.1007/978-3-319-16510-3_4
8. de Souza Machado AA, Kloas W, Zarfl C, Hempel S, Rillig MC (2018) Microplastics as an emerging threat to terrestrial ecosystems. Glob Chang Biol 24:1405–1416. https://doi.org/10.1111/gcb.14020
9. Andrady AL (2017) The plastic in microplastics: a review. Mar Pollut Bull 119:12–22. https://doi.org/10.1016/j.marpolbul.2017.01.082
10. Chae Y, An Y-J (2017) Effects of micro- and nanoplastics on aquatic ecosystems: current research trends and perspectives. Mar Pollut Bull 124:624–632. https://doi.org/10.1016/j.marpolbul.2017.01.070
11. Campanale C, Massarelli C, Savino I, Locaputo V, Uricchio VF (2020) A detailed review study on potential effects of microplastics and additives of concern on human health. Int J Environ Res Public Health 17:1212. https://doi.org/10.3390/ijerph17041212
12. Cózar A, Echevarría F, González-Gordillo JI, Irigoien X, Úbeda B, Hernández-León S, Palma ÁT, Navarro S, García-de-Lomas J, Ruiz A, Fernández-de-Puelles ML, Duarte CM (2014) Plastic debris in the open ocean. Proc Natl Acad Sci 111:10239–10244. https://doi.org/10.1073/pnas.1314705111
13. Eriksen M, Lebreton LCM, Carson HS, Thiel M, Moore CJ, Borerro JC, Galgani F, Ryan PG, Reisser J (2014) Plastic pollution in the World's oceans: more than 5 trillion plastic pieces weighing over 250,000 tons afloat at sea. PLoS One 9:e111913. https://doi.org/10.1371/journal.pone.0111913
14. Jambeck JR, Geyer R, Wilcox C, Siegler TR, Perryman M, Andrady A, Narayan R, Law KL (2015) Plastic waste inputs from land into the ocean. Science 347:768–771. https://doi.org/10.1126/science.1260352
15. Lebreton LCM, van der Zwet J, Damsteeg J-W, Slat B, Andrady A, Reisser J (2017) River plastic emissions to the world's oceans. Nat Commun 8:15611. https://doi.org/10.1038/ncomms15611
16. Lebreton L, Egger M, Slat B (2019) A global mass budget for positively buoyant macroplastic debris in the ocean. Sci Rep 9:12922. https://doi.org/10.1038/s41598-019-49413-5
17. Pingree RD, Pugh PR, Holligan PM, Forster GR (1975) Summer phytoplankton blooms and red tides along tidal fronts in the approaches to the English Channel. Nature 258:672–677. https://doi.org/10.1038/258672a0

18. Holligan PM, Swallow JC, Currie RI, Gill AE, Simpson JH (1981) Biological implications of fronts on the northwest European continental shelf. Philos Trans R Soc London Ser A 302:547–562. https://doi.org/10.1098/rsta.1981.0182
19. Fournier RO, Det MV, Hargreaves NB, Wilson JS, Clair TA, Ernst R (1984) Physical factors controlling summer distribution of chlorophyll a off southwestern Nova Scotia. Limnol Oceanogr 29:517–526. https://doi.org/10.4319/lo.1984.29.3.0517
20. Le Fèvre J (1987) Aspects of the biology of frontal systems. Adv Mar Biol 23:163–299. https://doi.org/10.1016/S0065-2881(08)60109-1
21. Sun Y-J, Cho Y-K (2010) Tidal front and its relation to the biological process in coastal water. Ocean Sci J 45:243–251. https://doi.org/10.1007/s12601-010-0022-3
22. Simpson JH, Pingree RD (1978) Shallow Sea fronts produced by tidal stirring. In: Oceanic fronts in coastal processes. Springer, Berlin, pp 29–42. https://doi.org/10.1007/978-3-642-66987-3_5
23. O'Donnell J, Marmorino GO, Trump CL (1998) Convergence and downwelling at a river plume front. J Phys Oceanogr 28:1481–1495. https://doi.org/10.1175/1520-0485(1998)028<1481:CADAAR>2.0.CO;2
24. Belkin IM, Cornillon PC, Sherman K (2009) Fronts in large marine ecosystems. Prog Oceanogr 81:223–236. https://doi.org/10.1016/j.pocean.2009.04.015
25. Belkin IM, O'Reilly JE (2009) An algorithm for oceanic front detection in chlorophyll and SST satellite imagery. J Mar Syst 78:319–326. https://doi.org/10.1016/j.jmarsys.2008.11.018
26. Kirches G, Paperin M, Klein H, Brockmann C, Stelzer K (2013) The KLIWAS climatology for sea surface temperature and ocean colour fronts in the North Sea. KLIWAS Schriftenreihe 23A. https://doi.org/10.5675/kliwas_climatology_northsea_A
27. Kirches G, Paperin M, Klein H, Brockmann C, Stelzer K (2013) The KLIWAS climatology for sea surface temperature and ocean colour fronts in the North Sea. KLIWAS Schriftenreihe 23B. https://doi.org/10.5675/Kliwas_Climatology_Northsea_B
28. Kirches G, Paperin M, Klein H, Brockmann C, Stelzer K (2013) The KLIWAS climatology for sea surface temperature and ocean colour fronts in the North Sea. KLIWAS Schriftenreihe 23C. https://doi.org/10.5675/Kliwas_Climatology_Northsea_C
29. Maximenko N, Hafner J, Kamachi M, MacFadyen A (2018) Numerical simulations of debris drift from the great Japan tsunami of 2011 and their verification with observational reports. Mar Pollut Bull 132:5–25. https://doi.org/10.1016/j.marpolbul.2018.03.056
30. Mountford AS, Morales Maqueda MA (2019) Eulerian modeling of the three-dimensional distribution of seven popular microplastic types in the global ocean. J Geophys Res Oceans 124:8558–8573. https://doi.org/10.1029/2019JC015050
31. Otto L, Zimmerman JTF, Furnes GK, Mork M, Saetre R, Becker G (1990) Review of the physical oceanography of the North Sea. Neth J Sea Res 26:161–238. https://doi.org/10.1016/0077-7579(90)90091-T
32. Huthnance JM (1991) Physical oceanography of the North Sea. Ocean Shoreline Manag 16:199–231. https://doi.org/10.1016/0951-8312(91)90005-M
33. Sündermann J, Pohlmann T (2011) A brief analysis of North Sea physics. Oceanologia 53:663–689. https://doi.org/10.5697/oc.53-3.663
34. Yanagi T (1987) Classification of "siome", streaks and fronts. J Oceanogr Soc Jpn 43:149–158. https://doi.org/10.1007/BF02109215
35. Sharples J, Simpson JH (2009) Shelf Sea and shelf slope fronts. In: Steele JH (ed) Encyclopedia of ocean sciences, 2nd edn. Academic Press, Oxford, pp 391–400. https://doi.org/10.1016/B978-012374473-9.00623-8
36. Tonani M, Sykes P, King RR, McConnell N, Péquignet A-C, O'Dea E, Graham JA, Polton J, Siddorn J (2019) The impact of a new high-resolution ocean model on the Met Office North-West European shelf forecasting system. Ocean Sci 15:1133–1158. https://doi.org/10.5194/os-15-1133-2019
37. Allen CM, Simpson JH, Carson RM (1980) Structure and variability of shelf sea fronts as observed by an undulating ctd system. Oceanol Acta 3:59–68

38. Simpson JH, Hunter JR (1974) Fronts in the Irish Sea. Nature 250:404–406. https://doi.org/10.1038/250404a0
39. Hearn CJ (1985) On the value of the mixing efficiency in the Simpson-Hunterh/u3 criterion. Deutsche Hydrographische Zeitschrift 38:133–145. https://doi.org/10.1007/BF02226285
40. Garrett CJR, Loder JW (1981) Circulation and fronts in continental shelf seas – dynamical aspects of shallow sea fronts. Philos Trans R Soc London Ser A 302:563–581. https://doi.org/10.1098/rsta.1981.0183
41. Lwiza KMM, Bowers DG, Simpson JH (1991) Residual and tidal flow at a tidal mixing front in the North Sea. Cont Shelf Res 11:1379–1395. https://doi.org/10.1016/0278-4343(91)90041-4
42. Hill AE, James ID, Linden PF, Matthews JP, Prandle D, Simpson JH, Gmitrowicz EM, Smeed DA, Lwiza KMM, Durazo R, Fox AD, Bowers DG (1993) Understanding the North Sea system – dynamics of tidal mixing fronts in the North Sea. Philos Trans Phys Sci Eng 343:431–446. https://doi.org/10.1098/rsta.1993.0057
43. Schrum C (1997) Thermohaline stratification and instabilities at tidal mixing fronts: results of an eddy resolving model for the German Bight. Cont Shelf Res 17:689–716. https://doi.org/10.1016/S0278-4343(96)00051-9
44. Pingree RD, Griffiths DK (1978) Tidal fronts on the shelf seas around the British Isles. J Geophys Res Oceans 83:4615–4622. https://doi.org/10.1029/JC083iC09p04615
45. Holt J, Umlauf L (2008) Modelling the tidal mixing fronts and seasonal stratification of the northwest European continental shelf. Cont Shelf Res 28:887–903. https://doi.org/10.1016/j.csr.2008.01.012
46. Dippner JW (1993) A frontal-resolving model for the German Bight. Cont Shelf Res 13:49–66. https://doi.org/10.1016/0278-4343(93)90035-V
47. Becker GA, Giese H, Isert K, König P, Langenberg H, Pohlmann T, Schrum C (1999) Mesoscale structures, fluxes and water mass variability in the German Bight as exemplified in the KUSTOS- experiments and numerical models. Deutsche Hydrographische Zeitschrift 51:155–179. https://doi.org/10.1007/BF02764173
48. Holt J, Proctor R (2008) The seasonal circulation and volume transport on the northwest European continental shelf: a fine-resolution model study. J Geophys Res Oceans 113: C06021. https://doi.org/10.1029/2006JC004034
49. Ricker M, Stanev EV (2020) Circulation of the European northwest shelf: a Lagrangian perspective. Ocean Sci 16:637–655. https://doi.org/10.5194/os-16-637-2020
50. Brumer SE, Garnier V, Redelsperger J-L, Bouin M-N, Ardhuin F, Accensi M (2020) Impacts of surface gravity waves on a tidal front: a coupled model perspective. Ocean Model 154:101677. https://doi.org/10.1016/j.ocemod.2020.101677
51. Wright WR (1976) The limits of shelf water south of Cape Cod, 1941 to 1972. J Mar Res 34:1–14
52. Condie SA (1993) Formation and stability of shelf break fronts. J Geophys Res Oceans 98:12405–12416. https://doi.org/10.1029/93JC00624
53. Hill AE (1996) Spin-down and the dynamics of dense pool gyres in shallow seas. J Mar Res 54:471–486. https://doi.org/10.1357/0022240963213538
54. Brown J, Hill AE, Fernand L, Horsburgh KJ (1999) Observations of a seasonal jet-like circulation at the Central North Sea Cold Pool Margin. Estuar Coast Shelf Sci 48:343–355. https://doi.org/10.1006/ecss.1999.0426
55. Brown J, Carrillo L, Fernand L, Horsburgh KJ, Hill AE, Young EF, Medler KJ (2003) Observations of the physical structure and seasonal jet-like circulation of the Celtic Sea and St. George's Channel of the Irish Sea. Cont Shelf Res 23:533–561. https://doi.org/10.1016/S0278-4343(03)00008-6
56. Hill AE, Brown J, Fernand L, Holt J, Horsburgh KJ, Proctor R, Raine R, Turrell WR (2008) Thermohaline circulation of shallow tidal seas. Geophys Res Lett 35:L11605. https://doi.org/10.1029/2008GL033459

57. Krause G, Budeus G, Gerdes D, Schaumann K, Hesse K (1986) Frontal systems in the German bight and their physical and biological effects. In: Nihoul JCJ (ed) Elsevier oceanography series. Elsevier, pp 119–140. https://doi.org/10.1016/S0422-9894(08)71042-0
58. McWilliams JC (2021) Oceanic frontogenesis. Ann Rev Mar Sci 13:227–253. https://doi.org/10.1146/annurev-marine-032320-120725
59. Kubota M (1994) A mechanism for the accumulation of floating marine debris North of Hawaii. J Phys Oceanogr 24:1059–1064. https://doi.org/10.1175/1520-0485(1994)024<1059:AMFTAO>2.0.CO;2
60. Lebreton LC-M, Greer SD, Borrero JC (2012) Numerical modelling of floating debris in the world's oceans. Mar Pollut Bull 64:653–661. https://doi.org/10.1016/j.marpolbul.2011.10.027
61. Maximenko N, Hafner J, Niiler P (2012) Pathways of marine debris derived from trajectories of Lagrangian drifters. Mar Pollut Bull 65:51–62. https://doi.org/10.1016/j.marpolbul.2011.04.016
62. van Sebille E, England MH, Froyland G (2012) Origin, dynamics and evolution of ocean garbage patches from observed surface drifters. Environ Res Lett 7:044040. https://doi.org/10.1088/1748-9326/7/4/044040
63. Nunes RA, Simpson JH (1985) Axial convergence in a well-mixed estuary. Estuar Coast Shelf Sci 20:637–649. https://doi.org/10.1016/0272-7714(85)90112-X
64. Brown J, Turrell WR, Simpson JH (1991) Aerial surveys of axial convergent fronts in UK estuaries and the implications for pollution. Mar Pollut Bull 22:397–400. https://doi.org/10.1016/0025-326X(91)90343-Q
65. Ricker M (2020) On the dynamics of floating marine litter in the North Sea. Doctoral dissertation, University of Oldenburg
66. Acha EM, Mianzan HW, Iribarne O, Gagliardini DA, Lasta C, Daleo P (2003) The role of the Río de la Plata bottom salinity front in accumulating debris. Mar Pollut Bull 46:197–202. https://doi.org/10.1016/S0025-326X(02)00356-9
67. Dris R, Tramoy R, Alligant S, Gasperi J, Tassin B (2020) Plastic debris flowing from rivers to oceans: the role of the estuaries as a complex and poorly understood key interface. In: Rocha-Santos T, Costa M, Mouneyrac C (eds) Handbook of microplastics in the environment. Springer, Cham, pp 1–28. https://doi.org/10.1007/978-3-030-10618-8_3-1
68. Franz H, Klein H (1986) Some results of a diffusion experiment at a river plume front. Deutsche Hydrographische Zeitschrift 39:91–112. https://doi.org/10.1007/BF02408133
69. Schöneich-Argent RI, Freund H (2020) Trashing our own "backyard" – investigating dispersal and accumulation of floating litter from coastal, riverine, and offshore sources in the German Bight using a citizen science-based wooden drifter recapture approach. Mar Environ Res 162:105115. https://doi.org/10.1016/j.marenvres.2020.105115
70. González-Fernández D, Cózar A, Hanke G, Viejo J, Morales-Caselles C, Bakiu R, Barceló D, Bessa F, Bruge A, Cabrera M, Castro-Jiménez J, Constant M, Crosti R, Galletti Y, Kideys AE, Machitadze N, Pereira de Brito J, Pogojeva M, Ratola N, Rigueira J, Rojo-Nieto E, Savenko O, Schöneich-Argent RI, Siedlewicz G, Suaria G, Tourgeli M (2021) Floating macrolitter leaked from Europe into the ocean. Nat Sustain 4:474–483. https://doi.org/10.1038/s41893-021-00722-6
71. Schöneich-Argent RI, Dau K, Freund H (2020) Wasting the North Sea? – a field-based assessment of anthropogenic macrolitter loads and emission rates of three German tributaries. Environ Pollut 263:114367. https://doi.org/10.1016/j.envpol.2020.114367
72. Vriend P, van Calcar C, Kooi M, Landman H, Pikaar R, van Emmerik T (2020) Rapid assessment of floating macroplastic transport in the Rhine. Front Mar Sci 7:10. https://doi.org/10.3389/fmars.2020.00010
73. Tramoy R, Gasperi J, Dris R, Colasse L, Fisson C, Sananes S, Rocher V, Tassin B (2019) Assessment of the plastic inputs from the Seine Basin to the sea using statistical and field approaches. Front Mar Sci 6:151. https://doi.org/10.3389/fmars.2019.00151

74. Liubartseva S, Coppini G, Lecci R, Creti S (2016) Regional approach to modeling the transport of floating plastic debris in the Adriatic Sea. Mar Pollut Bull 103:115–127. https://doi.org/10.1016/j.marpolbul.2015.12.031
75. Iwasaki S, Isobe A, Kako S, Uchida K, Tokai T (2017) Fate of microplastics and mesoplastics carried by surface currents and wind waves: a numerical model approach in the Sea of Japan. Mar Pollut Bull 121:85–96. https://doi.org/10.1016/j.marpolbul.2017.05.057
76. Politikos DV, Ioakeimidis C, Papatheodorou G, Tsiaras K (2017) Modeling the fate and distribution of floating litter particles in the Aegean Sea (E. Mediterranean). Frontiers in marine. Science 4:191. https://doi.org/10.3389/fmars.2017.00191
77. Stanev EV, Ricker M (2019) The fate of marine litter in semi-enclosed seas: a case study of the Black Sea. Front Mar Sci 6:660. https://doi.org/10.3389/fmars.2019.00660
78. Hajbane S, Calmanovici B, Reisser J, Jolly A, Summers V, Ferrari F, Ghadouani A, Pattiaratchi C (2021) Coastal garbage patches: fronts accumulate plastic films at Ashmore Reef Marine Park (Pulau Pasir), Australia. Front Mar Sci 8:379. https://doi.org/10.3389/fmars.2021.613399
79. Ruiz I, Basurko OC, Rubio A, Delpey M, Granado I, Declerck A, Mader J, Cózar A (2020) Litter windrows in the South-East Coast of the Bay of Biscay: an ocean process enabling effective active fishing for litter. Front Mar Sci 7:308. https://doi.org/10.3389/fmars.2020.00308
80. Cózar A, Aliani S, Basurko OC, Arias M, Isobe A, Topouzelis K, Rubio A, Morales-Caselles C (2021) Marine litter windrows: a strategic target to understand and manage the ocean plastic pollution. Front Mar Sci 8:98. https://doi.org/10.3389/fmars.2021.571796
81. Klein H (1986) On the diffusion of rhodamine B on both sides of a river plume front. Dt Hydrogr Z 39:225–234. https://doi.org/10.1007/BF02226333
82. Pogojeva M, Zhdanov I, Berezina A, Lapenkov A, Kosmach D, Osadchiev A, Hanke G, Semiletov I, Yakushev E (2021) Distribution of floating marine macro-litter in relation to oceanographic characteristics in the Russian Arctic Seas. Mar Pollut Bull 166:112201. https://doi.org/10.1016/j.marpolbul.2021.112201
83. Korshenko E, Zhurbas V, Osadchiev A, Belyakova P (2020) Fate of river-borne floating litter during the flooding event in the northeastern part of the Black Sea in October 2018. Mar Pollut Bull 160:111678. https://doi.org/10.1016/j.marpolbul.2020.111678
84. Atwood EC, Falcieri FM, Piehl S, Bochow M, Matthies M, Franke J, Carniel S, Sclavo M, Laforsch C, Siegert F (2019) Coastal accumulation of microplastic particles emitted from the Po River, Northern Italy: comparing remote sensing and hydrodynamic modelling with in situ sample collections. Mar Pollut Bull 138:561–574. https://doi.org/10.1016/j.marpolbul.2018.11.045
85. Pietrzak JD, de Boer GJ, Eleveld MA (2011) Mechanisms controlling the intra-annual mesoscale variability of SST and SPM in the southern North Sea. Cont Shelf Res 31:594–610. https://doi.org/10.1016/j.csr.2010.12.014
86. Lynch DR, Greenberg DA, Bilgili A, McGillicuddy J, Dennis J, Manning JP, Aretxabaleta AL (2014) Particles in the Coastal Ocean: theory and applications. Cambridge University Press, Cambridge. https://doi.org/10.1017/CBO9781107449336
87. Hardesty BD, Harari J, Isobe A, Lebreton L, Maximenko N, Potemra J, van Sebille E, Vethaak AD, Wilcox C (2017) Using numerical model simulations to improve the understanding of micro-plastic distribution and pathways in the marine environment. Front Mar Sci 4:30. https://doi.org/10.3389/fmars.2017.00030
88. Gutow L, Ricker M, Holstein JM, Dannheim J, Stanev EV, Wolff J-O (2018) Distribution and trajectories of floating and benthic marine macrolitter in the South-Eastern North Sea. Mar Pollut Bull 131(part A):763–772. https://doi.org/10.1016/j.marpolbul.2018.05.003
89. Kehl C, Fischer RPB, van Sebille E (2021) Practices, pitfalls and guidelines in visualising lagrangian ocean analyses. ISPRS Ann Photogramm Remote Sens Spatial Inf Sci V-4-2021:217–224. https://doi.org/10.5194/isprs-annals-V-4-2021-217-2021

90. Froyland G, Padberg K, England MH, Treguier AM (2007) Detection of coherent oceanic structures via transfer operators. Phys Rev Lett 98:224503. https://doi.org/10.1103/PhysRevLett.98.224503
91. Froyland G, Stuart RM, van Sebille E (2014) How well-connected is the surface of the global ocean? Chaos 24:033126. https://doi.org/10.1063/1.4892530
92. McAdam R, van Sebille E (2018) Surface connectivity and interocean exchanges from drifter-based transition matrices. J Geophys Res Oceans 123:514–532. https://doi.org/10.1002/2017JC013363
93. van der Mheen M, Pattiaratchi C, van Sebille E (2019) Role of Indian Ocean dynamics on accumulation of buoyant debris. J Geophys Res Oceans 124:2571–2590. https://doi.org/10.1029/2018JC014806
94. Duhec AV, Jeanne RF, Maximenko N, Hafner J (2015) Composition and potential origin of marine debris stranded in the Western Indian Ocean on remote Alphonse Island, Seychelles. Mar Pollut Bull 96:76–86. https://doi.org/10.1016/j.marpolbul.2015.05.042
95. Onink V, Wichmann D, Delandmeter P, van Sebille E (2019) The role of Ekman currents, geostrophy, and stokes drift in the accumulation of floating microplastic. J Geophys Res Oceans 124:1474–1490. https://doi.org/10.1029/2018JC014547
96. Mansui J, Molcard A, Ourmières Y (2015) Modelling the transport and accumulation of floating marine debris in the Mediterranean basin. Mar Pollut Bull 91:249–257. https://doi.org/10.1016/j.marpolbul.2014.11.037
97. Mansui J, Darmon G, Ballerini T, van Canneyt O, Ourmieres Y, Miaud C (2020) Predicting marine litter accumulation patterns in the Mediterranean basin: Spatio-temporal variability and comparison with empirical data. Prog Oceanogr 182:102268. https://doi.org/10.1016/j.pocean.2020.102268
98. Fossi MC, Romeo T, Baini M, Panti C, Marsili L, Campani T, Canese S, Galgani F, Druon J-N, Airoldi S, Taddei S, Fattorini M, Brandini C, Lapucci C (2017) Plastic debris occurrence, convergence areas and fin whales feeding ground in the Mediterranean marine protected area Pelagos sanctuary: a modeling approach. Front Mar Sci 4:167. https://doi.org/10.3389/fmars.2017.00167
99. Liubartseva S, Coppini G, Lecci R, Clementi E (2018) Tracking plastics in the Mediterranean: 2D Lagrangian model. Mar Pollut Bull 129:151–162. https://doi.org/10.1016/j.marpolbul.2018.02.019
100. Macias D, Cózar A, Garcia-Gorriz E, González-Fernández D, Stips A (2019) Surface water circulation develops seasonally changing patterns of floating litter accumulation in the Mediterranean Sea. A modelling approach. Mar Pollut Bull 149:110619. https://doi.org/10.1016/j.marpolbul.2019.110619
101. Carlson DF, Suaria G, Aliani S, Fredj E, Fortibuoni T, Griffa A, Russo A, Melli V (2017) Combining litter observations with a Regional Ocean model to identify sources and sinks of floating debris in a Semi-enclosed Basin: the Adriatic Sea. Front Mar Sci 4:78. https://doi.org/10.3389/fmars.2017.00078
102. Bagaev A, Mizyuk A, Khatmullina L, Isachenko I, Chubarenko I (2017) Anthropogenic fibres in the Baltic Sea water column: field data, laboratory and numerical testing of their motion. Sci Total Environ 599–600:560–571. https://doi.org/10.1016/j.scitotenv.2017.04.185
103. Schernewski G, Balciunas A, Gräwe D, Gräwe U, Klesse K, Schulz M, Wesnigk S, Fleet D, Haseler M, Möllman N, Werner S (2018) Beach macro-litter monitoring on southern Baltic beaches: results, experiences and recommendations. J Coast Conserv 22:5–25. https://doi.org/10.1007/s11852-016-0489-x
104. Delandmeter P, van Sebille E (2019) The parcels v2.0 Lagrangian framework: new field interpolation schemes. Geosci Model Dev 12:3571–3584. https://doi.org/10.5194/gmd-12-3571-2019
105. Dagestad K-F, Röhrs J, Breivik Ø, Ådlandsvik B (2018) OpenDrift v1.0: a generic framework for trajectory modelling. Geosci Model Dev 11:1405–1420. https://doi.org/10.5194/gmd-11-1405-2018

106. Callies U (2021) Sensitive dependence of trajectories on tracer seeding positions – coherent structures in German Bight backward drift simulations. Ocean Sci 17:527–541. https://doi.org/10.5194/os-17-527-2021
107. Hernández-Carrasco I, López C, Hernández-García E, Turiel A (2011) How reliable are finite-size Lyapunov exponents for the assessment of ocean dynamics? Ocean Model 36:208–218. https://doi.org/10.1016/j.ocemod.2010.12.006
108. Karrasch D, Haller G (2013) Do finite-size Lyapunov exponents detect coherent structures? Chaos 23:043126. https://doi.org/10.1063/1.4837075
109. Boffetta G, Lacorata G, Redaelli G, Vulpiani A (2001) Detecting barriers to transport: a review of different techniques. Phys D Nonlinear Phenom 159:58–70. https://doi.org/10.1016/S0167-2789(01)00330-X
110. Haller G, Yuan G (2000) Lagrangian coherent structures and mixing in two-dimensional turbulence. Physica D: Nonlinear Phenomena 147:352–370. https://doi.org/10.1016/S0167-2789(00)00142-1
111. Shadden SC, Lekien F, Marsden JE (2005) Definition and properties of Lagrangian coherent structures from finite-time Lyapunov exponents in two-dimensional aperiodic flows. Physica D: Nonlinear Phenomena 212:271–304. https://doi.org/10.1016/j.physd.2005.10.007
112. Peacock T, Dabiri J (2010) Introduction to focus issue: Lagrangian coherent structures. Chaos 20:017501. https://doi.org/10.1063/1.3278173
113. Haller G, Sapsis T (2011) Lagrangian coherent structures and the smallest finite-time Lyapunov exponent. Chaos 21:023115. https://doi.org/10.1063/1.3579597
114. Samelson RM (2013) Lagrangian motion, coherent structures, and lines of persistent material strain. Ann Rev Mar Sci 5:137–163. https://doi.org/10.1146/annurev-marine-120710-100819
115. Allshouse MR, Peacock T (2015) Lagrangian based methods for coherent structure detection. Chaos 25:097617. https://doi.org/10.1063/1.4922968
116. Haller G (2015) Lagrangian coherent structures. Annu Rev Fluid Mech 47:137–162. https://doi.org/10.1146/annurev-fluid-010313-141322
117. Hadjighasem A, Farazmand M, Blazevski D, Froyland G, Haller G (2017) A critical comparison of Lagrangian methods for coherent structure detection. Chaos 27:053104. https://doi.org/10.1063/1.4982720
118. Qi L, Chen H, Chen Y (2018) Tensor eigenvalues and their applications. Springer, Singapore. https://doi.org/10.1007/978-981-10-8058-6
119. Rypina II, Pratt LJ, Lozier MS (2011) Near-surface transport pathways in the North Atlantic Ocean: looking for throughput from the subtropical to the subpolar gyre. J Phys Oceanogr 41:911–925. https://doi.org/10.1175/2011JPO4498.1
120. Rypina II, Pratt LJ, Pullen J, Levin J, Gordon AL (2010) Chaotic advection in an archipelago. J Phys Oceanogr 40:1988–2006. https://doi.org/10.1175/2010JPO4336.1
121. Olascoaga MJ, Beron-Vera FJ, Brand LE, Koçak H (2008) Tracing the early development of harmful algal blooms on the West Florida shelf with the aid of Lagrangian coherent structures. J Geophys Res Oceans 113:C12014. https://doi.org/10.1029/2007JC004533
122. Prants SV, Budyansky MV, Uleysky MY (2014) Identifying Lagrangian fronts with favourable fishery conditions. Deep-Sea Res I Oceanogr Res Pap 90:27–35. https://doi.org/10.1016/j.dsr.2014.04.012
123. Ourmieres Y, Mansui J, Molcard A, Galgani F, Poitou I (2018) The boundary current role on the transport and stranding of floating marine litter: the French Riviera case. Cont Shelf Res 155:11–20. https://doi.org/10.1016/j.csr.2018.01.010
124. Mathur M, David MJ, Sharma R, Agarwal N (2019) Thermal fronts and attracting Lagrangian coherent structures in the north Bay of Bengal during December 2015–March 2016. Deep-Sea Res II Top Stud Oceanogr 168:104636. https://doi.org/10.1016/j.dsr2.2019.104636
125. Huntley HS, Lipphardt BL, Jacobs G, Kirwan AD (2015) Clusters, deformation, and dilation: diagnostics for material accumulation regions. J Geophys Res Oceans 120:6622–6636. https://doi.org/10.1002/2015JC011036

126. Jacobs GA, Huntley HS, Kirwan AD, Lipphardt BL, Campbell T, Smith T, Edwards K, Bartels B (2016) Ocean processes underlying surface clustering. J Geophys Res Oceans 121:180–197. https://doi.org/10.1002/2015JC011140
127. Hernández-Carrasco I, Orfila A, Rossi V, Garçon V (2018) Effect of small scale transport processes on phytoplankton distribution in coastal seas. Sci Rep 8:8613. https://doi.org/10.1038/s41598-018-26857-9
128. Kalda J, Soomere T, Giudici A (2014) On the finite-time compressibility of the surface currents in the Gulf of Finland, the Baltic Sea. J Mar Syst 129:56–65. https://doi.org/10.1016/j.jmarsys.2012.08.010
129. Duran R, Beron-Vera FJ, Olascoaga MJ (2018) Extracting quasi-steady Lagrangian transport patterns from the ocean circulation: An application to the Gulf of Mexico. Sci Rep 8:5218. https://doi.org/10.1038/s41598-018-23121-y
130. Rypina II, Scott SE, Pratt LJ, Brown MG (2011) Investigating the connection between complexity of isolated trajectories and Lagrangian coherent structures. Nonlinear Processes Geophys 18:977–987. https://doi.org/10.5194/npg-18-977-2011
131. Kirwan AD, Toner M, Kantha L (2003) Predictability, uncertainty, and hyperbolicity in the ocean. Int J Eng Sci 41:249–258. https://doi.org/10.1016/S0020-7225(02)00239-2
132. Mancho AM, Small D, Wiggins S (2004) Computation of hyperbolic trajectories and their stable and unstable manifolds for oceanographic flows represented as data sets. Nonlinear Processes Geophys 11:17–33. https://doi.org/10.5194/npg-11-17-2004
133. Mancho AM, Hernández-García E, Small D, Wiggins S, Fernández V (2008) Lagrangian transport through an ocean front in the Northwestern Mediterranean Sea. J Phys Oceanogr 38:1222–1237. https://doi.org/10.1175/2007JPO3677.1
134. Verma V, Sarkar S (2021) Lagrangian three-dimensional transport and dispersion by submesoscale currents at an upper-ocean front. Ocean Model 165:101844. https://doi.org/10.1016/j.ocemod.2021.101844
135. Callies U, Plüß A, Kappenberg J, Kapitza H (2011) Particle tracking in the vicinity of Helgoland, North Sea: a model comparison. Ocean Dyn 61:2121–2139. https://doi.org/10.1007/s10236-011-0474-8
136. Neumann D, Callies U, Matthies M (2014) Marine litter ensemble transport simulations in the southern North Sea. Mar Pollut Bull 86:219–228. https://doi.org/10.1016/j.marpolbul.2014.07.016
137. Hernández-Carrasco I, López C, Orfila A, Hernández-García E (2013) Lagrangian transport in a microtidal coastal area: the Bay of Palma, island of Mallorca, Spain. Nonlinear Processes Geophys 20:921–933. https://doi.org/10.5194/npg-20-921-2013
138. Ypma SL, van Sebille E, Kiss AE, Spence P (2016) The separation of the east Australian current: a Lagrangian approach to potential vorticity and upstream control. J Geophys Res Oceans 121:758–774. https://doi.org/10.1002/2015JC011133
139. Ypma SL, Georgiou S, Dugstad JS, Pietrzak JD, Katsman CA (2020) Pathways and water mass transformation along and across the Mohn-Knipovich ridge in the Nordic seas. J Geophys Res Oceans 125:e2020JC016075. https://doi.org/10.1029/2020JC016075
140. Hufnagl M, Payne M, Lacroix G, Bolle LJ, Daewel U, Dickey-Collas M, Gerkema T, Huret M, Janssen F, Kreus M, Pätsch J, Pohlmann T, Ruardij P, Schrum C, Skogen MD, Tiessen MCH, Petitgas P, van Beek JKL, van der Veer HW, Callies U (2017) Variation that can be expected when using particle tracking models in connectivity studies. J Sea Res 127:133–149. https://doi.org/10.1016/j.seares.2017.04.009
141. van der Molen J, García-García LM, Whomersley P, Callaway A, Posen PE, Hyder K (2018) Connectivity of larval stages of sedentary marine communities between hard substrates and offshore structures in the North Sea. Sci Rep 8:14772. https://doi.org/10.1038/s41598-018-32912-2
142. Wichmann D, Delandmeter P, van Sebille E (2019) Influence of near-surface currents on the global dispersal of marine microplastic. J Geophys Res Oceans 124:6086–6096. https://doi.org/10.1029/2019JC015328

143. Politikos DV, Tsiaras K, Papatheodorou G, Anastasopoulou A (2020) Modeling of floating marine litter originated from the eastern Ionian Sea: transport, residence time and connectivity. Mar Pollut Bull 150:110727. https://doi.org/10.1016/j.marpolbul.2019.110727
144. Rossi V, Ser-Giacomi E, López C, Hernández-García E (2014) Hydrodynamic provinces and oceanic connectivity from a transport network help designing marine reserves. Geophys Res Lett 41:2883–2891. https://doi.org/10.1002/2014GL059540
145. Reijnders D, van Leeuwen EJ, van Sebille E (2021) Ocean surface connectivity in the arctic: capabilities and caveats of community detection in Lagrangian flow networks. J Geophys Res Oceans 126:e2020JC016416. https://doi.org/10.1029/2020JC016416
146. Gawarkiewicz G, Monismith S, Largier J (2007) Observing larval transport processes affecting population connectivity: progress and challenges. Oceanography 20:40–53. https://doi.org/10.5670/oceanog.2007.28
147. Poje AC, Ozgokmen TM, Lipphardt BL, Haus BK, Ryan EH, Haza AC, Jacobs GA, Reniers AJHM, Olascoaga MJ, Novelli G, Griffa A, Beron-Vera FJ, Chen SS, Coelho E, Hogan PJ, Kirwan AD, Huntley HS, Mariano AJ (2014) Submesoscale dispersion in the vicinity of the deepwater horizon spill. Proc Natl Acad Sci 111:12693–12698. https://doi.org/10.1073/pnas.1402452111
148. Merlivat L, Boutin J, Antoine D (2015) Roles of biological and physical processes in driving seasonal air–sea CO 2 flux in the Southern Ocean: new insights from CARIOCA pCO 2. J Mar Syst 147:9–20. https://doi.org/10.1016/j.jmarsys.2014.04.015
149. Breivik Ø, Allen AA, Maisondieu C, Olagnon M (2013) Advances in search and rescue at sea. Ocean Dyn 63:83–88. https://doi.org/10.1007/s10236-012-0581-1
150. Serra M, Sathe P, Rypina I, Kirincich A, Ross SD, Lermusiaux P, Allen A, Peacock T, Haller G (2020) Search and rescue at sea aided by hidden flow structures. Nat Commun 11:2525. https://doi.org/10.1038/s41467-020-16281-x
151. Meyerjürgens J, Badewien TH, Garaba SP, Wolff J-O, Zielinski O (2019) A state-of-the-art compact surface drifter reveals pathways of floating marine litter in the German bight. Front Mar Sci 6:58. https://doi.org/10.3389/fmars.2019.00058
152. van der Mheen M, Pattiaratchi C, Cosoli S, Wandres M (2020) Depth-dependent correction for wind-driven drift current in particle tracking applications. Front Mar Sci 7:305. https://doi.org/10.3389/fmars.2020.00305
153. Lumpkin R, Özgökmen T, Centurioni L (2017) Advances in the application of surface drifters. Ann Rev Mar Sci 9:59–81. https://doi.org/10.1146/annurev-marine-010816-060641
154. van Sebille E, Zettler E, Wienders N, Amaral-Zettler L, Elipot S, Lumpkin R (2021) Dispersion of surface drifters in the tropical Atlantic. Front Mar Sci 7:1243. https://doi.org/10.3389/fmars.2020.607426
155. Meyerjürgens J, Ricker M, Schakau V, Badewien TH, Stanev EV (2020) Relative dispersion of surface drifters in the North Sea: the effect of tides on mesoscale diffusivity. J Geophys Res Oceans 125:e2019JC015925. https://doi.org/10.1029/2019JC015925
156. van Sebille E, Waterman S, Barthel A, Lumpkin R, Keating SR, Fogwill C, Turney C (2015) Pairwise surface drifter separation in the western Pacific sector of the Southern Ocean. J Geophys Res Oceans 120:6769–6781. https://doi.org/10.1002/2015JC010972
157. Dräger-Dietel J, Jochumsen K, Griesel A, Badin G (2018) Relative dispersion of surface drifters in the Benguela upwelling region. J Phys Oceanogr 48:2325–2341. https://doi.org/10.1175/JPO-D-18-0027.1
158. Corrado R, Lacorata G, Palatella L, Santoleri R, Zambianchi E (2017) General characteristics of relative dispersion in the ocean. Sci Rep 7:46291. https://doi.org/10.1038/srep46291
159. D'Asaro EA, Shcherbina AY, Klymak JM, Molemaker J, Novelli G, Guigand CM, Haza AC, Haus BK, Ryan EH, Jacobs GA, Huntley HS, Laxague NJM, Chen S, Judt F, McWilliams JC, Barkan R, Kirwan AD, Poje AC, Özgökmen TM (2018) Ocean convergence and the dispersion of flotsam. Proc Natl Acad Sci 115:1162–1167. https://doi.org/10.1073/pnas.1718453115
160. Meyerjürgens J, Badewien TH, Zielinski O, Ross L, Thölen C, Schmitz J, Braun A. Drifter trajectories North Sea July 2019, submitted to PANGAEA

161. Manning JP, Churchill JH (2006) Estimates of dispersion from clustered-drifter deployments on the southern flank of Georges Bank. Deep-Sea Res II Top Stud Oceanogr 53:2501–2519. https://doi.org/10.1016/j.dsr2.2006.08.004
162. Drinkwater KF, Loder JW (2001) Near-surface horizontal convergence and dispersion near the tidal-mixing front on Northeastern Georges Bank. Deep-Sea Res II Top Stud Oceanogr 48:311–339. https://doi.org/10.1016/S0967-0645(00)00084-9
163. Tarry DR, Essink S, Pascual A, Ruiz S, Poulain P-M, Özgökmen T, Centurioni LR, Farrar JT, Shcherbina A, Mahadevan A, D'Asaro E (2021) Frontal convergence and vertical velocity measured by drifters in the Alboran Sea. J Geophys Res Oceans 126:e2020JC016614. https://doi.org/10.1029/2020JC016614
164. Berta M, Griffa A, Haza AC, Horstmann J, Huntley HS, Ibrahim R, Lund B, Özgökmen TM, Poje AC (2020) Submesoscale kinematic properties in summer and winter surface flows in the northern Gulf of Mexico. J Geophys Res Oceans 125:e2020JC016085. https://doi.org/10.1029/2020JC016085
165. Berta M, Griffa A, Özgökmen TM, Poje AC (2016) Submesoscale evolution of surface drifter triads in the Gulf of Mexico. Geophys Res Lett 43:11,751–11,759. https://doi.org/10.1002/2016GL070357
166. Reed R (1971) An observation of divergence in the Alaskan Stream. J Phys Oceanogr 1:282–283. https://doi.org/10.1175/1520-0485(1971)001<0282:AOODIT>2.0.CO;2
167. Molinari R, Kirwan Jr A (1975) Calculations of differential kinematic properties from Lagrangian observations in the western Caribbean Sea. J Phys Oceanogr 5:483–491. https://doi.org/10.1175/1520-0485(1975)005<0483:CODKPF>2.0.CO;2
168. Lodise J, Özgökmen T, Gonçalves RC, Iskandarani M, Lund B, Horstmann J, Poulain P-M, Klymak J, Ryan EH, Guigand C (2020) Investigating the formation of submesoscale structures along mesoscale fronts and estimating kinematic quantities using Lagrangian drifters. Fluids 5:159. https://doi.org/10.3390/fluids5030159
169. Rascle N, Molemaker J, Marié L, Nouguier F, Chapron B, Lund B, Mouche A (2017) Intense deformation field at oceanic front inferred from directional sea surface roughness observations. Geophys Res Lett 44:5599–5608. https://doi.org/10.1002/2017GL073473
170. Nencioli F, d'Ovidio F, Doglioli AM, Petrenko AA (2011) Surface coastal circulation patterns by in-situ detection of Lagrangian coherent structures. Geophys Res Lett 38:L17604. https://doi.org/10.1029/2011GL048815
171. Nencioli F, d'Ovidio F, Doglioli AM, Petrenko AA (2013) In situ estimates of submesoscale horizontal eddy diffusivity across an ocean front. J Geophys Res Oceans 118:7066–7080. https://doi.org/10.1002/2013JC009252
172. Nencioli F, Petrenko AA, Doglioli AM (2016) Diagnosing cross-shelf transport along an ocean front: an observational case study in the Gulf of lion. J Geophys Res Oceans 121:7218–7243. https://doi.org/10.1002/2016JC011908
173. Hernández-Carrasco I, Orfila A (2018) The role of an intense front on the connectivity of the Western Mediterranean Sea: the Cartagena-Tenes front. J Geophys Res Oceans 123:4398–4422. https://doi.org/10.1029/2017JC013613
174. Huhn F, von Kameke A, Allen-Perkins S, Montero P, Venancio A, Pérez-Muñuzuri V (2012) Horizontal Lagrangian transport in a tidal-driven estuary – transport barriers attached to prominent coastal boundaries. Cont Shelf Res 39–40:1–13. https://doi.org/10.1016/j.csr.2012.03.005
175. Froyland G, Padberg-Gehle K (2015) A rough-and-ready cluster-based approach for extracting finite-time coherent sets from sparse and incomplete trajectory data. Chaos 25:087406. https://doi.org/10.1063/1.4926372
176. Beron-Vera FJ, LaCasce JH (2016) Statistics of simulated and observed pair separations in the Gulf of Mexico. J Phys Oceanogr 46:2183–2199. https://doi.org/10.1175/JPO-D-15-0127.1
177. Essink S, Hormann V, Centurioni LR, Mahadevan A (2019) Can we detect submesoscale motions in drifter pair dispersion? J Phys Oceanogr 49:2237–2254. https://doi.org/10.1175/JPO-D-18-0181.1

178. Gräwe U, Burchard H, Müller M, Schuttelaars HM (2014) Seasonal variability in M2 and M4 tidal constituents and its implications for the coastal residual sediment transport. Geophys Res Lett 41:5563–5570. https://doi.org/10.1002/2014GL060517
179. Stanev EV, Al-Nadhairi R, Staneva J, Schulz-Stellenfleth J, Valle-Levinson A (2014) Tidal wave transformations in the German Bight. Ocean Dyn 64:951–968. https://doi.org/10.1007/s10236-014-0733-6
180. Feng S, Cheng RT, Pangen X (1986) On tide-induced lagrangian residual current and residual transport: 1. Lagrangian residual current. Water Resour Res 22:1623–1634. https://doi.org/10.1029/WR022i012p01623
181. Sterl MF, Delandmeter P, van Sebille E (2020) Influence of barotropic tidal currents on transport and accumulation of floating microplastics in the global open ocean. J Geophys Res Oceans 125:e2019JC015583. https://doi.org/10.1029/2019JC015583
182. Stanev EV, Ricker M (2020) Interactions between barotropic tides and mesoscale processes in deep ocean and shelf regions. Ocean Dyn 70:713–728. https://doi.org/10.1007/s10236-020-01348-6
183. Callies U, Carrasco R, Floeter J, Horstmann J, Quante M (2019) Submesoscale dispersion of surface drifters in a coastal sea near offshore wind farms. Ocean Sci 15:865–889. https://doi.org/10.5194/os-15-865-2019
184. Haller G (2001) Distinguished material surfaces and coherent structures in three-dimensional fluid flows. Phys D Nonlinear Phenom 149:248–277. https://doi.org/10.1016/S0167-2789(00)00199-8
185. Green MA, Rowley CW, Haller G (2007) Detection of Lagrangian coherent structures in three-dimensional turbulence. J Fluid Mech 572:111–120. https://doi.org/10.1017/S0022112006003648
186. Bettencourt JH, López C, Hernández-García E (2012) Oceanic three-dimensional Lagrangian coherent structures: a study of a mesoscale eddy in the Benguela upwelling region. Ocean Model 51:73–83. https://doi.org/10.1016/j.ocemod.2012.04.004
187. Jacob B, Stanev EV (2017) Interactions between wind and tidally induced currents in coastal and shelf basins. Ocean Dyn 67:1263–1281. https://doi.org/10.1007/s10236-017-1093-9
188. North EW, Hood RR, Chao S-Y, Sanford LP (2006) Using a random displacement model to simulate turbulent particle motion in a baroclinic frontal zone: a new implementation scheme and model performance tests. J Mar Syst 60:365–380. https://doi.org/10.1016/j.jmarsys.2005.08.003
189. Garaba SP, Aitken J, Slat B, Dierssen HM, Lebreton L, Zielinski O, Reisser J (2018) Sensing Ocean plastics with an airborne hyperspectral shortwave infrared imager. Environ Sci Technol 52:11699–11707. https://doi.org/10.1021/acs.est.8b02855
190. Lilly JM, Pérez-Brunius P (2021) A gridded surface current product for the Gulf of Mexico from consolidated drifter measurements. Earth Syst Sci Data 13:645–669. https://doi.org/10.5194/essd-13-645-2021
191. Koszalka IM, LaCasce JH, Andersson M, Orvik KA, Mauritzen C (2011) Surface circulation in the Nordic Seas from clustered drifters. Deep-Sea Res I Oceanogr Res Pap 58:468–485. https://doi.org/10.1016/j.dsr.2011.01.007
192. Poulain P-M, Antonio B, Gerin R, Jungwirth R, Mauri E, Menna M, Notarstefano G (2013) Mediterranean surface currents measured with drifters: from basin to subinertial scales. Oceanography 26:38–47. https://doi.org/10.5670/oceanog.2013.03